RENEWABLE ENERGY: TRENDS AND PROSPECTS

BOOK PUBLICATIONS
of The Pennsylvania Academy of Science

1. **Renewable Energy: Trends and Prospects,** 2002. ISBN: 0-945809-17-4.
2. **Science, Technology and National Security,** 2002. ISBN: 0-945809-18-2.
3. **Ethics in Academia,** 2000. ISBN: 0-945809-16-6.
4. **The Ecology of Wetlands and Associated Systems,** 1998. ISBN: 0-945809-14-X.
5. **The Era of Materials,** 1998. ISBN: 0-945809-15-8.
6. **Forests– A Global Perspective,** 1996. ISBN: 0-945809-13-1.
7. **Environmental Contaminants, Ecosystems and Human Health,** 1995. ISBN: 0-945809-12-3.
8. **Medicine and Health Care Into the 21st Century,** 1995. ISBN: 0-945809-11-5.
9. **The Oceans: Physical-Chemical Dynamics and Human Impact,** 1994. ISBN: 0-945809-10-7.
10. **Biological Diversity: Problems and Challenges,** 1994. ISBN: 0-945809-09-3.
11. **Conservation and Resource Management,** 1993. ISBN: 0-945809-08-5.
12. **Global Climate Change: Implications, Challenges and Mitigation Measures,** 1992. ISBN: 0-945809-07-7.
13. **Natural and Technological Disasters: Causes, Effects and Preventive Measures,** 1992. ISBN: 0-945809-06-9.
14. **Air Pollution: Environmental Issues and Health Effects,** 1991. ISBN: 0-945809-05-0.
15. **Science Education in the United States: Issues, Crises, and Priorities,** 1991. ISBN: 0-945809-04-2.
16. **Environmental Radon: Occurrence, Control and Health Hazards,** 1990. ISBN: 0-945809-03-4.
17. **Water Resources in Pennsylvania: Availability, Quality and Management,** 1990. ISBN: 0-945809-02-6.
18. **Wetlands Ecology and Conservation: Emphasis in Pennsylvania,** 1989. ISBN: 0-945809-01-8.
19. **Management of Hazardous Materials and Wastes: Treatment, Minimization and Environmental Impacts,** 1989. ISBN: 0-9606670-9-1.
20. **Ecology and Restoration of The Delaware River Basin,** 1988. ISBN: 0-9606670-8-3.
21. **Contaminant Problems and Management of Living Chesapeake Bay Resources,** 1987. ISBN: 0-9606670-7-5.
22. **Environmental Consequences of Energy Production: Problems and Prospects,** 1987. ISBN: 0-9606670-6-7.
23. **Endangered and Threatened Species Programs in Pennsylvania and Other States: Causes, Issues and Management,** 1986. ISBN: 0-9606670-5-9.
24. **Management of Radioactive Materials and Wastes: Issues and Progress,** 1985. ISBN: 0-9606670-4-0.
25. **Solid and Liquid Wastes: Management, Methods and Socioeconomic Considerations,** 1984. ISBN: 0-9606670-3-2.
26. **Hazardous and Toxic Wastes: Technology, Management and Health Effects,** 1984. ISBN: 0-9606670-2-4.
27. **Pennsylvania Coal: Resources, Technology and Utilization,** 1983. ISBN: 0-9606670-1-6.
28. **Energy, Environment, and the Economy,** 1981. ISBN: 0-9606670-0-8.

Books and Journal of The Pennsylvania Academy of Science Publications
Editor: Shyamal K. Majumdar • Professor of Biology
Lafayette College, Easton, Pennsylvania 18042

RENEWABLE ENERGY: TRENDS AND PROSPECTS

EDITED BY

S.K. MAJUMDAR, Ph.D.
Kreider Professor of Biology
Lafayette College
Easton, PA 18042

E.W. MILLER, Ph.D.
Professor and Associate Dean (Emeritus)
Pennsylvania State University
University Park, PA 16802

A.I. PANAH, Ph.D.
Professor and Director of
Geology and Environmental Science
University of Pittsburgh at Bradford
Bradford, PA 16701

Founded on April 18, 1924

A Publication of
The Pennsylvania Academy of Science

Library of Congress Cataloging in Publication Data

Bibliography
Index
Majumdar, Shyamal K. 1938-, ed.

Library of Congress Catalog Card No.: 2002104916

ISBN-0-945809-17-4

A Publication of
The Pennsylvania Academy of Science, Easton, PA 18042

Printed in the United States of America by

The Typehouse
Bath, PA 18014

PREFACE

The huge reserve of coal and petroleum has provided most of the energy for the industrial development of the world. As the reserves are consumed, renewable energy must play an increasing role. With a growing population and industrial expansion, domestic requirements for energy have been doubling every ten years. The continuing demands for energy have resulted in substantial interests in developing and promoting renewable and alternative power generation technologies. This volume provides information on the availability and utilization of vast energy resources, covering a variety of renewable and alternative power sources and generation technologies, as well as the environmental impact, policies, and regulations pertinent to renewable energy.

The book consists of 31 chapters grouped in three parts. Part one contains seventeen chapters that cover renewable energy resources, technological development, and utilization. The major sources of renewable energy considered include bioenergy, solar energy, ocean power, hydroelectric power industry, wind energy, geothermal energy resources and uses, as well as hydrogen power production.

The second section considers a sustainable environment. Questions relating to environmental impacts and benefits with the application of resources, utilization, and energy production are analyzed in seven chapters. The role of various sustainable technologies such as, hydroelectric power, wind power, ethanol production, as well as impacts of renewable energy development on regional ecology, and climate change are examined. Additionally, sustainable agroecosystems and the viewpoints of different nations in their commitment to the utilization of renewable energy are evaluated.

The remaining seven chapters in the third section discuss policies and regulations concerning renewable energy. In these considerations there is a need to understand both the technological as well as the political rules and regulations. Several chapters have analyzed renewable energy policies and regulations and their impact on energy technologies and generation in competitive markets and at the state level. In addition, the integrated science and technology efforts, and trends and prospects of renewable energy programs in Nepal are examined in this section.

The editors express their appreciation to all authors for the quality of their work and cooperation during the preparation of this text. We trust that the volume will be a valuable resource book for all citizens and professionals who are concerned with energy and the environment. Each chapter has up-to-date information on the subject and complete references to the applicable topics.

The editors express their sincere gratitude to Lafayette College, the Pennsylvania State University, the University of Pittsburgh at Bradford, and the Pennsylvania Academy of Science for their support in the preparation of this book.

Shyamal K. Majumdar,
E. Willard Miller,
Assad I. Panah,
Editors
May, 2002

RENEWABLE ENERGY: TRENDS AND PROSPECTS

Table of Contents

ACKNOWLEDGEMENT

The publication of this book was aided
by contributions from the

E. Willard and Ruby S. Miller
book publication fund and the

Charles B. Reif
book publication fund.

Renewable Energy: Trends and Prospects. Edited by S.K. Majumdar, E.W. Miller and A.I. Panah

Chapter One

RENEWABLE ENERGY'S PLACE IN THE GLOBAL ENERGY MARKET

E. WILLARD MILLER

Department of Geography
The Pennsylvania State University
University Park, PA 16802

INTRODUCTION

The modern world economy is based on the availability of vast supplies of fossil fuels– coal, petroleum, and natural gas. The Industrial Revolution was, thus, primarily an Energy Revolution. The demands for energy continue to rise. During the 20th century, the global energy output has risen from a total of 911 million tons of coal equivalent in 1900 to 9,647 million tons of coal equivalent in 1997 (Table 1). The vast deposits of coal and petroleum are not evenly distributed. Asia is the largest producer of energy with an output of 4 million tons of coal equivalent or about 35 percent of the total. Africa and South America have only 5.8 and 5.2 percent, respectively, of the world's total (Table 2).

There has been not only an increase in output level but significant changes in the importance of each fuel in supplying the world's energy needs. In 1900 coal provided 5.5 percent of the total commercial energy and biomass materials 42 percent. Particularly significant in the 20th century has been the rise of petroleum and most recently natural gas. Coal and biomass materials have declined to 22 and 19 percent, respectively, while petroleum now supplies 30 percent and natural gas 23 percent of the total. Nuclear, once thought of as a potential source of energy, provides only about 6 percent of the world's annual energy production. At the present time, as demand for energy rises, particularly in the Third World countries that are poor in fossil fuels, renewable energy resources are receiving attention. Although the renewable energy resources - solar, wind, geothermal and hydropower - provide less than two percent of the present world's energy needs, the World Energy Council has predicted that this figure will rise to a minimum of 4 percent and could reach 12 percent with incentives.

This paper reviews the present status and some developmental trends of the renewable energy resources.

CLASSIFICATION OF RENEWABLE ENERGIES

The Department of Energy has classified renewable energy into three major groups:

Solar

a. Solar thermal electric facilities
b. Photovoltaics - solar cells
c. Wind/windmills
d. Ocean thermal energy
e. Hydroelectric

Geothermal

a. Heating and cooling of buildings, including hot water heating.
b. Agricultural and industrial process heating

Biomass

a. Plant matter, including wood and waste

FACTORS AFFECTING DEVELOPMENT OF RENEWABLE ENERGY

There are both positive and negative factors that influence the use of renewable energy.

TABLE 1
World Energy Consumption
(Millions of metric tons of coal equivalent)

	1900		1997	
Energy Sources	Total	Percent	Total	Percent
Coal	501	55	2,122	22
Crude oil	18	2	2,940	30
Natural Gas	9	1	2,173	23
Nuclear	0	0	579	6
Renewables[1]	383	42	1,833	19
Total	**911**	**100**	**9,647**	**100**

[1]Includes biomass, wind, solar and geothermal energy.

Sources: World Energy Council, 1995 Survey of Energy Resources, *Energy Statistics Yearbook*, 1995; NewYork, United Nations, 1995; *BP Statistical Review of World Energy*, London: Group Media and Publications, 1998.

CONSTRAINTS OF DEVELOPMENT

Inherent Constraints

There are a number of restrictions that have retarded development of renewable energy resources.

- Great regional variations in availability of renewable energy
- Consistency of supply - at night, no solar energy; great variations in wind velocities
- Storage requirement
- Private and public invesenent decisions have been biased because of high interest rate
- Reliability of renewable energy equipment
- Innovations slow to develop
- Huge reserves of low cost fossil fuels

Federal Policy Restraints

- Low taxes on reserves of fossil fuels
- Low taxes on uses of fossil fuels, such as gasoline
- Exclusion of social costs because competing fuels exclude such social costs as air, water, land, pollution impacts, national security risks, and health costs
- Subsidies for the development of competing fuels, particularly nuclear energy
- Limits on utility liability and similar programs that subsidize centralized technologies
- Marketing and other promotional endeavors for fossil fuels, such as rural electrification programs

TABLE 2
Production of Commercial Energy in 1995
(Millions of metric tons of coal equivalent)

Area	Total Energy	Coal	Petroleum Liquid Gases	Natural Gas	Hydro Electric
Africa	772	152	490	119	11
North America	3,246	949	947	950	399
South America	612	33	405	111	60
Asia	4,177	1,446	1,953	540	237
Europe	1, 711	473	448	363	429
Russian Federation	1,504	242	430	774	58
Oceania	257	195	46	38	8

Source : World Energy Council, 1995 Survey of Energy Resources, *Energy Statistics Yearbook*, 1995, New York: United Nations, 1995.

Economic Restraints

- Competition with low cost fossil fuels
- High cost of technological innovation
- Profit margins have been too low to fimd large-scale research and development programs
- Limited public funds
- The low energy density of renewable energy in market areas requiring a high energy density
- Insurance policies are not formulated for renewable energy development or use

Institutional Restraints

- Resistance to change of established practices by the building industry
- Lack of information reduces market appeal and prevents full assessment of renewable energy potential
- The legal uncertainties that ensure access to renewable energy, such as assuring access to sunlight for solar energy. The large number of windmills within a restricted area in order to provide electricity in sufficient amounts to serve an area

FACTORS ENCOURAGING DEVELOPMENT

There are a number of factors that will encourage the use of renewable energy in the future.

Environment - Physical

Renewable energy is essentially pollution free. Fossil fuel pollution may cause:

- Health problems - asthma, bronchitis, etc.
- Climatic change world wide
- Acid rain - destruction of vegetation and buildings

General Availability

- There is an unlimited supply of energy from most renewable energy sources
- A growing recognition that fossil fuel reserves are limited and that a new source of energy must be developed

Technology

- Since 1990 technological innovations have reduced the cost of renewable

energy
- Storage for larger quantities of energy are now available to assure a constant supply

Economic

- Declining cost per kwh
- Rising energy demand in many Third World countries with limited or no fossil fuels provide markets when costs are high.
- Fossil fuel costs are rising, particularly petroleum

Institutional

- The acceptance of new house designs in order to use renewable energy
- Trained construction workers to build renewable energy buildings

DEVELOPMENTS IN RENEWABLE ENERGY

From the earliest days of civilization, renewable energy, such as biomass, has been a source of heat. Renewable energy resources can be used directly to provide energy for heating and cooling or it can be converted into electricity. When renewable energy is converted to heat directly, it is, of course, used locally, but when converted to electricity, it may be transported for over 1,000 miles before consumed. To illustrate, the vast tropical deserts of the world have the potential to provide the energy consumed in the northern latitudes.

Biomass Energy

Biomass materials originate from the process of plant photosynthesis. The great mass of biomass material may be classified into three kinds – wood, plant residues, and animal wastes. For centuries biomass material was the only source of heat to cook food and heat homes. Although no longer supreme, biomass material still plays an important role in supplying heat to man.

Of all biomass sources, wood is most widely used. According to U.N. statistics, the world's largest firewood producers, in order, are India, Brazil, China, Indonesia, the United States, and Nigeria. The International Energy Agency (EA) reports that in most industrial nations the use of fuelwood could be expanded without depleting natural forests. On average, about 25 percent of the wood entering the timber industry is available for conversion to energy.

The United States provides fairly typical examples of the use of fuelwood by an industrial nation. In the U.S., industrial, commercial, and utility applications account for some two-thirds of the wood used for energy. The remainder provides 10 percent of the nation's residential heating, warming about 5 to 6 million homes exclusively and 21 million homes partially. Of the industrial uses, almost half of all wood fuel is consumed by the pulp and paper industry, which meets more than 55 percent of its energy needs from wood. The second largest market is the lumber industry. The fastest growing sector is the non-forest product industries. Users include electric utilities, industry giants such as Dow, Corning, and Proctor & Gamble, and small electric companies.

Biomass materials remain a significant source of energy in the Third World (Table 3). More than two-thirds of all Third World people rely on biomass material, primarily wood, for cooking, and heating. Rural dwellers obtain nearly 100 percent of their energy needs from wood. Even in oil rich nations, such as Nigeria, the rural, agricultural people do not possess sufficient funds to purchase expensive heating oil. As a consequence, families suffer when energy is lacking to cook foods.

In the Third World today, rural people are increasingly migrating to urban areas. This magnifies the ecological consequences of the firewood scarcity. Urban dwellers tend to rely on charcoal rather than wood because its light weight makes it more economical to transport from the rural areas. However, when wood is converted to charcoal, more than half of the primary wood energy is lost. Consequently, for every person who moves from a farm village to a city and uses charcoal rather than wood equals, in energy terms, two people.

The development of a plan for maintaining a supply of fuelwood has not been easy. In general, experts agree that such a plan to meet fuelwood needs for the Third World must include: increasing the productivity of national forests, making better use of wood now wasted, including logging residue, raising the efficiency with which wood is burned, and planting more trees.

During the 1970s and early 1980s, it was generally agreed that a "fuelwood crisis" loomed in the near future. This was based on the assumption that fuelwood

TABLE 3

Share of Total Energy Provided by Biomass Selected Countries

Country	Biomass share of total energy use
Nepal	94
Malawi	93
Sudan	74
Kenya	71
Paraguay	64
Nicaragua	50
Costa Rica	33
India	33
China	25
Brazil	20

Includes agricultural wastes and dung in addition to wood and charcoal
Source: Brown, Lester R. *et al*, 1988, *State of the World*, Washington DC: Worldwatch Institute., p. 88.

collection and deforestation were directly related and that meeting fuelwood demands would be greater than the replacement of the forests. Nevertheless, the development of planned forests in the Third World has encountered many problems. It was soon recognized that in the Third World the only source of labor to plant large areas of trees could come only from the large village population. In many areas, however, rural people were hesitant to participate in communal endeavors because the plans were usually vague as to how the benefits would be shared. Often their needs and ideas were not solicited. In addition, trees take years to mature before any fuelwood is available. Consequently, rural people in the Third World have rarely been motivated to plant trees exclusively for fuelwood. While a fuelwood crisis may exist for most rural dwellers in the Third World, the planting of trees for fruit, lumber, poles, and shade receive higher priorities.

Although energy planners in developing countries versus those in industrial nations face different challenges in the use of biomass, biomass energy remains important in each. In the Third World, biomass will remain one of the leading sources of energy to the average citizen. In the industrial world, biomass materials will increase as a commercial source of energy.

Hydropower

The earliest civilization used water power to grind grain to produce bread. In the early stages of industrial development, hydropower was used in factories to operate water wheels in order to run machinery. With the development of the turbine in the late 19th century, hydropower was converted to electricity becoming competitive with fossil fuels in most situations. The growth potential is, however, limited by the availability of suitable river sites. There must be a natural falls or the construction of a dam creating an artificial falls.

About one-quarter of the world's hydropower potential has been developed. During the past half century, there have been significant increases in the world's hydro capacity from about 44,000 megawatts in 1950 to 282,000 in 1970 and in the 1990s over 650,000 megawatts. Currently hydropower is the largest single renewable source of electricity. Hydroelectric power provides more than 20 percent of the world's electricity (Table 4).

Although hydroelectric capacity is found on all continents, the potential varies greatly depending on climatic conditions and, to a lesser degree, elevation above sea level. The United States and Canada are the major producers of hydropower with about 13 percent each of the world's total. The Russian Federation produces about 10 percent, Brazil 9 percent, and China has 5 percent. Norway, with over 100 inches of precipitation in the high mountains, gets 95 percent of its electricity from falling water and Sweden about 50 percent.

In the past, many dams were built on streams, not to produce electricity but to control flooding or for recreational purposes. In the United States, for example, fewer than 3 percent of the 67,000 existing dams currently produce electricity. Many of these dams could be converted to produce electricity, often at less expense than the construction of a new fossil fuel facility. The U.S. Department of Energy estimates that up to 10,000 megawatts could be produced by this type of dam conversion.

Most of the recently constructed dams are massive in size, such as those in China and Brazil. Although these dams produce thousands of megawatts of electricity and are built to develop the economy of the nation, many environmental problems have arisen. Vast areas are flooded, destroying agriculture and the habitat of natural life. Consequently, opponents to this type of dam have arisen, so it appears that not many huge dams will be built in the future.

Solar Energy

Experiments to use solar energy as a direct source of heat began as early as the 17th century. Solar panels were used to provide a source of heat for homes. Progress was, however, slow in developing these solar panels because of the availability of wood and fossil fuels.

The process to produce electricity directly from sunlight was invented in 1837. Technological progress was also slow but by the 1970s improved photovoltaic cells were available. By using a thin, silica wafer, sunlight could be converted directly into electricity. While the cost of producing electricity from these cells remains high, they are light weight and modular and thus, easily

TABLE 4
World Hydroelectric Generating Capacity
1950-1995

Year	Capacity (megawatts)
1950	44,596
1953	56,698
1956	114,916
1959	145,450
1962	180,206
1965	214,023
1968	251,249
1971	295,564
1974	339,271
1977	336,426
1980	466,938
1983	517,899
1986	574,275
1989	622,058
1992	643,511
1995	650,000 (prelim)

Source: United Nations, Department of International Economics and Social Affairs, *Energy Statistics Yearbook*, New York : United Nations, various years; British Petroleum, *BP Statistical Review of World Energy*, London: British Petroleum.

transported. Thus, more than half of the world's market for PVs is in the Third World countries where fossil fuel facilities to produce electricity are insufficient.

Although PV s are increasingly used, most governments have failed to explore the potential of solar cells. To illustrate, in 1991, U.S. governmental research and development in photovoltaic cells totaled $289 million, compared with $989 million spent on coal, and $4.49 billion for nuclear reactors. Solar energy funds amounted to 3.5 percent of total R&D funds. However, in the late 1990s it has risen a small amount. Some European governments are also increasing their support for PVs. Spain, in the 1990s, put 17 percent of its energy R&D budget in solar, Germany 16 percent, and Switzerland 14 percent.

Due to the growing market for PVs in the fossil fuel deficient Third World countries, the solar energy companies are increasing their solar R&D budgets, which will ultimately reduce costs.

The production of PV cells has grown steadily since 1971 from 0.1 megawatts to 46.5 megawatts in 1990 and 151.7 megawatts in 1998 (Table 5). The United States is the largest producer with an output of 54 megawatts in 1998. More than 70 percent of the PV cells produced in the United States are exported. Growth in the future will depend on partnerships between the federal government and the electric utilities, and the PV producing companies.

Japan is the second largest producer, shipping 49 megawatts in 1998. The United States and Japan account for over 90 percent of the world's trade in PVs. Europe produced 30 megawatts. Other countries with growing solar industries include Australia, India, China, and Taiwan, with a combined production of 19 megawatts.

The use of PVs has a number of environmental advantages. Their production of electricity generates no air pollution. Further, the production of standard silicon solar cells yields little waste materials. A few new cell technologies that are in

TABLE 5
World Photovoltaic Shipments
1971–1998

Year	Shipments (Megawatts)
1971	0.1
1975	1.8
1978	2.5
1981	7.8
1984	21.5
1987	29.2
1990	46.5
1993	60.1
1996	88.0
1998	151.7

Source : *Vital Signs 1999*, Washington DC; PV News, Various Issues.

the process of development generate some toxic materials. For example, some multi layer cells that are made of such materials as gallium arsenide or cadmium sulfide also produce some toxins.

Although panels require considerable space, their use of land is not excessive. Coal mining normally requires more space than solar panels. Further, solar panels are normally located on rooftops, along highways, or in the sun-rich empty deserts. PV cells have a versatility that permits their location in remote places where land is low-cost and available with easy transportation. Consequently, PVs can supply electricity to rural areas and other remote locations less expensively than most central power plants. In addition, sunshine is more widely distributed in the world than any other energy source.

Technological developments are reducing the cost of solar energy. To illustrate, the major effort is to achieve higher temperatures, up to 3,000 degrees. The simplest system, known as a solar pond, consists of lined cavities filled with fresh and salt water. The salt water, which is cleaner than fresh water, lies on the bottom absorbing heat while the fresh water traps the underlying heat. The largest solar panel is in southern California with a capacity of 48 megawatts. It provides electricity for some 48,000 homes. Another new system using thousands of mirrors tracks the sun and reflects the energy onto central receivers through which a fluid is heated to the exact degree for heating homes or other buildings or the heat can be transferred to operate a turbine for generating electricity.

Wind Energy

The windmill has been a source of power for centuries. The first windmill providing power to grind grain into flour was recorded in Persia over 3,000 years ago and similar reports are found throughout Asia and Europe. With the development of the turbine, wind power was converted to electricity. By the beginning of the 20th century, windmills provided a major source of energy in northern Europe. For example, Danish industry relied on wind power for a quarter of its energy in 1900.

With the growth of fossil fuel energy, the importance of wind power declined. However, in the 1970s, Danish engineers made major technological advances in the use of the windmill. The development of fiberglass blades, as long as 120 feet with electronically controlled variable speed drives and sophisticated micro controls, made wind power economically competitive with fossil fuel generated electricity.

In response, the world wind power generating capacity expanded rapidly in the 1980s from 10 megawatts in 1980 to 1,930 in 1990, and over 9,800 in 1998. In 1998 alone the world added 2,100 megawatts of new wind generating capacity (Table 6).

These wind turbines generated an estimated 17 billion kilowatt hours of electricity in 1999 worth $1.2 billion and enough to meet the electrical needs of 3.4 million homes. In 1998 the market for new windmills provided sales of about

$2 billion and created thousands of manufacturing jobs. This was an all-time record and 35 percent more than were added in 1997.

Wind power is now being developed throughout the world with major concentrations in a few countries. In Europe, Germany is the leader with a wind power capacity in 1998 of 2,800 megawatts. The present development of wind power began in Germany in 1991 and by 1998 wind was providing about one percent of the national electricity and had reached 14 percent in the North Sea coastal states of Schleswig and Holstein.

The United States is the second largest producer with capacity of about 1900 megawatts. In the United States, California is a major producer with wind farms concentrated in the high, windy mountain passes - Altamont, San Gorgonio and Tehcirhapi. Although initially wind power was concentrated on the West Coast, a much wider distribution is now occurring. The most recent developments are a 107 megawatt wind farm in Minnesota, one of 42 megawatts in Wyoming and one of 25 megawatts in Oregon. Government laws and regulation have been both favorable and unfavorable to the development of wind power. In certain countries, such as Germany and Spain, their dramatic increases in the use of wind energy have been due to favorable government policies. In the United States, the large electric facilities are now required to purchase any quantity of excess electricity produced by windmills. The dicontinuation of this practice in 1993 slowed growth. Many American companies producing turbines went bankrupt.

Wind power is still in its initial stage of modern development. While its importance will grow, it will continue to play a small role in supplying the energy needs of the world.

Geothermal Energy

Although it has long been known that there are vast quantities of heat beneath the earth's surface in certain regions, there has been little development of this

TABLE 6
World Wind Energy Generating Capacity

Year	Total (Megawatts)
1980	10
1983	210
1986	1,270
1989	1,730
1992	2,510
1995	4,820
1998	9,600

Source : International Wind Energy Development, World Market Update, World Watch Institute, Washington DC, *Vital Signs 1999.*

vast energy source. In 1950 only 239 megawatts of electrical energy were generated from geothermal sources. The utilization of this energy source was still small in 1980 when only 2,396 megawatts were produced. Since then the geothermal electrical capacity has increased to over 9,300 megawatts (Table 7). Development was encouraged in the United States by tax incentives, government research, and widespread exploration for new geothemmal energy sources. As a response, about one-half of the world development is in the United States with a concentration in Califomia.

The Philippine Islands, located on the western Pacific volcanic rim, possess a major supply of geothermal energy. This nation possesses the world's second largest capacity of geothermal energy and now supplies about 15 percent of its electrical needs of the nation. Mexico is the third largest producer with about 17 percent of the world's geothermal energy capacity. Italy is fourth with 9 percent followed by Japan.

In addition to electrical production, many countries are now using this geothermal energy for direct heating. In the 1990s, geothermal direct heating reached the equivalent of 11,730 electrical megawatts. About one quarter of this is in Japan. Other nations where geothermal heat is important include Iceland, China, Hungary, and the Russian Federation.

The future for the use of geothermal energy is bright. Technological advancements have made it competitive with the cost of electricity from fossil fuels. Electricity from the earth's heat may cost as little as 4.5 to 6 cents per kilowatt

TABLE 7

World Geothermal Electrical Generating Capacity

Year	Capacity (Megawatts)
1950	239
1953	242
1956	246
1959	362
1962	487
1965	558
1968	619
1971	834
1974	1,167
1977	1,314
1980	2,396
1983	3,383
1986	6,505
1988	7,279
1991	9,369

Source: United Nations, Department of International Economic and Social Affairs, *Energy Statistics Yearbook*, New York: United Nations, Various years.

compared to 5 to 7 cents per kilowatt from fossil fuels. However, geothermal energy costs vary considerably, depending upon the depth of the heat source within the earth. The depth of the source may vary from 300 to 10,000 feet. A major advantage in the use of geothermal energy to produce electricity, however, is that the source is constant and it is essentially pollution free. The use of geothermal energy remains in its initial stage. Technical advances will continue to reduce costs.

At the present time, most geothermal energy is secured from hot water or steam trapped in rocks. This heat is known as hydrothermal energy. In the future, two other sources of earth heat may be tapped. The first is dissolved methane gas, known as geopressed energy, located from 12,000 to 24,000 feet beneath the surface. Still more difficult to exploit is energy from hot rocks that is dry. The energy reserves are huge but technology does not exist for initially fracturing the deep rock, then injecting water into the formation and finally recovering the heat.

Ocean Energy

The oceans are a potential source of energy that is essentially untapped. Tidal energy uses the force of the incoming-outgoing flow of water in the same technique that hydroelectric plants use the flow of falling water. Each tidal site is specific and the engineering must reflect a design that ensures adequate water flow.

Wave energy technology is still in its infancy. In most systems, waves are forced through a channel to increase pressure. New designs also use floating rafts, buoys, or bobbing floats.

Although there are only a few tidal plants in operation, several exploratory facilities have been built. At La Rance, France, a 240-megawatt facility currently is in operation. Canada has built a plant on the large tidal flats located in the Bay of Fundy. In 1986, the Chinese government erected several small tidal plants, and completed a 10-megawatt plant in Zhejang Province.

Wave energy is most applicable to production of small quantities of electricity. Twelve wave energy conversion plants have been in operation - 11 for elevator power generation and one for fresh water production. Cost studies show that wave energy could be economical in the near future.

Research has begun in the United States and Japan in the program on the Ocean Thermal Energy Conservation (OTEC). It is now known that there are significant differences in temperature between the ocean surface water and those at 3,000 feet or more. To utilize these temperature contrasts, the deep cold water must be pumped to the surface to create a solar thermal cycle, or "heat engine." While scientists and engineers have investigated the concept of ocean thermal power for the past 50 years, no commercial plants are in operation. Research is needed to improve the technology and reduce costs, as well as resolve environmental problems.

Ocean energy will not be of any importance in supplying energy in the immediate future. Nevertheless, ocean energy research and technology deserves support as a future potential source of vast amounts of energy.

Hydrogen Energy

The technology is presently available for using hydrogen to produce electricity and as a fuel for motor vehicles. An inexhaustible supply of hydrogen could be made available and it can be transported by ocean carrier, pipelines, or trucks, so it could be available anywhere in the world. The conversion of the energy of H_2 to electrical energy is also essentially pollution free.

Currently, hydrogen can be produced from a number of existing processes. One of the simplest is the production of H_2 from the decomposition of biomass. However, since this process has a low conversion efficiency the biomass system would have to be extremely large to produce a commercial quantity of hydrogen. Nevertheless, if a huge supply of biomass is available, then this system is the least costly of any at the present time.

A number of techniques are at the experimental stage in producing hydrogen. One is the direct process of producing H_2 and O_2 from water. Perhaps the most practical process will be to use a photovoltaic system, coupled to electrolyzers to release hydrogen from its sources.

The motor vehicle industry now has the first vehicle that can be operated on hydrogen. However, no infrastructure exists, such as H_2 filling stations, and the motor industry will not invest billions of dollars until a mass market exists for hydrogen fueled cars. As a consequence, many technical and marketing problems must be solved before hydrogen becomes a common source of energy. Ultimately, however, it may become the major source of energy in the world, for water is inexhaustible and can be the major source of H_2 .

CONCLUSION

Of all energy resources, the renewable energy resources are most widely distributed. As the market for energy has grown, however, the small local capacity has been a major hardship for development. Of the fossil fuels, petroleum appears to be most likely to rise in cost and therefore most vulnerable. While the transition from one type of energy to another has occurred in the past– such as biomass to fossil fuel– energy demands today are so great that the problem of making changes is massive.

SELECTED BIBLIOGRAPHY

General

British Petroleum, *B P Statistical Review of World Energy*. London: British

Petroleum. Various issues.
Cannon, James S., 1995, *Harnessing Hydrogen: The Key to Sustainable Transportation*. New York: INFORM, Inc.
Dunn, Seth, September, October 1999. "King Coal's Weakening Grip on Power," *World Watch*. 12: 10-19.
Electric Power Research Institute, December 1997, Renewable Energy Technology *Characterization*, Palo Alto, CA: Electric Power Research Institute.
Flavin, Christopher and Nicholas Lenssen, 1990. *Beyond the Petroleum Age: Designing a Solar Economy*. World Watch Paper 100, Washington, DC: World Watch Institute.
Grubb, M.J., Sept. 1991, "The Integration of Renewable Electricity Sources," *Energy Policy*, 19: 670-688.
Grubler, Arnold, Alan McDonald, and Nebojsa Nakicengvie, Eds., 1998, *Global Energy Perspectives*, New York: Cambridge University Press.
Hankins, Mark, 1993. *Solar Rural Electrification in the Developing World: Four County Case Studies*. Washington, DC: Solar Electric Light Fund.
Harrigan, Alice, July/August 1997. "What on Earth is An Eccohome?" *The Environmental Magazine*, 8: 23-35.
Holdren, John P., et al. 1997, *Federal Energy Research and Development for the Challenges of the Twenty-First Century,* Report of the Energy Research and Development Panel of the President's Committee of Advisors on Science and Technology, Washington, DC.
Jackson, Tim, ed., Oct. 1991, "Renewable Series - Collection," *Energy Policy*, 19: 706-807.
Jain, B.C., September 1987. "Rural Energy Centers Based on Renewables– Case Study on an Effective and Viable Alternative, *IEEF Transactions on Energy Conservation*, 2: 329-335.
Kraft, Michael E. and Regina S. Axelrod, Dec. 1984, "Political Constraints on Development of Alternative Energy Sources: Lessons from the Reagan Administration," *Policy Studies Journal*, 13: 319-330.
McGowan, Francis, March 1991, "Controlling the Greenhouse Effect: The Role of Renewables," *Energy Policy*, 19: 110-118.
Prendergast, J., March 1990, "Tomorrow's Energy Today," *Civil Engineering (Am. Soc. Civ. Eng.),* 60: 70-79.
Reece, N.S. and G.R. Nuss, Sept. 1990, "Renewable Energy: Energy for Today and Tomorrow," *ASTM Standardization News* 18: 32-37.
Shea, Cynthia Pollock, 1988, *Renewable Energy: Today's Contribution. Tomorrow's Promise*. Washington, DC: World Watch Institute, Paper No. 81, 68 pp.
Smil, Vaclav, 1994. *Energy in World History*, Oxford, U.K.: Westview Press.
Williams, Susan *et al.* Spring 1990. "Renewing Renewable Energy," *Issues in Science and Technology*, 6: 64-70.
May 1987, "Alternative Energy," *IEE Proceedings*, Part A 134: 369-463.
July 1990, "Social and Private Costs of Alternative Energy Techniques,"

Contemporary Policy Issues 81: 1-307.
September 3, 1997, "The New Economies of Oil," *Business Week*, 144-148.

Biomass

Carioca, J.O.B. *et al.*, 1987, "Energy from Biomass," *Impact of Science on Society*, 37: 305-315.
Gowen, Marcia M., October 1989. "Biofuel v. Fuel Economics in Developing Countries," *Energy Policy,* 17:455-470.
Herendeen, R. and S. Brown, January 1987, "A Corporative Analysis of Net Energy From Woody Biomass," *Energy* 12: 75-84.
Reddy, A.K.N. and J. Goldenberg, September 1990, "Energy for the Developing World," *Scientific American*, 263: 110-118.
Strauss, C.H. and L.L. Wright, 1990, "Woody Biomass Production Costs in the United States: An Economic Survey of Commercial Populous Plantation Systems," *Solar Energy*, 42: 105-110.
Zerbe, John I., Winter 1988, "Biofuels: Production and Potential," *Forum for Applied Research and Public Policy*, 3: 38-44.
December, 1989. "Solid Fuels/Biomass: Products and Services," *Independent Energy*, 19: 93-100.

Hydroelectric Power

Lambert, Jeremiah D. and James R. O'Neill, February 4, 1988, "Privatization of Municipal Hydroelectric Facilities Under Current Law," *Public Utilities Fortnightly*, 121: 11-17.
Reisner, Marc. Spring 1985. "America's Newest Old Energy Sources: Hydro Power, Wherever Water Flows and Drops in Appreciable Distance, Someone Somewhere, Is Contemplating a Dam." *Amicus Journal,* 6: 42-52.
Swainson, Neil A., Spring 1986, "The Columbia River Treaty: Where Do We Go From Here," *Natural Resources Journal*, 26: 243-259.
Whittaker, M. Curtis, 1986, "The Federal Power Act and Hydropower Development: Rediscovering State Regulatory Power and Responsibilities," *Harvard Environmental Law Review*, 10: 135-182.

Solar Energy

Bluestone, Mimi, July 18, 1988, "Solar Power Alive, Well, and Almost Making Money," *Business Week*, 132-133.
Brodbrook, Adrian, Jr., 1989, "The Role of the Courts in Advancing the Use of Solar Energy," *Journal of Energy Law and Policy*," 9: 135-175.
Flavin, Chris and Nicholas Lenssen, March 1992, "Policies for a Solar Economy," *Foreign Policy* 20: 245-256.
Merritt, C. and R. Carter, October 1989, "Sun Sees its Future - and It's Bright."

Coal, 26, 34-47.
Thornton, J.P. *et al.*, 1990, "Photovoltaics - Today's Reality, Tomorrow's Power," *Energy Engineering*, 87: 65-79.
Weinberg, C.J. and R.H. Williams, September 1990, "Energy from the Sun," *Scientific American*, 263: 146-155.
October 1985, "Solar Thermal Power," *National Resources Journal,* Supplement, 25: 1097-1157, 4 articles.

Wind

BTM Consult, March 1997, *International Wind Energy Development*: World Market Update, 1996, Ringhobing, Denmark.
Caplan, Ruth, November/December 1992, "Losses and Gains, Energy Bill: Overview," *Power Line.*
Denniston, Derek, May/June 1993, "Europe Catches the Wind," *World Watch.*
Eide, David, Summer 1991, "Bright Prospects for Wind Energy," *Strategic Planning Energy and Environment,* 11: 40-52.
Hershey, Robert D.J., February 20, 1993, "Indirect Effects of the Energy Tax," *New York Times.*
Jarvis, Michaela, July 1989, "Power from Thin Air," *California Journal*, 20: 278-283.
Kellett, Jon, April 1990, "The Environmental Impact of Wind Energy Development," *Town Planning Review,* 61: 139-155.
Lenssen, Nicholas, July/August 1990, "California Wind Industry Takes Off," *World Watch.*
Shea, Cynthia Pollock, March/April 1988, "Harvesting the Wind: Launched in California, Wind Power is Now Spanning the Globe," *World Watch* 1: 12- 17.
Stans, Thomas A., 1988, Legislative Incentives and Energy Technologies: Government's Role in the Development of the California Wind Energy Industry," *Ecology Law Quarterly*, 115: 103-158.
U. S. Wind Power Monthly, (various issues).

Geothermal

Cataldi, R., 1987. "World Geothermal Development: The Present Situation and Opportunities for the Future," *Impact on Science and Society*, 37: 337-350.
Pasqualetti, M.J. and Mark Dellinger, Spring 1988. "Hazardous Waste from Geothermal Energy: A Case Study," *Journal of Energy and Development,* 13: 275-295.
Rummel, Fritz, 1985, "Geothermal Energy: An Alternative or Additional Potential Energy Source," *Applied Geography and Development*, 25: 7-24.

Ocean Power

Baker, A.C., May 1987, "Tidal Power," *IEE Proceedings* Part A, 134: 392-398

Kindl, John Warren. Winter 1984. "Ocean Thermal Energy Conversion," *Georgia Journal of International and Corporative Law*. 14: 1-27.

Penney, T.R. and D. Bharathan, January 1987, "Power from the Sea," *Scientific American* 256: 86-92.

Watts, S., December 7, 1988, "Marine Directorate to Investigate Energy from Oceans," *New Scientist* 120: 24-31.

Renewable Energy: Trends and Prospects. Edited by S.K. Majumdar, E.W. Miller and A.I. Panah

Chapter Two

BIOENERGY FROM BIOMASS: RENEWABLE ENERGY FROM TREES AND ORGANIC WASTE: A GLOBAL PERSPECTIVE

LYNDHURST COLLINS

Department of Geography
University of Edinburgh
Drummond Street
Edinburgh
EH8 9XP
Scotland

INTRODUCTION

Renewable energy from biomass is probably the most mundane and least exciting option for deriving alternative sources of energy discussed in this book. During the last quarter of this century, for example, most public and media attention has been focussed on so called "high-tech" applications of generating electricity as renewable energy from the sun, waves, tides, rivers and geothermal sources. Biomass, however, is the oldest source of renewable energy. Bioenergy from biomass is generated from both virgin energy crops such as trees and organic waste through the process of *incineration*. Fuel wood from trees along with animal dung, when burned on open fires for either cooking or heating, provided our earliest ancestors with the very first form of controlled renewable energy. Biomass is still used in exactly the same way by millions of people scattered across the earth's surface. Much later in human development, ancient people used wind power as a form of propulsion for their primitive sailing craft. Both water and wind power were subsequently adopted for turning wheels linked to mechanical machinery, first by belts and then by cogs for grinding grain. Later the technology was adapted to drive more sophisticated machinery for making cotton and woollen goods in new purpose built factories which symbolised the Industrial Revolution. This climaxed with the introduction of steam power, which again was first generated by renewable sources of wood before being replaced by finite but more efficient sources of coal.

The purpose of this chapter is to consider the possibilities of providing significant if not major sources of renewable energy from energy crops, especially trees and other forms of biomass in a global context. It is envisaged that no country will be able to rely on a single source of renewable energy though it can be noted that France is already 70 per cent dependent on Atomic power for generating its electricity. Most countries will have to consider strategies that will involve reliance on several sources of renewable energy. These strategies will of necessity be tempered by both the natural resource endowments and the particular geographical location and situations with respect to climatic and hydrological regimes.

The debate about the advantages and disadvantages of bioenergy needs, therefore, to be considered with some immediate reference to other sources of renewable energy. Possibilities of providing major sources of bioenergy will be influenced and constrained by a large number of factors. These include environmental, political, economic, cultural, aesthetic, social, medical and possibly ethical considerations. At the outset it must be emphasised that bioenergy is completely different from all other sources of renewable energy. The resources for the high-tech options are *already* freely available and are renewed continuously and naturally in the form of the sun, wind, waves, tides and rivers; in a more confined context almost "unlimited" sources of uranium are available for atomic power. In all these cases the problem is one of developing and applying efficient and reliable *technology*. Bioenergy is different. In general terms, for large scale applications the technology is already available but not the *resources*. Longer term strategies will need to be employed to provide a reliable and continuous source of energy crops. These will have to be nurtured, harvested and replaced continuously by human activity. Land for growing these crops will be a central issue in this debate.

Atomic energy though not strictly a renewable source, but because of the time scale with reference to its reserves (up to 5000 years), can be grouped with the high-tech sources of renewable energy. *All* sources of energy production, whether renewable or not have received increasing opposition from a number of sources and for a number of reasons.

ENVIRONMENTAL CONCERNS

These concerns recognise that *all* sources of energy production have an impact on the environment. Fossil fuels - coal, gas and oil - currently by far the most common source of energy are now attributed in the scientific world at large, to be the major contributors to global warming. Global warming is seen by many agencies and governments to be the planet's major environmental problem which is being accelerated through the burning of fossil fuels; these emit large quantities - approximately 6 Giga tonnes per year-of carbon dioxide (CO_2) into the earth's atmosphere. Fossil fuels when burned, also emit large quantities of sulphur dioxide (SO_2) and nitrogen oxides (NO_2) which contribute especially to acidification and ozone depletion respectively.

Acidification, which was seen by many as *the* major-environment problem during the late 70s and for most of the 80s causes "forest die back". This has been particularly evident among the northern coniferous softwood forests of eastern Canada where dieback is caused by the emissions from thermal power stations on the eastern, seaboard of the USA. Similarly forest dieback has occurred in northern Germany and in the Scandinavian countries of Sweden and Finland where it is caused by the emissions of thermal power stations in France, Germany itself and in the U.K. Prevailing winds carry the sulphur dioxide from these power stations over the affected forested areas where the sulphur dioxide mixed with water droplets fall as "acid rain". Acid rain not only has disastrous effects on the forests causing defoliage but also severely affects aquatic and especially fish life in the thousands of lakes in all these areas. The combination of anthropogenic activity and regional climatic regimes emphasised for the first time, on a large scale, that environmental pollution recognises no international borders; economic activity in one country can have disastrous environmental impacts in other, and not necessarily, neighbouring counties.

Ozone depletion, that is the creation of holes in the protective ozone layer high in the earth's atmosphere, especially over the polar ice caps, has been caused in part by the burning of fossil fuels but it is widely recognised that there are also often major contributors, such as for example the CSF gases that have been widely used until recently in the manufacture of refrigeration and air conditioning units. Depletion of the ozone layer allows dangerous ultra violet rays from the sun to penetrate through to the earth's surface where they can cause a whole range of severe cancerous skin problems. Because of these additional social and medical factors, reducing the volume of coal, gas and oil burned in situ for energy production in thermal power stations for generating electricity as well as reducing the use of oil in transportation, especially, automobiles is therefore now high on the world's political agenda. Global acceptance of these environmental issues was first recognised at the Earth Summit meeting in Rio de Janeiro in 1992. Global recognition of the need for international co-operation to solve the problem was re-emphasised five years later at the Earth Summit meeting in Kyoto (1997). At Kyoto strong positive commitments were made by many western governments, especially in the European Union (EU), and particularly by the UK, to reduce emissions of the so called "greenhouse gases" which are the major contributors of global warming. One major exception to this raft of support for reducing emissions was the USA, where powerful lobby groups, especially in the automotive and steel industries, were able to constrain any pledges that could have been made by the USA government. There are, therefore, very strong environmental imperatives to reduce the world's dependence on the burning of fossil fuels. But some governments also have other economic and political interests for reducing dependency on fossil fuels and for seeking alternative sources of energy.

ECONOMIC IMPERATIVES

Fossil fuels are finite sources of energy and some analysts estimate that known reserves of coal, oil and gas, at current levels of production, are expected to last for 250, 60 and 40 years respectively. (Fells, 1996; Elliott, 1997). Possible depletion of both oil and gas in approximately the first half of this century represents a major challenge to virtually all major national economies. At the beginning of the 20th century "King Coal" provided by far the largest share (over 80 per cent) of the world's primary energy but by 1997 its contribution of the world's annual primary energy consumption (8280 million tonnes oil equivalent or mtoe) had falled to 27 percent whereas the respective contributions of oil and gas and risen to 35 and 23 percent (BP, 1997); nuclear power, HEP and energy derived from other fuels such as wood supplied the remaining 11.5 per cent of global primary energy consumption in 1997. For longer-term economic interest therefore, there are very good reasons to develop alternative sources of energy and in particular sources of renewable energy.

Large scale hydro electric power schemes representing the first wave of the high-tech options for renewable energy began to be installed earlier in the last or in the 20th century in several northern hemisphere countries such as the USA, Canada, Sweden, Norway and Finland for sound economic reasons. Suitable sites for giant schemes such as the Hoover and Boulder dams are extremely limited but whatever criteria are used the building of these dams are considered as major feats of advanced engineering. For a long time HEP was regarded as a clean, safe and environmentally friendly source of renewable energy. But since 1900 more than 100,000 people have been killed by bursting dams; in 1979, for example, the bursting of the Guyurut dam in India killed 15,000 (Fells, 1996). Moreover, millions of people have been displaced from their traditional homelands for the creation of new reservoirs on major rivers in Africa, South America and Asia. Apart from these anthropological implications many of these very large reservoirs have irrevocably disturbed important ecological systems. There has been considerable concern for both these reasons on the overall environmental impact of the Three Gorges Dam on the Yangtze in China which when completed will generate 18,000 megawatts of electricity or seven times the output of the Aswan Dam on the Nile in Egypt. For further comparison, a large thermal power station in the UK will generate 2000-2500 megawatts. The growing awareness of the extent of environmental problems associated with these very large schemes has encouraged the World Bank in recent years, to resist further funding e.g. in Malaysia, for such schemes. Currently HEP provides approximately 20 per cent of the world's electricity - about the same as that provided by nuclear power.

Nuclear power likewise was viewed by many during the 1960s and 1970s as the obvious "alternative" source of producing cheap energy on a large scale. Initial economic assessments of nuclear power, however focussed only on actual production costs and little or no attention was given to the longer term and uncertain costs involved in the decommissioning of old plants and in the safe environmental storage of spent materials. Burial at sea in concrete laden drums was seen as a cheap and efficient way of getting rid of low-grade spent materials. Also, the true economic production costs of nuclear power have been disguised in an acceptable way by many governments because of the strategic desirability of owning sources of material that can be used for the manufacture of nuclear weapons.

During the 1980s the economic attractiveness of nuclear power came under increasing scrutiny especially in terms of issues relating to the environmental health and pollution. Major accidents such as those at Three Mile Island in the USA, at Chernobyl in the Ukraine and more recently in Japan in 1999 have generated very strong concerns among both the general public and governments about the safety, health and environmental credentials of nuclear installations. In the USA no nuclear plants have been constructed since the incident at Three Mile Island; in Switzerland, public protest groups have prevented the building of new plants along the Rhine; in Sweden a previous government pledged to phase out all its nuclear power-stations by the year 2010; and in Germany, in 2000, the government has pledged to phase out its 19 nuclear power stations. Nuclear-power, therefore, at the moment at least, is no longer seen as the panacea for solving the economic challenge that will emerge in the foreseeable future through the predicted depletion of fossil fuels.

POLITICAL AND STRATEGIC IMPERATIVES

Geology and geography dictate that very large reserves of fossil fuels are concentrated in a limited number of countries. Political control and dependency on major sources of energy are highly sensitive factors. Oil is the extreme example in the contemporary world since the Middle East accounts for 65 per cent (675 thousand million barrels) of the world's 1016 thousand million barrels of proven reserves at the end of 1996. Similarly, in the case of natural gas the former Soviet Union with 57.3 billion cubic metres and the Middle East with 45.8 billion cubic metres together accounted for over 70 per cent of the world's proven gas reserves; North America has only 3.5 billion cubic metres (6.1 per cent) and South and Central America has 4.2 per cent of the world's proven reserves of natural gas. This pattern of concentration has enormous strategic and political implications. Strategic implications were illustrated during the Second World War by the German thrust eastwards into the Ukraine and across North Africa towards Suez and the Middle East, both moves being propelled by the desperate need to secure supplies of oil. Economic implications were clearly illustrated by the first oil crisis of 1974, reflecting the political instability in the Middle East and the ability of a small number of large producers to strangle supplies, and to literally hold much of the world to ransom; more recently the Gulf War was fought to protect the world's major sources and arteries of oil. The oil crisis of 1974 provided the first, and economically motivated, impetus for a large number of countries, especially in the West, to consider developing 'alternative energies'. Apart from the purely economic implications, the oil wars have demonstrated that independent sources of renewable energy are *politically desirable* for self-sufficiency in times of conflict, (Collins, 1999).

Diversified sources of energy can be another important political factor within the framework of national boundaries. This was illustrated very clearly in several European countries in the post-war period when there was an increasing and deliberate relaxation of dependency for both economic and political reasons, on coal as the dominant source of energy especially in terms of generating electricity. This was especially true in the UK, where during the immediate post war period several governments were held to ransom by the might of one union– the National Union of Mineworkers (NUM). Threats and actual strikes by the NUM distorted pay bargaining because once higher rates had been secured by the miners, other industrial sectors followed suit with resulting inflationary trends. This climaxed in the 1970s with the "Winter of Discontent". Subsequently, increasing generating capacity of electricity by nuclear-power, oil and gas fired power stations reduced and undermined the power of the NUM. This was accelerated by the government's policy to allow imports of large quantities of coal from as far as Australia and China; large quantities of such coal was also stockpiled in coastal areas as a buffer to any domestic strikes of coal production. This new dependency on imported coal further accelerated the closure of hundreds of collieries throughout the UK. Before the closure of these collieries the UK had very substantial *accessible* reserves of coal but once coal pits are closed and flooded these reserves become totally inaccessible; they can never be recovered. Similar activity has occurred in other coal rich European countries, for example Germany.

For these environmental, long term economic, political and strategic reasons there is a growing desire, if not urgency, among many countries to encourage the development of alternative renewable sources of energy. It is a moot point to ask which of these reasons is really the most important. Currently, political forces, in the form of governments wishing to be identified as "green" by their voters are undoubtedly the major stimulus whereas economic factors reflected in the actual costs of producing alternative energy from renewable resources represent the major constraint to widespread innovation and implementation.

GOVERNMENT STIMULI OF RENEWABLE ENERGY

(i) High-tech Solutions

The oil crisis of 1974 was undoubtedly an economic and strategic factor that stimulated both the USA and the UK to establish a wide range of government funded research and development (R&D) projects for renewable energy (Elliot, 1997). In the UK the immediate response was the establishment of a Research and Development (R&D) Program (1974) under the guidance of the Energy

Technology Support Unit (ETSU) of the Department of Trade and Industry (DTI). Initial emphasis was given to the large scale harnessing of wave power since the UK, relative to its size, has some of the world's best deep sea wave and estuarine tidal energy resources. Some of the "giant" schemes envisaged, such as the proposal to build a tidal barrage across the Severn Estuary were too complex and too costly for the estimated returns from the very high levels of capital investment which would be required. There was also very strong opposition from the rapidly developing Environmental Pressure Groups (EPGs) about the effects such a giant scheme would have on both the visible environment and on estuary wildlife which is highly sensitive to variations in natural regimes.

Wavepower on a large scale was abandoned or at least set aside in the UK owing to the controversial cost estimates and attention was shifted instead to wind power (Elliot, 1994). Here again the geographical location of the UK provides the country with one of the most reliable sources of this form of energy compared with other countries of similar size. Although a number of wind farms have been established in the UK especially on the hilly/mountainous areas of Wales and Scotland the cost of providing electricity from wind turbines still remains uncompetitive vis a vis fossil fuels. The same economic disadvantage applied to solar energy. The annual solar energy falling on Europe is several hundred times Europe's annual commercial energy consumption, but photovoltaic energy is 5-6 times as expensive as conventional power generation (Grubb and Viogotti, 1997). Green technologies, therefore, have had to be given a secure financial framework within which they can achieve close to commercial viability in order to penetrate the market.

(ii) Biomass

The USA government was the first to take a step in this direction with the Public Utilities Regulatory Act (1978). This required electricity generating companies to purchase electricity from waste-to-energy plants at the price the plant would incur in producing the power. Waste-to-power energy plants were selected for special treatments in the USA not so much to provide an "alternative source of energy" per se but rather to solve another "green" or environmental issue. Rapid exhaustion of economically accessible land-fill sites in many parts of the USA meant that garbage had to be incinerated and efforts were made to do this economically. This issue is considered in more detail in the next section.

Ten years later, the British government introduced the Non Fossil Fuel Obligation (NFFO) as part of the Electricity Act (1989) which privatised the industry. The Electricity Act enabled the Secretary of State to require each public electricity supplier to secure a specific amount of electricity from non-fossil, including nuclear sources of energy (DTI, 1997). Indeed the whole emphasis of the NFFO, in the first instance, was to put nuclear power on the same economic platform as energy derived from commercial sources; nuclear power was seen as the obvious "alternative energy".

The objective of both schemes, in the USA and the UK, was to provide an "incubating mechanism" whereby the introduction of new technologies is given a contracted period of time, via subsidies, to enter the generating market on a commercial basis. In the UK the initial contract period was 8 years but has since been extended to 15 years. The NFFO has been implemented in a series of rounds or 'Orders' and the fifth, 'NFFO-5', was announced in November 1997. The Ministry of Science, Energy and Industry of the new Labour Government, which is pledged to strong environmental policies has embarked on a review of:

'...what would be necessary and practicable to achieve 10 per cent of the UK's electricity needs from renewables by the year 2010. NFFO-5 will therefore become the first step in our new and strong choice for renewables and it is expected that NFFO-5 will contribute substantially to reducing greenhouse gases and encouraging internationally competitive industries' (Department of Trade and Industry, 1997, p.2). The ministry acknowledges that, on the basis of work undertaken so far, in order to achieve its goal of 10 per cent by 2010 from renewables it will: '...require bringing forward technologies in addition to those proposed to be supported under NFFO-5, including offshore wind energy *and energy crops*' (DTI, 1997, p.2).

BIOENERGY FROM BIOMASS

Renewable energy from energy crops per se is only one part of a broader strategy being considered and implemented by several governments especially in the USA and Europe. Bioenergy from biomass describes this comprehensive strategy and is attractive for two important reasons - technological and environmental. Technologically, bioenergy is attractive because it is more versatile than the high-tech options which in general are concerned with generating only electricity. Environmentally, bioenergy is attractive for two separate reasons. First, it helps to solve the increasing problem of the diminishing number of landfill sites which represent a source of recycled biomass. Second, its ability to contribute directly to ameliorating the problem of global warming, by encouraging, in an economic framework, a programme of afforestation for increasing the global forest resource.

TECHNOLOGICAL ATTRACTION

Bioenergy can be derived in three main ways. Two of these involve the process of incineration, one from recycled biomass and the other from virgin energy crops.

(i) Recycled Biomass

Gas, mostly methane, is generated by the decomposition of organic material in Municipal Waste Landfills and Sewage Treatment Plants and can be extracted through a network of pipes inserted into the landfill as it is being dumped.

Alternatively, Municipal Solid Waste (MSW) can be burned in purpose-built incinerators, commonly referred to as waste-to-energy plants. These plants can be used to generate steam for local domestic heating, which tends to be the normal practice in European and British systems, or, if the MSW is burned in large installations at higher temperatures, to generate electricity that can be fed into the grid, as tends to happen in the USA. The impetus for the widespread proliferation of waste-to-energy plants, especially in North America and many European countries such as Germany and France, has come rather more from the rapidly declining number of available landfill sites than from pure environmental reasoning. In Germany, for example, the number of landfill sites in Bavaria declined from over 4000 in the early 1980s to less than 40 in the early 1990s. Similarly, in the USA approximately 3000 municipal landfills were closed between 1983 and 1988 and it is estimated that half of the remaining 8000 landfill sites in 1990 will have come to the end of their life by the year 2000; in New York State alone the number of landfill sites declined from 294 in 1986 to 76 in 1996 and it is estimated that only 13 will be left operational by the year 2006 (Collins, 1997). Fresh Kills, the world's largest landfill site in near New York is rapidly coming to the end of its life when it will then be over 500 ft above sea level making it one of the highest points on the East Coast south of Maine. Incineration for the generation of bioenergy, therefore, is almost the only option. Denmark and Switzerland already incinerate approximately 80 per cent of their MSW for energy recovery.

Before considering similar technological advantages of producing various types of power from energy crops it needs to be emphasized that some EPGs are strongly opposed to the principle of incineration for what are held to be sound environmental reasons. There is concern for example, that incinerators emit toxins into the atmosphere. Moreover, there is frequently strong public opposition to the building of incinerators close to or in major urban areas which are the main sources of MSW. Many incinerators, therefore, have been built in the face of fierce local opposition co-ordinated by protest groups which tend to adopt names with emotive acronyms, such as in the USA - Mothers Opposed to Mass Incineration (MOMI); Stop Incineration Now (SIN); and Citizens Reacting Against Pollution (XXXX) (Collins, 1997). In the UK one such group is called RATS (Gray, 1995). This kind of opposition is commonly referred to as NIMBYism - or "not in my back yard".

(ii) Energy Crops

Energy crops, like fossil fuels, can provide a continuous and steady supply of electricity. Unlike some other renewables, they do not suffer 'intermittency' due, for example, to darkness, cloud and calm weather. Energy crops also possess greater technological flexibility and adaptability in the production of different forms of energy (Patterson, 1994). Woody energy crops can produce electricity from direct burning in a designated power plant or they can be co-fired with coal in conventional thermal power stations, leading to reductions in the sulphur content of emissions. Woody energy crops can also be used, through the process of either gasification or pyrolysis of the biomass, for the production of liquid fuel for firing gas turbines or diesel engines. The other renewables, on the other hand, are used exclusively for the production of electricity. Finally, non-woody energy crops such as sugar cane can be refined to produce ethanol as a fuel for cars, which has happened in countries such as Brazil (Grubb, 1995) and is used also in the USA, France, Germany and Austria (Sims, 1999).

ENVIRONMENTAL ATTRACTION

Two main environmental attractions of bioenergy can be recognized. First, bioenergy derived from recycled biomass is helping to solve the landfill problem in many countries. This problem has in turn two facets. For a long time concern has been expressed about leachates seeping out from landfills following the decomposition of organic material which can contaminate both surface water courses as well as underground aquifers which are important sources of drinking water. In recent years however, the landfill problem has been more concerned with the rapidly diminishing number of economically accessible sites. In the USA, especially along the eastern seaboard where there has been strong opposition by protest groups to the building of incinerators, the increasing shortage of landfill space has spawned a new billion dollar private industry, largely unregulated by government, concerned with the transport of garbage to rural landfills (Collins, 1997). Many of these are located in the mid west, hundreds of miles from the eastern cities. Over 10 million tonnes of garbage are being moved across state boundaries each year. Most of this can be described as "back haulage" - trucks which would otherwise be empty on their return journeys to mid-western locations after delivering agricultural produce and other foodstuffs to the eastern seaboard for local consumption or export. Refrigerated vehicles have sometimes been used for this purpose. After tipping the garbage in mid-western landfills in Michigan, Ohio and Kentucky, at points up to 1300 km from source (though most sites are between 500-1000 km) the trucks are steam-cleaned before collecting more produce for delivery to the east.

The second environmental reason relates to the need for significantly increasing the earth's forest cover in an economic framework in order to mitigate the effects of global warming. Several major studies have confirmed the important role that global forests play in maintaining the finely balanced equilibrium in the earth's carbon cycle (Fig. 1.). Global warming occurs as a result of major disturbances of this cycle which over the last 200 years has been affected from two different anthropogenic activities namely the burning of fossil fuels and global deforestation.

(i) Fossil Fuels and the Carbon Cycle

Fossil fuels when burned emit carbon dioxide and other gases into the earth's atmosphere which produce a "greenhouse effect". Greenhouse gasses occur naturally in the atmosphere where they have the unique property of being transparent to incoming hot short wave-length solar radiation but of being opaque to outgoing cooler long wave-length radiation from the earth's surface. Unimpeded, the latter process would cool the earth's surface to about - 18°C but the 'one way trapdoor' provided by the greenhouse gases radiates much of the solar energy back to the earth's surface. This delicate balance between incoming and outgoing radiated energy maintained by natural greenhouse gases (mainly CO_2 and water) is, therefore, extremely important for maintaining global temperatures.

Unfortunately the large-scale burning of fossil fuels over the last 200 years has led to a greater concentration of various greenhouse gases in the atmosphere, thus making them more opaque so that more of the earth's radiation is trapped, leading to global warming. Carbon dioxide is currently increasing at 0.5% per annum in the atmosphere; again there are a range of estimates concerning the rate of global warming, and since the Industrial Revolution global mean temperatures have risen by between 0.3 and 0.6°C (DOE, 1994). Since 1965, however, the overall upward trend in atmospheric temperature has accelerated, and 10 of the 11 warmest years of this century have occurred since 1980. These are, in rank order: 1997 (0.43), 1995 (0.38), 1990 (0.35), 1991 (0.29), 1994 (0.26), 1988 (0.24), 1987 and 1983 (0.23), 1996 (0.22), and 1981 (0.20). Numbers in parentheses indicate temperatures in degrees centigrade above the average for 1961-1990. Prior to 1981, the hottest year this century was 1944 (0.21) (data from The UK Meterological Office).

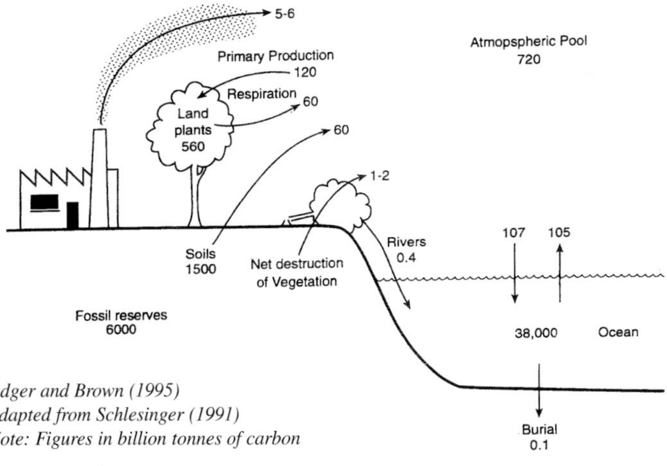

Source: Adger and Brown (1995)
Adapted from Schlesinger (1991)
Note: Figures in billion tonnes of carbon

FIGURE 1. The Global Carbon Cycle

(ii) Deforestation

Global deforestation is proceeding at an alarming rate. The vast plumes of smog sweeping across Singapore and Malaysia from Indonesia, several hundreds of miles away, where tropical rainforests are being burned and cleared for agricultural land is a visible and recent testament to this ongoing process. Although the mass burning of trees to clear the land for agriculture and ranching is the main contributor to deforestation, especially in the Tropics, some of the deforestation is attributable to the felling and logging of trees for specific purposes. Of these trees, approximately 54 per cent is used for fuel on use on open fires, 33 per cent as sawn wood for construction and furniture, and 13 per cent for pulp and paper. (Jaako Pöyry, 1996). There is some evidence to suggest however, that forest cover in temperate latitudes has actually increased as a result of major replanting schemes specially by pulp and paper companies (Kauppi, 1992; Sedjo, 1992; and Mather, 1990). Estimates concerning the current rate of global deforestation do vary but it is generally thought that since human intervention the global forest cover has been reduced by approximately a third from 6 billion to 4 billion ha (Furley, 1994); more specifically it is estimated that Latin America and Asia have lost two fifths of their former forest cover and Africa approximately one half (Furley, 1994). Most deforestation is occurring in tropical areas, and the Food and Agricultural Organization (FAO) of the United Nations (UN) estimated that in the early 1990s approximately 17 Mha of tropical forests were being lost annually, which is an area equivalent to more than half the size of Belgium (Lanly, 1991). Myers (1989) estimated that if the annual loss of approximately 1.8 per cent which occurred throughout the 1980s were to continue then the remaining 8 million km^2 of tropical rainforest would have a terminal lifespan of approximately 38 years; from time of writing that would give an end year of 2030 for tropical rainforests. Such a scenario would have obvious undesirable and irrevocable implications for anthropological, economic and biological reasons: such a scenario would also have major implications for global climatic change.

GLOBAL FORESTS AND CLIMATIC CHANGE

Over the last 10-15 years several scientific studies have confirmed the extremely important role that forests play as major "carbon sinks" in the global carbon cycle which influences climatic change (Nilsson and Schopfhauser, 1995; Houghton, 1990; Marland and Marland, 1992; and Nakicenovic, 1993). This concept has also been firmly accepted by the Intergovernmental Panel on Climate Change (IPCC) in formulating its strategties for the mitigation of global warming (Houghton *et al* 1992). Carbon is absorbed by trees as they grow to maturity over a period of 60 years and then absorption ceases (Hall and House, 1995). Most carbon absorption, however, occurs during the first 25-30 years when, on average one tree will absorb 1 tonne of carbon dioxide (Whyte, 1995). Subsequently, the level of absorption declines appreciably but the carbon remains stored in the tree and is referred to as "carbon sequestration". This implies that the oft-quoted phrase "tropical rainforests are the lungs of the earth" is not entirely accurate (Kimmins, 19922).

The IPCC recognizes that, in order to mitigate the effects of global warming, deforestation not only must be stopped but there will have to be large programs of reforestation to increase the global forest cover in order to increase the absorption capacity of the carbon sink. Such reforestation programs would not necessarily be confined to tropical areas since independent studies have demonstrated the much greater than expected effectiveness of temperate forests as carbon sinks (Tans et al, 1990; Kauppi *et al*, 1992; Sedjo, 1992). In New Zealand, for example, calculations have shown that fast-growing Radiata pine which are grown for lumber and the pulp and paper industries, were absorbing 70 per cent of New Zealand's fossil fuel emissions in the 1988-89 period (Holinger, 1993, Simms, 1999).

The IPCC estimated that the net increase in atmospheric CO_2 is approximately 3.8 GI-C/yr. To counter-balance this, several studies have suggested that at varying levels of growth rate it would require between 465 and 600 Mha of land to be reforested (Sedjo and Saloman, 1989; Kimmins, 1992). Estimates of the rates of reforestation that could be achieved vary quite significantly; Grainger (1988) for example suggested 3-5 million ha/yr would be possible but Trexler, (1991) is more cautious and envisaged that up to the year 2050 a realistic achievable rate would be only 0.7 million ha/yr in tropical regions. Variations of these estimates reflect the different data and the nature of the assumptions. Some assessments are based only on the physical availability of land and ignore ecological, social, economic and cultural barriers; some studies include land that is suitable whether or not it is available (Hall and House, 1994).

The underlying thesis of these studies is three fold. First, in order to mitigate the effects of global warming it is essential to increase the surface area of the global forest cover. Any large scale reforestation program, however, is expensive and only where an economic return is possible would such schemes be viable. Second, in order to achieve the desired levels of carbon sequestration, the new forests would have to be managed on a short rotation basis so that the trees are cut down when they reach maturity and before they cease to absorb carbon; the time periods will vary according to species. Third, once felled, the wood will have to be used for some economic purpose to pay for the reforestation programs. Economic return can be obtained in two ways. First, the wood can be used to make some long-lasting product such as buildings or furniture which continue to store carbon without releasing it into the atmosphere; there are of course extreme limits to this option.

Alternatively the wood can be incinerated on a commercial basis to produce energy in order to reduce dependence on burning fossil fuels which will reduce emission of greenhouse gases into the atmosphere. Energy derived from renewable sources of wood is potentially carbon-neutral since the carbon dioxide given off through incineration is more than absorbed by the new wood that it replaces (Van den Broek, *et al* 1996). The carbon absorption cycle, therefore, is sustainable. This concept was endorsed by scientists at the First International Conference on Carbon Dioxide who concluded that it was more efficient than to simply sequestrate CO_2 in forests (Bekkering, 1992).

GLOBAL CONSUMPTION OF BIOENERGY

In 1996, the world's primary energy consumption derived from commercially traded fuels, which excludes biomass, was approximately 350 million Giga Joules (GJ) or 350 Exa Joules (EJ), which is equivalent to 8380 million tonnes of oil (8380mtoe) (BP, 1997). This produced a 'final consumption' of almost 5866mtoe (70 per cent of total consumption); in the production and delivery of electricity, for example, over 65 per cent of the energy input is used. Currently, it is estimated that almost 15 per cent (50EJ) of the world's primary energy is obtained from biomass fuel which includes wood, plants and MSW; this means that biomass is already the world's fourth largest source of energy after oil, coal and gas (Hall and House, 1994; Schlamadinger and Marland, 1996). Bioenergy provides approximately 38 per cent of all primary energy in developing countries. These countries use 43EJ or 86 per cent of the total; industrialized countries use 7EJ which amounts to 3 per cent of the primary energy consumption. Most of the wood that is used in developing counties is burned inefficiently on open fires for cooking, with an energy yield of approximately 5-15 per cent (Hall and House, 1994). This degree of inefficiency has been generally assumed to be one of the prime causes of deforestation in Africa leading to the 'woodfuel crisis'. China derives over 25 per cent of its primary energy from biomass but this is less than in Brazil (31 per cent) and India (51 per cent); several African countries such as Uganda, Tanzania and Mozambique are over 95 per cent dependent on bioenergy (Kgathi and Zhou, 1995). In Western Europe, biomass accounted for 4 per cent (2EJ) of its total energy consumption in 1995 (57.8 EJ). The leading bioenergy countries in Europe are France (22% of total), Sweden (15%), and Finland (11%) (Bauen and Kaltschmitt, 1999).

POTENTIAL SCENARIOS OF BIOENERGY CONSUMPTION

Future scenarios of global bioenergy consumption will be fashioned by a wide range of variables including land availability, environmental considerations and biotechnological advances and by economic, political and social constraints.

(i) Land availability

Land availability is critical for increasing the supply of bioenergy. If the ultimate objective was the totally unrealistic scenario of producing all the world's primary consumption from trees at the current average yiels of 12t/ha/yr then approximately 10.4 per cent of the earth's land surface (13.05 million km2), or an area approximately one-and-a-half times as big as Brazil, would be required (Hall and House, 1994). To be self-sufficient, Europe would require 58 per cent of its landmass to be forested and North America just over 20 per cent; the developing countries, at present levels of consumption, would require less than 5 per cent. Land availability is more critical in developed countries, especially in Europe and to a lesser extent in North America. In both areas, however, the Green Revolution in agriculture has significantly transformed the potential for releasing land to forestry in recent years. In Europe, the introduction of genetically engineered high yielding crops and new methods of farming applied to larger scale farms resulting from a process of consolidation mean that more food can be produced on less land. To avoid the accumulation of food surpluses the Common Agricultural Policy (CAP) introduced the 'set aside' scheme whereby farmers are subsidised to use marginal land taken out of agricultural production for other purposes, such as forestry. Farmers will have to be convinced, however, that the strategy of planting energy crops, especially trees, will be economically viable in the long term (Patterson, 1994). Three independent studies (NSGP, 1992; Grassi and Bridgewater, 1993; Scurlock *et al.*, 1993) have suggested that, in the old EU12, between 30 and 50 million ha of the then existing 127 million ha of agricultural land could be 'set aside' for the production of energy crops; since its entry into the EU, Sweden plans to take 1 million ha of arable land out of production for new forest plantations.

Similar changing circumstances prevail in the USA, but there more 'agricultural' land is already available because of the different methods of subsidising farmers. In the USA almost the same amount of money has been used as in Europe to pay farmers to keep much of the available land out of production; in Europe, before the set-aside scheme, the same amount of money was used to pay farmers to produce more food than was required, hence the accumulation of the notorious food surpluses. In the USA, of the 386 million ha classified as farmland only 120-135 million ha are planted out each year; this implies that approximately 125-140 million ha are already available for energy crops (Turnbull, 1996). Moreover, as in Europe, the US Department of Agriculture has predicted that over the next 40 years it will be possible to release a further 40 million ha while producing the same amount of food through the introduction of genetically engineered high yielding crops.

(ii) Biotechnology and Environmental Constraints
(a) Biotechnology

Within the overall framework of Kondratieff's model of economic cycles it has been suggested that the fifth and next cycle extending to the year 2050 will be stimulated by biotechnology (Fig. 2) (Collins, 1994). Recent advances in biotechnology especially in terms of the "cloning of dolly" and possible future cloning of human cells, together with the rapid advances in the production of genetically modified organisms (GMOs) for farm food crops has thrown these issues to the forefront of debate by both the general public and by medical and government organizations. Environmental, medical and ethical considerations are being hotly debated in this framework.

Biotechnology has been applied for some time by the pulp and paper industry, both for accelerating and hence cheapening the pulping process of the delivered pulpwood with treatment by enzymes (Ratto *et al* 1993; Grant, 1993) and for increasing the yield of trees grown in plantations specifically for pulpwood. The same technology can be applied to trees grown as energy crops.

Two main methods prevail in the growing of trees as energy crops. In North America, in developing countries and in some European countries *single stem* trees are favoured. In North America, for example, there is already a close association between forest products industries, e.g. pulp and paper which use residual material from pulpwood. In developing countries in tropical areas single stem trees can achieve very high growth rates. In Brazil, for example, cloned eucalyptus pulpwood trees produce 20-40m^3/ha/yr compared with much less than half of that for most species grown in the northern temperate latitudes (McNabb, 1994) and ongoing experiments suggest that new genetically engineered species of eucalyptus may achieve growth rates of 80m^2/ha/yr.

In the UK and some of the other European countries the favoured method is coppicing. Coppicing is an ancient system of woodland management that was used widely to produce firewood as well as wood for specialized tasks such as fencing and basket making (ETSU, 1994). Coppicing involves repeatedly cutting back to just above the rootstock of each tree for continuous harvesting. The process encourages the growth of *multiple stems* which can be easily machine cut

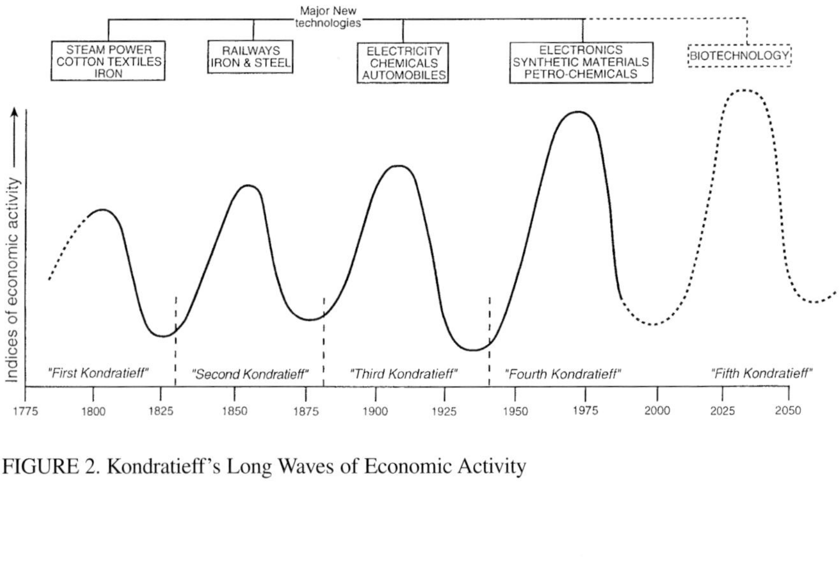

FIGURE 2. Kondratieff's Long Waves of Economic Activity

and harvested. Experiments in the UK with willow and poplar indicate that yields of 10-15 dry t/ha/yr are achievable; this would be equivalent to an annual production of 6 tonnes of oil (Department of Trande and Industry, 1994). Willow is the most favoured species because it is indigenous to most parts of Europe, is very easy to propagate and lends itself to genetic engineering for the development of fast-growing crops (Ledin, 1996). Short Rotation Coppicing (SRC) allows a harvesting cycle of approximately 2-3 years after the first initial cycle of 3-4 years. Approximately 10% (600,000 ha) of arable land in the UK is now available for alternative uses through the CAP set aside scheme. It is estimated that 5.5 million ha of land could be surplus to agricultural production by the year 2010. Chesworth and McDonald (1999) suggest that if this land were planted with SRC it would have the theoretical potential to meet over half the energy (electricity) needs for the UK.

The levels of yield for energy crops achieved through biotechnology will have significant results for national strategies especially in the USA and Europe where the land availability issue is perhaps more critical than in tropical areas. In its program for renewable energy, the US Department of the Environment has set a target of achieving yields of 22 dry t/ha/yr; at the moment most yields from larger stands are approximately 11-13 dry t/ha/yr; (Easterly and Burnham, 1996). Again, as in Europe, estimates of the potential energy that can be obtained in the US vary, but Trexler (1991), suggests that by 2020 biofuels could replace up to 25% of the US 1991 energy consumption. On the other hand, Wright *et al* (1992), point to studies at the Oak Ridge National Laboratories which estimate short rotation wood crops grown on 103 Mha could off set 80% of the total CO_2 released from US electricity production (Hall and House, 1994). Conservative estimates by the US Department of Environment (Easterly and Burnham, 1996) indicate that approximately 22,000 MW of bioenergy will be available and this will double by 2030. In its National Energy Strategy, the US Department of Energy estimates that energy crops could provide 17 EJ of energy to its future energy supplies as well as 5.3 EJ from agriculture and forest residues and 2.1 EJ from municipal solid waste. Altogether, this total of 24.7 EJ obtained from biomass would equal 28% of the US energy consumption of 1991 (Easterly and Burnham, 1996). If the US does proceed with developing its full bioenergy potential, it has been argued that three main types of agricultural land should be targeted in the first instance: highly erodable land; wetlands presently converted to agricultural uses; and marginal agricultural land in selected regions (Paine, 1996).

In Europe, on the other hand the spectre of large areas of the countryside being covered by plantations of rapidly growing cloned trees will generate significant environmental concerns and strong opposition.

(b) Environmental Constraints

Potential negative reactions will come from a general public that has been educated, especially in the UK, to believe that plantations are bad for the environment, both in terms of reducing diversity, which is not true in those cases where plantations use mixed species, and in terms of spoiling natural landscape aesthetics (Bell, 1994). Definitions of both natural landscapes and of what is aesthetically pleasing involve a high degree of subjectivity. The ancient or natural landscape of Scotland for example, was largely a pine forest (the 'Caledonian Forest'), not sheep-grazed moorland which today is regarded by many as the natural landscape. Is moorland more aesthetically pleasing than forest clad slopes and, if so, to whom? The moorlands of Scotland attract thousands of tourists each year but so does the Black Forest area of Southern Germany.

Other environmental concerns relate to the possibility of the rapid spread of pests and diseases through monoculture plantation (Evans, 1992). Measured against this, however, numerous studies have demonstrated that forestry projects on degraded or abandoned cropland can have numerous positive effects; improved soil structure; reduced run off and increased soil water storage; reduced flooding downstreams; increased local rainfall; modified local temperatures; increased biodiversity; creation of wind breaks, and reduced pressure on natural forests in terms of recreational activities. Also, the residual ash derived from the combusion of biomass has good agricultural value, unlike that remaining in coal-fuelled stations which on average is five to ten times greater in volume. Biomass-derived ash contains plant-growing nutrients such as potassium, calcium iron and silicon at high pH which can be used either as an agricultural or forestry fertilizer (Turnbull, 1996).

Transportation is another major environmental and technological issue of growing and using energy crops. (Collins, 1998a) both oil and gas are highly mobile through pipelines which can be buried under-ground, and coal is found in highly concentrated and localized reserves. Wood has a much lower energy content than coal and a much lower energy per unit of weight or cubic volume of fuel. Bituminous coal has approximately 30 kJ/kg compared with approximately 20 kJ/kg for wood on a dry basis. On a wet (green) basis, biomass has a moisture content of 40-50%, which reduces the energy value of green hardwoods to 10kJ/kg. In terms of volume, woods are generally in the range of 4.3-6.2 m^3 Mg (140-200 ft m3/tonne) whereas coal is in the range of 1.1-1.5m^3 Mg (35-50^3). Volume is, therefore, the main limiting factor in the transportation of biomass since most modes are capable of carrying greater weights than the volume allows (Angus-Hankin *et al*, 1995). For example, if the UK were to replace all its energy needs currently met by oil, gas and coal with imported wood carried by train through the Channel Tunnel, it would require a daily train 650 km long stretching from the Tunnel to the Lake District moving at 40 mph. In general terms, wood fuel can be transported to a power plant up to a distance of 120 km or 75 miles, within which transport will account for 20-40% of the total delivered fuel cost. Clearly a sensitive approach to the actual transportation infrastructure is required. Ongoing studies suggest that the problem of transporting woody biomass can be reduced by processing the fuelwood which can then be compacted for transport (Angus-Hankin *et al*, 1995) similarly, other ongoing trials are focusing on drying and storing comminuted wood for this purpose (Jirgis, 1995).

Current technology for burning wood favours small-scale power plants of approximately 50 MW capacity compared with the large 1000-2000 MW capacity of many coal-fired stations. Depending on the crop yields, which range from 12-15 dt/ha, and on the efficiency of the plant, then within an economically acceptable transportation radius of approximately 80 km, between 2 and 5% of the total land surface would be required for biomass crops, which could be grown in 100-200 parcels of land measuring approximately 1-2 km^2. Energy crop landscapes therefore, will be characterized by a dispersed pattern of a large number of relatively small generating stations. This will be in sharp contrast to the existing energy landscape which is typified by a concentrated pattern of few but large generators with capacities of approximately 2000 MW. Large numbers of such small plants, however, could generate an attitude of 'NIMBYism' in many rural localities. Weighed against this however, it needs to be recognized that 'energy crops' will be relatively labour intensive due to the need for planting, management, harvesting and transportation on a continuous basis. Currently, in the US, over 60,000 jobs can be attributed to the production of bioenergy and by 2010 it is estimated that this could increase to 250,000. The most important factor in this respect is that the jobs are being created in many declining rural areas.

(iii) Economic and Socio-Political Constraints
(a) Economic Constraints

Critics of the scenarios described above have maintained that bioenergy is uncompetitive with energy derived from fossil fuels and can survive only on the basis of subsidies already given indirectly to users of fossil fuels in terms of the very high *external costs* which are borne *not* by the producer but by the public at large in terms of health hazards resulting from pollution and global warming (Martin *et al*., 1992; Burniaux et al., 1995; Dean and Hoeller. 1992; Manne and Richels, 1992). The hidden subsidy argument is even stronger when directed at nuclear energy. These external costs or 'externalities', it is argued, should be allocated on the now familiar 'polluter pays' principle. The application of a carbon tax, for example, would completely re-balance the economics of energy production in favour of renewable sources, especially woody energy crops. Switzerland is seriously considering a carbon tax and in the USA the 1990 Clean Air Act Amendment (CAAA) has already been instrumental in encouraging many electricity generators in the Tennessee Valley Authority to reduce CO_2 and SO_2 emissions by developing co-firing wood-derived fuel and coal boilers (Gold and Tillman, 1996). Some advocates of the carbon tax see it as a means of transferring substantial financial support from industrialized countries, which are the main polluters, to developing countries where the money could be used for reforestation programmes to ameliorate global warming. At the end of 1999, for example, polluting industries in the USA have agreed to spend $5 billion on forestry programs in South America as a means of obtaining "green credits". Inevitably there will be strong and extended arguments against carbon taxes from powerful lobbies with vested interests in fossil fuels such as the automobile industry in the USA which was very vocal at the time of the Kyoto Summit. The actual economic framework for increasing the supply of bioenergy from biomass can, therefore, be recalibrated by political initiatives which will largely reflect social and medical attitudes to the continued burning of fossil fuels.

(b) Socio-political Constraints

Social attitudes and beliefs on behalf of the general public can have a powerful influence on the nature and direction of government policies. Strong arguments have been elaborated to show that there are now many major inconsistencies if not areas of conflict in the general environmental policies of these countries (Collins, 1996). This is particularly true for those policies concerned with the generation of energy from recycled biomass in many western countries. The general public have been educated to "believe" that recycling per se is good for the environment; Ackerman (1997) has described recycling in the USA as a "religion" since more people believe in the validity of recycling than go to church. In the UK, the Friends of the Earth concluded a poll which showed that 94% of people believe that recycling is good for the environment.

Recycling strategies have been built on a popular belief, started in the 1970s, which embraces the view that all recycling is good for the environment. Some measure of the strength of this belief has been shown in a major study carried out by the Centre for Social and Economic Research on the Global Environment (CSERGE). Their survey found that, on average, recyclers who made special journeys by car travelled a 3.66 km round trip from home to the recycling site, delivering on average 14 kg of recyclables. This suggests petrol-derived energy is needed for 261 km/tonnes. In addition, a further 161 km/tonne is required to transport the materials from the recycling sites to the storage depot; a total of 422 km/tonne (Powell and Craighill, 1996). While the belief that recycling is good for the environment may be true for products made from finite resources it is not necessarily true for paper products which are different for two important reasons. First, paper products are made from a renewable resource, and second, they have an alternative use as an agent for recovering energy. Recycling strategies, especially those related to wastepaper, have undoubtedly engendered a 'feel good' factor in society at large and initially were propelled forward by well-meaning slogans such as 'Save a Tree'.

Most European countries, under the guidance of the European Commission (EC), and the US have adopted strategies to promote higher rates of recycling activity, especially for waste paper; in the UK, for example, paper companies have received major-grants from the EC to subsidize the building of very large recycling installations such as the one at Aylesbury, in Kent approx 50 km from

London. In the US, MSW is typically composed of 40-50 per cent paper products, 30- per cent other organics and the rest is glass (9%), plastics (6%), ferrous metals (5%) and other inorganics including aluminum (4%). Strategies for recycling wastepaper are designed to divert as much of the paper content as possible from the waste stream. Unfortunately, apart from plastics, wastepaper is the most combustible component of MSW and has the highest calorific value for deriving energy. In order to burn out the pollutants from organic MSW, high temperatures are required at some time during the operation. The lower the paper content of the MSW the greater reliance there has to be on heat derived from fossil fuels to achieve the required temperatures. The higher the paper content the more energy is recovered from the incineration process.

The success of these recycling campaigns can be measured by the global consumption of wastepaper compared with the global production of paper and board products. Between 1970 and 1996 global paper and board consumption increased by more than 150 per cent (Rennel and McNutt, 1996) and between 1980 and 1996 the global average per capita consumption of paper and board products increased by over 37 per cent from 40 kg to 55 kg (Collins, 1998b). Earlier forecases (made in 1990) of a global consumption of 340 million tonnes by the year 2010 have been revised to reach 420 million tonnes by that year (McNutt and Rennel, 1996); by 1998 global consumption had exceeded 300 million tonnes. The global figures disguise huge national and regional variations since in the USA average annual per capita consumption rose from 272 kg (1980) to 336 kg in 1998; in the EU from 117 kg to 190 kg; in Japan from 153 kg to 237 kg and in China from 6 kg to 26 kg.

Alongside these trends between 1970 and 1980, global consumption of recycled wastepaper rose relatively slowly from 30 million tonnes (23 per cent of paper making fibre) to 41 million tonnes (24%). In the next 15 years, however, it increased by 250% to 104 million tonnes, or to 38% of global paper making (Collins, 1998b). Used paper is now a raw material that worldwide accounts for 2.5% of all industrial output (Pearce, 1997). Current forecasts indicate that recovered paper consumption is expected to increase to approximately 200 million tonnes by 2010 which would then account for almost 50% of the total making fibre (McNutt and Rennel, 1997).

Wastepaper recovery schemes in some countries, such as the US and Germany, appear to have been 'too successful' in that far more paper has been recovered than can be utilized by the respective paper-making industries. These surpluses have encouraged a significant new international trade in wastepaper which amounted to over 18.3 million tonnes in 1996 (Collins, 1997). The US, with the biggest surplus of approximately 6.5 million tonnes, exports most of it to Southeast Asia and Canada (2.1 million tonnes). Germany exported almost 3 million tonnes in 1996 and Sweden imported over 500,000 tonnes. Both Canada and Sweden are importing wastepaper to make newsprint with a 40% wastepaper content in order to meet the legal requirements of their respective markets, the US and Germany. This international trade in wastepaper requires significant amounts of energy derived from burning fossil fuels which contributes still further to global warming. Several independent studies have confirmed that, as far as global warming is concerned, the recycling of wastepaper is more damaging than its incineration (BNMA, 1995; Hamm and Gotsching, 1993; Virtanen and Nilsson, 1993; Leach *et al*, 1997; Pearce, 1997). The results of these studies depend on the assumptions on which they are based. Hekkert *et al* (1999) using specific assumptions for the Netherlands and Sweden conclude that maximum recycling of waste newspaper is better than its incineration but it is recognized that other definitions of their system boundaries would reverse their conclusion. More generally it is accepted that the incineration of wastepaper, like wood, is carbon neutral i.e. the cycle is sustainable.

Apart from the questions concerning the possible long term effects of recycling strategies on global warming, more immediate questions have been raised very recently about the use made of some recycled paper. In the UK in 1999, for example, the Ministry of Agriculture, Fisheries and Food has discovered chemicals leaching from recycled packaging materials into products such as breakfast cereals. The chemical - disopropyl naphthalene (DIPN) has in some cases contaminated the food through film overwrap. DIPN is a solvent used to make carbonless copy paper such as credit card receipts and it remains within recycled board used for packaging foods. Related concerns have been expressed for some time about the outwash from the de-inking process which releases heavy metals such as cadmium and mercury contained in many printing inks.

Recycling strategies are not designed to increase the use of the forests but simply to 'conserve' what is there already. Incineration of wastepaper however, will encourage if not force pulp and paper companies to re-engage in major reforestation schemes as they had done up to the 1970s, before recycled paper became a major source of fibre. It was up to the 1970s that there was evidence of an increase in the real extent of temperate forests which have been shown to be a greater than expected effective carbon sink. It is argued, therefore, that there is a need for new social-political initiatives to harmonize our overall objectives in generating more bioenergy from biomass including both recycled biomass and energy crops.

DISCUSSION

Heat for cooking food and boiling water derived from burning wood and recycled biomass (animal dung) is the oldest form of anthropogenically controlled renewable energy. Bioenergy, at the start of the new millennium, appears as the Cinderella option among the line up of high-tech alternatives using as their basic resource resource the sun, wind, waves, rivers, nuclear power and geothermal activity. Environmental concerns about the first five of these have been re-considered briefly in this chapter to emphasize that all forms of energy production will have some impact on the landscape and environment. All the negative impacts will have effects at the local, regional or national level though in the case of atomic power there are well founded fears after Chernobyl, that nuclear power can also have an international dimension.

It has been argued that these sub-national negative impacts reflect the environmental price we will have to pay in order to reduce our dependency on the continued burning of fossil fuels which are contributing at an accelerating rate, to global warming. Global warming is now firmly accepted by the world's scientific community to be planet earth's major environmental problem and which has a truly international dimension. National governments wishing to be seen "green" by their voters, especially in Europe, are rapidly developing strategies to stimulate and encourage the generation of renewable energy; in the UK a good example is the Non Fossil Fuel Obligation (NFFO) which in February 2000 was replaced by the Renewables Obligation. In some countries the quest for developing alternative sources of energy has been pushed also by economic imperatives reflecting the increasing shortage of fossil fuels within national boundaries and the political desire to be self-sufficient in energy production.

The evidence presented in this chapter has shown that global warming is accelerating as a result of two main factors: the burning of fossil fuels and deforestation. *Ceteris paribus*, the continued and increasing use of fossil fuels would in itself contribute to global warming through the accumulation of greenhouse gases, especially carbon dioxide, in the earth's atmosphere. This accumulation, however, is being accelerated by the declining capacity of the earth to absorb carbon dioxide even if it were being emitted at a constant rate. Again, it needs to be noted that the UN Population Division has revised downwards its earlier prediction of a global 10 billion people by the year 2050; this downward revision, however, is still predicting 9 billion or a 33% increase on the present 6 billion. The declining carbon absorption capacity is directly related to the ongoing and widespread decrease in the global forest cover which acts as one of the earth's important carbon sinks.

It is estimated that the global forest cover has decreased by one third since human intervention but most of this has occurred during the last 200 years or since the beginning of the Industrial Revolution which marked the start of using fossil fuels on a large scale. Global warming, therefore, has to be addressed on two fronts: reduce reliance on fossil fuels and increase the global forest cover.

Bioenergy, it has been argued, is perhaps the best alternative for addressing both issues simultaneously and effectively. The other alternatives can reduce the reliance on the burning of fossil fuels but their adoption would not in itself encourage reforestation on a global scale without the intervention of some taxation or green credit transfer system. At best the levies would be used to plant out new forests which if left to grow would then cease to continue absorbing carbon dioxide beyond their maturity. The forests would *sequester* carbon but the carbon cycle would not be sustainable at the required level. For that, forests need to be used in an economic way so that they are harvested on a regular basis to maximize carbon absorption during the first 50 years of a strategic framework in which extensive areas of the earth's surface can be reforested and harvested on a short rotation basis for providing the basic resource - fuel wood.

Bioenergy, therefore, is completely different from other alternative high-tech sources of renewable energy. All the high-tech sources require only the application of technology to existing sources of energy which are being replaced continuously by nature - the sun, waves, tides, wind and running water. All these sources, however, are subject to intermittency due to climatic variations involving cloud cover, drought, calm or periodicity for tides. Moreover the high-tech alternatives are limited to the production of one energy mode ie electricity.

Bioenergy from biomass, on the other hand, can be generated from two different sources, is not subject to intermittency of supply, and can produce a variety of energy modes. Bioenergy can be generated by incinerating, either recycled biomass (MSW) or virgin energy crops. MSW can be supplied to waste-to-energy plants located in or close to large urban areas on a continuous basis to produce either *steam* for local heating purposes or *electricity* which can be fed into the grid which is the preferred method in the USA. The supply of MSW will increase over the next 50 years in some proportion to the forecasted increase in world population from 6 to approximately 9 billion people. Bioenergy, in the form of *gas*, can also be derived at an increasing rate with growth in world population from either sewage plants or where they are available, landfill sites. The added attraction of bioenergy generated from the incineration of MSW is that as well as reducing the reliance on fossil fuels it is contributing to the solution of another major environmental problem by diverting MSW away from landfill sites which are increasingly scarce in some countries e.g. Denmark and Switzerland and the eastern seaboard of the USA. Even where landfill sites are available, there are strong concerns about the environmental hazards they present in the form of leachates.

The successful generation of bioenergy from recycled biomass highlights a major area of conflict in the Environmental Strategies of many western countries. The incineration of MSW has become a necessary economic solution to the landfill problem; to a large extent the electricity or steam that is produced can be seen as a useful bi-product which helps to make the process "more economic". Environmental strategies of many countries, however, have emphasized the "need" to recycle as much waste paper as possible. This "need" is based on a very popular and widely held belief promoted by EPGs that recycling is good for the environment. Several studies have argued and shown that in the case of waste-paper, which is made from a renewable source, this is not so. Recycling strategies from wastepaper have been extremely successful and in many countries as much as 70 per cent of paper and board products have been recovered from MSW in which paper and board products have typically comprised between 40-50% of the total. These strategies have no doubt engendered a "feel good" factor among the general public which has given strong support to this "green activity" promoted by respective national governments. Recycling strategies have been so successful that many western countries especially the USA and Germany now accumulate very large annual surpluses of wastepaper which are then internationally traded especially to countries in S.E. Asia. This international transport of wastepaper requires greater use of fossil fuels. Moreover the actual recycling process also requires very large inputs of energy derived from fossil fuels. Finally, the removal of the wastepaper from MSW makes it more difficult to incinerate the remaining components of MSW to the required temperature in order to generate steam or electricity. If wastepaper, which when burned is carbon neutral, were left in the MSW the dependency on fossil fuels for this process would be reduced. There is, therefore, a major conflict in the environmental strategies of governments who, on the one hand want to promote a "feel good" factor through recycling, and on the other want to solve the environmental problem of landfills by the incineration of MSW. Bioenergy from recycled biomass is similar to high-tech sources of renewable energy in that the resource is already available and is being supplied continuously on a daily basis as a result of human life.

Bioenergy from biomass in the form of virgin energy crops is different. For commercial purposes the resource is not yet available at least in most western countries. If energy crops, especially trees, are going to be used on a large scale within the next 50 years which is the forecasted time scale for the commercial exhaustion of oil and gas, then long term strategies for the extensive planting of fuel wood trees need to be implemented now as was recognized in the NFFO-5 in the UK. Cognisance has been taken of the various political, social and cultural constraints that such strategies will have to resolve. The most critical factor will be available land. It has been argued, however, that in the USA and Europe, at least, large scale afforestation for energy production can complement the ongoing agricultural policies in both sets of countries. In both areas there is widespread recognition that we are producing too much food and every effort is being made to withdraw more land from agriculture.

Within the framework of Kondratieff's model of long wave economic cycles (of approximately 50 years) it has been suggested that the next cycle of predicted growth will be driven by Biotechnology. Biotechnology is already making a very large imprint on the general public in the context of Genetically Modified Organisms (GMOs) for improving food production which again will reduce the amount of land required for agriculture. One attraction of GM crops, for example, is that they will not be susceptible to pests and other diseases so that greater certainty can be attached to the volume of harvest; once risk is removed farmers do not have to plant out such large areas of land as a hedge against a poor harvest. Surplus land, which will nearly always be marginal land, cannot be left idle. Economic returns can be obtained from this land through reforestation for either fuel wood or pulp wood. Biotechnology will or can be instrumental for increasing the yields and hence economic return for both purposes. Farmers would have to be assured, however, that the long term investment, which is an expensive operation, will provide a guaranteed return in the longer term. For this, government environmental policies will have to be clearly articulated and conflicts within existing policies will have to be resolved.

Single stemmed trees grown for pulpwood, energy production and furniture can be complementary activities. For pulpwood only the main part of the trunk is used; the main branches can be burned for energy production; the smaller branches can be chipped for garden compost; and the sawdust can be used to manufacture chipboard which when made into roofing, furniture, floors, etc. contributes to carbon sequestration. Nothing is wasted. Government reversal of existing recycling strategies would produce a significant increase in the global demand for pulpwood, especially in temperate locations where most pulp and paper production is located. The advent of the so called paperless office in the 1970s has increased rather than decreased the consumption of writing and printing paper in all western countries; at the same time the explosion in satellite TV channels and networking, many of which are dedicated to "news" has been paralleled by an increase in newspaper circulation in these countries.

With reference to the analyses of the global use of recycled wastepaper it has been demonstrated that a reversal of government environmental policies concerned with recycling wastepaper would have a significant impact on the demand for virgin pulpwood. Virgin fibre would provide a secure economic return for marginal land released from agriculture as a result of ongoing developments in biotechnology.

In conclusion, it has been argued that although Bioenergy from Biomass including recycled biomass (MSW) and virgin energy crops at first appears as the Cinderella among the high-tech options for producing renewable energy it has several attractions over those held by its more seemingly appealing sisters. Bioenergy from biomass has a greater ability to ameliorate the major problem of global warming by not only reducing our reliance on fossil fuels but also through encouraging reforestation. Second, bioenergy through reforestation and harvesting will unlike its high-tech sisters, generate a large labour force in declining rural areas. Third, bioenergy derived from recycled biomass helps to solve another environmental problem, that of landfills. Greater reliance on bioenergy, because of the technological constraints of transportation and optimum size of power-plant would mean a transformation of energy landscapes into a dispersed pattern of a large number of small production units. Other social and cultural constraints have been recognized. The main economic barrier which applies to all forms of renewable energy could be revalued by new political thinking about carbon taxes.

REFERENCES

Ackerman, F. 1997. *Why do we recycle? Markets, values and public policy.* Island Press.

Adger, W.N. and K. Brown. 1995. Policy implications of the missing global sink. *Area.* 27: 311-317.

Angus-Hankin, C., B. Stokes and A. Twaddle. The transportation of fuel wood from forest to facility. *Biomass and Bioenergy*, 9(1-5): 191-203.

Bauen, Ausilio and Martin Kaltschmitt. 1999. Contribution of biomass toward CO_2 reduction in Europe (EU) in R.P. Overend and E. Chanet (eds). *Biomass: A growth opportunity in green energy and value-added products. Vol. 1.* Proceedings of the 4th Biomass Conference of the Americas. Oxford: Elseview Science; 371-377.

Bekkering, T.D. 1992. Using Tropical Forests to Fix Atmospheric Carbon; The Potential in Theory and Practice. *Ambio*, 21: 414-419.

Bell, S. 1994. Energy forestry in the landscape. *Biomass and Bioenergy*, 6: 53-67. British Newsprint Manufactures Association (BNMA). 1995. *Recycle or Incinerate?: The future for used papers*, Swindon: BNMAS.

British Petroleum (BP). 1997. *Statistical review of world energy.*

Burniaux, J.M., J.P. Martin, J. Oliveria-Martins and D. Van der Mensbrughe. 1995. Carbon abatement, transfers, and energy efficiency, in Goldin, I. and Winters, A.L. (eds). *The economics of sustainable development.* Cambridge University Press. Chapter 2.

Chesworth, Andy and Adrian McDonald. 1999. Renewable energy from biomass in the UK. *Geography Review*, 13(1): 34-36.

Collins, Lyndhurst. 1994. Environmental performance and technological innovation: the pulp and paper industry as a case in point. *Technology in Society*. 16(4): 427-446.

Collins, Lyndhurst. 1996. Recycling and the environmental debate: a question of social conscience or scientific reason. *Journal of Environmental Planning and Management*, 39(3): 333-355.

Collins, Lyndhurst. 1997. The geography of waste: an international dimension. *Geography Review*, 11: 22-26.

Collins, Lyndhurst. 1998a. Renewable energy from wood and paper: technological and cultural implications. *Technology in Society*, 20: 159-177.

Collins, Lyndhurst. 1998b. Environmentalism and restructuring of the global pulp and paper industry. *Tijdschrift voor economische en Sociale Geografie*, 89(4): 401-415.

Collins, Lyndhurst. 1999. Renewable energy from energy crops. *Geography*, 84(2): 169-179.

Dean, A. and P. Hoeller. 1992. Cost of reducing CO_2 emissions: evidence from six global models. *OECD Economic Studies*, 19: 16-47.

Department of the Environment. 1994. *Climate change: the UK programme command 2427.* London: HMSO.

Department of Trade and Industry. 1994. Short rotation coppice production and the environment. *Technology Status Report 014*, Oxford, ETSU.

Department of Trade and Industry. 1997. *Renewable Energy Bulletin*, 7: November. Easterly, J.L. and M. Burnham. 1996. Overview of biomass and waste fuel resources for power production. *Biomass and Bioenergy*, 10(2-3): 79-92.

Elliott, D.A. UK Renewable Energy Strategy: The need for longer term of support. 1994. *Energy Policy*, 22(12): 1067-1074.

Elliott, D.A. 1997. *Energy, society and environment technology for a sustainable future*. London: Routledge.

Energy Technology Support Unit (ETSU). 1994. New and renewable energy: future prospects in the UK, *Energy Paper 62*, London, HMSO.

Evans, J. 1992. *Plantation forestry in the tropics*. Oxford: Clarendon Press.

Fells, I. Energy and Civilization. *Lloyds TSB Forum*. University of Edinburgh. Nov. 27th, 1996.

Furley, P. 1994. Tropical moist forests: transformation or conservation, in Roberts, N. (eds). *The changing global environment*. Oxford: Blackwells; 304-331.

Gold, B.A. and D. Tillman, 1996. Wood cofiring evaluation at TVA power plants. *Biomass and Bioenergy*, 10(2-3): 71-78.

Grainger, A. 1988. Estimating areas of degraded tropical lands requiring replenishment of forest cover. *International Tree Crop Journal*, 5: 31-61.

Grant, R. 1993. R&D Optimizes Enzyme Application. *Pulp and Paper International*, 35(9): 16-20.

Grassi, G. and A.B. Bridgewater. 1993. The opportunities for electricity production from biomass by advanced thermal conversion technologies. *Biomass and Bioenergy*, 4: 39-345.

Gray, M. 1995. Towards a geography of waste. *Geography Review*, 9(1): 2-6.

Grubb, M. 1995. *Renewable energy strategies for Europe foundation and context 1*. London: Earthscan.

Grubb M. with R. Vigotti. 1997. *Renewable energy strategies for Europe 2: Electricity system and primary electricity sources*. London: Earthscan.

Hall, D.O. and J.I. House. 1994. Trees and biomass energy: carbon storage and/or fossil fuel substitution. *Biomass and Bioenergy*, 6(1-2): 11-30.

Hall, D.O. and J.I. House. 1995. Biomass energy in Western Europe to 2050. *Land Use Policy*, 12(1): 37-48.

Hamm, U. and L. Gottsching. 1993. The CO_2 balance - how it is affected by the German pulp and paper industry. *Appita*, 46(3): 179-86.

Hekkert, Marko, Richard van den Broek, and Andre Faaij. 1999. Energy crops versus wastepaper: a system comparison of paper recycling and paper incineration on the basis of equal land-use, in R.P. Overend and E. Charnet (eds) op. cit: 487-493.

Hollinger, D.Y. 1993. 1 Carbon Sequestration by New Zealand's Plantation Forests. *New Zealand Journal of Forestry Science*, 23(2): 194-208.

Houghton, H.J.T., B.A. Callander, and S.K. Varney (eds). 1992. *Climate Change 1992: the supplementary report to the IPCC scientific assessment*. Cambridge University Press.

Houghton, R.A. 1990. The Future role of tropical forests in affecting the CO_2 concentration of the atmosphere. *Ambio*, 19: 204-209.

Jirgis, R. 1995. Storage and drying of wood fuel. *Biomass and Bioenergy*, 91(1-5): 181-190.

Kauppi, P.E., K. Mielikainen and K. Kausela. 1992. Biomass and carbon budget of European forests 1971-1990. *Science*, 256: 70-74.

Kgathi, D.L. and P. Zhou. 1995. Biofuel use assessments in Africa; implications for greenhouse emissions and mitigation strategies. *Environmental Monitoring and Assessment*, 38(2-3): 253-269.

Kimmins, H. 1992. *Balancing act: environmental issues in forestry*. Vancover: UBC Press.

Lanley, J.P. 1991. Tropical Forest resources. *FAO Forestry Paper*, 30. Rome: FAO.

Leach, M.A., A. Bauen, and N.J.D. Lucas. 1997. A systems approach to materials flow in sustainable cities; a case study of paper. *Journal of Environmental Planning and Management*, 40(6): 705-723.

Ledin, S. 1996. Willow wood properties, production and economy. *Biomass and Bioenergy*, 11(2-): 75-83.

McNabb, K. 1994. Silvicultural techniques for short rotation eucalyptus plantations in Brazil. Stokes, B.J. and McDonald, T.P. (eds). *Proceedings of the International Energy Agency, Task IX*. Mobile, AL.

McNutt, J. and J. Rennel. 1996. *Is a global long-term fibre shortage ahead? Fibre balance of the pulp and paper industry*. Helsinki: Pöyry, Consulting Oy.

Manne, A.S. and R.G. Richels. *Buying greenhouse insurance: the economic costs of CO_2 emission limits*. Cambridge, M.A.: MIT Press.

Marland, G. and S. Marland. 1992. Should we store carbon in trees? *Water, Air and Soil Pollution*. 64: 181-195.

Martin, J.P., J.-M. Burniaux, G. Nicoletti, and J. Oliveria-Martins. 1992. The costs of international agreements to reduce CO_2 emission: evidence from GREEN. *OECD Economic Studies*, 19: 93-121.

Mather, A.S. 1992. The forest transition. *Area*, 24(4): 367-379.

Myers, N. 1989. *Deforestation rates in tropical forests and their climatic implication*. London: Friends of the Earth.

Nakicenovic, N. 1993. Long term strategies for mitigating global warming. *Energy* (special issue) 18: 97-193.

Netherlands Scientific Council for Government Policy (NSCGP). 1992. *Ground for Choices: Four perspectives for the rural areas of the EC*. The Hague: NSCGP.

Nilsson, S. and W. Schopfhauser. 1995. The carbon sequestration potential of a global affestation program. *Climatic Change*, 30: 267-93.

Paine, L.K. 1996. Some ecological and socio economic considerations for biomass energy crop production. *Biomass and Bioenergy*, 10(4): 231-242.

Patterson, W. 1994. *Power from Plants: The global implications of new technologies for electricity from biomass. London: Earthscan.*

Pearce, F. 1997. Burn Me. *The New Scientist*, 2109 (Nov 22): 31-4.

Powell, J. and A. Craighill. 1996. Is recycling good for the environment? *Warmer Bulletin*, 51: 4-5.

Pöyry Oy, J. 1996. Competitiveness: global trends and perceptions. *Know-how Wire*, 2: 4-7.

Rättö, M., A. Kantelinen, M. Bailey and L. Viikan. 1993. Potential of Enzymes for wood debarking. *Tappi Journal*, 76(2): 125-128.

Schlamadinger, B. and G. Marland. 1996. The role of forest and bioenergy strategies in the global carbon cycle. *Biomass and Bioenergy* 10(5-6): 275-300.

Scurlock, J.M.O. 1993. Utilizing biomass crops as an energy source: a European perspective. *Water, Air and Soil Pollution*; 70: 139-159.

Sedjo, R.A. 1992. Temperate forest ecosystems in the global carbon cycle. *Ambio*, 21: 274-277.

Sedjo, R.A. and A.M. Soloman. 1989. Climate and Forests in Rosenberg, N.J. (ed). *Greenhouse warming; abatement and adaptation.* Washington, D.C.: Resources for the Future. 105-109.

Sims, Ralph E.E. 1999. Bioenergy and carbon cycling - the New Zealand way, in R.P. Overend and E. Charnet (eds). op. cit: 401-405.

Tans, P.P., Fung, I.Y. and Takahashi, T. 1990. Observational constraints on the global atmospheric CO_2 budget. *Science*, 247: 1431-1438.

Trexler, M.C. 1991. Estimating tropical biomass future: a tentative scenario, in Howlett, D. and C. Sargeant, (eds). *Proceedings of Technical Workshop to Explore Option for Global Forestry Management*, London: IEED.

Turnbull, J.H. 1996. Strategies for achieving a sustainable, clean and cost effective biomass resource. *Biomass and Bioenergy*, 10(2-3): 93-100.

Whyte, I. 1995. *Climatic change and human society*, London: Arnold.

Wright, L.L., R.L. Graham, A.F. Turnhollow and B.C. English. 1992. The potential impact of short rotation woody crops on carbon conservation, in Sampson, R.N. and Hair, D. (eds). *Forests and global change. Vol 1: Opportunities for increasing forest cover.* Washington, D.C. Chap. 8.

Van den Broek, R., A. Faaij, and A. Van Vijk. 1996. Biomass combustion for power generation. *Biomass and Bioenergy*, 11(4): 271-281.

Virtanen, Y. and S. Nilsson. 1993. *Environmental impacts of waste-paper recycling.* International Institute for Applied Systems Analysis. London: Earthscan.

Renewable Energy: Trends and Prospects. Edited by S.K. Majumdar, E.W. Miller and A.I. Panah

Chapter Three

SOLAR DRYING SYSTEMS FOR AGRICULTURAL PRODUCE

M.Y.H. OTHMAN, K. SOPIAN, B. YATIM, and W.R.W. DAUD

Solar Energy Research Group,
University Kebangsaan Malaysia
43600 Bangi
Selangor, Malaysia

ABSTRACT

The design and performance of two types of solar assisted drying systems are presented. The V-groove solar collector is used in the first dryer and the double-pass solar collector with porous media in the lower channel is used in the second dryer. The drying chambers are of the cabinet and flat bed types for the former and later respectively. Several agricultural produces have been used in the first dryer namely chilies and green tea. Ground oil palm fronds as used as animal feedstock is dried using the second dryer. The lower channel of the double-pass solar collector are filled up with porous media and acted as a storage medium for the system. Hence, the second solar dryer is suitable for higher temperature drying of agricultural produce. At a solar radiation level of 700 W/ m^2 and ambient temperature of 30°C, the outlet temperatures are 60°C and 80° C for the first and second dryers respectively.

KEYWORDS

Solar drying system, V-groove collector, double-pass solar collector, cabinet dryer, flat bed dryer.

INTRODUCTION

Solar drying provides an alternative to the use of fossil fuel. Solar drying of agricultural produce is one of the most promising applications of solar energy. However, there are problems associated with the intermittent natural of solar radiation and the low intensities of solar radiation. The former can be overcome by adapting heat storage devices, auxiliary energy sources and adequate control strategies. The second can be remedied by the use of heat storage, auxiliary energy source, control system and larger surface collector. However, this will result in a high capital investment. If the savings from the use of a solar assisted drying system is more than the cost of using fossil fuel than it is justifiable to use the solar drying system. Many other economic indicators, such as payback period, and internal rate of return can be used for the above purpose. The technical development of solar drying systems can proceed in two directions. Firstly, simple, low power, short life, and comparatively low efficiency-drying system. Secondly, high efficiency, high power, long life expensive drying system. The latter is characterized not only by an integrated structure but also integrating in an energy system involving processes other than drying, such as hot water when the drying system is not in use. In addition, hot water tanks of the farm can be used as heat storage of the solar drying system.

This chapter presents the design of two solar assisted drying systems for crop drying using different collector types,

(a) V-groove solar collector with cabinet dryer, and

(b) Double-pass solar collector with porous media in the lower channel and a flat bed drying chamber.

SOLAR DRYING PROCESS

Drying of material means removal of moisture from interior of material to the surface and then to remove this moisture from the surface of the drying material. Drying process is a complex heat and mass transfer process, which depends on external and internal variables. The external variables are temperature, humidity and velocity of drying air. The internal variables are a function of drying material and depends on parameters like surface characteristic (rough or smooth surface), chemical composition, physical structure, size and shape of the products.

The most important parameter is the saturation vapour pressure which is the maximum pressure of water vapour which depends on the temperature. In the drying process it is assumed that the material surface is saturated with water vapour. The ideal saturated absolute humidity is (Mujumdar, 1995):

$$Y_S = \frac{P_W^O}{P - P_W^O} \cdot \frac{M_W}{M_G} \qquad (1)$$

and the absolute humidity of drying air is:

$$Y = \frac{P_W}{P - P_W} \cdot \frac{M_W}{M_G} \tag{2}$$

Rate of evaporation can be expressed by a mass transfer coefficient and an air humidity gradient (Strumillo and Kudra, 1986):

$$W_D = k_g(Y_{st} - Y) \tag{3}$$

Y_{st} is absolute humidity at the surface, if the surface is saturated with water:

$$Y_{st} = Y_s \tag{4}$$

Y_s is the absolute humidity of air at the same temperature, so that W_D may be calculated. For drying process, one wants the greater difference between Y_s and Y. In solar drying Y_s depends on the temperature of the heating air. The heating air temperature is controlled by solar radiation. If solar radiation is low, the auxiliary heater is necessary to increase the air temperature.

SOLAR DRYING EFFICIENCY

Solar air heater received solar radiation, which is converted to heat and transferred to the air stream. Some heat is lost to the surrounding which have a lower temperature by the process of conduction, convection and radiation. Thermal efficiency is the ratio of the useful heat gain to the solar radiation incident on the plane of solar air heater:

$$\eta = \frac{mCp_a (T_o - T_i)}{A_C I_T} \tag{5}$$

The hot air that is produced by the solar air heater is used as a drying agent in the drying system. Drying efficiency of the drying system is the ratio of energy required by the moisture to heat a supply to the dryer (Brenndorfer, *et al.*, 1980), by the equation:

$$\eta_d = \frac{WL}{A_C I_T} \tag{6}$$

For a solar-assisted drying system, energy is used for heating the material to the drying temperature and then for heating air from ambient temperature to the drying temperature. The energy supplied to the system is energy from solar radiation, electric energy is used for blower and electric heater. Abdullah *et al.*(1994) define the drying efficiency of solar assisted drying system as:

$$\eta_d = \frac{m_p Cp_p (T_d - T_i) + m_w Cp_w (T_d - T_i) + m_v L}{A_C I_C + P_b + P_h} \qquad (7)$$

The outlet air temperature is not constant and depends on the intensity of solar radiation. To calculate the drying efficiency, the inlet air temperature and the initial temperature of dried material is considered equivalent to ambient temperature. Jannot and Coulibaly (1998) expressed the drying rate as evaporative capacity. The evaporative capacity is the weight of water that can be extracted by the air flowing over the products. The evaporative capacity can be expressed as:

$$E = m_{da}(X_{2m} - X_a) \qquad (8)$$

SOLAR ASSISTED DRYING SYSTEM WITH V-GROOVE COLLECTOR

The solar assisted drying system with V-groove collector has been designed to dry various types of agricultural produce such as chilies, green tea, anchovies, and grains. The system produces low air temperature up to about 50 ° C. An electric auxiliary heater was attached to the system in order to provide continuous heat as required by the drying commodity. The specification of the system is described in Table 1 (Othman *et al*, 1994).

TABLE 1.
Specification of the V-groove Solar Drying System

SYSTEM COMPONENTS	SPECIFICATIONS
Collector	
Type of absorber	V-groove back-pass
Material of absorber	Folded aluminum sheet SWG22 - standard size, 244cm x 122cm.
Angle of groove	49° and height 7.8 cm
Collector area	100 cm x 460 cm per unit collector
Total collector area	13.8 cm^2
Top cover	Glass: thickness 2.5 cm; one side tempered
Insulator	Fibreglass wool; 2.5 cm thickness, density 46.0 kgm^{-3}
Air Circulation	
Two unit axial fan	2700 rpm, 85 W. 230 V (AC), single phase motor, 18 cm outlet diameter.
Ducting	PVC pipe
Air flow rate	6.0 - 16.5 m^3 min^{-1}
Drying Chamber	
Type	Adjustable set of shelved frame
Size	1.0m x 3.0m x 3.0m

The system consists of a solar air collector, blower, auxiliary-heater and drying chamber. The entire system configuration is shown in Fig. 1 and the photograph in Fig.2. The solar collector is of the back-pass V-groove as shown in Fig. 3. The size of each collector put in 3 series is 4.6m x 1.0m x 0.15 m. The total area of the collector is 13.8 m^2. A 10 kW electric auxiliary heater is equipped with an on/off controller. The air temperature in the chamber can set according to required temperature. The size of the chamber is 1.0m x 2.5m x 2.9m. There are 3 layers of trays made of wire-mesh stainless steel inside the drying chamber as shown in the photograph in Fig. 4.

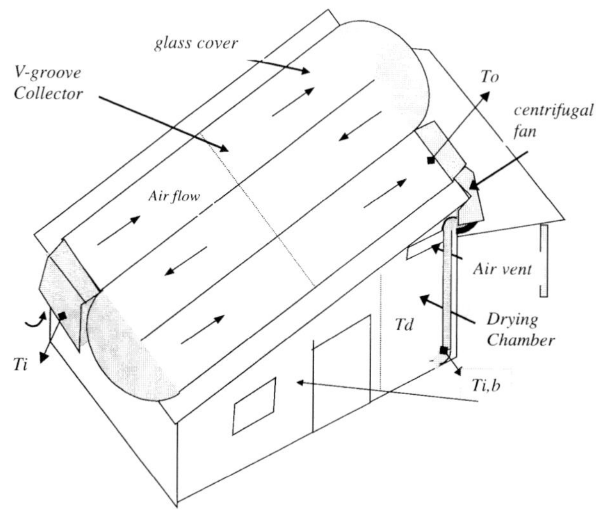

FIGURE 1. Configuration of the V-Groove solar drying system.

FIGURE 2. Photograph of the V-Groove solar drying system.

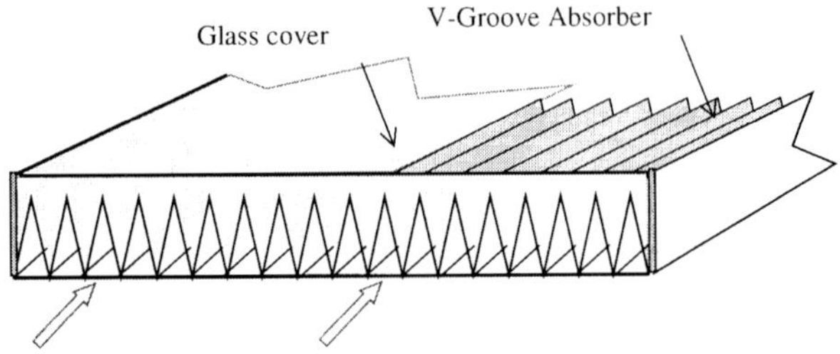

FIGURE 3. Cross section of the V-groove collector.

FIGURE 4. The cabinet dryer of the solar assisted drying system.

Figure 5 shows that at mean solar radiation of 700 Wm^{-2} and air flow rate of 6.1m^3 min^{-1} the mean efficiency of the system was 36.8%. The solar radiation was measured on the surface of the collector at an angle of 10° facing South. Table 2 shows the temperature gain for various volumetric air flow rates at different mean solar intensities. It can be seen in the table that air temperature in the chamber could achieve 55°C.

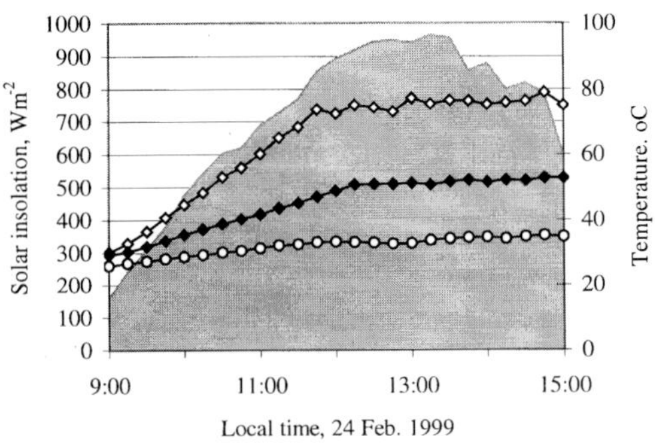

FIGURE 5. Solar drying system efficiency at 700 Wm^{-2}

TABLE 2
Performance of solar dyer with V-groove collector

Flow rate M^3s^{-1}	Intensity Wm^{-2}	Mean o/p Temp °C	Max. o/p Temp °C	Mean dryer Temp °C	Max.dryer Temp °C	Mean Eff. %
16.3	695.8	50.8	64.3	45.1	5.8	58.6
9.2	652.3	55.4	71.2	32.0	52.1	45.3
6.1	706.3	63.7	79.0	44.9	52.9	36.8
6.1	455.4	51.1	68.6	39.2	47.3	40.1
4.8	624.0	60.5	82.7	42.2	50.9	29.5
4.8	464.1	52.1	76.4	39.9	49.3	33.2

SOLAR ASSISTED DRYING SYSTEM WITH DOUBLE-PASS SOLAR COLLECTOR

The second type of solar drying system, namely solar assisted drying system with doublepass solar collector, produces much higher outlet temperatures (Sopian et al, 1999). The system consists of a solar air collector, blower, auxiliary-heater and drying chamber. The solar collector is of the double-pass type. The second or lower channel of the solar collector is filled up with porous media. This will increase the outlet temperature and the performance of the system. The auxiliary heater is equipped with an on/off controller. The set temperature is 50°C based on the temperature of the inlet to the drying chamber. The system is designed for drying of wet ground fronds from moisture content of 60% wet basis to 10% weight product and the drying time is not more than 10 hours. The collector width and length are 120 cm and 240 cm. The upper channel depth, dl, is 3.5 cm and the lower depth, d_2, is 10.5 cm. The schematic of the collector is shown in Fig. 6, the photograph of the solar assisted drying system is shown in Fig 7, and the schematics of the whole system is shown in Fig. 8.

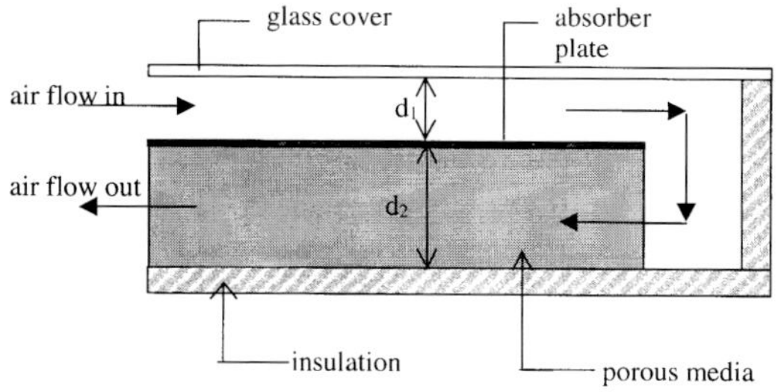

FIGURE 6. The Schematic of a double-pass solar collector with porous media in the second channel.

FIGURE 7. Photograph of the solar assisted drying system.

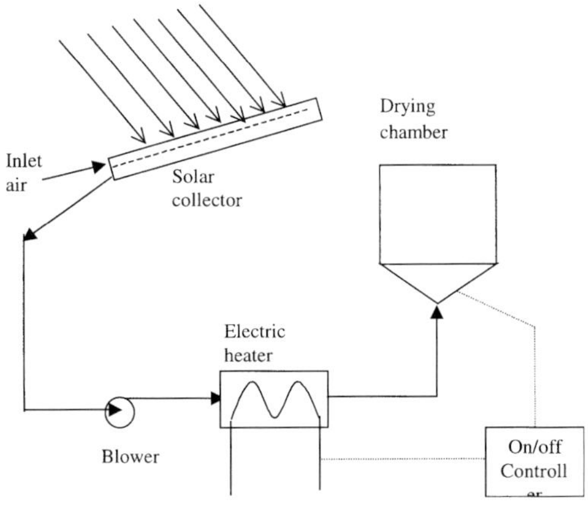

FIGURE 8. The Schematic diagram of the solar assisted drying system.

The solar air heater consists of 6 solar collectors, arranged as 2 banks of 3 collectors each in series as shown in Fig. 9, and the photograph of the flat bed drying chamber of the system is shown in Fig. 10. Air enters the inlet of the upper channel in the first collector and flows in the lower channel. Next, the air flows to the second collector. The lower channel of the third collector is the air outlet from the first bank. In the second bank, air flows in the upper channel of the sixth collector and the outlet air is the lower channel of the fourth collector. The outlet air from the third collector is mixed with the air outlet from the fourth collector. The centrifugal blower sucks in the hot air. Power rating the blower is 0.11 kW, 230 V with rotating 2520 RPM. The finned air electric heater of 5 kW is used as a heating unit for supply heat in the lower solar radiation. The heater is controlled with ON/OFF controller and sets at 50°C. The heater is on if the hot air temperature is lower than 50° C and off if higher than 50° C.

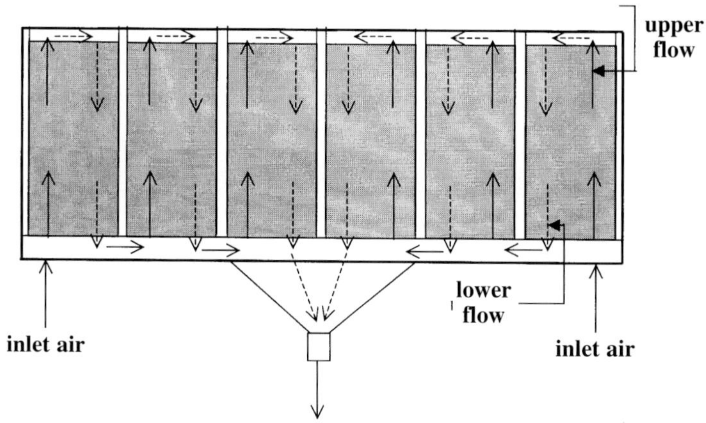

FIGURE 9. The collector array for the solar drying system.

FIGURE 10. The flat bed drying chamber of the solar assisted drying system.

The system has been tested under daily solar radiation. Chopped and ground palm oil frond has been use as the drying commodity. The ground fronds are dried from 09:30 a.m. to 05:30 p.m.. Solar radiation and various temperatures (dryer inlet and outlet, collector inlet and outlet) are show in Fig. 11., and the weight loss curve is show in Fig. 12. The time constant of the collector is about 21 minutes (Supranto,et al, 1999a), so that the solar radiation is averaged for every 30 minutes, $A_C I_T$ = 76121.3 Watts for drying time 7.5 hours. The drying efficiency of the system has been described elsewhere (Supranto, *et al*, 1999b).

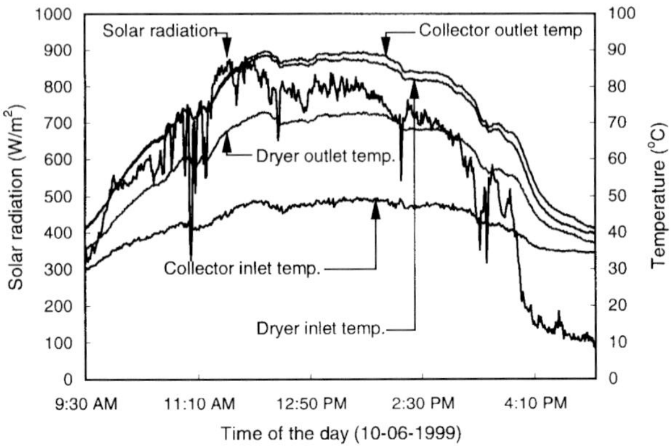

FIGURE 11. Variation of the solar radiation and various temperatures (dryer inlet and outlet, collector inlet and outlet).

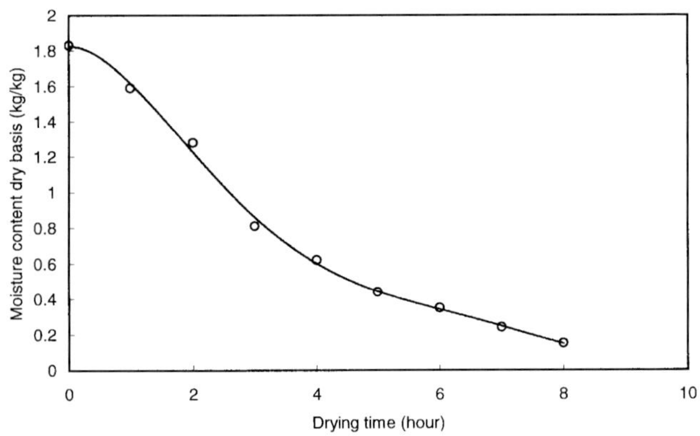

FIGURE 12. The drying curve of ground oil palm frond.

CONCLUSIONS

Two types of solar assisted drying systems have been developed for agricultural produce. Both systems consist of the solar collector, blower, auxiliary heater, and drying chamber. The first drying system used the V-groove type solar collector and the second drying system used the double-pass solar collector with porous media in the second channel. The drying chamber for the first dryer is the cabinet type and the drying chamber for the second dryer is a flat bed type. The first dryer is suitable for agricultural produce such as chilies, green tea, and dried fruits. The second dryer produces much higher outlet temperatures compared to the first drying system. The connections for the V-groove solar collector are straight forward and simple while the connections for the double-pass solar are a little complicated. The connections for the doublepass solar collector involve internal manifolding. The drying system is used to dry animal feedstock such as oil palm fronds and other higher temperature drying processes.

NOTATIONS

A_C = collector area, m^2
Cp_a = heat capacity of air, kJ/(kg)(°K)
Cp_p = heat capacity of drying product, kJ/(kg)(°K)
Cp_w = heat capacity of vapour, kJ/(kg)(°K)
E = evaporative capacity, kg/h
I_T = solar radiation, W/m^2
kg = mass transfer coefficient of humid gas, $kg/(m^2)(s)$
L = latent heat of evaporation of water at the temperature of the dryer, kJ/kg
m = air flow rate, kg
m_{da} = mass flow rate of dry air, kg/s
m_p = weight of drying product, kg
m_w = weight of produced water vapour, kg
m_v = evaporated water in the dryer, kg
M_G = molar mass of dry air, kg/mol
M_w = molar mass of moisture, kg/mol
P = total pressure, Pa
P_b = power to the blower, Watt
P_G = partial pressure of dry gas, Pa
P_h = power to heater, Watt
P_w = partial pressure of moisture vapor, Pa
P°_w = vapor pressure, Pa
T_d = the drying temperature, °K
T_i = inlet temperature, °K
T_o = outlet temperature, °K
η_d = drying efficiency
w_D = drying rate, $kg/(m^2)(s)$
W = weight of moisture evaporated, kg
X_{2m} = dryer outlet absolute humidity
X_a = ambient absolute humidity.
Y = gas humidity in the main gas stream (dry basis), kg/kg
Y_s = gas humidity on the surface (dry basis), kg/kg

ACKNOWLEDGEMENT

The authors would like to record our appreciation to the Ministry of Science, Technology and Environment, Malaysia and the Universiti Kebangsaan Malaysia for funding the project under IRPA 02-02-02-0001.

REFERENCES

Abdullah, K., Thamrin and Wenur, F., 1994, Drying of Black Pepper using Solar Energy, *Drying* '94, edited by V. Rudolph and R.B. Keey, Volume B. pp. 887-895.

Brenndorfer, B., Kennedy, L., Oswin-Bateman, C.O., Trim, D.S., Mrema, G.C. and Wereko-Brobby, C., 1990, Solar Dryers-The Role in Post-Harvest Processing, pp. 3235, Commonwealth Science Council, London.

Jannot Yves and Coulibaly, Y., 1998, The Evaporative Capacity as A Performance Index for A Solar-Dryer Air-Heater, *Solar Energy*, 63, 6, 387-391.

Mujumdar, A.S. and Menon, A.S., 1995, Drying of Solids: Principles, Classification and Selection of Dryers, in Handbook of Industrial Drying, Volume 1, pp. 6-7, edited by Arun S. Mujumdar, Marcel Dekker Inc., New York.

Othman M.Y., K. Sopian, M.M. Salleh, dan M.H. Roslan, 1994, Evaluation of a Force Convective Solar Dryer, *Sains Malaysiana* 23(2) pp 59-69 .

Sopian K., Supranto, W.R. W. Daud, B. Yatim and M.Y. Othman,, 1999, Thermal performance of the double-pass solar collector with and without porous media, *Renewable Energy*, 18/4, pp. 557 - 564.

Strumillo, C. and Kudra, T., 1986, Drying: Principles, Applications and Design, pp. 7072, Gordon and Breach Science Publisher, New York.

Supranto, K. Sopian, W. R. W. Daud, M.Y. Othman, and B. Yatim, 1999a, Performance Characteristics Curves of the Double-Pass Solar Collector with Porous Media, Proceeding of the World Renewable Energy Congress 1999 Malaysia, 8 -11 June 1999, Kuala Lumpur, pp. 147 - 152.

Supranto, K. Sopian, W. R. W. Daud, M.Y. Othman, and B. Yatim, 1999b, Performance of a Solar Assisted Drying System for Oil Palm Fronds, *Asia Australia Drying Conference* 1999, Bali, Indonesia, 24 - 26 October, pp. 492-500.

Renewable Energy: Trends and Prospects. Edited by S.K. Majumdar, E.W. Miller and A.I. Panah

Chapter Four

PERSPECTIVES OF CROP RESIDUE AS A RENEWABLE ENERGY SOURCE

APURBA SARKAR[1]

[1] Project Directorate for Cropping Systems Research
Modipuram, Meerut, U.P., India - 250110

INTRODUCTION

For time immemorial humans have been using renewable sources of energy. It turns towards non-renewable sources like fossil fuel due to population pressure. With dwindling reserve its prices are sky-rocketting. To cope with the needs of teeming billions by the turn of this century and beyond, we have to reduce our dependence of agriculture, especially for food supply, on such ever declining non-renewable energy sources. Assuming 2292 Kcal/head/day energy needed for human diet, the world needs 5019.48×10^{12} Kcal. of food energy per annum, of which 3841×10^{12} Kcal. must be supplied exclusively from food grains by the year 2000 A.D. In this respect organics show promise as a renewable source of energy. Moreover, soil often becomes sick because of continuous use of chemical fertilizers such as FYM, compost, green manure, sewage sludge etc. Organics are used to enhance the efficiency of chemical fertilizers, but most of them have one or the other limitations. A larger share of cowdung is used as domestic fuel, limiting the desired amount for farmyard manure (FYM) production. Transportation of bulky FYM to fields is an inherent problem. Production of green manuring crop costs a crop season land and labour which could have been utilized elsewhere. Production of good quality compost also requires a lot of effort. Sewage sludge and city refuse is available only in urban areas. A common problem in such waste is their foul smell. Presence of heavy metals like nickel, zinc, lead, cadmium, and copper in industrial sewage sludge pose health hazards. But crop residue is a unique renewable source of agro-energy. A part is used as food, feed and fodder leaving a greater chunk for its use as manure. In contrast to other organics, a crop residue has certain plus points. As they are produced on the farm itself, they require little or no transportation and there is no heavy metal problem, no health hazards, salinity, sodicity problem, or problem of nitrate toxicity. There is no cost in extra land area or effort for their production in an extra season because it is produced along with the crop and can be applied in situ. But, decomposition sometimes poses a problem for residues with high C:N ratio. Even so, it can do wonders if properly managed. We must know the total quantity for that matter, and the grain:straw ratio can give a good estimate of the amount globally available. But such ratio varies with the variation of weather, soil fertility, soil moisture, cultural practice etc. Moreover, crop residue is rarely weighed since these are bulky. In this backdrop, an effort has been made here to have a perspective view of its energetics, including environmental aspects.

GLOBAL QUANTUM OF ENERGY FROM CROP RESIDUE

Computation procedure

The Food and Agriculture Organization (FAO) of UNO publishes world crop yield data through its production yearbook. This does not report the residue production from different crops but FAO document no. 15 & 24 (1995) reported the grain to straw ratio or economic produce:residue ratio of different crops (Table 1). These ratios are used to compute the straw/stover yield of various crops from their grain or economic produce yield.

Note that since the world is very much concerned with the food for humans, only the principal food crops including sugarcane and potato have been included here for energy calculation. So other crops like fiber, fodder, fruits, many vegetables, medicinal and aromatic crops are excluded.

TABLE 1.

Grain:Straw or Grain:Stover or Economic Produce:residue ratio of various cultivated crops

Crop	Ratio	Crop	Ratio
Wheat	1:1.5	Sorghum	1.2
Rice	1:1.5	Millet	1:2
Total pulses	1:1	Maize	1:1.5
Pigeonpea	1:2.5	Sugarcane (Stripped cane)	1:0.1
Chickpea	1:1	Potato(Tuber)	1:0.5
Barley	1:1.5		
Oilseeds	1:2		
Groundnut	1:1.5		

For energy calculation the equivalent energy for different residues is used (Table 2)

TABLE 2.

Equivalent for different sources of energy

Source of Energy	Unit	Equivalent Energy(MJ)
1. Straw (Dry mass) (Rice, wheat, chickpea, other pulses, barley, oilseeds, groundnut (pod.))	Kg.	12.5
2. Crop residue (Dry mass) of pigeonpea, sorghum, pearlmillet, maize	Kg.	18.0
3. Crop residue (dry mass) of potato (tuber).	Kg.	10.0
4. Crop residue (dry mass) of stripped sugarcane	Kg.	16.1
5. Fertilizers N	Kg.	60.0
6. Fertilizers P_2O_5	Kg.	11.1
7. Fertilizers K_2O	Kg.	6.7

Source Mittal et al. (1985)

Total production of crop residue and its energy equivalence

The total production of crop residue from each crop is calculated by dividing the crop grain or economic produce yield by the corresponding grain: straw ratio. Adding all the residue yield of all the food crops, we get the global yield of food crop residue (Table 3). The energy equivalence figure is obtained empirically

TABLE 3.

Different cultivated crops, their total production of crop residue, and corresponding energy equivalence. Amount of available residue for recycling as manure and their corresponding energy equivalence.

Crop	Crop residue yield (million metric tonne)		Energy equivalence (10^{12} Kcal.)	
	Total	Available for recycling as manure	Total	Available for recycling as manure
Rice	843	281	2515	839
Wheat	885	295	2640	880
Maize	870	290	3738	1246
Millet	60	20	258	86
Sorghum	120	40	516	172
Barley	262	88	783	261
Total pulses	58	19	413	58
Oil crops	160	160	477	477
Sugarcane	104	104	400	400
Potato	145	145	346	346
Total	3507	1442	12086	4765

by multiplying the residue yield with the corresponding value of its energy equivalence as given in Table 2.

Note: 1. Total pulses include all the pulses like fieldpea, chickpea, lentil, blackgram, greengram, Bengal gram, horsegram, pigeonpea etc.
2. Oil crops include all oilseed crops like groundnut, rape and mustard, sesame, safflower, sunflower, soybean, niger seed, linseed, castorseed etc.
3. It is assumed that two-third of the residue quantum is used as feed for animal and other purpose and the other one-third is available for recycling as manure except oil crops, sugarcane and potato either as direct soil incorporation or through composting.

The total production of crop residue from food crops is estimated as 3507 million metric tonne (mmt), $3/4^{th}$ from rice + wheat + maize together and the rest from other food crops shown in Table 3. The total energy equivalence of this total crop residue comes to about 12086 x 10^{12} Kilo calorie, about three-fourth of which is contributed again by three prominent crops viz., rice, wheat and maize and rest by rest of the food crops mentioned above. The energy contribution of these food crop residues can be viewed from different angles. From cereal-non-cereal viewpoint, each category contributed equally (i.e., 50% each) to the equivalent energy. From cereal vs. millet angle, about half is contributed by cereals and 40% by millets. From physiological angle, C_3 food crops contributed 56% and 44% by C_4 food crops mentioned here. But all this energy is not available for recycling as manure.

Amount of food crop residue and its energy equivalence available for recycling as manure

A greater chunk of the crop residue is used as food, feed, fodder and fire and for other purposes. Thus, roughly one-third is available for recycling as manure for most food crops, except oil crops, sugarcane and potato. This available portion of crop residue amounts to 1442 mmt equivalent to 4765 x 10^{12} Kilo calorie. Here, also, the first three food crops, rice, wheat and maize together constitute the largest energy share (Table 3).

MAJOR PLANT NUTRIENTS AVAILABLE FROM CROP RESIDUE

Since a plant contains all its essential nutrients, crop residue can be considered as a complete nutrient packet for crop growth, at least for the three of its major nutrients, nitrogen, phosphorus and potassium. Based on the nutrient content of the respective crop residue, nutrient potential of crop residue available for recycling as manure comes to about 28.8 million metric tonnes ($N+P_2O_5+K_2O$), half of which is expected to be mineralized in the current season and thus can replace 9 to 10% of the total global fertilizer demand of 160 million metric tonne ($N+P_2O_5+K_2O$) and the other half will serve as a (Table 4) reserve for subsequent seasons. In a sense, quantitatively, crop residue can well meet up to one-fifth of the global fertilizer demand ($N+P_2O_5+K_2O$). But energetically, the picture is somewhat different. If recycled as manure, the available crop residue can replace 5% of the total 1553 x 10^{12} Kcal of global energy equivalence for fertilizer (160 mmt of $N+P_2O_5+K_2O$ fertilizers) required during the current season. And annually, we can expect a supply of 151.5 x 10^{12} Kcal of fertilizer energy from the crop residues available for recycling. This amounts to 10% of the global fertilizer energy requirement (Table 4).

FACTORS EFFECTING THE PRODUCTION OF CROP RESIDUE

All the factors that affect the production of a crop also influence its residue production, for example, meteorological factors, soil factors, management factors etc. Each crop flourishes best under an optimum combination of all these factors. But since it is beyond the purview of this topic, only the factors are enumerated here. *Meteorological factors*, influencing crop residue production are: temperature, humidity, sunshine-hours, wind velocity, rainfall etc. *Soil factors* are: soil temperature, soil aeration, soil moisture, its bulk density, texture, structure, soil nutrient status in respect of N, P, K, Ca, Mg, S and all essential micronutrients like Fe, Mn, Zn, Cu, Cl, Mo, B; its infiltration rate, permeability, pH, electrical conductivity etc. *Crop management factors* are: time of sowing, type of crop cultivar, method of sowing, level of fertilizer and manure, method of fertilizer application, sources of fertilizer nutrient, plant protection measures, time of harvesting, etc.

EFFECT OF CROP RESIDUE ON CROP PRODUCTION

Application of crop residue as soil incorporation improved the crop yield. Such effect on the rice-wheat cropping system has been reported by a number of workers like Gotah and Onikura (1991), Mishustin *et al* (1975), Hwang *et al* (1989), Will *et al* (1989), Wold *et al* (1989), Verma *et al* (1989) and Stott *et al* (1990), and on the maize-potato cropping system by Sarkar (1992). Experiments under an India coordinated research project on the cropping system showed that a part (25 to 50%) of the nitrogen level can be met through crop residue in rice-rice, rice-wheat, pearlmillet-wheat and sorghum-wheat systems (Anon 1998-1999). Thus, the incorporation of such residue is beneficial in the long run, particularly in combination with fertilizers (Vlek, 1990; Sarkar, 1996).

EFFECT OF CROP RESIDUE ON SOIL

Being organic in nature, crop residue benefits not only the crop but also the

TABLE 4.

Crop residue available from different crops for recycling, their nutrient content, nutrient potential for recycling as manure and fertilizer replacement value.

Crop	**Crop residue available for recycling as manure. (Million metric tonne) (a)**	**Nutrient content of crop residue (on oven dry basis) (%). (b)**			**Nutrient potential (million metric tonne)**		**Energy equivalent (10^{12} Kcal.)**	
		N	P_2O_5	K_2O	For recycling as manure ($N+P_2O_5+K_2O$)(c)	Fertilizer replacement value ($N+P_2O_5+K_2O$)(d)	For recycling as manure (e)	Fertilizer replacement value (f)
Rice	281	0.61	0.18	1.38	6.1	3.05	32.1	16.05
Wheat	295	0.48	0.16	1.18	5.4	2.70	27.1	13.95
Maize	290	0.52	0.18	1.35	6.0	3.00	29.2	14.60
Millet	20	0.45	0.16	1.14	0.4	0.20	1.7	0.85
Sorghum	40	0.52	0.23	1.34	0.9	0.45	4.1	2.05
Barley	88	0.52	0.18	1.30	1.8	0.90	8.8	4.40
Total pulses	19	1.29	0.36	1.64	0.6	0.30	4.2	2.10
Oil crops	160	0.80	0.21	0.93	3.1	1.55	21.6	10.80
Sugarcane	104	0.40	0.18	1.28	1.9	0.95	8.6	4.30
Potato	145	0.52	0.21	1.06	2.6	1.30	14.1	7.05
Total	1442				28.8	14.4	151.5	75.75

(a): See Table 4
(b): Source Tandon (1995)
(c): Based on (a) & (b)
(d): Assuming that half of N+P2O5+K2O in residue is mineralized in a season in most of the parts of the world including the tropics and subtropics except some part of the temperate region.
(e): Based on (c) and energy equivalent value in Table 2
(f): Based on (d) and energy equivalent value in Table 2

medium on which it grows. Different soil properties are influenced with the crop residue addition. This is presented briefly.

On soil physical properties: Bulk density

Cowpea or pearlmillet residue improved the soil as its bulk density is reduced from 1.78 to 1.6 g/cm^3 (Venkateswarlu, 1987, Hadimani, *et al,* 1982).

Soil structure

At molecular level, humus molecules are capable of binding clay particles and larger mineral particles at the point of contact existing in the soil matrix. But soil grains may be separated by greater distance in non-contact zones of cracks or micro-cracks where materials of molecular scale are ineffective. Under such situation, residue decomposition products like fungal hyphae or other metabolites do the binding work. Thus incorporation of crop residue helps the soil organism to overproduce metabolites that stabilize the soil structure and can easily improve the soil structure.

Infiltration and hydraulic conductivity

Wheat residue (5t/ha) doubled the infiltration rate of irrigation water with its incorporation (Mager, 1982). Hydraulic conductivity increased from 2.84 to 8cm/hr in loamy sand and from 4.19 to 10.53cm/hr in sandy loam soils on addition of cowpea or pearlmillet residue as reported by Venkateswarlu(1987) and Hadimani *et al*, (1982).

Soil aggregates

There was an increase in aggregates (of size more than 0.25 cm) from 62 to 87% in sandy soils on addition of pearlmillet or cowpea residue. Such residue addition also enhanced the mean weight diameter of both the types of soil as reported by the above investigators.

Miscellaneous physical properties

Residue mulch improves water retention when it is incorporated (Ramusseen & Smiley, 1989). It maintained soil texture and declined erosion. To get such benefit, about 35% of the crop residue should be left on the soil (Stout, 1982). Residue mulch also lowers the soil temperature, reduces evaporation rate, and increases the duration of soil drying in the beginning (Unger and McCalla). Canadian and U.S.A. farmers widely use cereal stubble and straw as protection against wind and water erosion for soil conservation.

On soil chemical properties

Addition of crop residue improves a number of chemical properties of soils. Nitrogen status: Applied residue has a residual effect on soil N. After integrated application of 157 Kg. mineral N along with sugarbeet tops, only 4% of N from residue was recovered by succeeding wheat crop and about 30% N remained as organic N. But with legume residue incorporation, 10 to 35% of N in legume became available to the following non-legume (Abshahi, *et al*, 1984). Likewise, 20-30% N in alfalfa-wheat system (Ladd, *et al*, 1983), and 12-33% N in wheat-sorghum system (Wagger, *et al*, 1987) was available to the following crops. Low recovery of the crop residue nitrogen is associated with the immobilization of N in microbial cell in soils. In a sense, the nitrogen recovery from crop residue of legume or non-legume origin is little. Similar information on phosphorus and potassium may be derived from other experiments. For simplicity and clarity, it is generally assumed that about 50% of N, P and K of the incorporated residue will be mineralized and are at the disposal for availability to the crop.

On biological properties of the soil

Based on his 50 years of field experiments at Sauborn, USA, Smith(1942) inferred that addition or removal of crop residue did not appear to influence the biochemical potential of microflora in the long run. But others (Crossley *et al*, 1984) had different views. Due to reduced water loss, control of temperature extreme, presence of relatively continuous substrate for decomposition, residue left over in no-tillage is congenial for soil and surface dwelling organisms. Knapp and coworkers (1983) showed through simulation modelling that the quantum of readily available carbon and nitrogen in wheat straw indicated the size of microbial biomass and consequently the degree of decomposition of the straw. Thus, crop residue addition enhanced the quantum of nutrients available to the microbes indicated by more ammonium nitrogen.

Effect of crop residue on sustainable agriculture

Agriculture can broadly be categorized into irrigated and rainfed. Maintenance of nutrient status, organic matter, soil moisture retention and microbial status in the rhizosphere are the major issues for sustaining irrigated agriculture. From the above discussion it is obvious that addition of crop residue has triple action on soil: nutritionally through its effects on major and micro-nutrients in soil and soil organic matter, physically though its effects on stability of aggregates, bulk density, infiltration rate, moisture retention, hydraulic conductivity, and biologically by enhancing microbial decomposition and the production of the soil biomass. Naturally, it is expected that an integration of crop residue with the mineral fertilizer would be able to sustain the irrigated agricultural production. Major issues under rainfed agriculture are nutrient and moisture availability. Residue mulch reduces evaporation from land surface, thereby helping retain more moisture, keeping optimum temperature, aeration and associated qualities of soil. Crop residue incorporation has a nutritional benefit. Thus, a judicious management of crop residue can well meet up the sustainability needs of rainfed agriculture too.

ENVIRONMENTAL ASPECTS OF CROP RESIDUE:

Both for construction work and fuel, demand for wood gradually increased with the growth of teeming millions. Naturally, animal dung became the principal fuel in rural areas. Globally we burn 400 millions tonnes of fuel wood annually. Without getting a proper avenue for disposal of crop residue, a part of it is usually burnt. During burning, oxygen, carbon, hydrogen and nitrogen escape as gases, leaving the plant ash as mineral containing calcium, magnesium, manganese, zinc, iron, copper, boron, potassium, phosphorus, silicon, cobalt etc. Important gases like carbon dioxide, carbon monoxide, nitrogen oxides, various hydrocarbons evolve, including non-gaseous substances like particulates (minute solid particles freely floating in air) and ash. A number of such products of residue burning have environmental consequences briefly put hereunder.

Emission of Carbon Dioxide

The global importance of carbon dioxide emission is well known. In a nutshell its thoughts and theories can be summed up as follows: Carbon dioxide is added to the atmosphere due to material activity in the kingdom of plants and animals. It takes a pollution dimension when too much is produced in our surroundings because of fossil fuel burning and allied human activities. Fossil fuel burning itself accounts for 5 to 6 billion tonnes carbon dioxide per annum (Chahal, 1991) with a net burning growth of 2-4.5% per year. A modest estimate shows 11.5 billion tonnes of carbon dioxide produced in the biosphere and transferred to the atmosphere through normal burning of fuel from all sources like all types of wood like broadleaved, wetland, coastland, conifers and all types of non-wood like gram, crop residues, shrubs, including non-wood sources of the Tundra. Crop residue burning contributes 16% of the total carbon dioxide production (Perry, 1982). So atmospheric carbon dioxide concentration is now 339 ppm and is increasing @ 1.5 ppm yearly. Due to green house effects of carbon dioxide, there will be an increase in atmospheric temperature causing the melting of ice caps with the resultant increase in the level of seawater and the predictable dangerous consequences. Recent estimate shows that at present rate of carbon dioxide emission, the atmospheric temperature will be increased by one third of a degree and the level of seawater 6cm per decade. But contrary to the general belief, with the increase of carbon dioxide level, crop yield is likely to be increased. But certainly there is global warming with increased emission of carbon dioxide. On the other hand, we know that carbon dioxide plays an important role in temperature control of the earth. Some part of the sunlight coming to the earth is converted to heat. Normally all this heat would have radiated in the space had there been no carbon dioxide in the atmosphere. But carbon dioxide absorbs a part of the heat and thus keeps the mother earth optimally warm, otherwise, it would have been as cool as the moon having no atmosphere.

Emission of Carbon Monoxide

It is a poisonous gas as such. Incomplete burning of organic materials results in carbon monoxide production. It has 200 times more affinity for haemoglobin of human blood than that of oxygen, which can be fatal sometimes. Crop residue burning produces carbon monoxide from 33 kg of this gas for each tonne sugarcane residue to 93 kg per tonne of red bean residue burning. For other crops, the production of carbon monoxide are 74, 64, 53, 68, 72, 41 & 56 for pea, wheat, alfalfa, oat, sorghum, safflower, rice and maize respectively (EPA, USA, 1976).

Production of particulate on crop residue burning

Minute solid particles of dust, hydrocarbon and plastic floating or suspending in air replacing some other constituents of it are called particulate matter. These particles are capable of reflecting back a part of sunlight reaching the earth, causing a cooling effect. With the increase of such matter in the atmosphere the earth may be possibly so cool that it may revert us back to ice age. These are all speculation indicating that particulates have dire environmental consequences if we do not take care of them. Quantitatively, particulate production from crop residue burning may vary from 4 Kg to 22 Kg out of one tonne of rice and oat crop residue. Figures for other crops are 11, 9, 94, 22, 16 for maize, soybean, sugarcane, redbean, and pea respectively (EPA, USA, 1976).

Production of ash on crop residue burning

The ash becomes a waste after residue burning although it contains a lot of minerals. About one third carbon in ash remains on soil surface. Burnt residue absorbs a part of herbicides (Ramusseen & Smiley, 1989). Burning of crop residue produces 1.8 tonne ash waste per hectare in case of alfalfa residue, 3.8, 6.7, 5.6, 3.6, 4.3 & 9.4 tonne per hectare in case of barley, grass, sorghum, sugarcane, maize and wheat crop residue. It becomes a huge handling problem for such a huge quantity of ash (EPA,USA, 1976).

Nitrogen oxides and sulphur dioxide from crop residue burning

Apart from the evolution of carbon dioxide, carbon monoxide, water vapour and hydrocarbons, some other gases are lost during burning of crop residue. These are nitrogen oxides and sulphur dioxide resulting into further nutritional loss of the residue. Note that nitrous oxide is involved in the photochemical destruction of the ozone shield of the upper atmosphere leading to the well-known allied problems. Acid rain results due to the presence of nitric oxide, nitrogen dioxide and sulphur dioxides. The exact quantification of production of these gases through residue burning is yet to be estimated (Ramussen & Smiley, 1989).

Open burning of crop residue

Open burning of crop residue on surface ground is a general practice in many countries. Here are given some of the important factors that affect the open burning of crop residue: (I) Temperature: More burning under higher temperature (ii) Composition of crop residue: Better burning or rapid burning with better quality of crop residue. (iii) Wind: More burning under speedy wind velocity. (iv) Residue compaction: Slower burning under more compact residue heap. (v) Percentage moisture of the residue: Slower burning of moist residue heap. It is generally accepted that with low temperature, more moisture of the residue encourages more amount of carbon monoxide, hydrocarbon and particulate matter production with the simultaneous suppression of nitrogen oxide evolution (EPA, USA, 1976).

Acetic acid production on crop residue burning

A number of organic acids are produced during the burning of crop residue. These acids are phytotoxic in nature. Acetic acid, butyric acid from corn residue; butyric, propionic and acetic acid from green manure and wheat crop residue; lactic, valeric, formic and acetic acid from rice straw residue are produced during residue burning. Acetic acid and some fatty acids are found to be phytotoxic. In the presence of nitrate nitrogen, the production of these acids declines since nitrate nitrogen is thought to act as an electron acceptor and fatty acid as a carbon source during decomposition process as reported by Harper and Lynch (1981).

Problems associated with maintaining crop residue on surface soil

(I) More proliferation of weed, pest, etc. (ii) Chemical weed control becomes essential. (iii) Placement of fertilizer becomes difficult. (iv) Crop stand becomes poorer because of difficult seed placements vis-a-vis problem of pathogen. (v) Surface soil becomes a bit acidic resulting into deactivation of some of the soil applied chemicals. (vi) Traffic lane is more compacted.

Denitrification of added residue

It is well-known that application of crop residue under unaerobic soil conditions encourages denitrification, leading to loss of nitrogen from soil although its dynamics are yet to be explored. Decomposition of cellulose and hemi-cellulose imparting major portion of crop residue is more important here.

DISCUSSION AND CONCLUSION

Globally we have a potential of 3507 million metric tonne of crop residue production from principal food crop, 41% of which is available for recycling as manure either through direct soil incorporation or compost production followed by soil application, rest consumed as food, feed and fodder for animals etc. This available portion of crop residue (1442 million metric tonne) is equivalent to 4765 x 10^{12} Kilo calorie energy, two-third of which is imparted by 3 food crops, rice, wheat and maize and the rest by other food crops like millet, sorghum, barley, total pulses, oil crops, sugarcane and potato. Recycling of crop residue in soil has a lot of benefit. It has a nutrient potential of 28.8 million metric tonne ($N+P_2O_5+K_2O$) half of which is expected to be mineralized in the current season in most of the parts of the world. And this can replace about 10% of the global demand of 160 million metric tonne of fertilizer ($N+P_2O_5+K_2O$). The other half will serve as a reserve for the subsequent seasons. In a sense, quantitatively, crop residue can meet 1/5th of the global fertilizer demand. Energetically, of course, the picture is somewhat different. Crop residue can substitute during the current season 5% of the total 1553 x 10^{12} Kcal. of global energy equivalence for fertilizer. Thus, annually it can account for a supply of 151.5 x 10^{12}amounting to 10% of the global fertilizer energy requirement. Apart from chemical angle, crop residue benefits the soil both physically and biologically and ultimately increases the crop yield. But it is not an unmixed blessing. Decomposition of crop residue under anaerobic condition leads to nitrogen loss from soil through denitrification and evolution of acetic acid, butyric acid and other phytotoxic acids. Presently, in the absence of a proper avenue for its disposal, a part of residue is usually burnt. This produces a lot of gaseous and non-gaseous matter affecting our environment directly or indirectly. 1.83 billion tonnes of CO_2 emission from residue burning causes green house effect and the associated problems. Incomplete burning of crop residue produces the poisonous carbon monoxide causing health hazards like suffocation and even death. Particulate production ranging between 4 to 22 Kg per tonne burning of crop residue reflects back the incoming solar radiation probably having a cooling effect on the atmosphere. Ash production ranging between 1.8 to 9.4 tonnes per hectares of residue burning becomes a handling problem, moreover it absorbs herbicide and pesticide sprays making these ineffective. Residue burning also produces evolution of nitrous oxide implicated in the photochemical destruction of ozone layer. Acid rain results due to evolution of nitric oxide, nitrogen dioxide and sulphur dioxide during residue burning. Maintaining a thick layer of crop residue on soil surface instead of its incorporation makes fertilizer and seed placement and weed control difficult, encourages pathogen growth, absorbs part of herbicides and pesticides or nutrient sprays. Thus, open burning of crop residue is wasteful as well as harmful. Even if we use crop residue to fuel the thermal power plant, it is not renewable. But recycling the crop residue as manure either through direct soil incorporation or by composting followed by soil application is a much more rational and renewable approach in this age of energy crisis since crop residue begets crop residue in addition to crop production.

FUTURE THRUST

a) Crop residue is a complete nutrient packet. Its total nutritional impact on crop production from a micronutrient point of view is yet to be ascertained for different crops and cropping systems.

b) A systematic agro-ecosystem-wise detailed study on the effect of crop residue on soil flora and fauna would be a well-appreciated line of research.

c) The below ground part of crop residue including root, rootlets and stubbles and its effect on soil properties and in turn on the following crop in a cropping system and their energetics are yet to be worked out fully.

d) Energetics of residues based cropping systems will be a right approach towards renewable energy research.

REFERENCES

1. Abshahi, A., Mills, F.J. and Broadbent, F.E. 1984. Nitrogen utilization by wheat from residual sugarbeet fertilizers and soil incorporated sugar crops. *Agron J.* 76: 954-958.
2. Anonymous 1998-99. Long term integrated nutrient supply system in cereal-based crop sequences. *Ann. Rep.* PDCSR, Modipuram, India, pp. 59-64.
3. Crossley, D.A. Jr., House, G.J., Snider, R.M., Snider, R.J. and Stinner, B.R., 1984. The positive interaction in agroecosystems pp. 73-81. In R.R. Larse, B.R. Sinner and G.H. House (Ed). Agricultural ecosystems-unifying concepts-J*ohn Willey & Sons, New York.*
4. Gotah, S. and Onikuru, Y. 1991. Organic acid in a flooded soil amended with rice straw and their effect on the growth of rice. *Soil Sc. and Plant Nutrition* 17(1):1-8
5. Hadimani, A.S., Hedge, B.R. and Satyanarayana, T., 1982. Management of red soils. Review of soil research in India, Part-II, and *12th international congress of soil science.* February, 8-166, 1982, New Delhi. pp. 686-702.
6. Harper, S.H. and Lynch, J.M. (1981). The kinetics of straw decomposition in relation to its potential to produce the phytotoxic acetic acid. *J. Soil Sc.* 32: 627-637.
7. Hwang, S.W., Ryu, I.S. and Park, J.K. 1989. The effect of application of rice straw and gypsum on the chemical properties of the soil and rice growth in saline soil. *Res. Rep. of Rural Dev. Adm. Soils and Fert.* 31(1): 37-50.
8. Knapp, F.D., Elillett, L.F. and Campbell, G.S. 1983. Carbon, N and microbial biomass interrelationship during the decomposition of wheat straw. A mechanistic simulation model. *Soil Biol. and Biochemistry* 15: 455-461.
9. Mager, S.S. 1982. Soil-water management aspects related to crop production in black cotton soils (Vertisols). In: *Proc. of the workshop in water Resources Planning and Management*, Jan. 22-24, 1982. At Walmi, Aurangabad, Maharashtra, India p. 5-18.

10. Mishustin, E.W., Vestrov, I.S., Delgikh, Yu. R., Rao, V.R. and Sidorenko, O.D. 1975. The use of rice straw as an organic manure for rice. *Agrokhimiya* No. 7: 80-87.
11. Mittal, V.K., Mittal, J.P. and Dhawan, K.C. 1985. Research digest on energy requirements in agricultural sector (1971-82). Coordinating Cell. All India Coordinated Research Project on Energy Requirements in Agricultural Sector, Punjab Agricultural University, Ludhiana, pp. 259.
12. Ramusseen, P.E., and Smiley, R.W., 1989. Long term management effect on soil productivity and crop yield in semi-arid region of eastern Oregeon. *Station Bull 675. Columbia Basin. Agril Res. Stn., Pendleton, Oregeon*, pp. 57.
13. Sakar, A. 1992. Studies on integrated nutrient management in maize-potato system under upland condition. Proceedings of the 79th *session of Indian Science Congress Association, Section X, Agricusltural Science Article* No. 18 pp. 11.
14. Sakar, A. 1996. Energy use pattern in sub-tropical rice-wheat cropping under short-term application of crop residue and fertilizers. *Agri. Ecosystems & Environ. (Netherlands)*. 61: 59-67.
15. Smith, G.E. 1942. Sauborn field: fifty years experiments with crop rotations, manures and fertilizers. *Missouri Agri. Exp. Stn. Bull* 458.
16. Stott, D.F., Stroo, H.F., Elliet, L.P., Papendick, R.I. and Unger, P.N. 1990. Wheat residue from field under no-till management. *Agron, J*., 54(1):92-98.
17. Stout, E.A. 1982. Agriculture biomass for fuel experimentation. 38: 145-151.
18. Tandon, H.L.S., 1995. Fertilizers, organic manure, recyclable waste and biofertilizers (Ed.). *Fertilizer Development & Consultation Organization*, New Delhi pp. 36-51.
19. Unger, P.W. and McCalla, T.M. 1980. Conservation tillage System. *Adv. Agron* 33:2-53.
20. Verma, U.N., Srivastava, V.C. and Prasad, R.S. 1989. Effect of zero tillage on wheat (*Triticum aestivium*) with rice (*Oryza sativa*) stubble mulching. *Indian J., Agric. Sc.* 59(10):669-671.
21. Venkateshwarlu, J. 1987. Efficient resource management systems for dryland of India. *Adv. Soil Sc*. 7:166-221.
22. Vlek, P.L.G. 1990. The role of fertilizers in sustaining agriculture in Sub-Saharan, Africa. *Fert. Res*. 26:327-339.
23. Wagger, M.G., Kissel, D.E. and Smith, S.J. 1987. Mineralization of N from N^{15} labelled crop residue under field conditions. *Soil Sc. Soc. A., J*. 49:1220-1226.
24. Will, B.R., Norman, R.J. and Helms, R.S. 1989. The effect of fertilization of wheat when wheat follows rice in rotation. *Res. Series Arkansus Agri. Expt. Stn*. No. 385:35-37.
25. Wold, J.D., Rhodes, G.N., Groveel, J.G., Gosaver, B.M., Arvin, M.K. and Churck, R.L. 1989. Activity of imazaguin in soil solution as affected by incorporated wheat (*Triticum Aestivum*) straw. *Weed Science*. 37(2):254-258.

Renewable Energy: Trends and Prospects. Edited by S.K. Majumdar, E.W. Miller and A.I. Panah

Chapter Five

RENEWABLE ENERGY RESOURCES FROM THE OCEANS

B. BHATTACHARYA[1] and S.K. SARKAR[2]

[1]Department of Metallurgical Engineering
Jadavpur University
Calcutta 700 032, India

[2]Department of Marine Science
University of Calcutta
35, B.C. Road
Calcutta 700 019, India

INTRODUCTION

Energy is the capacity for doing work and can be restored in other different forms such as gravitational, kinetic, heat, elastic, electrical, chemical, radiant, nuclear and mass energy. So-called energy resources are stores of convertible energy. The oceans receive, store and dissipate energy through various physical processes. Abundant ocean energy sources include tidal, wave, current, thermal and salinity gradient energy. Although the total energy of these renewable resources is large, only a fractional component of their potential is likely to be generated due to the following reasons: (i) ocean energy is spread diffusely over a wide area, requiring large and costly plants for its collection and (ii) the energy is often available in areas remote from centers of consumption.

The United Nations conference on Environment and Development (UNCED) held in Rio de Jenerio, Brazil in June 1992 addressed the challenges of achieving worldwide sustainable development. The goal of sustainable development cannot be realized without major changes in the world's energy system. Accordingly, Agenda 21, which was adopted by UNCED, called for "new policies or programs as appropriate, to increase the contribution of environmentally safe and sound and cost effective energy system, particularly new and renewable ones, through less polluting more efficient energy production, transmission, distribution and use".

Renewable, nonpolluting sources of energy can be generated from the motions and patterns of heat distribution in the atmosphere and oceans which include: (i) kinetic energy of the winds (ii) heat stored in the oceans (iii) potential and kinetic energy of tides and currents and (iv) potential and kinetic energy of waves.

There are about eight ingenious methods from which energy can be generated from the oceans. These are:

(a) tidal energy
(b) wave energy
(c) ocean thermal energy conversion
(d) salinity gradient
(e) biomass conversion
(f) ocean winds
(g) ocean currents
(h) ocean geothermal

TYPES OF OCEAN ENERGY SOURCES

(a) *Tidal Energy*

The tides are periodic rise and fall of sea level over a given time interval, and they are caused by the interaction between the gravitational attraction of the moon and the sun on the earth, and the centrifugal force resulting from the rotation of the earth and the moon. The ocean tides generated by this gravitational attraction of the sun and the moon on the mass of the ocean affect every particle of water from the surface to the deepest part of the ocean basin. Hence the tide obviously has a more pronounced effect on ocean phenomena than one can observe at the ocean surface.

The earth has a diameter of 12,682 km. The diameter of the moon is 3,478 km which is 0.27 that of the earth. The diameter of the sun is 1,392,000 km which is 109 times the diameter of the earth. These relative distances seem to be a major factor in determining their tide-generating effect on the earth where m_1 and m_2 are the masses and r is the distance between the masses.

$$\text{Tide-generating force} \qquad \frac{m_1 m_2}{r^3}$$

Based on relative masses, the sun is 27 million times more larger than the moon and has 27 million times the tide-generating effect. But, since the sun is 390 times further than the moon from the earth, the tide generating effect is reduced by 59 million times.

Tidal power may be regarded as a technically proven, dependable, clean and long-lived source of electric power. The exploitation of wave energy, by comparison, is still in its primary stage. Small shoreline and near shore devices are likely to be developed first, but their applicability and potential is restricted. A negative consideration involves the periodicity of the tides. Power would be generated throughout only a portion of a 24 hour day unless special design features were included in the construction of the facility. Because the tides operate on a lunar period and our energy demand operates on a solar period, the energy available through generating power from the tides could only be properly utilized whenever the availability of lunar tide generating force coincides with human working hours. Extraction of tidal energy is considered practical only when the tides are large and suitable sites for tidal plant construction can be found.

Tidal energy which involves the estuarine barrages at sites having high tidal ranges, offers the best prospects in the short to medium term (Cavanagh, Clarke and Price, 1993). The components are also commercially available for installing a tidal plant. A tidal energy scheme consists of a barrage or dam to be constructed across an estuary, being provided with a series of gated sluices. The construction method usually proposed for the main barrage structure involves the use of caissons (Bernshein, 1986: Bay of Fundy Tidal Power Review Board, 1997; U.K. Energy paper, 1981, 1989; Energy Technology Support Unit, 1988, 1990).

Two types of turbine generator are widely used:

(i) conventional bulb turbine and
(ii) rim-generator turbine

The technical set-up in conventional bulb turbine is generally used for larger machines, such as those at La Rance and those proposed for the Severn (UK Energy Paper, 1989). For smaller machines (as in the case of the Mersey and the Wyre in the United Kingdom), a geared bulb turbine in a pit configuration may be cheaper (Energy Technology Support Units, 1991). A large prototype of the rim-generator turbine has been successfully tested at Annapolis, Nova Scotia, in Canada and is favored for future tidal installations in Canada's Bay of Fundy (Fundy tidal power workshop papers, 1990).

The world's first large tidal power station was built across the estuary at La Rance, France and is operating satisfactorily. The estuary has a surface area of approximately 23 km^2 and the tidal range at La Rance reaches a maximum of 13.5 m. A barrier was built across the estuary a little over 3 km upstream, where it is 760 m wide to protect it from storm waves. The deepest water ranges from just over 12 m at low tide to more than 25 m at high tide. Sluices (artificial channels) have been built into the barrier to allow water to flow through the barrier where the generator units are shut down. It can produce 240 million watts of power- enough for a city of 300,000 people. As the tide falls, water is kept stored at high tide level inside the barrage. When the difference in water levels is about 3 meter, water is allowed to flow out of the barrage to the sea, flowing through 24 huge turbines which drive electricity generators. As the tide rises again, water is allowed to flow through the barrage to fill the estuary, until it reaches high tide level and the processes begins again in repetitive cycles. The La Rance barrage was originally operated in two-way generation i.e., extracting energy from both ebb and flow tides (La Rance 20th anniversary colloquium papers, 1986). Pumping is increasingly favored because of its ability to increase energy output. Studies in the United Kingdom on a number of tidal-energy schemes indicate that the energy gain achieved in such pumping (i.e., operating the turbines in reverse to act as pumps), as compared with simple ebb generation, is small but useful, typically ranging from 5 to 15 per cent (UK Energy Paper, 1989).

Tidal energy offers significant environmental impacts. It is nonpolluting and can replace coal and hydrocarbon fuels. The most significant benefit of using the tides to generate electrical power would be in reduced operating costs as compared to the conventional thermal or nuclear power plants that require fossil fuel or radioactive isotopes. Besides prevention of a huge amount of carbon dioxide emissions, a barrage also protects vulnerable coastlines against storm-surge tide. Still, tidal-energy projects may change the surrounding estuarine ecosystem; site-specific environmental impact assessments are therefore needed to identify these changes and determine their acceptability before construction begins.

The variability in tidal power generation is fully predictable. Such variability could be evened out by servo-controlled variable pitch-bladed turbines or using captive power plants (like gas turbines) to support during short falls.

TABLE 1

Existing Tidal Energy Plants in the World

Name of the station	Mean tidal range (meters)	Basin area Km^2	Installed capacity (MW)
LaRance, France	8.0	17	230
Kislays Guba, Russia	2.4	02	0.4
Jiangxia, China	7.1	02	3.2
Annapolis, Canada	6.4	06	17.8
Potential Sites in India			
Gulf of Kutch	900 MW		
Gulf of Cambay	7000 MW		
Sunderbans (West Bengal)	50 MW		

(b) *Wave Energy*

Ocean waves, created by the interaction of winds with the sea surface, contain both kinetic and potential energy, the later is a function of the amount of water displaced from the mean sea level. Ocean waves represent a sizeable renewable energy resource and for which a number of countries, including Japan, India, China, Denmark and others, have undertaken wave - energy research. The highest concentration of wave energy occurs between the latitudes 40° and 60° in each hemisphere because of the presence of strong winds. Wave energy exhibits localizations both in scale (frequency) and physical space (time), which implies the energy intermittency that energy at a given frequency may not be evenly distributed over the whole time span. Especially the local peak and the corresponding peak frequency of the local wavelet energy spectrum have significant fluctuations as time evolves. Such fluctuations imply that wind waves may have the instaneously remarkable energy, which is much more than the average over this time interval determined by the dominant wave component and may cause great damages to the offshore structures. The west coast of Europe and the United States, and the coast of New Zealand and Japan are particularly suitable for wave energy extraction (Leishman and Scobie, 1976).

Wave Activated Turbine Generators (WATG) have been in use for several decades but each unit produces a very small amount of electricity. An Oscillating Water Column (OWC) system is comprised of a chamber exposed to wave action through an entrance along the side. Due to wave action, the air inside the chamber expands under pressure and drives an air turbine at the top. The National Engineering Laboratory, Glasgow, UK first developed a conventional axial flow air turbine for power generation. Prof. Wells of Belfast University, UK developed a turbine which will rotate in the same direction with the pulsating air flow in and out of the system. This is an excellent turbine not requiring any valve. Many wave-power devices are required to be deployed offshore to produce considerable amounts of energy. It has been suggested that mass production might produce a decrease in the cost of each unit by 15 to 25 percent for each doubling of the cumulative production volume (Fisher, 1974). In India, research on wave energy was commenced in 1983, with the formation of an inter-disciplinary group at the Indian Institute of Ocean Engineering, IIT, Chennai was established, financed by Department of Ocean Development, New Delhi. The fishing harbor at Vizhinjam, near Trivandrum, offered most of the infrastructural facilities required for the construction of the wave energy plant. The wave energy plant operating in India demonstrates a clean and renewable conversion process from waves to electricity (Qasim, 1998).

Wave energy can reduce greenhouse-gas emissions and atmospheric pollution since this can partly replace energy from fossil fuels. The installation of a wave-power station would modify the local wave climate in the coastal environment and would create a potential hazard for ships because of their low height above the water, the devices would be relatively invisible either by sight or by radar

(Cavanagh, Clarke and Price, 1993). Uses of antifouling paints and their adverse effects on the biota would also have to be taken into consideration.

(c) *Ocean Thermal Energy Conversion*

In the tropical and subtropical oceans of the world, a natural temperature difference exists between the surface waters and those at depth. The concept of ocean thermal energy conversion (OTEC) exploits this temperature difference to drive power plants to produce electricity. Ocean thermal energy is a process developed to use the difference in temperature between warm tropical surface water and the cold water below the thermocline to generate electricity and ammonia. The upper layer of the tropical oceans are a vast reservoir of warm water that is held at a temperature near 27°C by a balance between the absorption of heat from the sun and the loss of heat by evaporation, convection and long-wavelength radiation. A practical and economical technology for converting even 0.01 % of the absorbed solar energy into electricity or fuel in a form suitable for delivery to consumers on land could have a profound impact on world energy availability and economics. The technology called thermal ocean energy conversion (OTEC) fulfills the technical and economic requirements for productive use of the solar energy continually absorbed by the oceans. Unlike other solar technologies, however, a reliable OTEC plant would be able to generate electricity continuously because the temperature difference lasts 24 hours a day.

The OTEC system operates according to a simple principle, which has been described as nothing more than sophisticated plumbing (Lavi et al., 1973). The system works in the opposite way from a typical refrigeration system and on a much larger scale. The main components of an OTEC system are; (a) Boiler: Warm surface water is used to evaporate a low-boiling point fluid (b) Turbine: The expanding gas is used to drive a turbine which generates electricity and (c) Condenser: The gas is condensed back into a liquid by using cold water pumped up from depths of about 1000 m. A 'working fluid' like ammonia and freon with a low boiling point is pumped into a closed tube exposed to the warm water zone. The heat of the warm water vaporizes the working fluid. The fluid is vaporized and passes through a turbine that drives an electrical generator. After passing through the turbine, the fluid is condensed by cold water that has been pumped up from the deep ocean. The condensed liquid is pumped into a boiler-evaporator to start the cycle afresh. An OTEC station capable of generating about 100 megawatts of electricity must pump about 450 m^3 of both warm and cold water through its heat exchanges per second (Grickspoor and Van der Pot, 1979).

No doubt, the power plant of this type would need sea-worthy marine platforms, moorings etc., on the one hand and energy transmission problems to the coastal areas.

The world's first commercial plant, constructed and installed on the island of Nauru in the Pacific, remained functional for some time. In India, Tamil Nadu is the only state which has recognized that OTEC could probably be an ideal substitute for conventional methods of energy production. Temperature and salinity profile along many stretches of the coastline of this state in different seasons along with a detailed knowledge of weather conditions are to be studied precisely.

This seems to be difficult to assess the environmental effects for the operation of an OTEC plant. The large flows hot and cold water might impose changes in local or even global weather patterns to a considerable extent. However, the carbon dioxide released would only be about one-fifteenth that released by oil or one twenty-fifth that released by coal in a plant of equivalent electrical output (World Energy Conference, 1992).

(d) *Salinity Gradient*

This is a very simple technique using the principle of osmosis. If a semipermeable membrane is placed between two salt solutions of different concentrations, then there is a natural movement of ions through the membrane until both sides attain equal concentrations. There exists a great difference in the osmotic pressure between freshwater and seawater which is utilized for generating electricity. A number of techniques (e.g., reverse dialysis etc.,) has been evolved for generating the energy in salinity gradients but none of them is still considered promising. A pilot salinity power station was carried out in Sweden using the salinity gradient mechanism. The generating potential of such a station was 200 MW.

(e) *Biomass Conversion*

Seaweeds which convert solar energy by photosynthesis can be processed to yield fuel as well as energy. The easily harvestable seaweeds like Sargassum spp., etc., could be cultured in suitable coastal waters adopting some conventional methods. The seaweeds grow to enormous size and these can be utilized for bioconversion.

(f) *Ocean Winds*

The winds that transform solar energy into ocean currents have for centuries been used by society to drive its machines. Wind velocity is generally high near the sea. Moreover, there are certain regions where the velocity of wind is high practically throughout the year. This provides enormous scope for generating power. Westerly winds off New England represent a near-shore wind supply that is reasonably sustained and contains a huge amount of energy. Offshore windpower systems (OWPS) could provide electricity to meet the needs of large sections of the North Atlantic coasts of the United States.

(g) *Ocean Currents*

The designs to generate power from the ocean currents are simple but the main problem is their low-energy density. Many have considered the great amount of energy in the Florida-Gulf Stream current system and dreamed of generating it to do the work of society. A group of scientists and engineers estimated that about 2000 megawatts (MW) of electricity could be recovered along the east coast of southern Florida. In 1973 William J. Moulton of Tulane University proposed the coriolis program. It is estimated that an array of 242 coriolis hydroturbines anchored in Gulf Stream current could provide 10% of the present electricity needs of Florida.

(h) *Ocean Geothermal*

Geothermal energy represents a potential and inexhaustible source of energy, which has been utilized in many countries for a long time. In the zone of tectonic activity surrounding the Pacific Ocean, molten rock lying beneath the Earth's surface (magma) may occur at depths of only a few kilometers (Qasim, 1998). The heated water may rise to the surface as "hot springs" or stream and energy can be derived from the continuous flow of this hot water or steam.

CONCLUSION

The renewable, non-polluting energy sources from the oceans have a profound impact on the future world economy. The mechanical motions such as waves and mean currents constitute a larger ocean energy source. Gravitational tide motion at sea is a possible source. Turbines mounted in dams or on moorings can capture some energy in elevated sea levels. Surface wave energy can be used. Coastal areas with narrow channels and large amplitude tides can be dammed with turbines. Persistent and strong currents like the Gulf Stream could be used for useful energy. All of the modern devices using these sources are coupled with electric generators. Large-scale demonstration plants are to be built as forerunners to commercial plants that will be economically feasible and environmentally benign. The start of the new millennium signifies another chapter in human endeavor and with it a clear dependence on the planet's oceans. We recognize that the oceans not only play a crucial role in the planetary processes that sustain life on earth, but also yield an increasingly diverse set of resources to support global economic prosperity. This creates a difficult balance between exploitation and sustainability.

REFERENCES

Bay of Fundy Tidal Power Review Board, 1977. Reassessment of Fundy tidal power, report of the Bay of Fundy tidal power Review Board and Management Committee.

Bernshein, L.B. 1986. Tidal power engineering in the USSR, Water Power and Dam Construction, March 37-41.

Cavanagh, J.E., Clarke, J.H. and Price, R. 1993. Ocean energy systems, In "Renewable Energy - sources for fuels and electricity", edited by T.B. Johanson, H. Kelly, A.K.N. Reddy and Robery H. Williams, pp. 513-547, Earthscan Publications Limited, London.

Energy Technology Support Unit (ETSU), 1988. Tidal power from the Mersey: a feasibility study, stage I, ETSU TID 4047.

Energy Technology Support Unit (ETSU), 1990. Conwy estuary feasibility study of tidal power, ETSU TID 4075.

Fisher, J.C. 1974. The energy crisis in perspective, Wiley, New York.

Grickspoor, W. and Van der Pot, B. 199979. OTEC - principles, problems and progress. Offshore Engineer September.

Lavi, A. et al., 1973. Plumbing the ocean depths: a new source of power IEEE Spectrum October.

Leishman, J.M. and Scobie, G. 1976. U.K. National Engineering Laboratory Report EAU M25.

Qasim, S.Z. 1998. Glimpses of the Indian Ocean, Universities Press (India) Limited, pp. 206.

U.K. Energy paper, 1981. Tidal power from the Severn Barrage, U.K. energy paper 46.

U.K. Energy paper, 1989. The Severn Barrage Project general report, U.K. energy paper 57.

World Energy Conference, 1992. Ocean energy, chapter 3 in World Energy Conference committee on renewable energy sources, opportunities and constraints 1990-2020.

Renewable Energy: Trends and Prospects. Edited by S.K. Majumdar, E.W. Miller and A.I. Panah

Chapter Six

OCEAN ENERGY CONVERSION – A REALITY

ANUPAM SARKAR

National Institute of Oceanography
Dona Paula, Goa 403004, India

INTRODUCTION

In order to cope with the ever-increasing global demand for supply of energy, the whole world is in search of renewable sources of energy. Most importantly, the demand is more for the industrialized countries, as they require more energy than the developing countries. In all physical and chemical changes, energy is neither created nor destroyed, but may be converted from one form to another. Most significantly, fossil fuels, coal, oil and natural gas have been the fuels of choice for performing these tasks. Fossil fuels are the most prominent among the renewables in the sense that they are finite resources, i.e. there is certain amount preserved under the earth and, once used, they will run out unless there is any unique technology evolved for their regeneration. The renewable energy, on the other hand, refers to fuel sources more consistently available than their fossilized counterparts. Sources for this energy commonly fall under five categories: i) Biomass (organic matter), ii) Geothermal (heat from under the earth), iii) Solar, iv) Water and v) Wind.

These sources are abundantly available in nature. As long as the earth revolves around the sun, the earth will continue to receive enormous amounts of harvestable energy from the sun. Heat energy from the sun creates atmospheric conditions conducive to wind and water production, although not in even quantities throughout the world or with scientific precision in any one particular location. Finally, the sun produces the light necessary for growing plants and trees, which constitute the biomass category. The use of natural resources was in practice to serve as energy resources long before the beginning of the industrial revolution. This is clearly evident from the historical records of watermill and windmill use, which dates back to ancient China (Liao, et al. 1986), Greece and Rome (Claude, 1930).

During the renaissance, a large number of innovative machines were built using the available energy for various purposes like grinding of grains, textile engineering and development of metal technology. Gradually more advanced technology has replaced the old technology for the same machines to meet the energy needs of a larger population and operate on similar principles, utilizing the energy of the world around us.

SIGNIFICANCE OF RENEWABLE ENERGY

The most significant part of renewable energy is perhaps the cleanliness. While fossil fuels require a combustion process (the use of an engine or turbine) to convert energy into a form capable of performing the task at hand, renewable energy sources do not require any such materials. No combustion means no emissions of byproducts that are likely to cause the most common types of air pollution today—acid rain, smog and climate change. There's probably no single answer, however, consider the following two examples related to solar (photovoltaic systems) and wind energy.

According to an estimate, about 830 kilowatt-hours of electricity are being used by an average U.S. household per month. Interestingly, the advantage of producing 1000 KWh of electricity with solar power is that it significantly reduces emissions by nearly 8 pounds of sulfur dioxide, 5 pounds of nitrogen oxides, and more than 1,400 pounds of carbon dioxide (Ditmars and Paddock, 1979; Green and Guenther, 1989; Keeling, 1968). It was projected that during the 28 years of clean energy production, for a rooftop system with 2-year payback, meeting half of a household's electricity use would be good enough to avoid conventional electrical plant emissions of more than half a ton of sulfur dioxide, one-third a ton of nitrogen oxides, and 100 tons of carbon dioxide (Ditmars and Paddock, 1979; Dunbar, 1981).

The American Wind Energy Association claimed that, in the wind power area, "A single 660-Kw wind turbine will displace emissions of 1,100 tons of carbon dioxide (the leading greenhouse gas), 6 tons of sulfur dioxide (the leading component of acid rain), and 4 tons of nitrogen oxides (the leading component of smog) every year." It is estimated on the basis of the U.S. average utility fuel mix that 375 acres (more than half a square mile) of forest would be required to absorb the same amount of CO_2.

In fact, the global interest for the conservation of energy started vigorously during the 1970s and early 1980s, primarily due to the exorbitant increase in the price of oil, which was induced by reductions in oil supplies as a result of the OPEC oil embargo in 1973 and the Iranian hostage crisis in 1979. The high price of oil spurred the whole world for development of renewable energy sources such as solar power, wind, geothermal, and biomass. In the late 1980s, however, the urge for the renewable energy waned as the price of oil plummeted. The government and consumers were not interested in investing in more costly renewable energy sources and programs when nonrenewable fossil fuels were so inexpensive and easily accessible.

It has been observed that, in recent years, there has been a greater interest in the issue of renewable energy, due to *environmental* concerns, especially the burning of fossil fuels, which are believed to contribute significantly to acid rain and global warming. Another factor contributing to the interest in energy issues is the realization of the developed countries like the United States of their increasing dependence on foreign oil. This became prominent during the Persian Gulf war. Moreover, in the changing scenario the public policy concerning the issues involving energy have tremendous economic implications (Penney and Daniel, 1989). In the present context of increasing demand for alternate sources of renewable energy it is of utmost importance to explore the natural resources like oceans, thermal energy, wind energy, tidal energy, wave energy, etc.

OCEAN THERMAL ENERGY CONVERSION

An enormous amount of energy is preserved in the oceans, which cover a little more than 70 percent of the Earth's surface. On an average day, 60 million square kilometers (23 million square miles) of tropical seas absorb an amount of solar radiation equal in heat content to about 250 billion barrels of oil. If less than 0.1% of this stored solar energy could be converted into electric power, it would supply more than 20 times the total amount of electricity consumed in a country like the USA on any given day.

Ocean thermal energy conversion is commonly known as OTEC. It is an energy technology that converts solar radiation to electric power. OTEC systems use the ocean's natural thermal gradient—the fact that the ocean's layers of water have different temperatures—to drive a power producing cycle. As long as the temperature between the warm surface water and the cold deep water differs by about 20°C (36°F), an OTEC system can produce a significant amount of power. The temperature difference between the surface and a depth of 1000 meters at different locations all around the globe is shown in figure 1. The oceans are thus a vast renewable energy resource, with the potential to produce billions of watts of electric power. This potential, according to some experts, is estimated to be about 10^{13} watts of base load power generation. The cold, deep seawater used in the OTEC process is also rich in nutrients, and it can be used to culture both marine organisms and plant life near the shore or on land.

OTEC is very promising and useful as an alternative energy resource for tropical island communities that rely heavily on imported fuel. OTEC plants in these markets could provide islanders with much-needed power, as well as desalinated water and a variety of mariculture products (Reed, 1982).

THERMODYNAMIC PRINCIPLES

According to the second law of thermodynamics, "all spontaneous processes are irreversible" or it can be said that "all spontaneous processes tend to equilibrium." As illustrated by Clausius (Taylor, 1924), "heat of itself, will not pass from a lower to a higher temperature." It is impossible for a self-acting machine, unaided by any external agency, to convey heat from a body at a low temperature to one at a high temperature. In order to convert heat into work, certain conditions are to be fulfilled. In fact, without the aid of an engine the conversion of heat into work is impossible. Further, the engine must work in a reversible cyclic process and operate between two temperatures. Then it will take up heat at a higher temperature, convert a portion of it into work, and release the rest of heat to a body at a lower temperature.

When heat flows from a reservoir at a higher temperature (called the source) to a reservoir at lower temperature (called the sink), it can be transformed into work with the intervention of an engine. The engine would pick up Q1 calories of heat from the source, transform only a portion of it into work W, and return the rest of the heat *Q2* to the sink.

According to the first law of thermodynamics,

$$W = Q_1 - Q_2$$

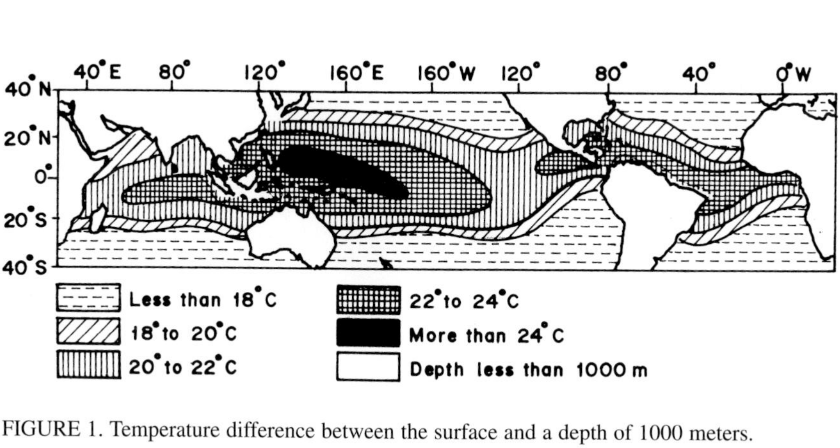

FIGURE 1. Temperature difference between the surface and a depth of 1000 meters.

The amount of work to be done from a given amount of heat will depend on the nature of the engine and the temperatures of the source and the sink. Thus the efficiency of an engine can be shown as follows:

$$\eta = \frac{\text{Work produced}}{\text{heat supplied}} = \frac{W}{Q} = \frac{Q_1 - Q_2}{Q_1}$$

$$\eta = 1 - \frac{Q_2}{Q_1}$$

Its utility, in practice, can be realized by considering a cyclic process. According to Carnot's theorem (Taylor, 1924), a reversible engine is more efficient than an irreversible one and all reversible engines are equally efficient.

The typical Carnot cycle consists of four successive operations using one gm mole of a perfect gas as the working substance. The gas taken is enclosed in a cylinder fitted with a frictionless piston. To start with, the cylinder containing the gas is kept in a large thermostat at the higher temperature (source) and suppose the volume of the gas is V_1.

Operation 1:

The gas is allowed to expand isothermally and reversibly from volume V_1 to V_2.

The heat absorbed by the gas = Q_1 (say)
The work done by the gas = $RT_1 \ In \ V_2 / V_1$,
Since the gas is ideal, we know $Q_1 = RT_1 \ In \ V_2 / V_1$

Operation 2:

The cylinder is next taken out of the thermostat and kept in a thermally insulated enclosure. The gas is allowed to expand further from volume V_2 to volume V_3 adiabatically and reversibly until the temperature falls down to that of the sink T_2.

The heat absorbed by the gas = nil
The work done by the gas = $C_v (T_1 - T_2)$ [C_v is the heat capacity of the gas]

Operation 3:

The cylinder is then placed in a thermostat at temperature T_2 (sink); the gas is compressed isothermally and reversibly until the volume changes to V_4 from V_3.

The heat given out by the gas = Q_2 (say)
The work done by the gas = $RT_2 \ In \ V_4/V_3$

[Since V_4 is smaller than V_3 the work is negative which means work is done on the gas.]

Operation 4:

The cylinder is now thermally isolated as in operation 2. The gas is then adiabatically and reversibly compressed to its initial volume V_1 and its original temperature T_1 is attained. The gas thus comes back to its initial state.

The heat absorbed by the gas = nil
The work done by the gas = $-\ C_v (T_1 - T_2)$

Thus the net work done by the gas in the complete cycle,

$$W = RT_1 \ In\ V_2/V_1 + C_v\ (T_1 - T_2) + RT_2\ In\ V_4/V_3 - C_v\ (T_1 - T_2)$$
$$= RT_1\ In\ V_2/V_1 - RT_2\ In\ V_3/V_4$$

But, considering the adiabatic change in operations 2 and 4, we have

$$T_1V_2^{\gamma-1} = T_2V_3^{\gamma-1}$$
$$T_1V_1^{\gamma-1} = T_2V_4^{\gamma-1}$$

Where $\gamma = C_p/C_v$ ratio of heat capacities at constant pressure and at constant volume

Thus $V_2/V_1 = V_3/V_4$

Therefore the net work done,

$$W = RT_1\ In\ V_2/V_1 - RT_2\ In\ V_2/V_1 = R(T_1 - T_2)\ In\ V_2/V_1$$

The efficiency of the process is given by the equation

$$\eta = W/Q_1 = R(T_1 - T_2)\ In\ V_2/V_1/RT_1\ In\ V_2/V_1 = T_1 - T_2/T_1 = 1 - T_2/T_1$$

and

$$W = Q_1(T_1 - T_2)/T_1 = Q_1\ \Delta T/T_1$$

This relation expresses the maximum amount of work obtainable from the heat flowing from T_1 to T_2.

It is thus amply clear that work can be done by a heat engine so long as a temperature difference exists between two heat reservoirs (in this case warm and cold seawater).

The maximum thermodynamic efficiency of such system is

$$\eta = \frac{T_1 - T_2}{T_1}$$

where T_1 is the temperature of the hot reservoir and T_2 is the temperature of the cold reservoir in absolute degrees. Using a closed cycle, a suitable fluid with good heat transfer characteristics and a high vapor pressure at ambient temperature can generate power. Among the fluids that are generally used are ammonia and propane.

APPLICATIONS AND ACHIEVEMENTS IN OTEC TECHNOLOGY

It was a French physicist, Jacques Arsene d'Arsonval (1881), who first proposed to utilize the thermal energy of the ocean. An open-cycle OTEC system was built in 1930 for an experiment at Matanzas Bay, Cuba (Claude, 1930). The system produced 22 kilowatts (KW) of electricity by using a low-pressure turbine. Claude constructed another open-cycle plant aboard a 10,000-ton cargo vessel moored off the coast of Brazil in 1935. Unfortunately, both plants were destroyed by weather and waves.

In 1974, the Natural Energy Laboratory of Hawaii (NELHA, formerly NELH) was established at Keahole Point on the Kona coast of the island of Hawaii to explore the ocean thermal energy (Harms, 1981). It has become the world's foremost laboratory and test facility for OTEC technologies. The first 50 kilowatt (KW) closed-cycle OTEC demonstration plant went up at NELHA and was known as "Mini-OTEC." The plant was mounted on a converted U.S. Navy barge moored approximately 2 kilometers off Keahole Point (Harms, 1981). A cold-water pipe was used in the plant to produce 52 KW of gross power and 15 KW net power.

In 1980, OTEC-1, a test site for closed-cycle OTEC heat exchangers was built by the U.S. Department of Energy (DOE) and installed on board a converted U.S. Navy tanker. It was tested and the results obtained were utilized to identify the methods for designing commercial-scale heat exchangers and they proved that OTEC systems can be operated from slowly moving ships with little effect on the marine environment (Fujioka, et al.,1980). A new design for suspended cold-water pipes was validated at that test site. In order to promote the commercial development of OTEC technology, two laws were enacted in 1980: the Ocean Thermal Energy Conversion Act, Public Law (PL) 96-320, later modified by PL 98-623, and the Ocean Thermal Energy (Block and Valenzuela, 1985).

The desalinated water was produced by using the open-cycle process at Hawaii's Seacoast Test Facility, which was established as a joint project of the State of Hawaii and DOE. In order to demonstrate how large polyethylene cold-water pipes can be used in an OTEC system, a 1-meter-diameter cold seawater/0.7-meter-diameter warm-seawater supply system was deployed at the test site. Japan has developed a shore-based, 100-KW closed-cycle plant in the Republic of Nauru in the Pacific Ocean (Iso, et al., 1986). This plant has used cold-water pipe laid on the seabed to a depth of 580 meters. The working fluid and the heat exchanger used in this plant were *Freon* and a *titanium* shell-and tube respectively. The plant has exhibited excellent results beyond the expectations by producing 31.5 KW of net power during continuous operating tests. Later, aluminum alloy was found to be more useful in place of more expensive titanium to make large heat exchangers for OTEC systems, especially because of its ability to control biofouling and corrosion (Larsen-Basse, 1984). Biofouling does not appear to be a problem in cold seawater systems. However, in warm seawater systems, it can be controlled with a small amount of intermittent chlorination (Panchal, et al.,1984).

A vertical-spout evaporator was developed to convert warm seawater into low-pressure steam for open-cycle plants by the scientists at the Solar Energy Research Institute (SERI, presently known as the National Renewable Energy Laboratory), in 1984. It was observed that energy conversion efficiencies as high as 97% can be achieved. Moreover, direct-contact condensers were also shown to be an efficient way to dispose of steam (Francis, et al., 1980).

The most significant achievement in regard to OTEC technology was the production of a record 50,000 watts of electricity during a net power-producing experiment with an open-cycle OTEC plant at Keahole Point, Hawaii (Harms, 1981, NELH, 1988). Thus scientists from different parts of the world focused their attention to develop new, cost-effective, state-of-the-art turbines for open-cycle OTEC systems (Nihous and Vega, 1988). In this regard, it is of immense importance for the researchers to ponder the following points:

- Collect data on appropriately sized demonstration plants for the operation of an OTEC plant
- Create a database of information on materials, design, deployment, and installation to develop and characterize cold-water pipe technology
- Conduct extensive research on the heat exchanger systems to improve heat transfer performance and decrease costs
- Develop innovative turbine concepts for the large machines required for open-cycle systems
- Evaluate and identify advanced concepts for ocean thermal energy extraction

Besides these points, there are several environmental as well as economic parameters that would be required for an in-depth analysis to set up of a prototype plant.

- Environmental parameters for site analysis - currents, wave, wind, tide, geothermal, (other sources of warm water), water quality, bottom survey, water temperature (time and spatial variation)
- Economic parameters for site analysis - site availability, market for product, social benefit, construction cost, maintenance and operation cost, environmental impact, salvage value.

In order to observe the suitability of a site for a plant facility it is necessary to look into the following four factors:

- Maximum gradients of water temperature with high volumes of water supply
- The suitability of multiple use of plant output, i.e. power, freshwater, aquaculture
- Social benefits
- Proximity of market for plant products

BENEFITS OF OTEC

It is of great significance to evaluate the use of an ocean thermal energy conversion (OTEC) plant and the continued OTEC development from the point of view of the economic and non-economic benefits. The economic benefits from the use of OTEC plant are that it:

- generates baseload electrical energy
- facilitates the production of fuels such as hydrogen, ammonia, and methanol
- is useful as a source for on-shore and near-shore mariculture operations
- provides moderate-temperature refrigeration
- supplies necessary power for air-conditioning
- has great potential to supply clean and cost-effective electricity for the future
- produces desalinated water for agricultural, industrial, and residential purposes

The noneconomic benefits of the use of OTEC to achieve global environmental goals are that it:

- reinforces energy independence and energy security
- accelerates competitiveness and international trade
- helps alleviate greenhouse gas emissions resulting from burning fossil fuels
- helps pursue international sociopolitical stability

The most significant usefulness of OTEC is that it is very beneficial in small island nations, as it provides self-sufficiency, minimal environmental impacts, and improved sanitation and nutrition, which result from the greater availability of desalinated water and mariculture products (Bardach, 1981).

WIND ENERGY

Wind energy is another type of energy abundantly available on this planet. Wind is produced by the movement of air due to uneven heating of the earth's surface by the sun. Wind power has been used for thousands of years to convert the wind's kinetic (motion) energy into mechanical energy for grinding grain or pumping water, sailing ships, etc. Modern science has used wind machines increasingly to produce electricity.

There are mainly two most common types of wind machines, *horizontal* and *vertical*, used for producing electricity. Horizontal machines have blades that look like airplane propellers. Vertical machines look like giant eggbeaters. It is quite easy to maintain the vertical machines and they can accept wind from any direction. Moreover, it does require protective features to guard against high winds. However, horizontal machines produce more electricity, and, for this reason, are used far more than their vertical counterparts.

It may be noted that most electricity production systems are based on large wind farms. Interestingly, most of the wind farms are not owned by public utility companies. Instead, independent producers, who operate the farms, sell electricity back to utility companies for distribution. The Public Utility Regulatory Policies Act (PURPA) requires utility companies to purchase electricity from independent energy producers at fair and nondiscriminatory rates. It was reported that wind energy provided the United States with about 10% of its total electricity in 1990, with California producing 98% of this amount. It is believed with great optimism that wind energy will provide much more of our future electrical production. There are some advantages and disadvantages of the use of wind energy as enunciated below:

Advantages of Wind Energy	***Disadvantages of Wind Energy***
Produces no water or air pollution	Wind farms require significant amounts of land
It is a "free" source of energy	Requires constant and significant amounts of wind
Relatively inexpensive to build wind farms	Can have a significant visual impact on landscapes
Can make use of the lands around wind farms	

TIDAL ENERGY

The word "tides" is a generic term used to define the alternating rise and fall in sea level with respect to the land that is produced by the gravitational attraction of the moon and the sun. There are some nonastronomical factors such as configuration of the coastline, local depth of the water, ocean-floor topography, and other hydrographic and meteorological parameters which are likely to play an important role in altering the range, interval between high and low water, and the times of arrival of the tides.

The most familiar evidence of the tides along the seashores is the observed recurrence of high and low water—usually, but not always, twice daily (NOAA, 1989). The term tide can be correctly referred to as a relatively short period, astronomically induced vertical change in the height of the sea surface (exclusive of wind-actuated waves and swell). But the expression of tidal current relates to accompanying periodic horizontal movement of the ocean water, both near the coast and offshore (but as distinct from the continuous, stream-flow type of ocean current).

The most striking point is that the major influx of energy to this planet, solar energy, is converted by natural processes, i.e. through wind generation, to the energy associated with waves. The energy thus contained is significant, in favored latitudes with values of around 70 MW/km of wave frontage (Burdette and Gordon, 1983). It is believed that wave energy could contribute significantly to world energy supply if an innovative technology could be made to extract this energy economically. Similarly, the tides, a result of the gravitational interaction of the Sun, Moon and Earth and the centrifugal force of the Earth-Moon system, carry enough energy to enable generation of 400 GW worldwide, if a tidal range of four meters is considered as the minimum for efficient extraction (Burdette and Gordon, 1983).

The most useful part of tidal energy is that given the right coastal environment, resonance can occur, leading to far greater than average tidal ranges, usually at the entrances to large estuaries, which could relatively conveniently be blocked off. Such situations are generally found in the Bay of Fundy, Canada, with a mean tidal range of 10.8 meters, the largest in the world, or in the Severn Estuary in Britain, with a mean range of 8.8 meters, making large scale projects at both these locations verge on the economical. The advantage of the use of tidal energy is that it can be applied with conventional technology in small scale.

IMPACT OF ASTRONOMICAL FORCES ON TIDE

High tides are produced in the ocean waters by the "heaping" action resulting from the horizontal flow of water toward two regions of the earth, which represents positions of maximum attraction of combined lunar and solar gravitational forces. Low tides are created by a compensating maximum withdrawal of water from regions around the earth midway between these two humps. The alternation of high and low tides is caused by the daily (or diurnal) rotation of the earth with respect to these two tidal humps and two tidal depressions. The changing arrival time of any two successive high or low tides at any one location is the result of numerous factors.

The tide-raising forces at the earth's surface thus result from a combination of basic forces: (1) the force of gravitation exerted by the moon (and sun) upon the earth; and (2) centrifugal forces produced by the revolutions of the earth and moon (and earth and sun) around their common centers-of-gravity (mass).

With respect to this center-of-mass of the earth-moon system (known as the barycenter), the above two forces always remain in balance (i.e., equal and opposite). However, at local points on, above, or within the earth, these two forces are not in equilibrium, and oceanic, atmospheric, and earth tides are the result.

The center of revolution of this motion of the earth and moon around their common center-of-mass lies at a point approximately 1,068 miles beneath the earth's surface, on the side toward the moon, and along a line connecting the individual centers-of-mass of the earth and moon.

It is noteworthy that noticeable differences can be observed in the successive time of occurrence and the heights of the tides. In the open ocean, the actual rise of the tidally induced wave crest is only one to a few feet. But considerable variations in the height of sea level can be detected only when the tidal crests and troughs move into shallow water, against land masses and into confining channels. The observed variations in the heights of the tides and in their times of occurrence are greatly influenced by the following factors.

i) Since the ocean waters possess a definite, although small, internal viscosity, this property prevents their absolute free flow, and somewhat retards the overall movement of the tides.

ii) The ocean waters follow the principle of traveling waves in a fluid. As the depth of the water shallows, the speed of forward movement of a traveling wave is retarded, as deducted from dynamic considerations. In shoaling situations, therefore, the advance of tidal waters is slowed.

iii) A certain relatively small amount of friction between the water and the ocean floor affect the movement of the tides, particularly as they move inshore. Further internal friction (or viscosity) between tidally induced currents and contiguous current in the oceans, especially where they are flowing in opposite directions— also affect the tidal movement.

iv) The presence of land masses imposes a barrier to the progress of the tidal waters. Where continents interpose, tidal movements are confined to separate, nearly closed oceanic basins and the sweeps of the tides around the world is not continuous.

v) Topography on the ocean floor can also provide a restraint to the forward movement of tidal waters - or create sources of local-basin response to the tides. As these waters enter confined bays, estuaries and harbors, restrictions to the advance of tidal waters imposed both by shoaling depths and the sidewalls of the channel can further considerably alter the speed of their onshore passage.

vi) In such particularly confined bodies of water, so-called "resonance effects" between the free-period of oscillation of the traveling, tidally induced wave and that of the confining basin may cause a surging rise of the water in a phenomenon basically similar to the action of water caused to "slosh" over the sides of a wash basin by repeatedly tilting the basin and matching wave crests reflected from the opposite side of the basin.

All of the above, and other less important influences, can combine to create a considerable variety in the observed range and phase sequence of the tides as well as variations in the times of their arrival at any location.

DATA NECESSARY FOR PRACTICAL APPLICATION OF TIDAL POWER

The detailed knowledge of the heights, times, and extent of inflow and outflow of tidal waters is of immense importance in a wide range of practical applications as stated below:

i) Navigation through intracoastal waterways, and within estuaries, bays and harbors; work on harbor engineering projects, such as the construction of bridges, docks, breakwaters, and deep-water channels.
ii) Establishment of standard chart datums for hydrography and for demarcation of a base line or "legal coastline" for fixing offshore territorial limits both on the sea surface and on the submerged lands of the Continental Shelf.
iii) Provision of information necessary for underwater demolition activities and other military engineering uses.
iv) Furnishing of data indispensable for fishing, boating, surfing, and a considerable variety of related water sport activities.

PROSPECTS OF TIDAL POWER PLANTS

The prospects for tidal power are similar, if not worse. Even though there is obviously a tremendous amount of energy available in the tides of the oceans, only a few sites with exceptionally high tidal oscillations due to resonances within a bay are likely to be economical, notably the Bay of Fundy in Canada and the Severn Estuary in Britain. But even there the situations are not favorable and so far only a few relatively small scale (up to 240 MW) power plants are in operation.

The development of wave power has virtually been stopped. In Britain, one of the theoretically most favored countries in the world for its climatic condition, a research group had been set up in view of the apparent energy crises in 1976, but the research program was abandoned in 1983, when it became apparent that large scale plants would not be economical in the near future.

Interestingly, in Norway, development of tidal power plants was taken a step further, concentrating on small applications on remote islands and the like, and for quite a while a small power plant (500 KW) operated successfully in Toftestallen until it was swept away by the sea. And while wave energy is used successfully in very small scale applications, such as powering lighthouses or navigation buoys, its short term prospects as a major contributor to large scale energy production seem to be almost economically ruled out.

Tidal electricity generation involves the construction of a barrage across an estuary to block the incoming and outgoing tide. The head of water is then used to drive turbines to generate electricity, similar to hydroelectric dams. Barrages can be designed to generate electricity on the ebb side, or flood side, or both. Tidal range may vary over a wide range (4.5-12.4 m) from site to site. A tidal range of at least 7 m is required for economical operation and for sufficient head of water for the turbines.

The only major tidal power stations in operation today are the 240 MW facility in St. Malo, France, finished in 1966 and comprising a large range of experimental facilities, a 20 MW unit in the Bay of Fundy, Canada, dating from 1984, and a 100 MW facility in China built in 1987. This compares to 28 GW and 7-10 GW thought possible in the Bay of Fundy and the Severn Estuary respectively.

There have been proposals for new facilities, such as on the Severn Estuary (UK), which might produce 8.2 GW. Tidal energy schemes are characterized by low capacity factors, usually in the range of 20-35%. There is a high capital cost for a tidal energy project, with a possible 10-year construction period. Therefore, the electricity cost is very sensitive to the discount rate.

WAVE ENERGY CONVERSION

The most striking form of ocean energy is the surface wave. A considerable amount of energy is present in the ocean waves pounding against a breakwater. However, it is not from the breaking waves that the energy can best be extracted. Waves are generated by the force of the wind blowing over the ocean's surface. The regular breakers seen on most beaches originate at sea and can come from a variety of storms. The water's surface acts like a great conveyor belt, delivering power from great distances. Waves vary in several ways. They vary from location to location and also from season to season. Where the winds are steady the waves persist for long periods of time. Waves at deep sea can have a shape that is close to a sinusoidal wave. The sources of wave energy are the following four phenomena: (1) bodies moving on or near the surface causing relatively low period waves of low energy; (2) winds generating seas and swells; (3) seismic disturbances causing the tidal wave; (4) the lunar and solar gravitational fields causing the largest waves, the tides (the actual tidal waves).

The two measurable properties of water waves are the height and the period. These properties are measured by using an assortment of wave gauges. In practice, the observed wave heights did not correspond to the average wave height, but more nearly to the average of the one-third highest waves. This statistically averaged wave is called the significant wave. It should be noted that there is no mathematical theory that exactly describes the behavior of the waves. The various wave theories are simply to approximate, to some degree, the actual phenomena.

THE LINEAR WAVE

The swell is a wave of relatively great length and small height. When a storm rages at sea, waves of various heights and periods are generated; however, as these waves travel away from the storm region, the waves of low period disappear leaving only waves of high period (i.e. swell). The swell is the most frequently observed wave and, as such, is of primary interest in wave energy conversion. Although the *wave height* H is much smaller than the *wavelength* λ, the relative energy of the swell is high. The wave height is the vertical distance between a wave crest and an adjacent trough; the wave period is the time it takes two successive crests to pass a fixed point; and the wavelength is the horizontal distance between two crests.

For a wave height to wavelength ratio H/λ of 1/50 or less, the linear theory can be used with excellent accuracy in predicting the kinematic properties of waves. The mathematical expressions for the *free-surface displacement* and the *wave period* are illustrated below.

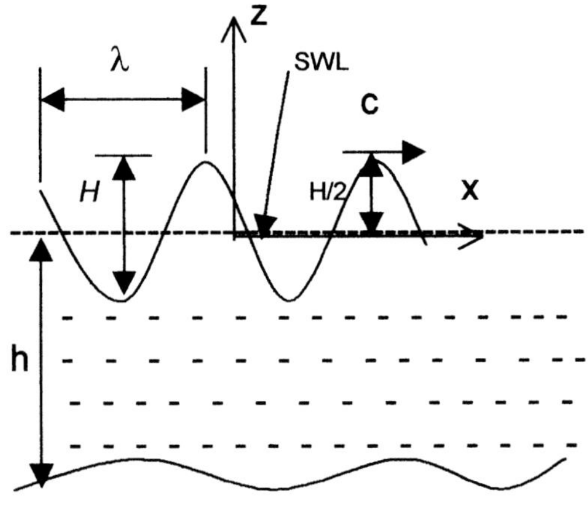

FIGURE 2. A linear wave – having a sinusoidal profile.

$$\eta = \frac{H}{2} \, Cos \left(\frac{2\pi x}{\lambda} - \frac{2\pi t}{T} \right) \tag{1.}$$

$$T = 2\pi \left[\frac{2\pi g}{\lambda} \tan h \left(\frac{2\pi h}{\lambda} \right) \right]^{-1/2}$$

$$= \frac{1}{f}$$

$$= \frac{2\pi}{\omega} \tag{2.}$$

where tan h $(2\pi h/\lambda)$ is the hyperbolic tangent, f is the wave frequency, ω is the circular wave frequency $(2\pi f)$, g is the gravitational constant (9.81 m/sec.) and h is the water depth. The period T is normally considered to be invariant with time t and depth h. However, this is not true over long travel distances.

Equation (2) can be rearranged to the expression for the wavelength:

$$\lambda = \frac{gT^2}{2\pi} \tan h \left(\frac{2\pi h}{\lambda} \right) \tag{3.}$$

Individual waves travel at a phase velocity C. This velocity is described by the equation

$$C = \frac{\lambda}{T} = \frac{g\,T}{2\pi} \tan h\,(kh) \tag{4.}$$

where k is the wave number defined by $k = \dfrac{2\pi}{\lambda}$ 5.

The kinematics properties of linear waves are functions of period and depth only. The wave energy conversion appears to be feasible in deep water. For engineering purposes, deep water is defined where h/λ 2≥1/2, or $kh \geq \pi$

In this case, equations (3) and (4) respectively are approximated by

$$\lambda = g\, T^2 / 2\pi \qquad 6.$$

and

$$c = g\, T / 2\pi \qquad 7.$$

Theoretically, both λ and c approach zero as h approaches zero. The water particles within the wave travel with horizontal and vertical velocity components of

$$u = \frac{\pi H}{T} \frac{\cos h\, [k\,(z + h)]}{\sin h\, (k\, h)} \operatorname{Cos}(k\, x - \omega\, t) \qquad 8.$$

and

$$w = \frac{\pi H}{T} \frac{\sin h\, [k\,(z+h)]}{\sin h\, (k\, h)} \sin(k\, x - \omega\, t) \qquad 9.$$

respectively. The deep water approximations of equations (8) and (9) are

$$u = \frac{\pi H}{T} e^{kz} \cos(k\, x - \omega\, t) \qquad 10.$$

and

$$\omega = \frac{\pi H}{T} e^{kz} \sin(k\, x - \omega\, t) \qquad 11.$$

In shallow water, where $h/\lambda < 1/20$; the approximate expressions are

$$u = \frac{H}{2} \sqrt{\frac{g}{h}} \cos(k\, x - \omega\, t) \qquad 12.$$

and

$$\omega = \frac{\pi H}{T} \frac{(z + h)}{h} \sin(k\, x - \omega\, t) \qquad 13.$$

Equations (8) and (9) describe water particle motions in water of intermediate depth, where 1/2 <h/λ<1 /20.

The total energy in a wave derived by McCormick (1973, 1978) is as given below:

$$E = E_p + E_k = \frac{\rho g H^2 \lambda b}{8} \qquad 14.$$

where b is the width of the crest, ρ is the mass density of water. The total energy in deep water in a wave described by the linear theory is equally composed of potential energy E_p and kinetic energy E_k; thus

$$E_p = E_k = \frac{\rho g H^2 \lambda b}{16} \qquad 15.$$

The potential energy is exhibited by the wave height H, whereas the kinetic energy is dependent on the motions of the particles.

The transfer of wave energy from point to point in the direction of wave travel is characterized by the energy flux or, more commonly, wave power:

$$P = \frac{\rho g H^2 c_g b}{8} \qquad 16.$$

where c_g is called the group velocity and is represented by

$$C_g = \frac{c}{2}\left\{1 + \frac{2kh}{\sin h\,(2kh)}\right\} = n\,c \qquad 17.$$

In the case of a deep water wave patch or group (where $h \geq \lambda/2$) the relationship between the phase velocity and the group velocity is

$$C_g = c/2 \text{ (deep water)} \qquad 18.$$

In shallow water ($h \leq \lambda/20$), however, the waves remain stationary with respect to the group boundaries; thus

$$C_g = c \text{ (shallow water)} \qquad 19.$$

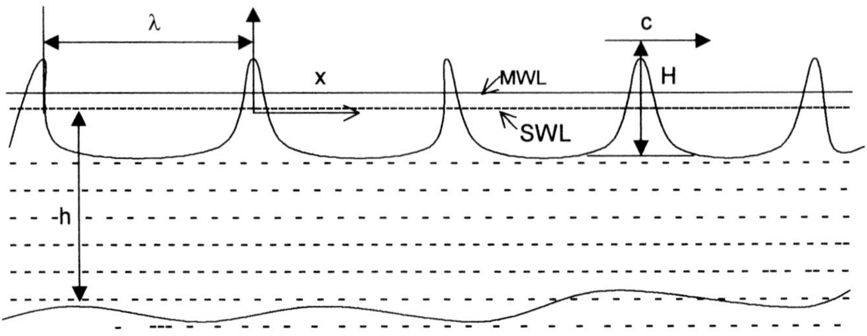

FIGURE 3. Sketch of a nonlinear wave profile.

THE NON-LINEAR WAVE

The failure of the linear wave theory is because it always predicts a sinusoidal profile. A deep water swell having a low value of H/λ will be well approximated by this profile; however, as the wave begins to shoal, that is to be affected by the seafloor, the wave profile will begin to change to one with a narrow crest and broad trough (as shown in the sketch above). This profile is said to be nonlinear (Fig. 3). The still water level (SWL) and the mean water level (MWL) coincide for a linear wave, however, the SWL will be below the MWL for the nonlinear wave since the MWL is defined as being half the distance from trough to crest.

Stokes (1847, 1880) introduces an irrotational water wave theory that utilizes a series of wave properties. The accuracy of the theory depends on the number of terms contained in the series. Stokes' first order theory is identical to the linear theory. His second order theory improves the accuracy in determining the wave profile and, in addition, the mass transport convection velocity and the breaking condition of the wave. Stokes' third and higher order theories successively add to the accuracy of the wave profile prediction. In the case of shallow water the wave *profile* predicted by Stokes' second order theory is given below:

$$\eta = \frac{H}{2} \, Cos\,(k\,x - \omega\,t) + \frac{3}{16}\,\frac{H^2}{K^2\,h^3}\, Cos\,[2(k\,x - \omega\,t)] \qquad 20.$$

where k is the wave number of the equation (5) and ω is the circular frequency $2\pi f$. The expression for the horizontal water particle velocity in shallow water according to the linear theory is as follows:

$$u = \frac{\omega H}{2kh} \cos\,(k\,x - \omega\,t) \qquad 21.$$

whereas that according to Stokes' second order theory is

$$u = \frac{\omega H}{2\,kh} \cos(k\,x - \omega\,t) + \frac{3}{16}\,\frac{\omega H^2}{k^3 h^4} \cos[2(k\,x - \omega\,t)] \qquad 22.$$

The expressions for the total wave energy and the wave power in shallow water according Stokes' second order theory are, respectively,

$$E = \frac{\rho\, g\, H^2 \lambda\, b}{8} [1 + \frac{9}{64}\,\frac{H^2}{k^4 h^6}] \qquad 23.$$

and

$$P = \frac{\rho\, g\, H^2 c_g\, b}{8} [1 + \frac{9}{64}\,\frac{H^2}{k^4 h^6}] \qquad 24.$$

where b is the creast width and the group velocity c_g is equal to the phase velocity c in shallow water, from equation (19).

ECONOMIC VIABILITY OF TIDAL AND WAVE POWER GENERATION

Since both methods suffer from the low concentration of energy it has become almost impossible to install huge power stations for large scale extraction, and making such use, even if not environmentally, then economically viable. In view of the economic viability of tidal and wave power generation the following assessments are made:

a) The maximum concentration of wave power can be obtained in the areas of the strongest winds, i.e. between latitudes 40° S and 60° S in both the northern and southern hemispheres on the eastern sides of the oceans. In this regard, the United Kingdom is found to be one of the world's most suitable locations for the extraction of wave power, but even here large scale projects have been put aside for the dark future, as they were deemed uneconomical compared to traditional power generating systems.

b) Designs for wave power stations, on the other hand, tend to be more revolutionary. Usually only a few of the components have already been tested in other applications, typically the air or water turbines that do get used. Furthermore, the aim of most of the original large-scale designs are mainly focused for use out in the sea rather than on the coastline, which results in major mooring difficulties and thus leads to an area having no experience other than through the employment of oil rigs.

c) There are some disadvantages of wave power stations. A wave power unit will probably not have much more than three times the output of a single wind turbine.
d) The construction costs are likely to be quite high due to mooring problems, the bulkiness and comparative complexity of the whole structure and the water-based location.
e) The only comparative bonus, the fact that they use up and deface less land, will prevail over economic considerations.

PROBLEMS FACED BY TIDE AND WAVE ENERGY CONVERSION

There are several problems faced by tide and wave energy conversion due to orientation in the sense that neither method is strictly economical on a large scale in comparison with conventional power sources.

In addition, none of them will produce electricity at a steady rate and thus not necessarily at times of peak demand, so unless a way can be found of storing energy efficiently, any storage devices currently available could incur a considerable loss. On the contrary, it will only allow them to run at a lower rating for a certain amount of the time.

As a result, such a power station should be considered as an insurance policy against future times when fossil or nuclear fuels run out and decrease the rate of fuel consumption, rather than as a replacement for other power stations.

Wave power stations suffer even more from these problems since their rate of production is unreliable. They can be used as a contributor to our energy supply in the future and can be restricted to small scale applications in isolated locations, where other forms of energy are hard to come by.

While ideas for the exploitation of wave and tide power have been around for hundreds of years, an organized effort to find alternate renewable energy sources has not.

ENVIRONMENTAL AWARENESS

A different scenario has arisen under which renewable energy sources might be desirable. Over the past decade ecologic awareness has steadily risen amongst the population in the West, and after the accident in Chernobyl, which proved a set back for most nuclear energy programs and compelled some governments to abandon nuclear power completely, it is now recognized that the greenhouse effect will force a decrease in the dependence on fossil power.

The major contribution (roughly 50%) to the greenhouse effect is carbon dioxide emission (Green and Guenther, 1989), which, in turn, is largely a result of the combustion of fossil fuel. It has been concluded that the temperature on this planet during this century has steadily increased with the enhanced concentration of carbon dioxide, at a rate in accordance with theoretical prediction and this effect would continue, with the temperature rising for another 75 years even if carbon dioxide emission was stopped today (Keeling, 1968). Because of such an unwanted situation, the National Research Council (1983) insisted on a decrease of carbon dioxide emission by 1% each year, with a special emphasis on industrialized countries as Third World countries are expected to continue having a steeply rising energy demand.

Tidal power generation may change the sedimentation and erosion patterns in the estuary (Quinby-Hunt, et al., 1984). Pollutants discharged into the rivers upstream from the barrages may accumulate in the estuary (Quinby-Hunt, et al.,1980).

So until the cost of maintaining the present rate of carbon dioxide emission is taken into account when building new power stations and a policy is adopted that depends less rigorously on market forces, the likelihood of tidal or wave power playing a major part in the energy supply of western industrialized nations, even in the medium term future, is extremely small.

ACKNOWLEDGEMENT

The author is thankful to Dr. Ehrlich Desa, Director, N.I.O. and Dr. S.W.A. Naqvi, head, COD, N.I.O. for their constant inspiration and valuable suggestions. Thanks are due to Mr. K.G. Chhitari, for reprographic work and Dr. P.V. Narvekar for his valuable help during computation of the data.

REFERENCES

Bardach, J.E. 1981. "Oil, Fish and the Sun and the Wind." *Proceedings, Asian Development Bank International Center for Living Aquatic Resources Management Workshop on Appropriate Technology for Alternative Energy Sources in Fisheries*; February 21-26, 1981, Manila, Philippines. pp. 1 -14.

Block, D., & J. Valenzuela. 1985. *Thermoeconomic Optimization of OC-OTEC Electricity and Water Production Plants.* SERI/STR-251-2603. Golden, CO: Solar Energy Research Institute.

Claude, G. 1930. "Power from the Tropical Seas." *Mechanical Engineering*. Vol. 52, No.12, pp.1039-1044.

d'Arsonval, A. 1881. "Utilization des forces naturelles. Avenir de l'electricite." *Revue Scientifique*. Vol.17, pp. 370-372.

Ditmars, J.D., & R.A. Paddock. 1979. "OTEC Physical and Climatic Environmental Impacts." Edited by G.L. Dugger. *Preprints of Volume II of the Sixth Ocean Thermal Energy Conversion Conference*; June 19-22, 1979. Ocean Systems Branch, Division of Central Solar Technology. U.S. Department of Energy. CONF-790631 /2. Paper 6A-3/1.

Dunbar, L.E. 1981. "Potential For Ocean Thermal Energy Conversion as a

Renewable Energy Source For Developing Nations." Science Applications International Corporation.
Francis, E.J., Avery, W.H., & G.L. Dugger. 1980. "Comparison of Cost Estimates, Sharing Potentials, Subsidies, and Uses for OTEC Facilities and Plantships." *Proceedings, Seventh Ocean Energy Conference*, June 1980; Washington, D.C.: U.S. Department of Energy; Sec.13.1, pp.1-8.
Fujioka, R.S., Loh, P.C., & L.S. Lau. 1980. Survival of Human Enteroviruses in the Hawaiian Ocean Environment: Evidence for Virus-inactivating Macroorganisms."*Applied and Environmental Microbiology*. 39:6; pp. 1105-1110.
Green, H.J., & P.R. Guenther. 1989. "Carbon Dioxide Release from OTEC Cycles." Presented at the American Society of Civil Engineers International Conference on Ocean Energy Recovery, Honolulu, HI, 28-30 November 1989. New York: ASCE.
Harms, V.W. 1981. "Current Measures off Keahole Point Hawaii; Application to OTEC-1 Ocean Test Platform." Proceedings of the Eighth Ocean Energy Conference, Vol. 1. Pub. No. DOE/Conf-810622-EXC. Washington, D.C.: U.S. Department of Energy.
High Technology Development Corporation. August 1985. *Final Environmental Impact Statement, Development Plan for the Hawaii Ocean Science and Technology Park and Expansion of the Natural Energy Laboratory of Hawaii*. Prepared by The Traverse Group, Inc., Honolulu, Hawaii.
Iso, S., et al. 1986. "Installation of OTEC Cold Water Pipe in Tokunoshima by Semi-Floating Type Submarine Method." *Proceedings, Fifth International Offshore Mechanics and Arctic Engineering Symposium April 13-18, 1986; Tokyo, Japan. New York: American Society of Mechanical Engineers*, pp. 599-605.
Keeling, C.D. 1968. "Carbon Dioxide in Surface Ocean Waters." IV. *Global Distribution Journal of Geophysical Research* 73:14; pp. 4,543-4,553.
Larsen-Basse, J. 1984. *OTEC Heat Transfer, Biofouling, and Corrosion Studies*. Honolulu, HI: Research Corporation of the University of Hawaii.
Liao, T., et al. 1986. "Feasibility and Concept Design Studies for OTEC Plants along the East Coast of Taiwan, Republic of China." *Proceedings, Fifth International Offshore Mechanics and Arctic Engineering Symposium; April 13-18, 1986; Tokyo, Japan. New York: American Society of Mechanical Engineers*; pp. 618-625.
McCormick, M. E. 1973. Ocean Engineering wave mechanics, Wiley Interscience, New York.
McCormick, M. E. 1978. Wind-wave power available to a wave energy converter array. *Ocean Engineering*, Vol. 5, no. 2, pp. 67-74.
National Oceanic and Atmospheric Administration. 1989. Tide Table 1990, *High and Low Water Predictions, West Coast of North and South America Including the Hawaiian Islands*. Washington, D.C.: U.S. Department of Commerce.
Natural Energy Laboratory of Hawaii. 1988. The Natural Energy Laboratory of

Hawaii: *1988 Annual Report*. Keahole Point, Kailua-Kona, Hawaii.

National Research Council. 1983. *Changing Climate: Report of the Carbon Dioxide Assessment Committee*. Washington, D.C.: National Academy Press. pp. 496.

Neighbor Island Consultants. 1976. *Keahole Master Plan: Phase 1*. Prepared for Research Corp. of the University of Hawaii.

Nihous, G., & L. Vega. April 1988. *Bottom-Mounted OTEC Seawater Systems Workshop—Summary Report*. Pacific International Center for High Technology Research, Kailua-Kona, Hawaii. Golden, CO: Solar Energy Research Institute.

Panchal, C., Larsen-Basse, J., & B. Little. August 1984. “Biofouling Control for Exchangers Using Intermittent Chlorination.” *Proceedings, 22nd ASME-AlChE National Heat Transfer Conference; August 1984, Niagara Falls, New York*. New York: American Society of Mechanical Engineers.

Penney, T.R., & T.H. Daniel. 1989. “Energy From the Ocean: A Resource for the Future.” *Science and the Future: 1989 Year Book*. Encyclopedia Britannica, Inc.

Quinby-Hunt, M.S., Wilde, P., & A.T. Dengler. 1985. *Potential Environmental Impacts of Open-Cycle Ocean Thermal Energy Conversion*. Environmental Impact Assessment.

Quinby-Hunt, M.S., Wilde, P., & C. Case. 1984. “Environmental Considerations: Ocean Thermal Energy Conversion on Pacific Islands.” *Proceedings of the Pacific Congress on Marine Technology (PACON)*. Honolulu, HI: Marine Technology Society.

Reed, S.A. August 1982. *Desalting Seawater and Brackish Waters: 1981 Cost Update*. ORNL/TM-8191. Oak Ridge, TN: Oak Ridge National Laboratory. Prepared for the Office of Water Research and Technology, U.S. Department of the Interior, Washington, D.C.

Stokes, G.G. 1847. On the theory of oscillatory waves. Transactions, *Cambridge Philosophical Society*. vol. 8 and paper 1, no. 197.

Stokes, G.G. 1880. Supplement to the 1847 paper under the same title, paper, *Cambridge Philosophical Society*, 1, No. 314

Taylor, H.S. 1924. The First and Second Laws of thermodynamics. In: A Treatise of physical chemistry, vol. 1, Taylor, H.S. and Glasstone, S. (eds) pp, 437-510.

Renewable Energy: Trends and Prospects. Edited by S.K. Majumdar, E.W. Miller and A.I. Panah

Chapter Seven

OCEAN POWER FROM WAVE TO WIRE

R. CURRAN

School of Aeronautical Engineering
The Queen's University of Belfast
David Keir Building
Stranmillis Road
Belfast, UK, BT9 5AG

INTRODUCTION

The following is concerned with the harnessing of ocean wave energy as a renewable power resource. A brief generic overview of wave energy conversion techniques to date will be given whilst also highlighting the present state of the art. Various components in the conversion chain will be described and some conclusions to their future relevancy made. Subsequently, a case study will be presented to elucidate some of the more significant decisions to be faced, in which an Oscillating Water Column will be coupled to an air turbine; being the system that is now closest to realizing full commercial viability. Throughout the text, the author intends to highlight the future relevancy and feasibility of wave energy as a resource whilst also making the reader aware of the associated issues connected to its advancement. In the new millennium, wave power will feature increasingly in the energy strategies of coastal nations, with the present state of the art being primed to act as a springboard for that expansion.

WAVE TO WIRE CONVERSION

Two-thirds of the earth's surface is comprised of oceans that derive their energetic state from the sun and moon; giving rise to thermal, wave and tidal energy. It has been estimated that over a four-day period the thermal and wave input derived from the sun alone is equivalent to the sum of the world's oil reserves. In the case of ocean waves, energy is imparted through the viscous interaction between the wind and sea as the air travels across thermal differentials. Consequently, the air velocity, duration, and distance or fetch give rise to the magnitude of the generated waves. These waves contain both kinetic and potential energy while the most energetic occur in unbroken deep water, and are characterized as fast moving with long wavelengths. The power is measured as the rate at which energy is transferred across a metre width section and therefore, is quoted per metre of wave front and may have an average value as high as 75kW/m at a reasonable offshore site.

As one of the most concentrated and regular forms of renewable energy, the potential for electrical generation is not to be overlooked. Indeed, even apart from relieving fossil-fuel dependency in a non-pollutant and renewable manner, wave energy offers a unique range of attractive features. It is largely unobtrusive; suited to isolated communities; highly marketable in modular form; provides sheltered waters for fish farming etc., protects coastlines and integrates into breakwaters; has a global significance; can be used to pump water or for desalination plants; and will support offshore wind turbines and solar arrays. However, it is also worth noting that the resource is greatest in deep and often hostile waters; tends to be irregular in direction, duration and size; may give rise to expensive transmission and short-term storage costs; and may interfere with shipping traffic. Notwithstanding, over the past couple of decades, many types of wave power converter have been championed and a few of the more lucent systems have been taken to the prototype stage and successfully connected to national electricity grids (Count, 1980).

WAVE POWER

It has been estimated that 2-3 terawatts (10^{12} W) of wave power is incident on the world's coastlines (WEC, 1995). Annually, this sums to approximately 22 petawatt-hours (10^{15} Whr) or 79 exajoules (10^{18} J), where 1 watt-hour (Whr) is equivalent to 3.6 kilojoules (kJ). In context, global energy consumption in 1993 was estimated by UNSTAT (1995) at 326 exajoules and is increasing at a rate of nearly 50 percent every 20 years. Consequently, the wave power resource is significant, although all coastal nations are not in a position to exploit it. Wave energy tends to be most prevalent between the latitudes of 30° and 60° in both the northern and southern hemispheres where winds are strongest. This is illustrated in Fig. 1 where the earth's less energetic regions have been shaded to highlight those that are more feasible. In addition, countries such as Ireland have a high coastline-to-population ratio with large fetch distances and therefore, can expect much greater returns through the exploitation of their wave resource. Mollison (1982) has carried out detailed wave investigations around the Irish coast and estimates that the annual resource is approximately 25 gigawatts (10^9W). Annually, this provides 790 petajoules of energy as compared to the

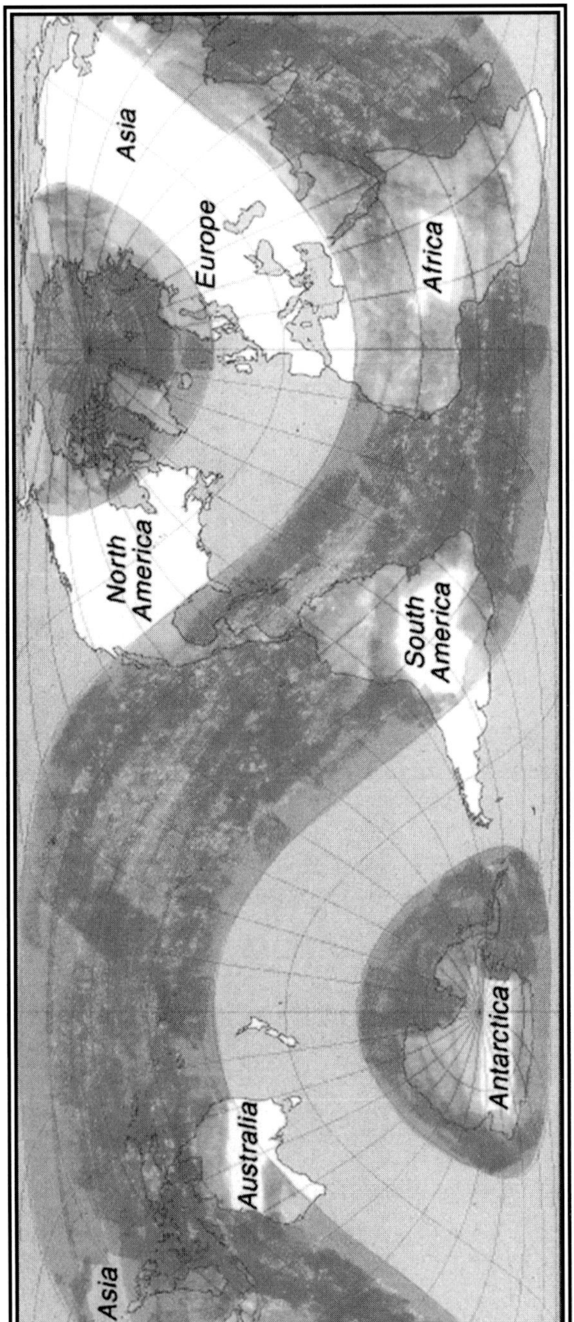

FIGURE 1. Feasible wave energy zones (not shaded).

TABLE 1
Typical levels of power (kW/m) recorded off the Northern Irish coast

	1978	*1979*	*1980*	*1981*	*Monthly Avg*
Jan		67.8	54.3	51.4	57.8
Feb		87.8	93.4	81.7	87.6
Mar		125.0	70.7	54.3	83.3
Apr		44.4	33.6	18.8	32.3
May		40.0	7.3	15.6	21.0
Jun		21.9	20.7	18.6	20.4
Jul		21.8	9.7	21.3	17.6
Aug		24.6	14.4	14.8	17.9
Sep		57.0	36.0	24.0	39.0
Oct	70.8	57.1	53.2		60.4
Nov	121.2	127.2	51.4		99.9
Dec	85.2	139.5	133.5		119.4
Annual Avg		64.0	55.3	44.9	

country's consumption of only 152 petajoules. Therefore, a conversion efficiency of less than 20 percent would provide Ireland's entire power requirement, although this assertion ignores the indicative short and long-term fluctuations. It was estimated that one quarter of Ireland's power consumption could be supplied from wave energy, while also noting that the seasonal variations tracked customer demand.

Table 1 presents 3 years of data from Malin Head, 45km off the Northern Irish coastline. The variations highlight the difficulty in determining longer-term productivity, the month of May being a particular case in point. The figures return an overall average of 57kW/m, being very similar to the estimated mean along the Western Coast, while the winter, summer and April-September averages were estimated to be 89.3kW/m, 19.2kW/m and 35.6kW/m respectively. Furthermore, the wave period varied ±5s about a 10s mean, and the direction ±48 degrees from West at 278 degrees.

CONVERSION TECHNOLOGY

Capture

The component that first converts wave motion into a more readily exploitable form is known as the capture device (Salter, 1988). The majority of these can be categorized as one of the following and are simply described through Fig. 2:

a) Oscillating Water Column (OWC)
b) Heaving and/or pitching bodies

c) Wave-hydro schemes
d) Pressure transmission via flexible membrane

An OWC device comprises a damped chamber with a submerged opening wherein an entrained water mass oscillates vertically according to the incident wave action. The piston-like motion of the water mass is used to force the air trapped above its surface out of the chamber via an exit duct that vents to atmosphere, thereby providing pneumatic power according to the damping applied by the turbine. The chamber may be located on the coastline or the seabed, or can be an anchored floating structure. It is designed to resonate within an optimal period band and may also utilize harbor focusing, for a given cross-sectional shape and inclination. Alternatively, the classic tapered channel is a simple shoreline wave-hydro device that involves heavy civil engineering works and is site-specific (Malmo and Reitan, 1985). Converging walls are used to optimally focus the waves up through an inclined channel into an elevated reservoir that stores potential energy for cost-effective generation. However, a floating type is also envisaged that channels the waves up over a front ramp into a water tank encased within a floating structure. However, the captured seawater continuously drains

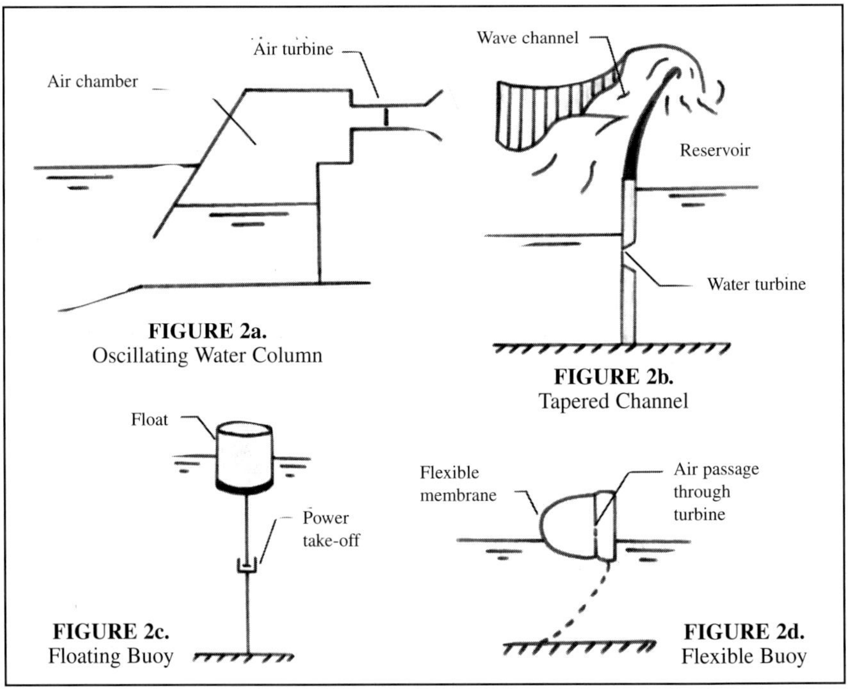

FIGURE 2a. Oscillating Water Column
FIGURE 2b. Tapered Channel
FIGURE 2c. Floating Buoy
FIGURE 2d. Flexible Buoy

through a floor duct that houses a conventional water turbine. The flexible membrane principle typically incorporates a number of floating chambers with elastic walls that effect air passage from one chamber to another as the outer membrane walls are depressed by the action of the waves, thereby providing pneumatic power. This arrangement has fewer advocates and as an extraction technique, poses more practical problems than the remaining category that uses the reaction forces of several bodies. The latter may all be identical and connected by hydraulic power take-off systems or one/several may float freely relative to a datum platform. However, the most simplistic configuration is the heaving and/or pitching float that is either fixed to the seabed or to an anchored base-plate, utilizing the reactive forces (Budal and Falnes, 1975). An alternate method uses the rotation of a body about a fixed axis that is perpendicular to the incident wave where, for example, a teardrop shaped floating device is induced to pivot about a stiff shaft located through the bulbous aft (Salter, 1988). The power take-off is facilitated by the arc-like motion described by the narrow leading edge, which is pointed towards the incident wave. Yet another subcategory consists of submerged devices (Evans, 1979), some of which may act as submerged floats or pendulous flaps.

Power Take-Off

The power take off mechanisms associated with the above devices are of either hydraulic or pneumatic nature and fit into one of the following categories:

- Closed-system high-pressure hydraulic fluid motor
- Open-system seawater turbine
- Low pressure pneumatic turbine

The simple closed-system oil machine constitutes a singular hydraulic ram while more complex setups utilize radially arranged cylinders with pistons that operate on an eccentric (Salter and Rampen, 1993). Each cylinder would have two valves that connect high and low pressure manifolds so that hydraulic fluid can be transferred in either direction for the fluid to drive pistons and produce rotational power as a generator, or vice-versa when operating as a pump. A seawater system is likely to use a standard water turbine such as a high-pressure Pelton or Kaplan turbine. Consequently, these must incorporate flow rectification to ensure that the water travels through the turbines in a single direction, this being already inherent to the tapered channel.

The final and most expansive category constitutes turbomachinery that converts pneumatic power to mechanical power (Horlock, 1966). The two most common types of low-pressure pneumatic machine are the axial-flow impulse turbine and axial-flow Wells turbine. Impulse turbines tend to have a low rotational speed and high flowrate characteristic, while guide vanes are required to improve the efficiency over a favorably large operational flow range

(Setoguchi *et al*, 1988). The performance of the Wells turbine is aerodynamically determined by the pressure differential caused by the symmetrical blades set around the hub at 90 degrees stagger, as shown through Fig. 3 (Abbott and von Doenhoff, 1949; Gato and Falcao, 1989; Raghunathan, 1995; Curran *et al*, 1998). The turbine was developed at The Queen's University of Belfast (QUB) for its equivalent performance in either flow direction, and it achieves a relatively high peak efficiency of 80 percent. Three contemporary variants that significantly enhance the basic set-up are as follows:

- Variable pitch monoplane
- Guide vane monoplane
- Contra-rotating biplane

The variable pitch monoplane (Gato and Falcao, 1989) allows blade rotation in order to optimize the angle of attack, in a similar way to wind turbines. Essentially, when the air velocity increases to a point where the angle of attack leads to separation from the blade surface, the aerofoil is pitched into the flow to decrease the angle. Alternatively, the other two designs both recover the swirl energy component (Curran and Gato, 1997). The use of guide vanes maximizes the conversion efficiency over a flow range as the entry vanes direct

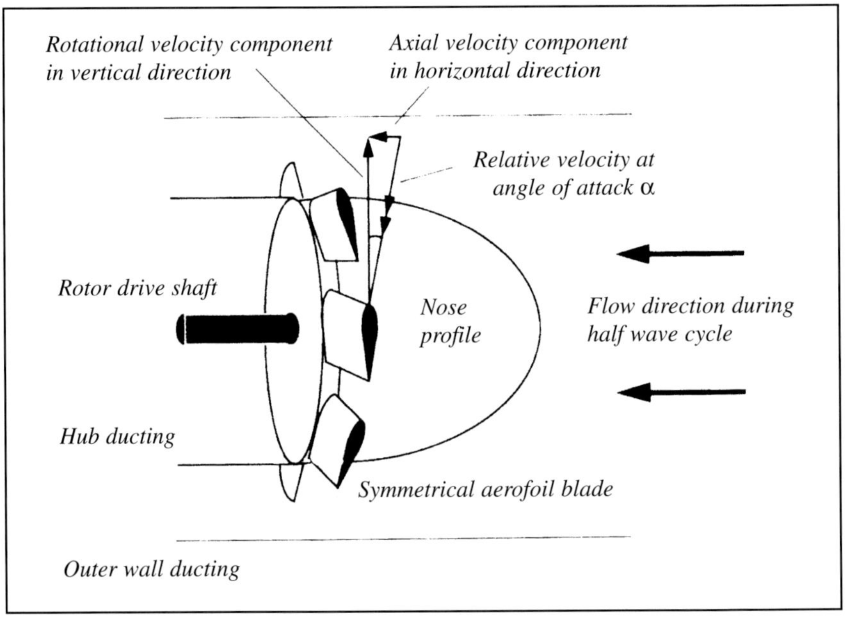

FIGURE 3. Schematic diagram of a horizontally mounted Wells air turbine.

flow unto the rotor and the exit vanes subsequently 'straighten' it in order to minimize swirl energy losses. However, fixed vanes have limited range and induce turbulence when operating outside of this. The final variant also recovers swirl energy but by counter-rotating the flow with a second downstream rotor in order to return it to an axial path with minimal losses (Curran et al, 1998). The additional attractions of this technique include the increased pressure drop accommodated by the two rotor planes and their favorable aerodynamic interaction that ensures gentle stall. In summary, the variable pitch turbine is complex while the guide-vane version sacrifices cyclic efficiency due to its narrow flow range and therefore, the author advocates the contra-rotating turbine as a simple, reliable machine with relatively high cyclic efficiency.

Electrical Supply

The final link in the wave-to-wire conversion chain is the electrical machine, which tends to induction so that it can operate as a generator or motor. It supplies power during the short periods when the turbine's speed tends to decrease, due to the lack of converted power relative to running losses, thereafter switching to generator mode for export. Further to that, wave power systems also require some form of integrated power smoothing, whether from inertial storage or electrical manipulation (Beattie at el. 1995). Presently, power engineers are favoring a joint system where a semiconductor converter is used in conjunction with a flywheel, while the control strategy aims to lop off the power peaks in order to supplement export as required. Further to this, there has also been complimentary research into the active control of the airflow, whether smoothly through a variable-rate blow-off valve; crudely through a turbulent valve; or by completely shutting off the flow at certain times through 'latching' (Sarmento et al. 1990; Justino et al., 1993).

STATE OF THE ART

Capture devices from each of the four generic categories mentioned have been taken to small-scale prototype stage and all have shown merit and commercial potential. However, the present state-of-the-art seems to continue favoring OWCs and floating devices. The author concludes that the relative success of these is largely a consequence of the ease with which each could be developed to full-scale; not least in terms of finance, technology, complexity, manufacture and maintenance. The OWC system is most favored and prototypes have ranged from low wattage lighting buoys for shipping navigation, to grid-connected conversion plants. Apart from the hundreds of buoys in Japanese waters, notable examples of larger systems include the Japanese 60kW Kamei project (1977-1985); 500kW Norwegian plant (1985-1988); 70kW Islay plant in Scotland (1991-1998); and the current 150kW Indian plant (1991-). Presently, a fresh wave of more rigorously designed OWC plants are in construction, highlighting renewed interest after the successful pioneering research of plants such as the QUB Islay unit (Whittaker et al., 1997; Curran et al. 1997). Notwithstanding, certain aspects still need attention, such as investment cost; production cost; reliability; environmental and socio-economic impact; and economies-of-scale. Consequently, it is important to gather complete information that relates to all aspects of commercial viability from the current generation of OWC plants. The Norwegians built the first standard shoreline tapered channel device in 1988 and are now negotiating a 2nd commission on Java, Indonesia. Alternatively, a Danish group has been operating a 45kW floating buoy for some years and lately, commercial interest in the system has again emerged. Also, several new techniques have emerged such as an American project looking at pizo electric plastics that produce electricity when stretched or the Japanese 'pendular' system that comprises a hanging plate that is displaced by the waves. However, all of the main contemporary projects and initiatives are summarized below and are listed under country for convenience:

- ***UK:*** 500kW 'LIMPET' OWC: QUB-Wavegen. Notably: concrete construction was aided by excavating and building behind a protective bund of coastal rock; survivability is dependent on front-face wave loadings; the contra-rotating turbine is controlled by an electrical converter; and it should prove to be reliable, although costs need to be reduced and the technology moved offshore. 1MW 'OSPREY' OWC: WaveGen. Notably, the first near-shore version was compromised in severe weather during deployment, underpinning the need for speedy operation; the OWC was manufactured in steel at a shipyard; the potential for modular construction and mass economies-of-scale is high; and a wind turbine was also to be accommodated on the plant. 375kW 'PELAMIS' floating device: Ocean Power Delivery. Notably, it will be deployed in 20m of water and consists of logtitudinal sections that lie in the incident direction; reactive forces from hydraulic couplings on adjacent sections facilitate power take-off; the design is highly modular; the principle is simple although power take-off is unproven; maintenance will be expensive and hindered; and anchoring and shipping regulation may prove problematic. 400kW and 1200kW floating channel devices: Sea Power International. Notably, a water depth of 50-80m has been identified; the front 'lip' pitches to optimize performance through water intake; the technology is simple and well established; but the device is material intensive and vulnerable to damage.
- ***Portugal:*** 500kW shoreline OWC: Instituto Superior Technico. Notably, it is being built on the Azores and is of a similar design to the 75kW Islay prototype; it will also operate as a test bed for turbines (guide-vane and variable-pitch Wells); and the construction involved intensive civil works at a site-specific location requiring three substantial cofferdams. 2MW submerged 'AWS' device: Teamwork Technology. Notably, it will operate in 50m of water; is a submerged oscillating float that is connected to a fixed base housing a

linear generator; and economies-of-scale from large 'wave-farm' configurations is high, although maintenance and installation costs will also be high.

- ***India:*** 10x110kW shoreline OWCs: National Institute of Ocean Technology (NIOT). Notably, the OWC-caissons will be optimally spaced in an array, each with two turbines in a 0.5km long breakwater; and the intensive civil engineering approach is cost-effective in the 3rd-world.
- ***Japan:*** 120kW floating 'MIGHTY WHALE' OWC: JAMSTECH. Notably, the floating device will have several power take-off units; the objective is two-fold in also providing calm waters for fish farming (as the wave climate is quite tame); and the team is very experienced with previous prototyping and various power take-off mechanisms.
- ***Australia:*** 500kW shoreline OWC: EnergeTech. Notably, the plant is free-standing with a walkway connecting it to an existing breakwater; a parabolic-shaped reflector focuses the energy into the OWC at the locus; a new highly efficient pitching air turbine is used; and combining it into the breakwater reduces overall supply costs. 20kW floating buoy: Ocean Power Technology. Notably, the testing of the single unit is the first stage in a program that is to be expanded to a 1 MW and then 10_MW wave-farm; and economies of scale should be large.
- ***Further activities:*** A probabilistic wave energy computer atlas has been developed through a European Program while there has been many developments with other devices including the Japanese 'PENDULAR' device; the Norwegian 'ConWEC'; the Swedish 'IPS' buoy and 'HOSEPUMP' device; and the 40m-long 'McCABE WAVE PUMP'. The Wave Pump has been successfully tested in County Clare, Ireland and is to be replicated at 400kW-scale. In a manner not unlike the aforementioned Pelamis device, it lies parallel to the incident wave direction and hydraulic power take-off is subject to the relative motion of its constituent sections, however, it comprises only three parts while the central of those is a neutral highly damped static member. It is known for its ability to pump water, for example, freshening lagoon water or supplying 'Reverse-Osmosis' plants.

DESIGN CASE: THE AUSTRALIAN OWC

Wave Resource

Port Kembla, Sydney was identified as a favorable site for the building of the first Australian wave energy device, even although the energy climate there is moderate. The main incentives included easy access from Sydney, the proximity of Australia's largest industrial area with an extremely stiff grid, and the existing breakwater. Wave data was supplied by the Sydney Port Authority and had been gathered using a wave-rider buoy located 100m offshore in water 22m deep. The action of the buoy was recorded over 219677 20-minute scans from the period

24/08/87 to 30/06/97, with each scan being averaged according to significant wave height H_s and peak spectral period T_p . The data is presented graphically in Fig. 4 where the entire range of recorded wave heights and periods has been subdivided into a 21x25 matrix of 0.25m and 1 s bands respectively, for the number of times n that each sea state occurred. It was found that the average values of significant wave height and peak spectral period were 1.12m and 9.32s respectfully, the values having been weighted according to occurrence.

The wave power associated with each of one of the 525 sea states was calculated and used in conjunction with its occurrence in order to estimate the average annual power level. The power calculation is described below through Eq. 1.

$$W_I = \tfrac{1}{16}\rho_w g H_s^2 C_g \tag{1}$$

where ρ_w is water density, g is gravitational acceleration and C_g is group velocity; which is a lengthy function of period and water depth. The contribution to annual incident wave power is presented in Fig. 5 where the cumulative incident wave energy E_I was calculated to be 667MWhr per annum, according to Eq. 2, giving an annual average of 7.61kW/m.

$$E_I = \sum_{i=0}^{i=525} 8766 p(i) W_I d \tag{2}$$

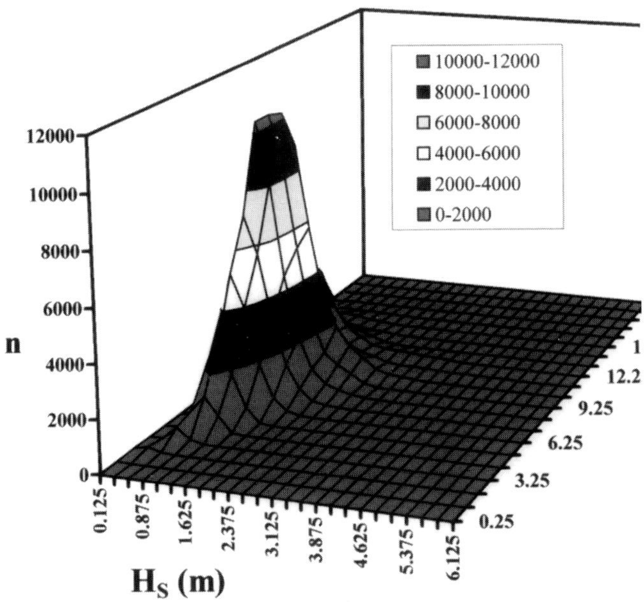

FIGURE 4. Port Kembla wave climate.

where i represents each of the 525 sea states, 8766 is the number of hours contained in one year, $p(i)$ is the probability of any sea occurring and d is the chamber width. The largest amount of energy, 18.9 percent, was generated in the 1.125m band while 24 percent was generated in the 10.25s period-band. Furthermore, 86.2 percent of the energy occurred between wave heights of 0.5m and 2.5m while 95 percent of the energy occurred between periods of 5.75s and 13.75s. In comparison to Fig. 4, it is noticeable that the surface plot has tended to populate a greater portion of Fig. 5.

Hydraulic Performance

The incident power results shown in Fig. 5 were used as input data for the calculation of pneumatic power output from the collector device (Mei, 1976). Here performance results from the University of New South Wales were used in conjunction with previous typical-case OWC tests (Curran et al, 1997) in order to estimate performance. However, certain influential factors have not yet been investigated, such as: the wave direction relative to the parabola's axis of symmetry; having a nonuniform seabed that may cause wave breaking; having very irregular incident waves; and the diffraction effects at the harbor mouth. Notwithstanding, the variation in hydraulic performance will be influenced in the main by the wave height and period while the tidal swell is less important farther

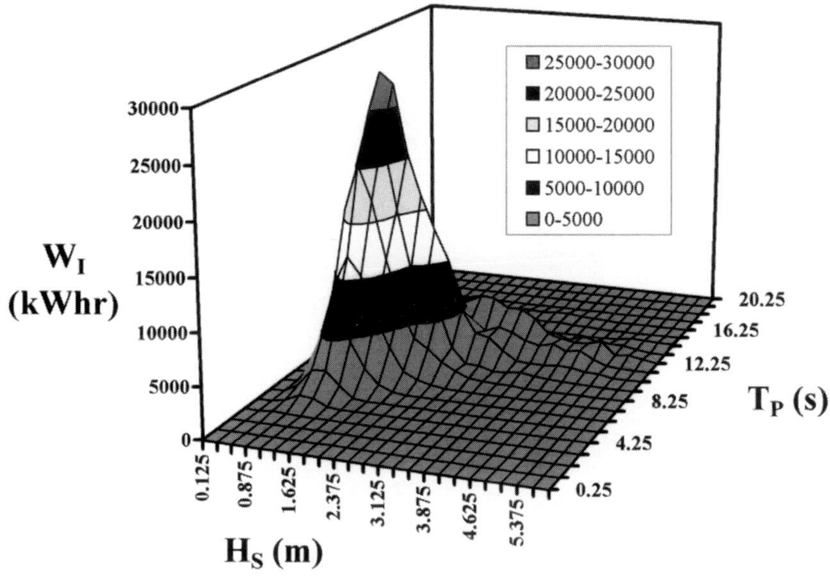

FIGURE 5. Annual contribution to incident power from each sea category.

from the shoreline. Consequently, the performance was related to these by the capture factor *CF*; by convention being the ratio of output pneumatic power to input wave power incident on the OWC width. The transfer function used to model the performance is detailed through the response surface presented in Fig. 6. The maximum efficiency ratio of the OWC η_{owc} was 2 and was associated with wave heights and periods of approximately 1.12m and 9.3s respectively, as shown by the distribution.

Pneumatic power was calculated from the product of incident wave power and efficiency while the cumulative pneumatic energy output E_p was calculated through Eq. 3, in a similar manner to Eq. 2.

$$E_P = \sum_{i=0}^{i=525} 8766 p(i) W_1 d \eta_{owc} \tag{3}$$

The shape of the distribution was very similar to that of Fig. 5 although the pneumatic energy output was considerably higher at 1207MWhr per annum. In terms of annual average, the pneumatic power rating of the plant was 137.7kW and the overall average hydraulic efficiency ratio was 1.81, although this is again relative to the width of the OWC. It is also interesting to note that the most

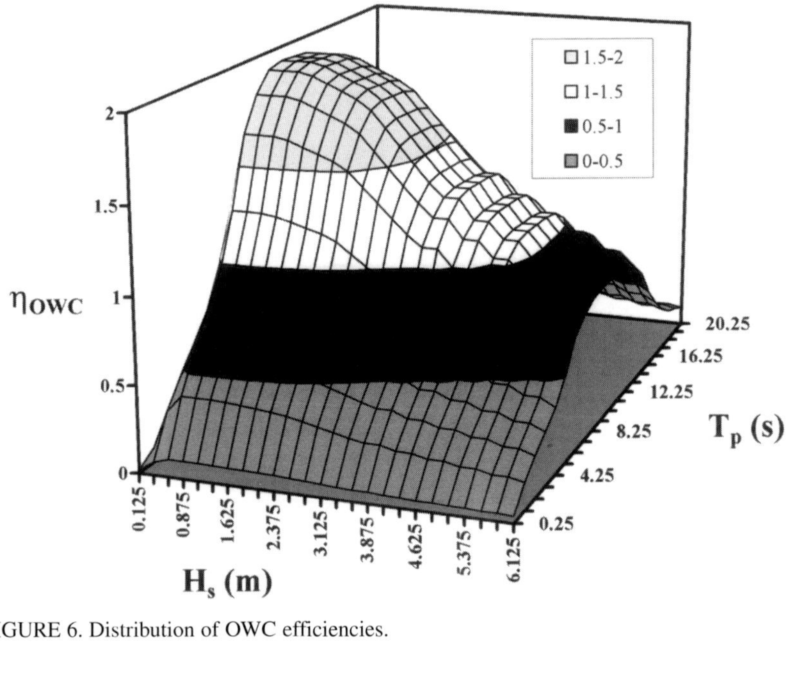

FIGURE 6. Distribution of OWC efficiencies.

energy intensive band is now 1.375m rather than 1.125m, the distribution remaining similar in terms of period.

Performance Coupling

The hydraulic performance has to be matched to that of the turbine for the predicted converted power. Essentially, the pneumatic power output from the OWC must be defined in terms of volume airflow through the turbine duct. This is achieved by considering the force-velocity relation known as applied damping (Mei, 1976), with reference to the mechanical mass-spring-damper analogy, where the turbine damping influences the motion of the water mass entrained inside the OWC. For statistical analysis, the turbine damping is expressed by Eq. 4, if mass flow is conserved and air incompressibility is assumed; Mach number being restricted to less than 0.3.

$$B_A = P \frac{A_C^2}{A_A V_A} \tag{4}$$

where P is total pressure drop, A_C and A_A are the cross-sectional areas at the water column surface and annular turbine duct respectively, and V_C and V_A are the respective air velocities. With reference to the turbine's fundamental performance variables as expressed later through Eqs. 7 and 8, the above can be further manipulated to yield Eq.5, highlighting the main design parameters.

$$B_A = 4\rho_A \left(\frac{A_C^2}{A_A} \right) U_t \left(\frac{P^*}{\emptyset} \right) \tag{5}$$

where ρ_A is air density, U_t is tip rotational velocity, P^* is non-dimensional pressure drop and $\emptyset$ is non-dimensional flowrate. It will also be seen that the pressure-flow ratio is approximately constant for the fixed speed turbine under consideration and therefore, that the applied damping remains constant. The pneumatic power is the product of pressure and flowrate and therefore, Eq. 5 can be further manipulated to yield Eq. 6.

$$V_A = \sqrt{\frac{W_P A_R^2}{B_A}} \tag{6}$$

where A_R is the area ratio between the water column surface and annular turbine duct. Consequently, Eq. 6 can be used to provide the axial air velocity that corresponds to any value of pneumatic power generated by the OWC. The velocity is then inputted to a turbine performance model for calculation of the torque and ultimately, power conversion. However, one further use of Eq. 5 is that the applied damping has direct influence on the hydraulic performance of the OWC, although the relationship is relatively insensitive; the pneumatic requirement being much more influential.

Power Productivity

A new type of axial flow air turbine was utilized instead of the more commonly used Wells turbine or impulse turbine. Notwithstanding, in a similar manner to the variable pitch Wells turbine, it is aerodynamic in principle with aerofoil blades that pitch in order to provide an optimal angle of attack. However, the blades are located on the periphery of the rotor hub in a neutral position that is parallel to the axial direction, rather than tangential to the direction of rotation. The turbine's predicted cyclic performance is presented in Fig. 7 where each of the variables are presented in terms of non-dimensional flow rate Ø, pressure P^*, and efficiency η_{Tuurb}, and are defined in Eqs. 7 through 9 respectively.

$$Ø = \frac{V_A}{U_t}, \quad P^* = \frac{\Delta P}{\rho_A \omega^2 D_t^2} \quad \& \quad \eta_{Turb} = \frac{T\omega}{PQ}, \qquad (7\text{-}9)$$

where ω is angular velocity, D_t is turbine tip diameter, T is torque, and Q is volume airflow. It can be seen from Fig. 7 that the pressure-flow ratio is approximately linear for the majority of the flow range like the Wells turbine. Consequently, the ratio can be easily substituted into Eq. 6 in order to calculate the air velocity while the corresponding efficiencies can be taken from Fig. 7, and used to predict the turbine's output through Eq. 10, where W_P is the averaged pneumatic power for each sea state.

$$E_C = \sum_{i=0}^{i=525} 8766 p(i) W_P \eta_{Turb} \qquad (10)$$

It was found that a diameter of 0.73m maximized output and that the overall value of turbine efficiency was 64 percent, although the full distribution of efficiencies relating to each sea state is presented in the response surface of Fig. 8. The peculiar horseshoe shape is best explained by considering the pneumatic power levels associated with each category. The highest efficiencies corresponded to seas that had pneumatic power levels that were well matched to the turbine, through the magnitude of air velocity, while alternatively, the least energetic and most energetic seas resulted in poor efficiencies. Furthermore, Fig. 9 shows that the highest pneumatic levels occurred at the largest wave heights

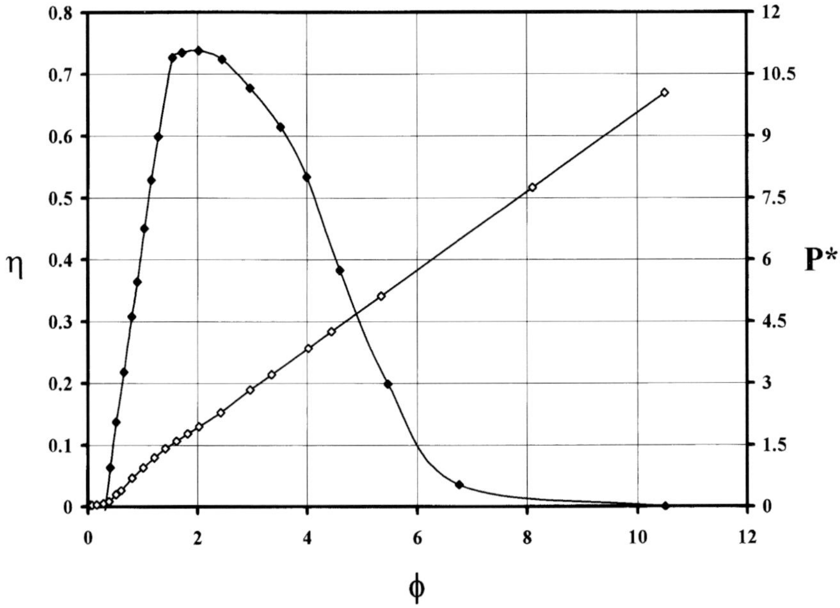

FIGURE 7. Fundamental turbine characteristics.

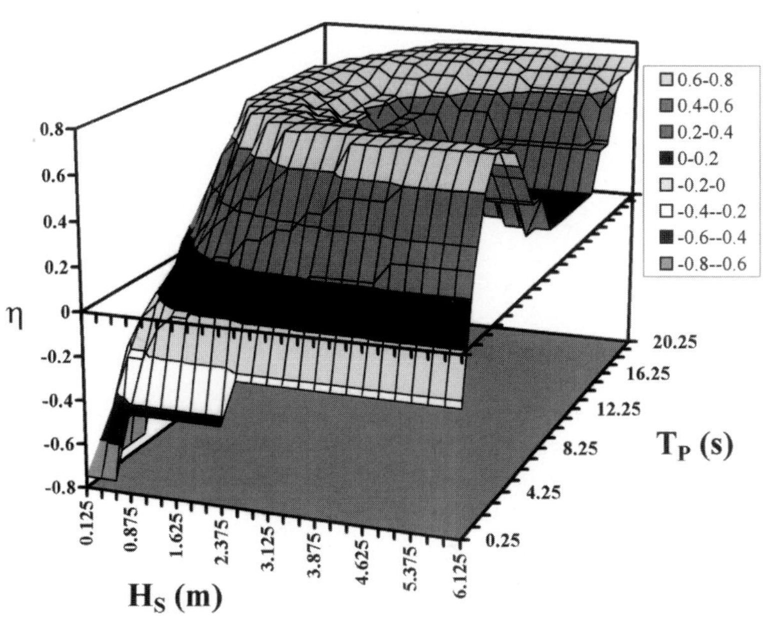

FIGURE 8. Distribution of turbine efficiencies.

with periods of approximately 10s. Therefore, with reference to Fig. 6, it can be seen that these are seas in which the OWC operates more efficiently, while the remaining seas have limited pneumatic power supply due to the unsuitable period restricting the efficient band-width response. The annual distribution of contributions to converted power is plotted in Fig 10, where the cumulative output is estimated at 751MWhr: giving an annual average power rating of 85.7kW. It is interesting to note that all of the sea states that skirted the energy intensive region resulted in power loss as the converted power was not enough to overcome the turbine drag losses incurred at the lowest flowrates, with reference to Fig. 7. Furthermore, it was found that the most energy intensive height was again in the 1.125m band, with a percentage of 22.3, while 94 percent of the converted energy was generated in the 0.5-2.5m range, and 97 percent between periods of 5.75s and 13.75s. Therefore, it is noted that there has been a refocusing due to the band-width response, both hydraulically and aerodynamically, when comparing the figures with those of the incident wave power; having increased from 86.2 percent in the 0.5-2.5m range. Finally, the distribution of average converted power associated with each of the sea states is presented in Fig. 11. Evidently, without further electrical control, the installed capacity of the plant would be 500kW, allowing the generator to operate within range for most of the sea states experienced annually. Finally, it is worth noting that for the more favorable sites to be populated in the future, the plant's installed capacity will range. from 1 - 1.5MW.

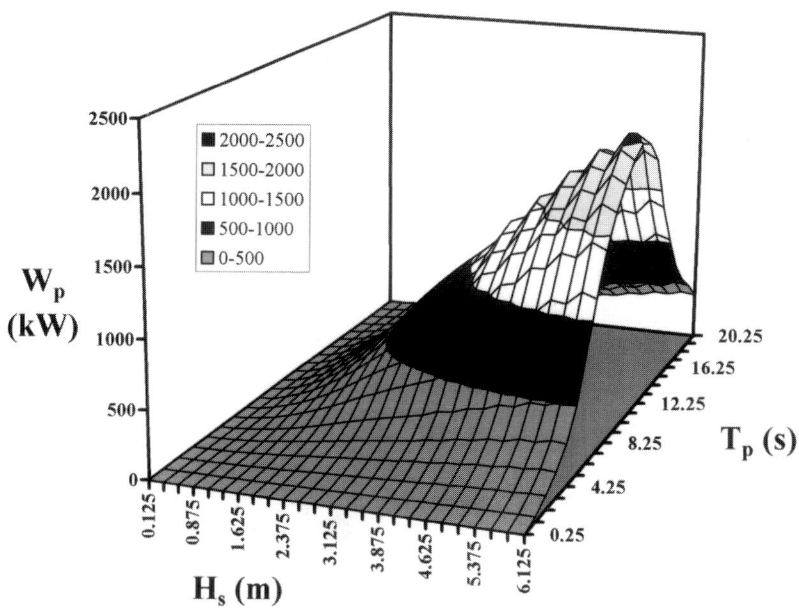

FIGURE 9. Distribution of pneumatic powers associated with each sea state..

COMMERCIAL STATUS, PRIORITIES AND PROSPECTS

Wave power exploitation is only now shifting into the hands of commercial groups who are able to utilize and profit from the experience of research teams that have nurtured the sector since the false boom of the late seventies and early eighties. It is readily evident that the technology is now in place but that the industrial growth will only be realized, as necessary, under the full anger of market forces and their synonymous streamlining. Presently, political and development grants are necessary but those companies in place will have a massive advantage in the ensuing expansive industry where time-to-market, credibility and experience will be crucial. However, to be viable, there is a strong need for baseline data on aspects such as investment cost, production cost, reliability, environmental and socio-economic impact, and economies-of-scale. Therefore, it is important that these aspects, pertaining to the current generation of wave energy devices, be defined, quantified and documented. Other institutional factors include planning and consent; grid connection, strengthening and distribution; design, health and safety codes; and other such political aspects that may arise. However, the overriding requirement at present is the credibility and inherent survivability of plants that will raise political and commercial confidence. Notwithstanding, one defining drive that will become ever more prevalent is the shift in emphasis from unclean finite fossil fuels to green renewable energies.

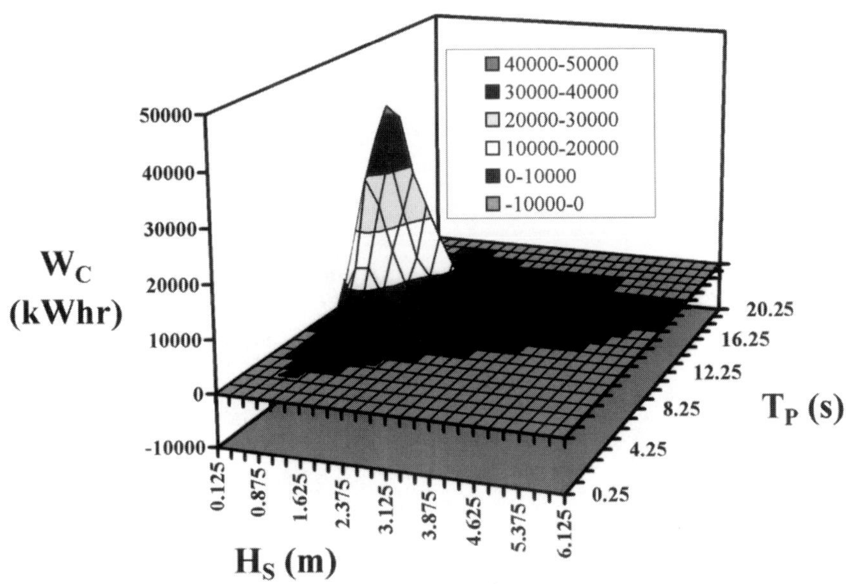

FIGURE 10. Annual contribution to converted power by the turbine.

In line with global climate policy and levies, sustainability, and competitiveness, the European Commission White Paper on renewables has proposed that from present, double the energy requirement (12 percent) should be met by renewables by 2010. The World Energy Council scenario analysis indicates that the cumulative investment in renewables will be between \$250 billion and \$600 billion between 2000 and 2010 while Shell predicts that 40 percent of global energy requirements will be renewable by the middle of the next century. Similarly, UK government analysis proposes that it is feasible to meet 50 percent of energy requirements by 2025 with renewables at an average cost of less than 5c/kWh, at time of publishing. Therein, it was estimated that wave energy could provide 10 percent of the energy at 6c/kWh and 20 per cent 7c/kWh. It is further encouragement that the longer established wind energy sector has been growing at 40 percent per annum over the past decade while nuclear growth has decreased to only 1 percent and coal to zero. Therefore, in conjunction with export opportunity, further design improvements, integrated energy policies, and increasingly constrictive restraints on fossil fuels, the above point to the firm market establishment of wave energy within next 10 years.

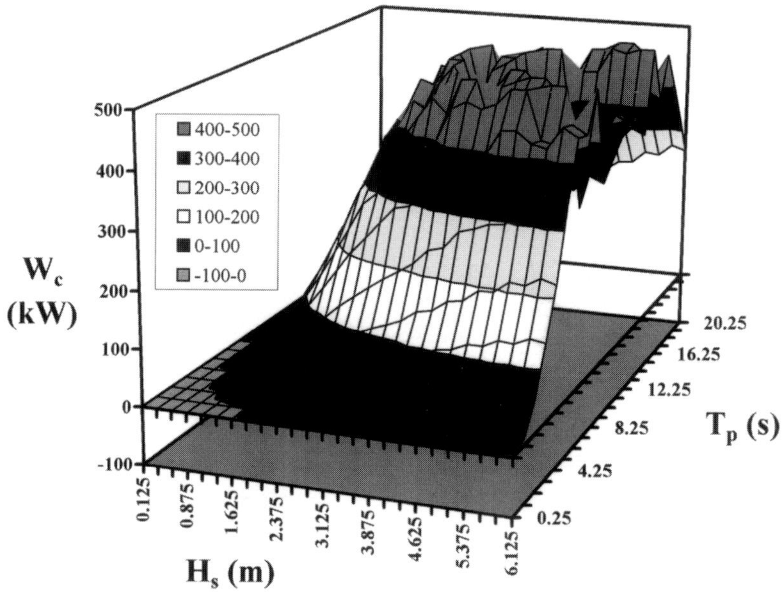

FIGURE 11. Distribution of converted powers associated with each sea state.

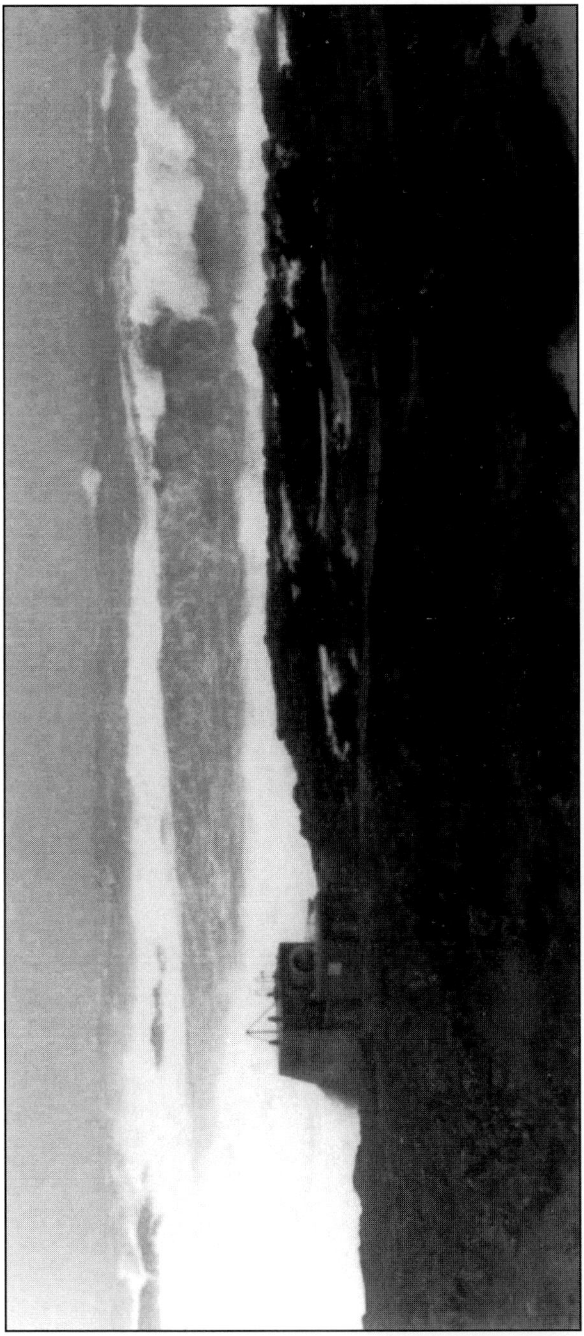

PLATE 1. The Islay 75kW prototype OWC.

PLATE 2. The 2MW AWS system.

PLATE 3. The 120kW Mighty Whale OWC.

PLATE 4. The 500kW EnergeTech OWC

CONCLUSIONS

Wave energy conversion is now technically feasible, ready for commercial exploitation and will feature increasingly in the integrated energy strategy of the 21st century. The ensuing growth will be evolutionary, with plants moving from shoreline, to near shore, to offshore, and from the research realm to the global market. Presently, the balance of power is passing into the commercial sector, with full commercial realization to follow within the next 10 years. The pioneering companies that are at present entering the sector are gaining a secure footing in a forthcoming industry that will realize a turnover of Millions within the short-term, with the full longer-term gains and integrated role to be established according to the global energy policies of the new millennium.

REFERENCES

Abbott, I.H., and von Doenhoff, A.E. 1949. *Theory of Wing Sections.* Dover Publications.

Beattie, W.C., Peters, G., and Alcorn, R.G. 1995. Strategies for the Control of a Wave Power Station with a Semiconductor Converter, *Proc 2nd Europ Wave Energy Conf*, Lisbon, 273-280.

Budal, K., and Falnes, J. 1975. A Resonant Point Absorber of Ocean Wave Power. *Nature*, **256:** 478-489

Count, B. 1980. *Power from Sea Waves*, Academic Press, New York, ISBN 0-12-193550-7.

Curran, R., and Gato, L.C. 1997. The Energy Conversion Performance of Several Types of Wells Turbine Designs. *Jnl. Power Energy*, **211:** 133-145.

Curran, R., Stewart, T.P., and Whittaker, T.J.T. 1997. Design Synthesis of OWC Wave Energy Converters: Performance Matching. *Jnl. Power and Energy*, **211:** 471-482.

Curran, R., Whittaker, T.J.T., Raghunathan, S., and Beattie, W.C. 1998. Performance Prediction of the Contrarotating Wells Turbine for Wave Energy Converters. *Jnl. Energy Engineering*, ASCE, **124:** 35-53.

Evans, D.V. 1979. Submerged Cylinder Wave Energy Device: Theory and Experiment. *Applied Ocean Research*, **1:** 3- 12.

Gato, L.M.C., and Falcao, A.F.deO. 1989. Aerodynamics of the Wells Turbine: Control by Swinging Rotor-Blades. *Int. Jnl. Mech. Sci.*, **31:** 425-434.

Horlock, J.H. 1966. *Axial Flow Turbines. Fluid Mechanics and Thermodynamics.* Butterworths.

Justino, P.A.P. Nichols, N.K. and Falcao, A.F.de.O. 1993. Optimal Phase Control of OWC's. *Proc lsd Europ Wave Energy Conf*, Edinburgh, 145-149.

Malmo, O., and Reitan, A. 1985. Wave Power Absorption by an Oscillating Water Column in a Channel. *Jnl. Fluid Mechanics,* **158:** 153-164.

Mei, C.C. 1976. Power Extraction from Water Waves. *Jnl. Ship Research*, **20:** 63-66.

Mollison, D. 1982. *Ireland 's Wave Power Resource*. Nat. Board. Sci. Tech., ISBN 0-86282-023-5.

Raghunathan, S. 1995. A Methodology for Wells Turbine Design for Wave Energy Conversion. *Jnl Power Energy*, **209:** 221-232.

Salter, S.H. 1988. World Progress in Wave Energy–1988. *Int. Jnl. Ambient Energy*, **10** (1): 3-24.

Salter, S.H. and Rampen, W.H.S. 1993. Wedding Cake Multi-Eccentric Radial Piston Hydraulic Machine. *Proc. 10th Int. Conf: On Fluid Power*, Brugge, 47-64.

Sarmento, A.J.N.A., Gato, L.M.C. and Falcao, A.F.de.O. 1990. Turbine Controlled Wave Energy Absorption by Oscillating Water Column Devices, *Ocean Engineering*, **17:** 481-497.

Setoguchi, T., Kaneko, K., Maeda, H., Kim, T.W., and Inoue, M. 1988. Impulse Turbine with Self Controlled Blades for Wave Energy Conversion. *Int. Jnl. Rotating Machinery*, **3** (4): 233-238.

UNSTAT, 1995. *1993 Energy Statistics Yearbook.* UN Statistical Division, New York.

WEC, 1995. Survey of Energy Resources. *World Energy Council*, London.

Whittaker, T.J.T., Beattie, W.C. Raghunathan, S., Thompson, A., Stewart, T. and Curran, R. 1997. The Islay Wave Power Project: an Engineering Perspective. *Water Maritime and Energy*, 189-201.

Renewable Energy: Trends and Prospects. Edited by S.K. Majumdar, E.W. Miller and A.I. Panah

Chapter Eight

THE U.S. HYDROELECTRIC POWER INDUSTRY: TRENDS AND PERSPECTIVES

DMITRY V. MESYANZHINOV
and
DAVID E. DISMUKES
Center for Energy Studies
Louisiana State University
Baton Rouge, LA 70803

INTRODUCTION

The purpose of this chapter is to provide a broad overview of the U.S. hydroelectric power industry, its current condition, past developments, and future trends. The first sections explains the role played by hydroelectricity in the U.S. power industry. Next two sections briefly describe hydroelectric technology and history of hydropower development in the U.S. Then follows an analysis of the current status of the industry. The chapter concludes with a discussion of the future of the U.S. hydroelectric power industry.

THE ROLE OF THE HYDROELECTRIC INDUSTRY

Hydroelectric generation occupies a special place in the U.S. power industry. This uniqueness is due primarily to a combination of several features, that distinguishes hydroelectric generation from other power generation technologies, including limited geographic placement, free fuel costs, relatively high capital costs, and multiple uses/purposes of hydroelectric projects. It is easy to think of hydroelectric power generation as a unique renewable energy resource that should be developed for its positive environmental benefits. Yet in order to promote a technology/resource, it is important to understand the context in which hydroelectric generating facilities operate and are used in modern bulk electric power systems.

Electric power systems are usually composed of three vertically integrated sectors: generation, transmission, and distribution. Generation, usually powered by either water, natural gas, coal, oil, or uranium, is an important sector within the power industry amounting to approximately 50 percent of all total plant in service and close to 75 percent of total operation and maintenance expense.[1] Thus, the operating characteristics of an individual generating unit, play a large role in any given utility's costs as well as its ability to provide electric power.

In addition to its relative size, generation is the production process center of the power industry. This production process is multi-faceted and starts with the conversion of primary energy, such as fossil fuels, uranium, or the kinetic energy of water, to electrical energy. The process by which this primary energy is converted to electricity varies depending upon the prime mover of the power generator. Mainstream generation technologies include hydroelectric facilities, internal combustion, or combustion turbine facilities, and steam generation facilities.

Hydroelectric facilities are one type of very unique electric generation facility that uses the kinetic energy of falling water to turn a water turbine to create electricity. These types of facilities usually have limited technical applicability and are located in geographic areas that meet certain elevation, water level, and stream flow requirements. The advantage of hydroelectric facilities is their free fuel costs. However, a disadvantage, in addition to their limited geographic applicability, is that they have relatively high capital costs.

In the past, electric generation planning consisted of developing and maintaining a portfolio of resources to meet the various types of electricity loads that occurred in any given hour, across any given day, in any given season. Since load varies considerably across hour, day, and year, utilities have traditionally segmented their generation facilities into three categories: baseload generation, intermediate or cycling generation, and peaking generation. Baseload generators are typically steam generation facilities that are used to service minimum system load and as such are run continuously. While these units are the most efficient to operate, they usually have high fixed capital costs and are costly to start up from a cold shut down, therefore, they are usually run at a near-constant rate. Running these plants at near constant rates reduces overall average costs and variable operating costs associated with ramping up very large generating facilities.

Other types of power generating facilities include intermediate load plants and peaking units. Intermediate units can be thought of as units that bridge the dispatch of baseload and peaking units during periods of high demand. These units can be older and less efficient than baseload units. Peaking units are typically combustion turbines that have the ability to come on line immediately and serve temporary spikes in demand, such as during a heat wave when surging residential and commercial air conditioning use begins to overload the system.

In the past, electric utilities dispatched generating units to meet demand on a lowest- to highest-cost basis. This form of dispatch is commonly referred to as "economic dispatch." The marginal or incremental cost of dispatching units is traditionally the benchmark used to rank order available generators. These marginal costs, in the very short run, are typically associated with changes in fuel costs and other variable operating and maintenance (O&M) costs. Large baseload units typically have the lowest incremental costs and are dispatched first to meet the load. As load increases during the day, or across seasons, less efficient, intermediate or cycling units, which generate electricity at slightly higher costs, are brought on line. Higher cost peaking units would be the last type of units brought on line under an economic dispatch regime.

Hydroelectric facilities can be dispatched in a number of different ways including serving as baseload units, intermediate load units, and "peakers." Usually, very large hydroelectric facilities are designated as baseload units. These units are typically "must-run" facilities and are operated at a near constant rate for a number of reasons. First, capital costs for these large units tend to be relatively high. Thus, operating at a continuous rate drives down average costs. Second, since hydroelectric generation does not involve the use of conventional fuels, marginal or incremental costs are close to zero. Despite the traditional application of hydroelectric facilities as baseload units, "pumped storage" hydroelectric facilities are often used to meet intermediate and peak loads.

HYDROELECTRIC TECHNOLOGY

Hydroelectric technology, or the technology of converting kinetic energy of water into electric energy, can be viewed as an extension of an old practice of using the force of falling water to rotate a water wheel, which in turn drives various machinery, such as grist and textile mills. The principle difference, however, is that in hydroelectric technology the falling water is used to spin a turbine, which is in turn connected to a generator that produces electrical rather than mechanical power.

The amount of energy that can be generated in a hydroelectric facility depends on the head and the flow of water. The head is the vertical distance which water travels through the system; in other words, the vertical distance between the intake and discharge of water. The flow is the speed at which water passes through the system, like water flowing in a pipe. Generally, a combination of high head with low flow is the most efficient; and it is typically achieved in hydroelectric facilities belonging to an impoundment type.[2] Impoundment facilities use dams to block the free flow of water in the river. After the dam is installed, the water begins to accumulate in the reservoir thus moving the intake level up and thereby increasing the head. Once the desired head is reached, the water may be passed though the turbines to create a required flow. The amount of water passing through the turbines as well the amounts discharged idly can be changed either to adjust electricity generation or as water (flood) control measures or both. Due to such versatility, impoundment facilities can be used for base load and peak generation; however, base load generation is more common for hydroelectric plants in general because of the combination of free fuel and high construction costs. Besides generation and flood control, impoundment, also called "storage," hydroelectric plants are also commonly used for water supply, irrigation, and recreation.

Facilities of another type called "run-of-river" use pipes with turbines placed directly in the water stream. Because the flow of water through such a system is identical to the natural water flow of the river, little or no impoundment is required.[3] While this is an environmentally less intrusive method of hydroelectric power generation, run-of-river facilities tend to have relatively small generation capacity.

"Diversion" or "conduit" facilities channel a portion of the flow into a bypass with turbines at the end of it. After the water passes through turbines, it is discharged back into the river at a lower point downstream. Some diversion facilities, however, use dams to increase the head. Diversion plants often exist as part of larger irrigation or water supply projects.

Finally, there are "pumped storage" facilities which do not require a constant stream of water. These facilities consist of two reservoirs positioned on different levels and separated by a dam equipped with turbines and pumps. During off-peak hours, when the price of electricity is relatively low, the pumps move the water from a lower to an upper reservoir. During peak hours, such a system functions as a regular impoundment hydroelectric facility. Pumped storage facilities are net consumers of electricity and generation for them is usually reported with a negative sign. However, off- and on-peak price differentials for electric energy make pumped storage facilities feasible.

Three types of turbines are employed in hydroelectric facilities. The smallest and the simplest type is Pelton wheel. Large hydroelectric plants primarily use complex devices known as Kaplan and Francis turbines.[4]

HISTORY OF HYDROELECTRIC POWER DEVELOPMENT

The history of hydroelectric power development in the U.S. began in 1880 in Grand Rapids, Michigan where a dynamo driven by a water turbine generated enough electricity to light 16 lamps. Interestingly enough, this predates Thomas Edison's Pearl Street Station by two years. However, it was not until 1895, with the inauguration of the 5,000 horsepower (37 kW) generators at Niagara Falls, New York that the hydroelectric power industry became an important part of the growing U.S. economy.[5] A rapid growth of hydroelectric power projects ensued in the following years.

From its earliest days, the development of hydroelectric power in the U.S. was closely associated with the Federal government.[6] The Reclamation Act of 1902 established the Reclamation Service, later renamed into the U.S. Bureau of Reclamation, which received an authority to develop the hydroelectric potential of reclamation projects. The Federal Water Power Acts of 1901 and 1910 created a legislative base for Federal involvement in hydropower development. The first large scale Federal hydroelectric project was built during the WWI in Mussell Shoals, Alabama. However, the peak of the Federal activity occurred during the Great Depression, when the construction of extremely large hydro facilities provided a short-term employment solution for many thousands of workers in addition to bringing long-term regional economic development benefits in the form of cheap and abundant electricity and water control. Besides the Bureau of Reclamation and the U.S. Army Corps of Engineers, two special entities were created to administer federal hydroelectric projects: the Tennessee Valley Authority (1933) in the Appalachia, and Bonneville Power Administration (1937) in the Northwest. During the mid-1930s, hydroelectricity accounted for a third of the total U.S. generation (Figure 1), and although hydroelectric generation had increased thereafter, its relative share of total generation had been declining.

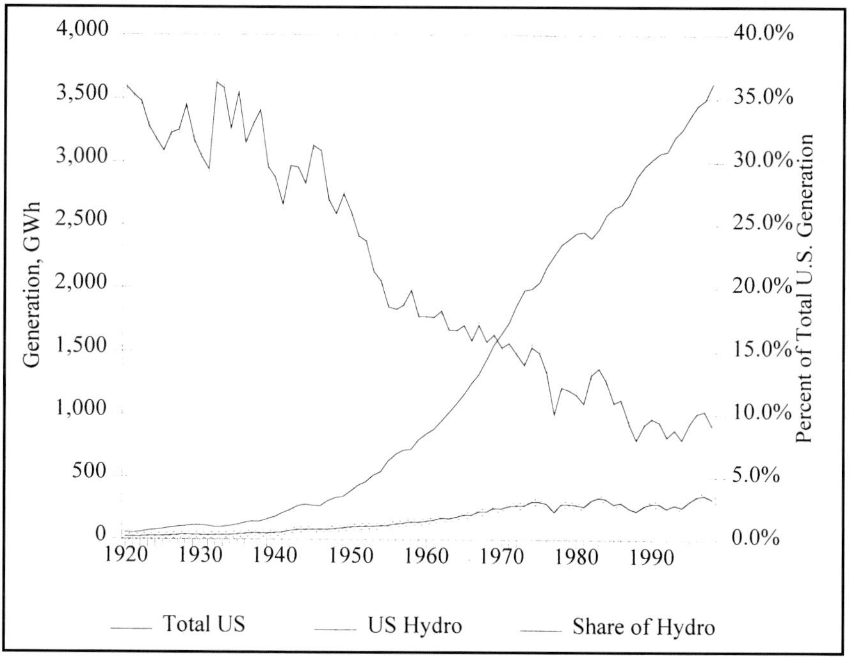

FIGURE 1. U.S. Generation Capacity, Total and Hydro

Through the 1970s, the installation of new hydroelectric capacity continued (Figure 2), but at a much smaller scale, compared to nuclear and fossil fuel based plants, hundreds of which were built to meet the demands of the booming post WWII economy. The oil crises of the 1970s stimulated development of peak-managing pumped storage hydro projects, like the enormous 4.2 GW of combined pumped storage capacity installed in 1973 alone. A significant effect on small-scale hydroelectric development was generated by the Public Utility Regulatory Policy Act (PURPA) of 1978. PURPA exempted "qualifying" facilities (QFs) from certain State and Federal regulation and required utilities to buy power from QF at the utility's "avoided," or incremental, cost of generation. Small facilities using renewable energy resources were classified in PURPA as "qualifying," consequently stimulating development of small-scale hydroelectric projects. However, this development was heavily constrained by the growing public awareness of negative environmental impacts associated with hydropower such as landscape destruction, obstruction of fish deoxidization, and mercury contamination.[7] In the late 1980s and throughout 1990s, largely due to environmental controversy, a very small number of new hydroelectric facilities were constructed.

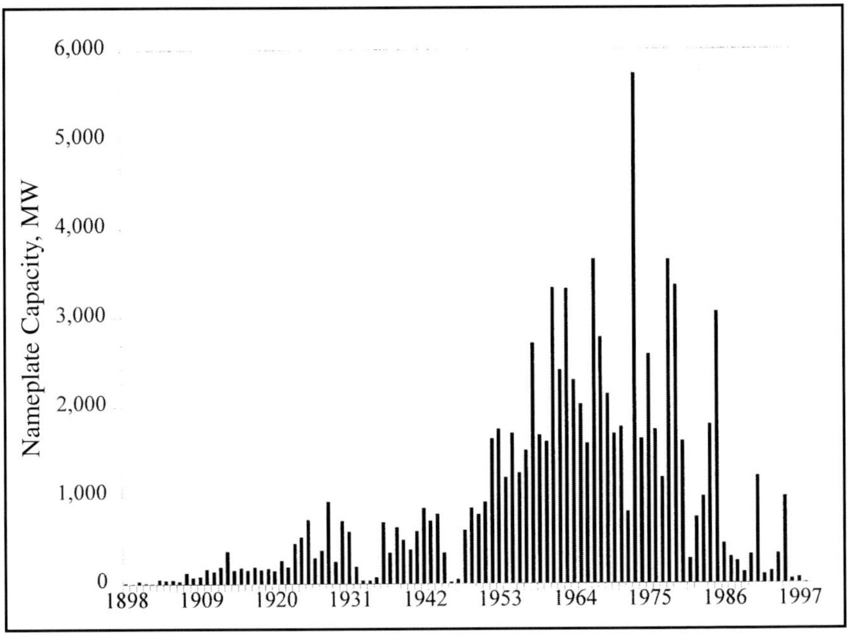

FIGURE 2. Annual Installation of the U.S. Hydroelectric Capacity, 1898-1998

CURRENT STATUS OF THE INDUSTRY

In 1998, total U.S. hydroelectric generation capacity was approximately 92 GW, of which 80 percent is conventional hydroelectricity and the remaining 20 percent is pumped storage. Currently, hydroelectricity accounts for about ten percent of total U.S. generation. There are approximately 2,300 hydroelectric plants with 4,600 generating units. The Federal government continues to be a major player in the industry, controlling 44 percent of the total U.S. nameplate hydroelectric capacity. Private, or investor-owned utilities, control 35 percent, while the remainder belongs to cooperatives, municipal and non-utilities.

Since hydroelectric development can occur only at sites with adequate water resources and appropriate terrain, distribution of hydroelectric facilities across the country primarily is a function of physical geography. However, despite the fact that the distribution of hydroelectric capacity across the states is highly uneven (Table 1), almost every state has some facilities listed in the 1998 data.[8]

Washington state has the largest installed hydroelectric capacity in the country, 20.6 GW or 22 percent of the U.S. total. The three western states combined–Washington, California, and Oregon–account for almost half of the nation's nameplate capacity. The map in Figure 2 shows the location of individual plants. There are four major areas with significant concentrations of hydroelectric facilities: South Appalachia (TVA); Upstate New and Central New England; Sacramento/San Joaquin rivers in California; and Columbia/Snake rivers in the Northwest (BPA). Of the 19 super-large (over 1 GW capacity) hydroelectric U.S. plants, eight are located in the Northwest, including the largest U.S. (and third ranked in the world) hydroelectric plant Grand Coulee, which boasts 6.5 GW capacity (Table 3). While overall U.S. hydroelectric capacity is concentrated within a few large units, the overwhelming number of facilities are of very small size (Figure 4). For instance, the top quarter of all plants accounts for 96 percent of the total U.S. hydro generation capacity; and 36 largest facilities (just 1.5 percent of the total number) possess a half of the total capacity.

THE FUTURE OF HYDROELECTRICITY IN A COMPETITIVE POWER MARKET

The electric power industry is currently undergoing a dramatic change. For the past 75 years, most utilities have been regulated entities with limited forms of competition. However, over the past several decades, a number of the economic, technical, and public policy advantages that have preserved monopoly power for electric generation have started to erode. Today, many states have passed legislation allowing customers to choose their own power providers. This increased ability to choose, however, will have significant implications for the

TABLE 1.
Hydroelectric Capacity, Plants, and Generating Units, by State, 1998

Rank	State	Nameplate Capacity (kW)	Number of Hydroelectric Plants	Number of Generating Units
1	WA	20,625,585	90	298
2	CA	12,445 765	397	594
3	OR	8,822,982	100	232
4	NY	5,323,928	214	425
5	TN	3,730,660	30	88
6	SC	3,429,632	43	138
7	VA	3,084,061	48	96
8	AL	2,886,570	22	93
9	GA	2,403,020	43	120
10	MI	2,399,590	115	293
11	MT	2,399,354	34	98
12	NV	2,091,925	10	30
13	NC	2,045,832	79	141
14	PA	1,942,510	33	90
15	AZ	1,934,100	13	33
16	SD	1,741,308	8	31
17	MA	1,733,598	88	129
18	ID	1,676,807	110	178
19	AR	1,195,730	14	43
20	CO	1,096,926	56	74
21	OK	1,027,678	12	39
22	MO	991,600	8	25
23	KY	747,150	8	31
24	TX	737,590	28	52
25	ME	708,141	124	218
26	ND	517,750	1	5
27	WI	507,074	111	245
28	AK	506,081	38	64
29	MD	494,550	8	20
30	VT	410,234	96	143
31	NJ	404,444	8	10
32	NH	365,746	113	143
33	UT	281,199	64	97
34	WY	270,737	16	30
35	WV	241,770	11	24
36	MN	220,680	33	67
37	NE	183,930	12	21
38	CT	154,890	36	56
39	IA	130,035	5	24
40	OH	127,000	5	9
41	IN	89,552	8	24
42	NM	66,220	6	9
43	FL	42,310	2	6
44	IL	30,598	8	18
45	HI	24,072	16	19
46	RI	7,528	9	9
47	KS	3,860	2	8
48	DE	500	1	1

TABLE 2.
Largest U.S. Hydroelectric Plants, 1998

Plant Name	State	Type*	Nameplate Capacity (kW)	Generators
Grand Coulee	WA	HY	6,548,500	28
Chief Joseph	WA	HY	2,457,320	27
Moses Niagara	NY	HY	2,200,000	13
Bath County	VA	PS	2,100,600	6
John Day	OR	HY	2,025,000	15
Ludington	MI	PS	1,978,800	6
The Dalles	OR	HY	1,713,700	21
Raccoon Mountain	TN	PS	1,530,000	4
Glen Canyon	AZ	HY	1,296,000	8
Rocky Reach	WA	HY	1,279,650	11
Castaic	CA	PS	1,275,000	6
Bad Creek	SC	PS	1,065,000	4
Helms Pumped Storage	CA	PS	1,053,000	3
Hoover	AZ	HY	1,041,800	11
Boundary	WA	HY	1,039,800	6
Wanapum	WA	HY	1,038,000	10
Hoover	NV	HY	1,037,000	8
Bonneville	OR	HY	1,025,899	18
Blenheim-Gilboa	NY	PS	1,000,000	4

* HY - Conventional Hyd PS - Pumped Storage

future development of hydroelectric power.

In the short run, it would appear unlikely that the existing level of hydroelectric power generation will be negatively affected by the restructuring of the electric power industry. In fact, given the cost and operating characteristics of the technology/resource, many hydroelectric operators stand to gain considerably from higher prices for electric energy. Understanding the nature of market clearing prices in a competitive power market helps to clarify this opportunity.

The illustrative supply curve presented in Figure 5 shows the economic dispatch of each generating unit in a given regional power market. The supply curve slopes upwards as more and more expensive generating units are brought on-line to meet electrical loads. The demand curve is simply the market summation of all individual electricity demands in any given hour. As seen in this graph, the market clearing price is determined by the last dispatched unit to meet demand. In this case, the 20 MW unit dispatched at $30 dollars sets the market clearing price. All generators that were dispatched prior to the marginal unit are paid this $30 price for their output. Thus, assume that the first dispatched plant in this example is a hydroelectric facility that has very low marginal costs since its fuel is essentially free. The cost for dispatching the plant is $15 while the market clearing price is $30; therefore, the hydroelectric facility makes a profit of $15 for every dispatched MW at that given hour. Since hydroelectric facilities are usually constantly run, they can stand to gain considerable profits in a competitive market.

The possibilities of hydroelectric power taking advantage of competitive

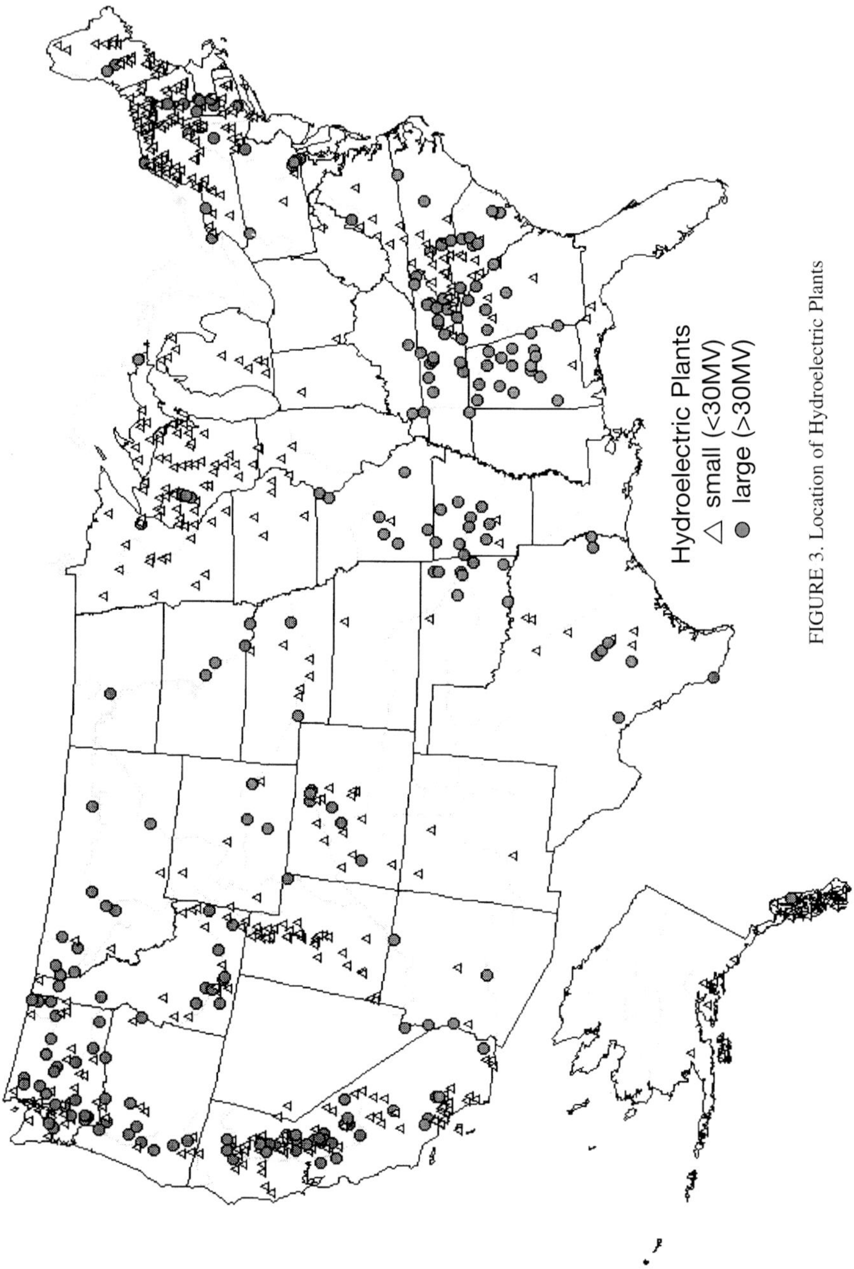

FIGURE 3. Location of Hydroelectric Plants

power market conditions has a number of significant industry stakeholders concerned, particularly in the Pacific Northwest, where the proportion of hydroelectric generation is relatively high to the total. In competitive markets, the large regional price differentials that have existed over the past two decades should start to erode as power moves freely from low-cost to other regions where prices may be higher. This can result in a regional balancing effect until the market reaches a new equilibrium. For customers in the Northwest, it is possible that their electricity prices may increase as a result of cheap hydroelectric power seeking greater profit opportunities. A recent study conducted by the U.S. Department of Energy provides one such scenario.[9]

These competitive opportunities for cheap hydroelectric power raise an additional concern that has potentially greater ramifications than changes in regional power price increases. A large number of hydroelectric facilities are owned by public institutions, including the federal government. These facilities have been financed and supported through taxpayer money and constructed under "public power" premises: providing cost effective, non-profit power to rural areas. Thus, if these public institutions take advantage of competitive market opportunities, they run the risk of displacing other competitive, non-governmental subsidized, power providers. In fact, the largest problem associated with hydro-

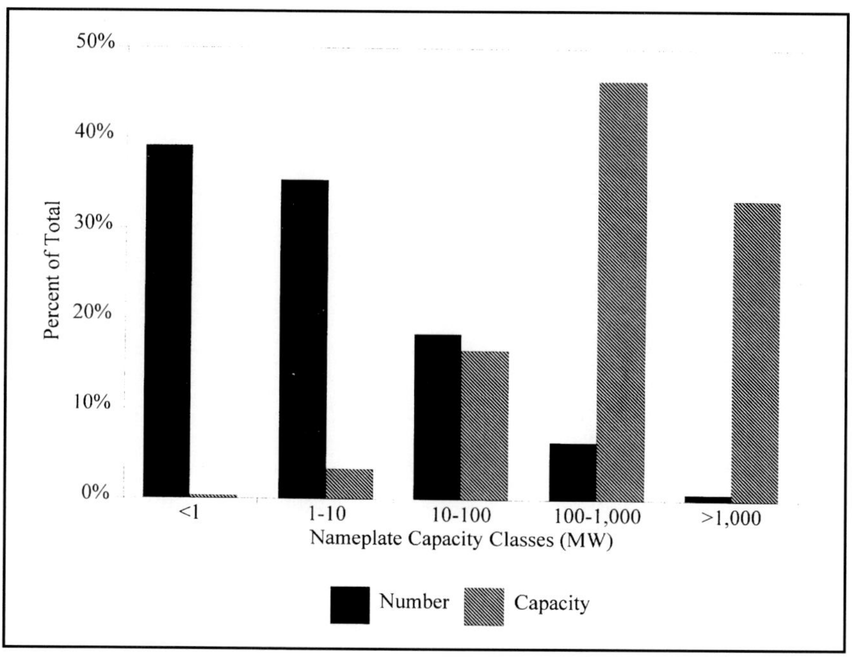

FIGURE 4. Distribution of Hydroelectric Plants by Number and Capacity

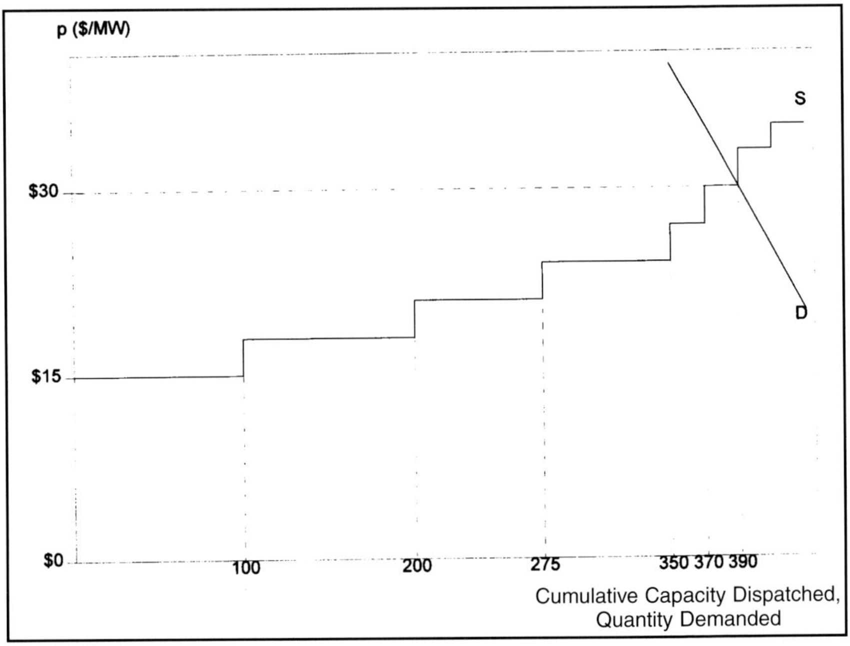

FIGURE 5. Hypothetical Supply and Demand Curves for Electric Power

electric power today rests not with the technology, but with its ownership characteristics. Hydroelectric facilities are so economical, on a marginal cost basis, that they essentially stifle traditional fossil fuel and nuclear power generation. It is this concern that has a number of investor-owned utilities calling for the privatization of large hydroelectric generation facilities of the Bonneville Power Administration and the Tennessee Valley Authority.

CONCLUSION

The future of hydroelectric power is relatively myopic: the very near future for the technology is relatively clear, but becomes rapidly unclear as one moves out further in time. In the short run, hydroelectric power will probably maintain its status as a significant regional source of electricity. According to a forecast prepared by the Energy Information Administration of the U.S. Department of Energy, both generation and capacity of conventional hydropower are expected to remain at their current levels at least until 2010.[10] Despite changes in regional power markets and the competitive implications of government financed hydroelectric facilities, most of these power providers have guaranteed long term "preferred" contracts with regional municipal utilities that will secure their sales base in the short run. This secured sales base is another issue which concerns the private sector of the industry. The actions of public hydroelectric entities will continue to be a thorn in the side of the privately owned sector of the power industry, but it seems unlikely that any large consensus for privatization will develop.

As the future unfolds, the opportunities for hydroelectric power become more uncertain and dismal. First of all, there is an issue of physical constraint on new hydroelectric plants: half of the U.S. hydropower potential is developed and the best sites have already been taken.[11] However, there is little doubt that existing hydroelectric facilities, provided they comply with the tough federal licensing standards, would continue to operate into the near future. From a market perspective, it is possible that several small power hydropower facilities (those less than 100 MW) will be constructed since these will not be seen as competitive threats by privately-owned power producers.

Then, there is an issue of economic feasibility: greater environmental costs as well as pressures from private power producers and the environmental lobby will most likely prevent large-scale hydroelectric development and will force the gradual retirement of older facilities that will be unable, or find it exceedingly difficult, to use public money to finance facility upgrades. Even if some hydroelectric plants were retired, the extremely long construction duration for hydropower facilities creates such an exceptional financial risk that it seems unlikely that any entity except a public institution would care to enter into such an endeavor to replace the retired units. The construction period for a hydroelectric generation facilitiy of any meaningful size is at least seven to eight years, while new state of the art natural gas fired turbines can be constructed within two years. These long construction durations make projects that appear to be attractive today, obsolete within a matter of years.

An exception to this dismal forecast rests with the future of public power entities in a restructured power industry. If public policies are developed that at least reconcile some of the more controversial disagreements between public and private power institutions, then publicly financed hydroelectric power may have a continued future in the fuel mix of the U.S. generation portfolio. However, it seems unlikely that all differences between public and private power generators will be eliminated, and the lack of compromise may result in an effective moratorium on new hydroelectric generating facilities, especially those that offer sizable generating capacities.

In conclusion, large hydroelectric generating facilities, like the regulated monopoly utility, may become a phenomenon of the 20th century alone. Competitive pressures and environmental concerns appear to be the undoing of a technology that, at least from an economic and engineering perspective, ought to have continued opportunities. Until these conflicts are resolved, the industry will continue to contribute an increasingly smaller proportion to overall U.S. electric generation.

REFERENCES

1. U.S. Department of Energy, Energy Information Administration. 1998. Financial Statistics of Privately-Owned Electric Utilities. U.S. Department of Energy, Washington DC.
2. Carless, Jennifer. 1993. Renewable Energy: A Concise Guide to Green Alternatives. Walker and Company, New York, NY.
3. Gulliver, John S. and Roger E. A. Arndt (editors). 1991. Hydropower Engineering Handbook. McGraw-Hill, New York, NY.
4. Warwick, C.C. 1984. Hydropower Engineering. Prentice Hall, Englewood Cliffs, NJ.
5. Hay, Duncan. 1991. Hydroelectric Development in the United States, 1880-1940. Edison Electric Institute, Washington DC.
6. Hausman, William J., and John L. Neufeld. 1999. Falling Water: The Origins of Direct Federal Participation in the U.S. Electric Utility Industry, 1902-1933. Annals of Public and Cooperative Economics, 70: 49-74.
7. Rosenberg, D.M. R.A. Bodaly, and P. J. Usher. 1995. Environmental and Social Impacts of Large Scale Hydroelectric Development: Who is Listening? Global Environmental Change, 5: 127-148.
8. Energy Information Administration Form 860, 1998; and Renewable Electric Plant Information System (REPiS), National Renewable Energy Laboratory, http://www.eren.doe.gov/repis/repis.html.
9. U.S. Department of Energy, Energy Information Administration. 1997. Electricity Prices in a Competitive Environment: Marginal Cost Pricing of Generation Service and Financial Status of Electric Utilities, A Preliminary Analysis Through 2015. U.S. Department of Energy, Washington DC..
10. U.S. Department of Energy, Energy Information Administration. 1993. Renewable Resources in the U.S. Electricity Supply. U.S. Department of Energy, Washington DC.
11. Federal Energy Regulatory Commission. 1992. Hydroelectric Power Resources of the United States, Developed and Undeveloped. Federal Energy Regulatory Commission, Washington, DC.

Renewable Energy: Trends and Prospects. Edited by S.K. Majumdar, E.W. Miller and A.I. Panah

Chapter Nine

WIND POWER IN VIEW

MARTIN J. PASQUALETTI

Department of Geography
Arizona State University
Tempe, AZ 85287-0104

THE PROBLEM WE FACE

Although none of us should shirk the global challenges to resource quality, most people think little about such large-scale problems; instead, we live in our "neighborhoods," in the sense that most of what attracts our attention is immediate and nearby rather than long-term and distant. All of humanity has the same localized field of reference, although poorer people perhaps more so because they have fewer cultural buffers between themselves and the natural world. Regardless of the scale or the immediacy of such perceptions, most of the major environmental concerns of importance have the same root causes: energy demand, population growth, environmental health. In spite of what occupies our attention minute by minute, these are the basics of life on Earth, and especially where they converge, we need to keep track of them.

Even a quick examination sees the first two concerns expanding and the third shrinking, trends with neither deviation nor end in sight. For example, we have been hearing for many years that we need the oil believed to be lying under the wilderness permafrost of the Arctic National Wildlife Refuge in northeastern Alaska, even though there have been strong warnings that development activities would despoil an area little touched by the human hand, our "last pork chop," as Edward Abbey has called it.[1] Eight thousand miles away, Ukraine officials were reluctant to turn off the undamaged reactors at Chernobyl because of the predicted impacts such an action would create for the regional population and its economy, despite the continued health risks of further environmental contamination. East of Ukraine, in already-polluted and overcrowded China, most coal is being burned without pollution abatement because clean-coal technologies are costly and would put the brakes on the accelerating economic engine. In the booming American southwest, the need for substantial new electrical generation stations is stronger than ever, despite the additional pollution such additional facilities would pour into air usually considered the nation's cleanest. Most tragic and conflicting of all, the chance that the two billion people living in less developed countries will break from poverty will remain low without access to electricity, yet if they get it, their great number will drive pollution levels up on a planetary scale.

Some measures have successfully slowed the growth of population, and there has even been success in arresting the downward spiral of environmental quality. But, inexorably, the demand for energy continues to rise, and there is no way to totally avoid further accumulation of the impacts of its use. How should we address this clutch of problems? How can we meet the conflicting demands of energy, environment, and overpopulation?

There has been a constant, if relatively minor, call for many years for more vigorous development of renewable energy resources. The Carter Administration in the late 1970s and the Clinton Administration in the 1990s pursued this goal with unusual enthusiasm and some stuttering success, arguing that renewable energy would provide resource diversity, sustainability, environmental protection, and reduced fuel-price risk.[2] The consistent goal of increasing renewable energy availability has been to satisfy more demand with little additional environmental harm. One of the elements retarding progress on this front has always been the sheer number of possible paths that might yield significant results, whether the research and development effort should stress geothermal, solar electric, solar thermal, ocean thermal gradients, or other possible contributors. Lately, public and private attention turned like never before to wind power, the renewable resource attracting the most attention in recent years. In this chapter I examine the development of wind energy for the generation of electricity at one of the sites of its earliest and most famous installations, Palm Springs, California.

THE RENEWABLE ENERGY OF THE WIND

In 1978, Congress passed the Public Utility Regulatory Policies Act (PURPA) guaranteeing the purchase of any electricity generated by small, nonutility-owned cogeneration and renewable energy facilities. Today about 10 percent of total U.S. generating capacity is contributed by these generators. Around the same time, state regulators started requiring utility companies to include a wider variety of supply-side means of generation, energy efficiency and life-cycle analysis when they were projecting their energy needs and the means they would employ to meet this demand. This was called integrated resource planning, or IRP.

With PURPA and IRP policies encouraging and supporting more vigorous development of renewable energy resources, recent years have witnessed wind energy's move into a leadership position in the renewable energy realm. In 1999 winds generated more than 10 times the kWh of solar electric (3,626 million kWh to 323 million kWh for solar). While this is a relatively small amount compared to hydroelectric generators (322,698 million kWh), the speed of expansion was impressive.

And, this expansion attracted a lot of attention. Wind power caught the contemporary imagination and entrepreneurial spirit like no other renewable energy resource, and it has achieved a position of importance and visibility in the United States and other countries. For example, the World Watch Institute, among several public interest organizations, has put its full weight behind the future contribution of wind power. The speed with which Germany, Denmark, the United States and several other countries are erecting wind turbines has made it the fastest-growing renewable energy resource in the world (Figure 1; Table 1). The present worldwide installed capacity of 14,000 megawatts is an increase of more than 37 percent over the previous year. Still, it is only a small part of the worldwide potential now recognized from Outer Mongolia to North Dakota. In just this single northern Plains state, wind energy could satisfy at least 36 percent of the electrical needs of the United States (Figure 2).

Wind power has several recognized natural shortcomings, among them its low energy intensity, and this means that any appreciable increase in wind's contribution will have to be on a grand and visually obvious scale. Whether such a presence, if multiplied toward its full potential, would amount to a commitment

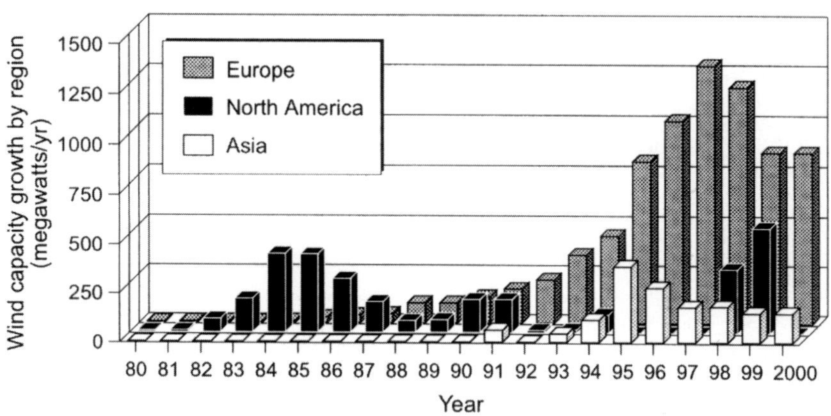

FIGURE 1. Growth of world wind power.
Courtesy of Paul Gipe

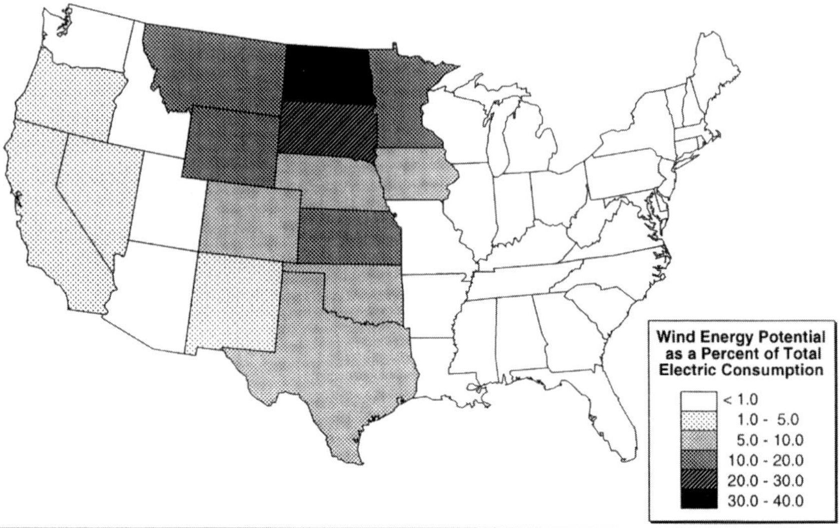

FIGURE 2. Wind electric potential as a percentage of contiguous U.S. 1990 total electric consumption. Source D L Elliott and M N. Schwartz, 1993 *Wind Energy Potential in the United States,* September 1993 PNL-SA-23 109 Richland, WA Pacific Northwest Laboratory NTIS no DE94001667.

TABLE 1
Wind Generating Capacity

Top Wind Energy Markets	1998 Additions	1998 Year End Total	1999* Additions	1999 Year End Total
Germany	793	2872	1200	4072
United States	193	1770	732	2502
Denmark	310	1433	300	1733
Spain	368	822	650	1722
India	82	1015	62	1077
United Kingdom	10	334	18	534
Netherlands	50	375	53	428
China	55	224	76	300
Italy	94	199	50	249
Sweden	54	176	40	216

Note: * Values for U.S., UK, China, India and Netherlands are confirmed and final. Year-end, confirmed estimates are made for Germany, Spain and Denmark. Italy and Sweden are confirmed estimates. Additions only include projects that have been installed and are operating in the calendar year. The year end total figure for the U.S. is net of retired projects. Source: American Wind Energy Association.

unacceptable to the public is a matter of debate, but that there is substantial potential is well known. In the United States, Class 4 winds (at least 6m/s) could annually produce 11.65 million kWh/km^2 on lands that total approximately 460,000 square kilometers, or about 6% of the total land area in the contiguous United States. The potential average power from areas with class 4 and higher is estimated at 500,000 MW. If future generation technology is utilized to take advantage of areas with wind resource class 3 and higher, then the amount of windy land available is more than 1,000,000 square kilometers, or almost 14% of the land area in the contiguous United States. A group of 12 states in the middle part of the country has enough wind energy potential to produce nearly four times the amount of electricity consumed by the nation in 1990[3].

In addition to its potential generating capacity, wind energy entrains several advantages over conventional sources of electricity. For example, wind turbines can be erected and removed quickly, with little trace or contamination. The generation of electricity from wind energy does not pollute air, land, or water. Wind generators can, like other forms of solar power, be erected in places that have no electrical infrastructure and can remain off-grid and isolated. Their modularity allows them to be installed in small increments, unlike the large power plants built between the 1960s and the 1990s that were designed to capture large economies of scale. The attributes of wind power make it well suited for electrical generation in many locations, including industrialized countries, as well as poorer, less developed lands.

The other growing attraction of wind power is its improving economic viability. During most of the 1980s, various incentives supported wind power development, even when it was mechanically inefficient, irresponsibly sited, or poorly operated. Although the most enticing economic incentives have lapsed, some smaller ones remain. For example, the federal government still encourages electricity production from wind farms with a 1.5-cent per kilowatt/hour tax credit. At the state level, California also offers incentives through the Existing Renewables Program ($70.2 million for Tier 2, wind) and the New Renewables Program. These economic aids, and more mature wind energy engineering, are reflected in the continuing investment and deployment of wind turbines in California, Iowa, Minnesota, Texas, and several other states.

At present, wind economics seem attractive. In a recent study, Wiser and Kahn estimated that the costs of generating electricity could be as low as 3.5 cents/kWh.[4] Such amounts are subject to many assumptions, and so are easily challenged, but the lowering installation costs also indicate an improving financial condition for wind power. In constant 1995 dollars, the installed cost per kilowatt in California in 1999 of $ 1322 is approximately one-third of what it was in 1981 ($4251) and one half of what it was in 1982 ($2926). In Germany the cost per kilowatt has dropped from $2776 in 1988 to $1,123 in 1996 (in $1995).[5]

Like any resource, wind energy has uneven distribution and availability. For this reason, it must be developed where it is found or not at all. Once this initial development is accomplished, the similarities with fossil and nuclear fuels thin substantially.

Unlike wind, for example, coal removed from the ground can be transported anywhere to be burned to make electricity. Modern wind development lacks such flexibility. The entire fuel chain common with fossil and nuclear fuels, is compressed with wind power into one location. Only the final product, electricity, has mobility. In effect, the fuel cycle is spatially constrained to locations where the wind happens to be strong and consistent. This generally means places where the wind exceeds 5 m/s (11 mph) for grid-connected development and 6 m/s (13 mph) in a windfarm.

A second constraint of wind power is found in its physics. The power that can be extracted from the wind rises with the cube of wind speed. This means that a site with 5 m/s will produce nearly twice as much power as a wind turbine at a location where the wind averages 4 m/s. This puts a steep premium on the windier sites, thereby concentrating developmental efforts there.

Wind's third constraint stems from its imposition on other land use qualities. Coal, oil, and uranium are routinely and usually removed from the location of natural occurrence for further processing and conversion to electricity elsewhere. With wind power, everything must to done in a single location. Sometimes this requirement is of little significance–as when wind turbines are installed out of sight–but not always.

In some locations wind developments interfere with land values such as wilderness, scenery, recreation, urban residences, or animal migration. In such places, public complaints are sometimes higher, counterbalancing some of the other advantages wind power can offer. In the more bucolic landscapes of the United Kingdom, for example, wind generators have been called "lavatory brushes in the air." Such rhetoric has contributed to the present "standstill" in Britain, severely inhibiting plans to accelerate the use of wind power to help meet some of the carbon reductions agreed in Kyoto. Complaints about visual intrusion have occurred in many other countries, including Sweden, Germany, and the United States. Denmark's response has lately involved erecting wind turbines in the North Sea in an effort to minimize intrusion on the countryside and the farmlands near rural communities.

The development of wind energy cannot help but attract attention. The generators necessarily require large areas, and their nature is to be in motion, albeit inconsistently. In addition to their visual presence, wind generations are sometimes hazards to migrating birds, although there is some accumulating evidence that this problem has been overstated.[6] Wind energy is otherwise considered to be environmentally benign and economically viable. These two qualities may be enough to overcome its perceived shortcomings.

THE WINDS OF PALM SPRINGS

Wind has been a part of the Palm Springs area from the early days of the Cahuilla Indians to the present day of golf resorts (Figure 3). Early Anglo settlers heard stories about the wind, and they could also see an abundance of silent evidence of its power. There were rocks sanded smooth, creosote bushes bent leeward, dunes piled high. Although many of these signs have been erased, graded and covered over by houses and hotels, still the wind blows, especially in the northern part of the city, away from the shelter of Mt. San Jacinto, which reaches heights of almost 11,000 just west of downtown. When the wind funnels through the narrow gap of San Gorgonio Pass, as it does especially in spring and summer, the sand it picks up still pits windshields and stings the skin, windbreaks are still needed to protect houses and railroad tracks, and blowing dust clouds still occasionally close the interstate highway. San Gorgonio Pass is a windy place, and it has always been so.

To most of those who live in the area, wind is an annoyance. To those developing renewable energy, wind is a godsend, and there have been many attempts to convert it into marketable electricity. The modern era of wind power began in the early 1980s. It accelerated with the aid of PURPA, stabilized and then shrank slightly in the 1990s, and is now experiencing a revival. By the end of 1999, there were 2528 wind turbines of various styles in San Gorgonio Pass, with a generating capacity of 353 MW. These turbines produced about 800 gigawatt-hours of electricity (enough for about 40,000 private homes). "Repowering" of the wind fields continues, a process which is aimed at the consolidation, removal and replacement with better turbines, increasing the efficiency of the entire system. (For comparison, 1986 saw 4155 turbines, with a capacity of 295 MW, generating 373 gigawatt-hours.)[7]

The wind power industry in San Gorgonio Pass today is a smoother operation than it was during its first major spurt of growth. Indeed, it was in Palm Springs that many of the most strident anti-wind public reactions germinated and grew, presaging those that the industry would face again and again, both domestically

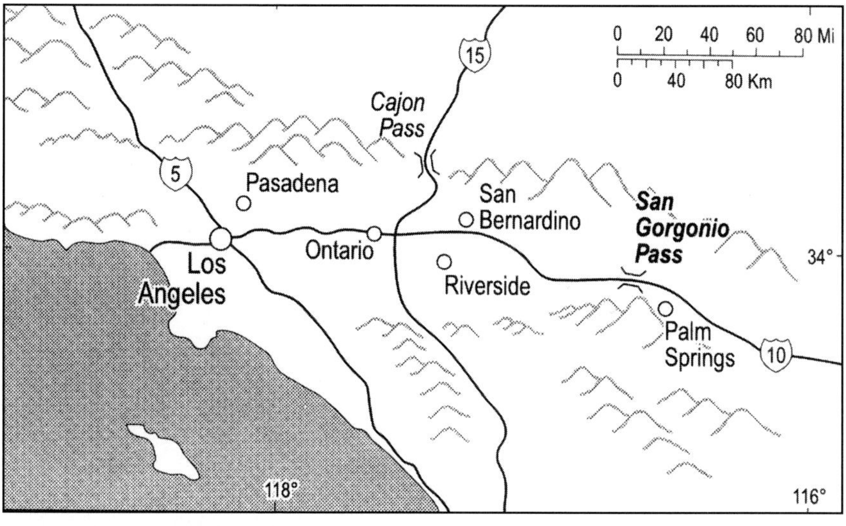

FIGURE 3. Palm Springs and its relationship the other sites in southern California.

and abroad. The most vocal initial public responses about wind development near Palm Springs objected to turbine noise, interference with television reception, possible bird mortality, and the lack of government control and oversight. People were particularly put off by the generous tax incentives that the early developers received to erect the turbines in the first place, a valid grievance given the poor performance of many of the early projects.

Most of all, however, complaints focused on wind power's visual intrusion (Figure 4). While there can be no argument that the installation of wind turbines quickly changes landscapes, it has been especially true in San Gorgonio Pass where there is no vegetation to offer camouflage, where majestic snow-covered desert mountains are a local and admired scenic attraction, and where travelers on the nearby interstate highway are fenced in by wind turbines every time they travel through the pass.

There was no preemptive measure that could have reduced the impact the turbines sufficiently to avoid the intense backlash of criticism they attracted, but this was not on the minds of the developers of the mid-1980s. They were interested only in getting the wind turbines in place, and they were not concerned with whether anyone liked them or even if they produced much electricity. As a consequence, environmental mitigation was little more than a nominal concern, deployment was chaotic, and turbines seemed to work infrequently. The public viewed power in those early days as having high costs and low benefits. Palm Springs, in a well-publicized action, responded to public displeasure by suing the

FIGURE 4. Wind generators can interfere with the view of Mt. San Gorgonio, the northern border of San Gorgonio Pass. *Photo courtesy of Paul Gipe.*

relevant jurisdictions of the US Bureau of Land Management and Riverside County over the wind turbines, contending that the "wind farms" degraded the entryway to their resort community. In one form or another, this same complaint has dogged wind power development ever since, despite all the adjustments that have been made by the wind industry in response. There seems to be a sequence, however, from initial enthusiasm for the development of renewable energy, to quick public backlash, to final public acceptance. That is, at least, what has happened in Palm Springs, with the last phase presumably the result of the implementation of planning requirements.

THE PLANNING RESPONSE

Various administrative units have jurisdiction over the land in San Gorgonio Pass, including the City of Palm Springs, the County of Riverside, and the Bureau of Land Management. Each has responded to the public's concerns about wind power in similar ways, although zoning ordinances have been the most popular mechanism. Palm Springs zoning ordinances now restrict turbines to four zones, namely W, 0-5, E-1, and M-2 (watercourse zone, open land zone, energy industrial zone, and manufacturing zone), and there are many restrictions that apply within these zones. The noise levels for the Wind Energy Conversion Systems (WECS) must be 55 dB(A) or less, unless it is determined that a "pure tone" will be generated. All parcels in the vicinity of the commercial WECS project property which are actually used or designated for residential, hospital, school, library, or nursing home purposes must be identified. A commercial WECS or WECS array cannot exceed 65 dB(A). A commercial WECS or WECS array must not be operated so that impulsive sound below 20 Hz adversely affects the habitability or use of any dwelling unit, hospital, school, library, or nursing home. All noise measurements and noise projections must be made in accordance with the technical specifications and criteria developed by the County Health Department and adopted by resolution of the Board of Supervisors. No perceptible vibrations of dwelling structural elements are allowed over a range of frequencies from 0.1 Hertz to 100 Hertz

Various setbacks have also been mandated. The center of each individual Wind Turbine Generator (WTG) must be installed at least 1.25 times the total WECS height from above ground electrical transmission lines of more than 12kV, 1.25 times height from a public highway or road, railroad or off-site building, 1.25 times height from any lot line (at least 200 feet from any lot line which contains a dwelling). A five-rotor diameter setback is required from any lot line that is perpendicular to and downwind of, or within 45 degrees of perpendicular to and downwind of, the dominant wind direction. For purposes of preserving some of the area's scenic assets, specific restrictions have been applied: No WTG can be erected within 1/4 mile of Highway 62; 500 feet of Indian Avenue and Interstate-10; 2/3 mile of Highway 111; 1/25 times total height from Dillon Road; nor south of Highway 111 (Figure 5).

Several other requirements address safety, security, birds, and culpability. In stark departure from the wind developments in England, which welcome the curious and sauntering tourist, people are kept out by fencing and patrols. The rotor at its lowest sweep must be at least 25 feet from the ground, and height limits are at 200 feet (to be amended to 300 feet; the same as the County). Dead birds within 500 feet of a wind generator must be reported, and bonds are required by the Palm Springs Municipal Code at $2,000 per acre.

In reference to aesthetics and public convenience, the stockpiling of any WECS or any other activity or structure during construction and life of the WECS is prohibited outside of temporary and permanent disturbance areas, there shall be no television interference, all failures must be reported in writing, all collector cables must be placed underground, and use of neutral 'environmental' paint colors (including samples for approval) is mandatory (with colors required to weather within a six-month period to a lusterless condition). These requirements provide a generalized summary of the form and themes that are being used to govern wind development in other places in the US and in Europe.

THE COMMUNITY TURNS AROUND

The early days of modern wind development near Palm Springs were controversial and well publicized, but the measures and controls that were instituted in San Gorgonio Pass sufficiently mollified public disdain for the

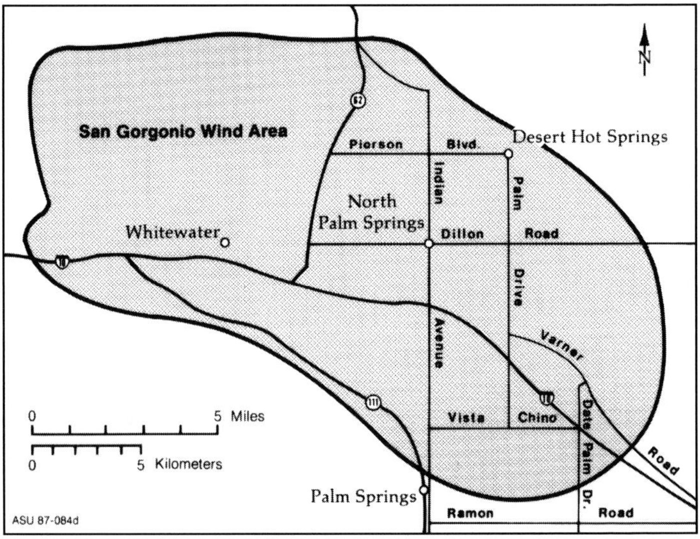

FIGURE 5. Palm Springs wind area.

wind projects that recent years have seen few examples of the early dissension and opposition. In addition to the establishment of codes and ordinances, part of the turnaround came with changes in the attitude of city government. It was proposed at the time that former entertainer Sonny Bono was mayor of Palm Springs, that a large area north of the municipal boundaries be annexed as a way to sweep further tax revenue into city coffers. As part of the proposed annexation, Stanley R. Hoffman Associates evaluated the potential benefits that would accrue from wind development, providing a general template for other jurisdictions. The property totaled 92,300 sq. feet (8,640 acres or 13.5 sq. mi), with 190,900 sq. fit of existing commercial and industrial uses and would bring in new employees totaling 388, including 139 from wind. Estimated revenue generation would come from property tax, property transfer tax, sales and use tax, transient occupancy tax, franchise fees, charges for current services, fines and forfeitures, and interest income. Costs of the annexation would include services by police, animal control, fire, street maintenance, signalized intersections, storm drainage, and citywide overhead. The estimated revenues in the initial year were $338,800 revenues vs. $164,400 costs, for a net surplus of $173,400 and a revenue/cost ratio of 2.05. These estimates did not include job creation or tourism value, something that unexpectedly developed as the wind farms became popular for films, advertisements, and the private tours that were promoted by the nearby tourist offices and resorts.

These economic benefits are specific to Palm Springs and are only applicable in a general way elsewhere; in other locations the various revenues might be greater or less. Were the lands of San Gorgonio Pass suitable for ranching, for example, landowners would be able to reap the additional revenues of the wind installations and make their land less susceptible to residential or commercial development. Lease payments can range from 2 percent to 10 percent or more of gross revenues on an annual basis. According to an estimate by the National Wind Coordinating Committee,[8] a 20-megawatt wind project operating with a 25 percent capacity factor that receives an average energy payment of 5 cents per kilowatt hour will produce more than $2 million in gross revenue annually. The benefit to the land owner if the lease terms specify 2 percent of gross revenue is approximately $43,800 per year. The Union of Concerned Scientists estimates that the farmer could increase the return on his or her land anywhere from 30 percent to more than 100 percent by leasing a small fraction for wind turbines.[9]

A RECENT EXAMPLE

Palm Springs continues to receive requests for planning approval for new wind developments. The details of such a request can provide some sense of how the planning process might be employed elsewhere. Recently, Wind Energy

Conversion Systems (WECS) 28 and 67 have been proposed for a site west of Indian Ave and north of Interstate-10. The proposal calls for the installation of eight turbines in WECS 28, which currently contains 34 Micon and Nordtank 65kW turbines. Nine are to be installed on WECS 67, which is currently vacant. Seventeen Zond Z750-KW wind turbine generators, towers, foundations, appurtenances and a facility yard are planned for this 126 acres of land. The turbines have a hub height of 65 meters (213 fit), an overall height of 90 meters (296 ft), and a rotor diameter of 50 meters (164 ft), and they generate up to 750 kW at 26 mph, with a start up at approximately 4.0 m/s (9 mph) with a cut out speed (for 5 sec) of 29.1 m/s (65 mph). They are rated for a wind speed 11.6 m/s (26.0 mph). Foundation for each WTG will be approximately 4 feet in depth, and a 160-foot utility easement is needed in the southwest portion of WECS 28. There was no request for a change in existing zoning: WECS 28 - E-I (energy industrial), WECS 67 - M-2 (manufacturing).

In response to this application, the City planning staff recommended that the plan meet all requirements of Section 9402.00.H.8 of the Zoning Ordinance with the exception of the height limit for new WTGs of 200 feet and the safety setback and wind access setback from property lines on both sites. Each individual WTG requires a safety setback of at least 1.25 times the total WECS height (296 x 1.25 or 370 feet) from any property line for safety purposes and 5 times the rotor diameter from any downwind property line for wind access purposes. In order to meet the proper property setback, the applicant proposed to install the new WTGs approximately 1.1 times the total height (approx. 326 feet) from the closest lot lines on both WECS sites. There was to also be a 326 foot setback, equating to a wind access setback of just under two rotor diameters, significantly less than the 5 rotor diameters (or 820-foot in this case) setback required in the Zoning Ordinance, and therefore requiring a variance. The use was considered an extension of the existing use already established, and the density of WTGs per acre was consistent with surrounding projects. The height of the existing WTGs in this area varies from 101 feet to 197 feet. The Planning Commission had already approved two Conditional Use Permit applications with related variances to install WTGs of approximately 233 feet in height on two WECS sites west and southwest of the subject sites. The proposed height variation would align with the subject property more closely with established projects in the Riverside County and BLM properties, where turbines to a maximum of 300 feet in height are currently allowed. Given the unique shape and ownership patterns of the two WECS on their adjacent lands, and the limitations on WECS development between the two sites, the proposal to reduce the required wind access setbacks was deemed "justifiable." The advantages of the turbines, as identified by staff reviewers, included that they would be more efficient energy producers, that they would include the latest safety measures, that they would make more efficient use of the terrain, that there would be lower noise levels and slower rotational speeds, and that their installation would reduce overall visual impacts because fewer machines will be needed to generate similar levels of energy.

The detail of the zoning ordinances that have been put into effect, as well as the recommendation of Palm Springs planning staff, illustrates a current approach to wind development that is substantially different from that existing in the mid 1980s. So great has the public mood changed toward wind power, that the city is even promoting future installations of WTGs in this area: "The primary function of the E-I Zone is to encourage the development of alternative energy systems that take advantage of the local meteorological conditions and capture wind energy in an efficient manner." The intended future of this area would be primarily for WECS.

THE REAL POWER OF WIND ENERGY LANDSCAPES

The rich experience of San Gorgonio Pass wind developments, complete with its controversy, visibility, notoriety, and regulatory controls, is an appropriate place to yield lessons transferrable to other sites. The most important lesson develops from the simple observation that, despite the tumult and growing pains that have been experienced, the wind turbines have become a part of the landscape. They are still there. That wind energy is not only growing but is being locally promoted by political jurisdictions further suggests that despite many public questions about the wisdom of its spread, wind generators are with us for the foreseeable future.

Among the several ingredients that can explain wind energy's current status near Palm Springs is that it is simultaneously profitable and environmentally sustainable. But more noteworthy is that the Palm Springs wind developments have overcome their inescapable public visibility, including some embarrassing episodes during the early years of deployment. Unlike the environmental damage that can result from energy projects in distant and isolated places, wind energy's immobility and unavoidable visual presence has stimulated more immediate, local, and satisfying problem recognition and redress. Admittedly, a higher environmental price might have precluded such accommodation, but their concrete, neighborhood presence has inspired us to learn more about them and what we must do to properly adjust and guide their development. In Palm Springs, we seem to have come to the conclusion that wind energy's nonpolluting and sustainable nature more than outweighs its visual presence, a conclusion I suspect will be reached in many other locations as well. The greatest power of wind energy, rather than its greatest weakness, may be that it is impossible to hide.

ENDNOTES

1. Edward Abbey. 1984. Gather at the river, in *Beyond the Wall:* Essays from the Outside, New York: Henry Holt and Company, Inc.

2. Nancy A. Rader and Ryan H. Wiser. 1999. *Strategies for Supporting Wind Energy: A Review and Analysis of State Policy Options,* Washington, D.C.: National Conference of State Legislatures.
3. These estimates are taken from D.L. Elliott and M..N. Schwartz, 1993. *Wind Energy Potential in the United States,* September 1993. PNL-SA-23109. Richland, WA: Pacific Northwest Laboratory. NTIS no. DE94001667. The authors point out that the although the wind potential of the country is large, only part of it can be exploited economically, and that it will vary from one utility to another, and according to several environmental and jurisdictional restrictions.
4. Ryan Wiser and Edward Kahn, 1996, "Alternative Windpower Ownership Structures: Financing Terms and Project Costs,", *Wind Energy Weekly*, No 709, 12 August 1996. Based on a 50-MW wind farm with an installed cost of $1,000/ kW, a 30% capacity factor, and operations and maintenance (O&M) expenses of 0.65 cents/kWh., if an investor-owned utility owned and financed the facility instead of a wind developer.
5. Paul Gipe, May 24, 2000, personal communication.
6. Richard Anderson, Michael Morrison, Karin Sinclair, Dale Strickland. 1999, *Studying Wind Energy/Bird Interactions: A Guidance Document,* prepared for the Avian Subcommittee and the National Wind Coordinating Committee, Washington, D.C. (*www.nationalwind.org*)
7. Paul Gipe, May 24, 2000, personal communication.
8. http:*//www.nationalwind.org/pubs/wes/wes05.htm*
9. M. Brower, M. Tennis, E. Denzler and M. Kaplan, 1993. *Powering the Midwest: Renewable Electricity for the Economy and the Environment,* Union of Concerned Scientists.

Renewable Energy: Trends and Prospects. Edited by S.K. Majumdar, E.W. Miller and A.I. Panah

Chapter Ten

MODELLING AND MEASUREMENTS OF COASTAL WIND SPEEDS

R. J. BARTHELMIE

Department of Wind Energy and Atmospheric Physics
Riso National Laboratory
4000 Roskilde, Denmark
(Also affiliated with the Atmospheric Sciences Program
Indiana University, Bloomington, IN 47405)

INTRODUCTION

In order to reduce emissions of greenhouse and other polluting gases and increase security of supply, the European Union (EU) White Paper on Renewable Energy [1] proposes that 12% of energy should be supplied from renewable energy sources by 2010. Within this framework the EU target is 40 GW of installed wind capacity. A significant part of this wind energy capacity will be generated by offshore wind energy installations. Offshore wind farms are an attractive proposition due to the potential for large developments, which pressure for land and other socio-economic factors will not permit on land in many European countries.

By the end of 1999, approximately 30 MW of installed capacity of offshore wind turbines will have been installed in Europe. Developments planned for the next decade include at least ten individual wind farms of up to 100 turbines each rated between 1 and 2 MW [2]. The first installations of large offshore wind farms are planned in the region 5-15 km from the coast which reduces visual intrusion and optimizes the wind resource while keeping transmission cable costs relatively low. However, in terms of predicting the wind energy resource, this region is the most difficult to model due to the complex interactions between heat and momentum fluxes as air flows over the coastal discontinuity. This chapter describes the modelling and measurement initiatives which are being employed to reduce uncertainties in resource prediction and to provide information regarding wind and turbulence climates to improve design of offshore wind turbines and wind farms. Although many European countries have plans for exploitation of offshore wind energy [2] plans are furthest advanced in Denmark and so this county will provide the focus of much of the research described here.

Wind energy in the coastal zone

Owing to lower surface roughness offshore, wind speeds over the sea are typically higher than over land (except where wind speeds on land are enhanced by topography), and the variation of wind speed with height (wind shear) is typically lower. Higher wind speeds and lower vertical wind speed variations across the turbine blades increase power output and lead to reduced stress and fatigue on blades, hubs and towers offshore. The negative implications of the reduction in turbulence offshore associated with the lower roughness of the sea surface is that wake effects (i.e. the disturbance of the wind flow experienced by a wind turbine caused by the other wind turbines located 'up stream') will be larger. Therefore optimum wind farm design (e.g. turbine spacing) may be different from land-based wind farms.

It is difficult to accurately predict wind energy resources at offshore sites which are influenced by the coast where wind speeds and turbulence are adjusting to changes in roughness, temperature and humidity which distort the shape of the wind speed profile. Research has demonstrated that the influence of land significantly modifies the wind speed distribution over 20 km from the shore [3] and the coastal discontinuity can still be detected in meteorological observations at 50 km from the coast. The propagation of the influence of land as air is advected offshore is especially important under stable conditions offshore [4] when the internal boundary grows very slowly [5].

The change of wind speed with height in near-neutral conditions is described by the logarithmic wind speed profile (equation 1):

$$U = \frac{u_*}{\kappa} \ln\left(\frac{z}{z_0}\right) \qquad (1)$$

where U is the wind speed at height z, u_* is the friction velocity, κ is the von Karman constant and z_0 is the roughness length.

In non-neutral conditions the diabatic (stability-corrected) profile is used:

$$U = \frac{u_*}{\kappa} \ln\left(\frac{z}{z_0}\right) - \psi\left(\frac{z}{L}\right) \qquad (2)$$

Where ψ (z/L) is defined differently for unstable conditions and for stable conditions. Higher wind speeds offshore are typically associated with near-neutral stability conditions. However, previous analyses suggest that stable conditions can persist under relatively high wind speed regimes given large differences between land and sea surface temperatures [6]. Roughness of the sea surface also varies with wind speed [7]. Hence, accurately predicting predicting wind speed profiles responses to changing surface conditions offshore (stability, roughness and wave heights) is complex.

Although stability conditions offshore are often assumed to be near-neutral, short-term variations of stability conditions (hours to months) occur. Examples of the effects caused by stability influences are:

• Inverted diurnal cycles: Sea surface temperatures change slowly on time scales of days (except in very shallow water) whereas the temperature of adverting air in the coastal zone varies in response to daytime solar heating of the land surface. In addition to mesoscale sea-breeze effects [8] this can cause an inverted diurnal cycle to form in offshore areas producing wind speeds which are higher at night than they are during the day [9].

• Non-logarithmic wind speed profiles (see 'Vertical profiles of wind speed in the coastal zone'): Changes in sea surface temperatures typically lag changes in mean air temperatures by about one month. This means that in spring atmospheric conditions in coastal areas are frequently stable (warm air from land blowing over relatively cool seas) while in autumn unstable conditions are more frequent [10], [11].

• Lower wind speeds offshore than those in adjacent land areas: Increased land roughness causes increased production of turbulence which more than compensates for enhanced frictional losses giving small acceleration of flow close to the coastline on land [12].

• Development of internal boundary layers (IBL): When air is advected over a roughness change an IBL develops in which the atmosphere is adjusted to the new surface layer. Above this layer, flow continues to respond to the previous surface. In between these two layers, an adjustment layer exists. Thus the wind speed profile cannot be predicted using a logarithmic profile. If sea surface temperatures are cooler than the advecting air temperature this IBL grows slowly.

• Low-level jets: Low-level jets are caused by a number of different mechanisms but those which have been observed to occur frequently in the Baltic Sea in spring (about two thirds of the time) [13] are related to the occurrence of stable conditions [14]. Since the observed low-level jets occur below 200 m height they are estimated to make a significant impact on wind resources (6-15%) [15], [16].

• Horizontal roll vortices: Horizontal roll vortices have been observed to occur in unstable conditions in the Baltic Sea and are likely to increase loading on wind turbine blades [17].

To effectively model both thermal and roughness influences, a meso-scale model is required. However, applying such a model is an expensive and time-consuming task, thus for most wind resource estimates simpler models are used. Although this approximation typically neglects thermal (and hence stability) effects this is frequently a reasonable assumption for prediction of long-term annual average wind speeds [18].

THE DANISH OFFSHORE WIND MONITORING PROGRAM

As part of the Danish Energy program up to 4000 MW of wind capacity will be installed offshore by 2030 which, together with an additional 1500 MW onshore, will produce more than 50% of total electricity requirements in Denmark [19].

The main purpose of the monitoring program described here is to collect data for wind resource estimates but it is also a unique data set which is being used for development of models to provide improved tools for offshore wind energy siting and design. In 1999 two meteorological masts were installed by the ELSAM utility at Horns Rev and Laesø Syd (Figure 1) which are not described further here. Data from the monitoring sites in eastern Denmark which are operated by the utility ELKRAFT and managed by Risø National Laboratory are the focus of this chapter. The program incorporates meteorological monitoring at the Vindeby offshore wind farm which has been ongoing since 1993 (Figure 1). Four new masts were installed offshore (Rødsand, Omø Stålgrunde and Gedser Rev in 1996, Middelgrunden in 1997). One additional coastal mast was installed at

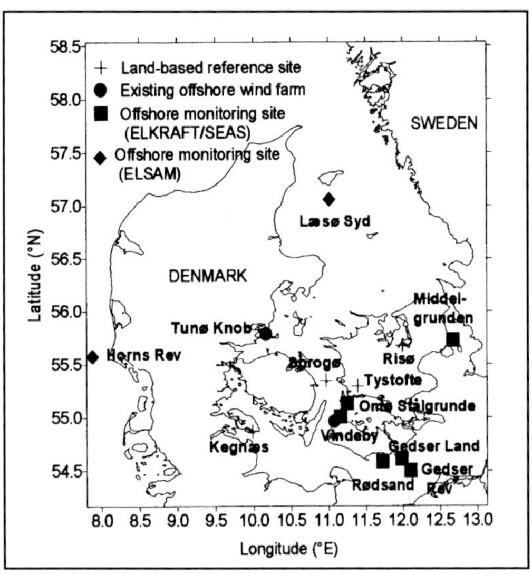

FIGURE 1. Location of monitoring and reference sites in Denmark. The main areas for developing large wind farms are Gedser Rev, Rodsand, Horns Rev, Laesø Syd and Omø Stålgrunde. A 40 MW wind farm was erected during 2000 at Middelgrunden, just outside of Copenhagen harbour.

Gedser in 1996. The masts and data acquisition are fully described in [20]. Despite the difficulties inherent in offshore measurements, data recovery rates in 1999 range between 73 and 95%.

ANALYSES OF DATA COLLECTED IN THE DANISH OFFSHORE WIND MONITORING PROJECT

Development of climatological time series for offshore sites

Meteorological conditions vary on many time-scales and so short-term measurements, such as those from the offshore masts described above, can not capture the full-range of variability of wind conditions at the sites. Thus a number of methods have been used to assess the wind resource at the offshore measurement sites, here Vindeby, based on longer-term (climatological) data sets [21]:

1. Weibull correction method [22].
2. WAsP analysis [23].
3. Measure-correlate-predict (MCP) method.

The climatological data series of wind speeds employed is from the 39 m level at Tystofte (see Figure 1) from the period 1984-1997. As shown in Table 1, all three methods give similar predictions of the long-term wind speed at Vindeby (50 m) and hence give similar energy and potential power predictions. (Potential power output is based on the characteristics of the Offs 1500 turbine [24]). Using different methods to establish the relationship between the measured time series offshore and a reliable long-term series helps to reduce the uncertainties in the wind resource prediction. The MCP method typically predicts lower wind speeds than the other two methods; this is most likely due to a degradation in spatial coherence as the distance between the predictor and predictand site increases. Uncertainty at Vindeby is estimated as ± 4% based on seven years of measurements. At the remaining sites with data coverage of one to four years, uncertainties in the resource predictions are estimated as between ± 8 and ± 10% [25], [26].

Vertical profiles of wind speed in the coastal zone

TABLE 1.
Results of predicting wind resources for 48 m height at Vindeby SMW

Prediction method	Mean wind speed (m/s)	Weibull scale parameter (m/s)	Weibull shape parameter	Mean energy density (W/m^2)	Annual power production (Gwh/y)
Weibull	7.9	8.9	2.1	540	4.6
WAsP	8.1	9.2	2.4	550	4.8
Correlation	7.9	8.9	2.2	540	4.8
Observed	8.1	9.1	2.3	-	-

Since wind speed measurements are rarely available at the desired turbine hub-height, vertical profile of wind speeds profiles must be extrapolated either from measured or modelled data. If conditions are near-neutral on average the logarithmic wind speed profile is used (equation 1). Changes in the state of the sea surface do not lead to distortions of the profile away from logarithmic predictions but cause changes in roughness. Roughness length can be related to friction velocity using the Charnock equation [7]. The fetch dependence of sea state which is not accounted for by [7] has been investigated by e.g. [27].

At all of the offshore sites in Denmark, the observed wind speeds at 50 m are higher than those predicted using 10 m wind speeds and the logarithmic profile with constant roughness of 0.0002 m [25]. Using a higher roughness length gives better predictions but the wind speed profiles at the offshore masts do not conform to a logarithmic profile. Three possible reasons for this are [21]:

1. Mast shadow effects are more important offshore due to a lower turbulence [28]: Evaluation of mast shadow corrections for the offshore masts gave small changes to the directional wind speed distribution rather than a significant adjustment to the mean wind speed or the wind speed profile [26].
2. Stability corrections do not average out over the course of a year [29].
3. The profile is impacted by the presence of internal boundary layers.

In the following sections the second two points are examined and data corrected where possible to improve vertical profile predictions.

Stability corrections

To demonstrate how using a stability correction improves the predictions of the wind speed profile, a single example of application of stability corrections to the wind profile (at Rodsand) is presented. The Monin Obukhov length is estimated based on the wind speed and temperature profile based on [30] for the

TABLE 2.

Observed and predicted wind speed profiles at Rødsand based on data from August 1996 to July 1999 (m/s). Stability corrections are as in equation 2 and discussed in [31].

Height (m)	Observed wind speed	Wind speed predicted using the logarithmic profile	Wind speed predicted using the diabatic profile
10	7.76	7.76	7.76
15		8.04	8.12
20		8.24	8.39
30	8.89	8.52	8.81
38		8.69	9.07
43		8.77	9.22
50	9.61	8.85	9.35

period August 1996 to July 1999. The predicted wind speed profiles using measured data, the logarithmic profile and adding the stability correction calculated from the Monin-Obukhov length are given in Table 2. Here the logarithmic profile (equation 1) is used with a constant roughness of 0.0002 m and the diabatic (stability corrected) profile is defined in equation 2.

All predictions are based on observations at 10 m height. The mean Monin-Obukhov length is 510 m indicating that conditions are slightly stable on average. The occurrence of different stability classes is given in Table 3 showing that stable and near-neutral conditions are the most frequent at this site. As shown in Table 2, using a logarithmic profile the mean wind speed at 50 m is underestimated by 9 % and has a root mean square error of 1.5 m/s while using a stability correction improves the prediction to within 3 % of the observed and reduces the root mean square error to 1.2 m/s.

Internal Boundary Layers

Internal boundary layers develop in response to discontinuities in surface conditions (roughness, temperature and humidity changes). Atmospheric properties within the layer reflect the influence of the new surface while those above the layer are still responding to those of the old surface. In between these two layers an adjustment zone exists.

The offshore fetch (distance to the coast) at Rodsand varies between 11 km and 110 km. For each observation an internal boundary layer (IBL) height can be calculated based on the fetch distance (x) and the Monin-Obukhov length as given by [12]:

$$z_{IBL} = 0.2\boldsymbol{x}^{(0.78\text{-}0.33z\ /L)} \qquad (3)$$

Figure 2 shows average IBL heights plotted by fetch for different stability conditions. The IBL grows most slowly in stable conditions when its development can be expected to have the most impact on near-surface wind speed profiles. Mean IBL heights range between 300 and 800 m suggesting that on average the IBL is not a major influence on the wind speed profile at this site. However,

TABLE 3.

Frequency of different stability classes at Rødsand

Stability class	Monin-Obukhov length (m)	Frequency (%)
Very stable	0 to 200	30
Slightly stable	200 to 1000	26
Near-neutral	<-1000 or >1000	40
Slightly unstable	-200 to -1000	5
Very unstable	0 to -200	1

minimum IBL heights in strongly stable conditions can be as low as 20 m height.

Fetch and stability have a strong influence on the wind speed profile which can be identified by calculating the ratios of wind speed at 50 m to those at 10 m. Figure 3 illustrates these relationships by fetch distance at Rødsand. In unstable conditions, the wind speed profile adjusts most rapidly to the new surface conditions and the increase of the ratio with fetch reaches an equilibrium value much more rapidly. Conversely in stable conditions the ratio continues to increase with fetch to well-beyond 50 km. Hence the near-surface wind speed profile is continuing to adjust even at considerable distances to the coast. This has significant implications for wind resource assessment.

Turbulence intensity

Turbulence intensity (T.I.) is critical to determining wake propagation and turbine lifetimes and typically lower offshore than onshore. T.I. is defined here as the ratio between the standard deviation of wind speed and the wind speed. As shown in Figure 4 T.I. is high at low wind speeds then decreases as wind speeds increase to about 12 m/s and then begins to increase in response to increasing wave height. For wind speeds lower than 20 m/s the relationship between wind speed and T.I. offshore can be modelled using an empirical fit to the 50 m data as a 6th degree polynomial. The form of this figure reflects a transition from predominantly thermally generated turbulence at lower wind speeds, to predominantly mechanically driven turbulence at higher wind speeds.

Figure 5 shows the mean T.I. by direction for Vindeby SMW. At SMW highest T.I. are found in the north east sectors owing to the influence of the wind farm

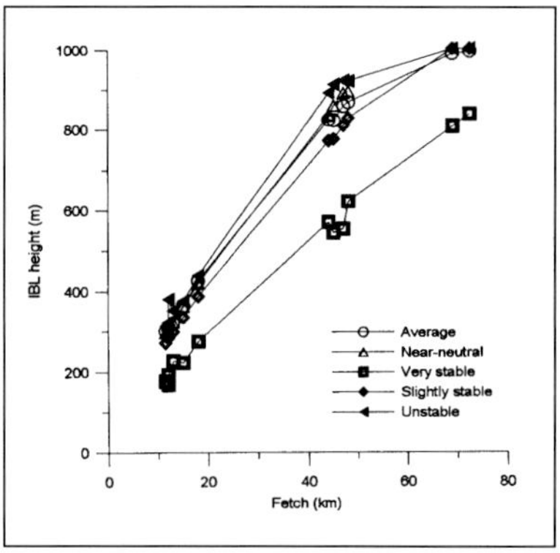

FIGURE 2. Average IBL heights (m) at Rødsand by fetch distance (km). Stability classes are as defined in Table 3.

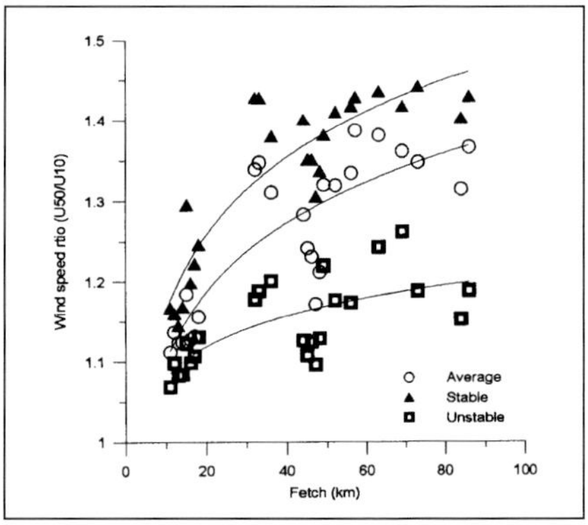

FIGURE 3. Ratios of wind speeds at 50 m to those at 10 m at Rødsand averaged by fetch distance and different stability classes (here unstable = Monin-Obukhov length <0 m and stable = Monin-Obukhov length >0 m). An exponential has been fitted to the data for each stability class.

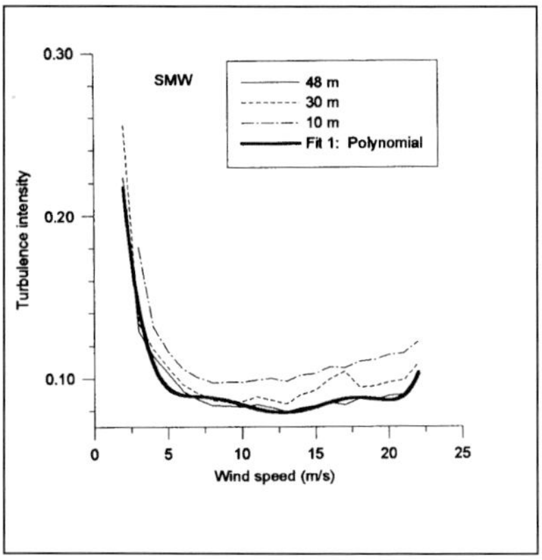

FIGURE 4. Variation of T.I. with wind speed and measurement height at Vindeby SMW

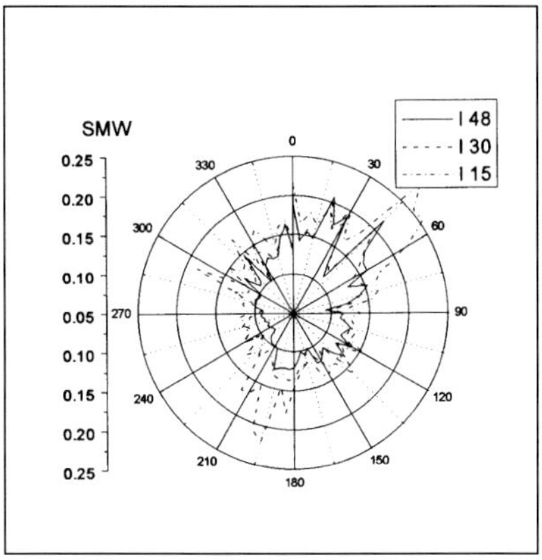

FIGURE 5. Mean T.I. by direction and measurement height at Vindeby SMW

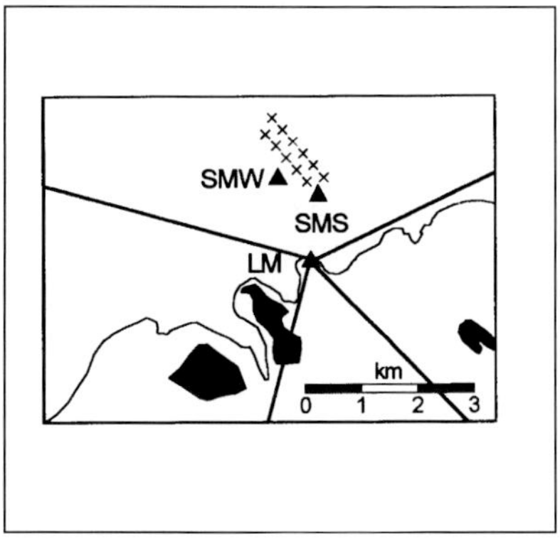

FIGURE 6. Location of the masts (triangles) and wind farm (crosses) at Vindeby.

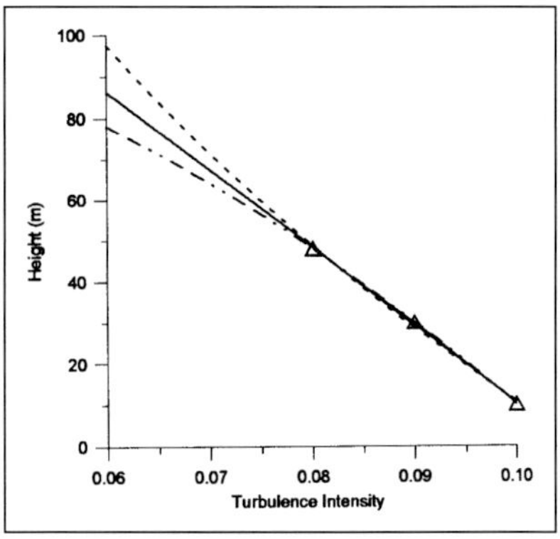

FIGURE 7. Vertical profile of T.I. offshore - mean profile from Gedser (sea sectors), Gedser Rev and Omø Stålgrunde.

and at the lower heights in the south south west sectors presumably relating to the presence of land in the fetch (see Figure 6). Lowest T.I. is observed in the sectors with the longest sea fetch (south west). Mean T.I. at Gedser Rev, Omø Stålgrunde and Gedser Land (sea sectors) is 0.10, 0.09 and 0.08 at 10, 30 and 50 m heights, respectively (Figure 7). The mean vertical T.I. profile from these sites suggests that T.I. decreases approximately linearly with increasing height.

Figure 8 shows sector mean T.I. from Vindeby SMW and Rodsand plotted by mean sector fetch distance (sectors containing the wind farm at SMW and mast influences are excluded). As shown, there is considerable scatter in the data but T.I. decreases with increasing distance from the coast and a log-linear fit suggests that after 20 km or so fetch turbulence intensity reaches a level (around 0.08 at 50 m height) beyond which it will not decrease substantially with increasing fetch.

WIND SPEED MODELLING

To capture the complexity of thermally driven flows in coastal areas, a mesoscale model should be applied to examine wind speed variability [32]. However these models have large input requirements, are computationally expensive and require extensive set-up and analysis. Hence for most wind energy applications more straightforward models are applied. In this chapter we will focus on the use of a model with simplified stability corrections (the WAsP model) [24] as a tool for predicting wind speeds in the coastal areas and examine its performance for different applications. The ability of WAsP to estimate wind speed and power prediction for long averaging periods in the coastal zone is demonstrated and the major uncertainties of applying WAsP to coastal predictions are discussed.

Application of WAsP in coastal areas: Default parameters

In brief, the methodology of the Wind Analysis and Application Program (WAsP) [24] is to create wind atlas files which provide the major characteristics of the wind climate on the micro- to meso-scale. WAsP corrects local wind speed and direction measurements for orography, roughness and obstacles to create wind atlas files for the surrounding region. The wind atlas describes the wind speed distribution in terms of Weibull shape and scale parameters [24]. An inversion of the same methodology is then used to create a wind climatology for a nearby site from a wind atlas file. WAsP is well suited to far offshore applications [33] since there is no complex orography offshore and there are no abrupt roughness changes (although there is a roughness dependence on wind speed.)

Although WAsP assumes near-neutral conditions on average, corrections are made for stability which are slightly different for onshore and offshore applications. Adjustments of stability parameters in the coastal zone are described in [34].

Application of WAsP in the coastal zone: Mean wind speed parameters

Because of the intensive monitoring conducted at Vindeby since 1993, comprehensive databases exist describing wind characteristics for the site. There are three 48 m meteorological masts one at the coast, the Land Mast (LM), one to the south of the wind farm, Sea Mast South (SMS) and one to the west of the wind farm (Seas Mast West) (Figure 6). The minimum over-water distance

TABLE 4.
Predicted and observed wind speeds and Weibull factors at Vindeby SMW

Prediction method/ Observations	Mean wind speed (m/s)	Weibull scale parameter (m/s)	Weibull shape parameter
WAsP-using data from LM	8.2	-	-
WAsP using data from Tystofte	8.1	9.2	2.4
Observed wind speeds at SMW	8.1	9.1	2.3

between the LM and SMW is 1.6 km. WAsP gives good estimates of the wind climate if the model is applied based on data from a long-term onshore reference site (Tystofte; shown in Figure 1) as in [25]. WAsP has also been applied using observations from LM and is found to slightly over-predict the mean wind speed [35]. Table 4 shows a comparison between observed and predicted wind speeds and Weibull factors. WAsP predicts the major adjustment of wind speeds moving offshore will occur within about 3 km of the coast [36] and [37]. This assumption may lead to slight over-prediction of near-surface wind speeds close to the coast.

Application of WAsP: Prediction of vertical profiles

Accurate prediction of vertical wind speed profiles offshore is essential for assessment of the wind resource, since measurements are not typically made to turbine hub-heights and rarely to tip heights. Applying the logarithmic wind speed profile to predict wind speeds above measurement heights should give a reasonably accurate prediction if conditions are close to neutral on average. WAsP has been shown to predict wind speed profiles in good agreement with selected observations from tall masts [33].

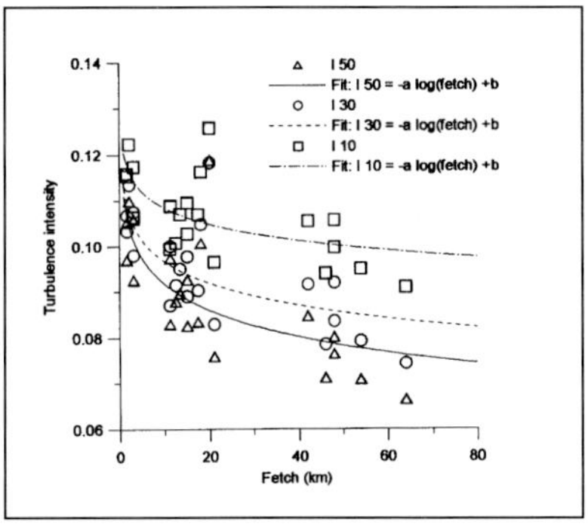

FIGURE 8. Relationship between T.I. and fetch for Vindeby SMW and Rodsand.

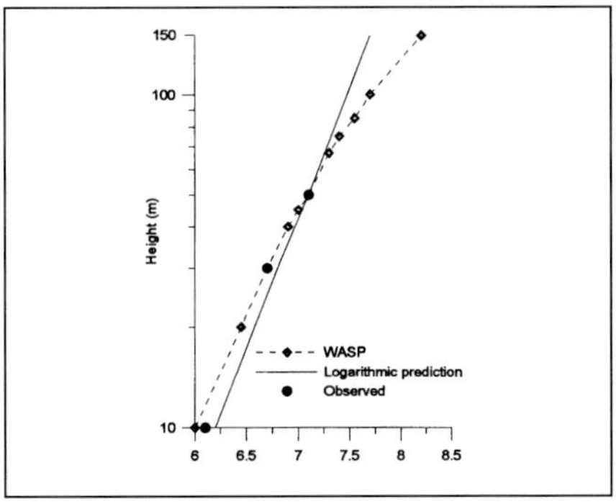

FIGURE 9. Predicted and observed (corrected for mast shadow effects) vertical profiles at Middelgrunden (October 1997-July 1999)

The Middelgrunden site is unusual in that it is frequently impacted by flow from Copenhagen with relatively high turbulence since the prevailing wind is south-westerly and the fetch is just over 2 km. Atmospheric conditions are slightly stable on average (Monin-Obukhov length 400 m). Figure 9 shows the average wind speed profile predicted by WAsP compared with the observed wind speeds (October 1997 to July 1999) and the profile predicted using a logarithmic profile with roughness of 0.0002 m.

All predictions are based on wind speeds at 50 m height. WAsP captures the wind speed profile below 50 m and predicts a stable profile in accord with the analysis of [38]. However, it is not possible to evaluate whether the predicted increases in wind speed compared with the logarithmic prediction above 50 m are accurate and these will have a significant impact on the predicted power output of turbines with hub-heights above 75 m.

SUMMARY

Offshore wind energy will contribute significantly to the power supply of Europe in the first decades of the new millennium. This paper describes the sites and data collected by the Danish Offshore Wind Monitoring Network. Three methods of estimating offshore wind climatologies from short-term data records have been compared and are shown to give results which are in reasonable accord (± 4 % for wind speed at sites with long data records). The length of the observation record is the main factor in the uncertainty of the predicted wind speed. In addition to the wind resource predictions, new analyses have been conducted on the data sets focussing on vertical profiles of wind speeds and turbulence. The effect of the coastal discontinuity persists in wind speed and turbulence characteristics at 50 m height for at least 20 km offshore.

The WAsP model has been applied in coastal areas where the wind climate is known to be difficult to predict due to the presence of a physical discontinuity, the importance of thermal effects and the non-equilibrium status of the wind flow. WAsP predicts both horizontal and vertical variability of wind speeds in reasonably good accord with observations. Comparisons with observations at sites in the Danish Offshore Wind Monitoring Network indicate that WAsP slightly over-predicts the increase in near-surface wind speeds in the first few kilometers offshore.

More research using existing and new measurement techniques and model development is also required for prediction of wind speeds, turbulence and wake effects offshore, particularly in complex coastal areas for long time scales (resource estimation, extreme events) and for short-term prediction (forecasting, maintenance). Since the relationship between wind and waves (in terms of separate and combined loads on foundations and towers) is not yet well understood, this area also requires major research initiatives. Continued monitoring and detailed investigation of these effects at the first large offshore wind farms will provide invaluable data for use in better evaluating and harnessing the offshore wind resource. Current work is focusing on development of the Coastal Discontinuity Model which accounts for stability impacts on coastal wind resources as part of a European Commission project (Predicting Offshore Wind Energy Resources).

ACKNOWLEDGEMENTS

This paper brings together results from a number of different projects including the European Commission 'Cost optimization of large wind turbines' and 'Predicting Offshore Wind Energy Resources' projects 'Offshore wind resources' and 'Wind resources at Middelgrunden' funded by the Danish Energistyrelsens Udviklings-programmet for Vedvarende Energi. Monitoring is continuing through support from SEAS/ELKRAFT and particular thanks are due to Jan Svenson, Frank Olsen and Per Vølund.

The technical staff of the Wind Energy and Atmospheric Physics Department have made a significant contribution to this research and thanks are also due to Jorgen Hojstrup and Met Support. Bernhard Lange, Morten Nielsen, Anna Maria Semprevia and Ole Rathmann contributed to the data analysis. Sara Pryor of Indiana University is acknowledged for helpful discussions on the analyses and manuscript.

REFERENCES

1. European Commission, 1997. *Energy for the future: renewable sources of energy,*. 1997, European Commission, Brussels.
2. Barthelmie, R.J., I. Bryden, J.C. Coelingh and S.C. Pryor. 1999. *Energy from the oceans - wind, wave and tidal, in Seas at the Millennium*, C. Sheppard, Editor, Elsevier: Chapter 29, pp. 303-321.
3. Pryor, S.C. and R.J. Barthelmie. 1999. *Wind speed distributions in the coastal zone. In Proceedings of Offshore Wind Energy in Mediterranean and Other European Seas*. Sicily April 2000: ATENA: pp. 29-43.
4. Pryor, S.C. and R.J. Barthelmie, 1998. Analysis of the effect of the coastal discontinuity on near-surface flow. *Annales Geophysicae*, **16**, 882-888.
5. Garratt, J.R. 1987. The stably stratified internal boundary layer for steady and diurnally varying offshore flow. *Boundary-Layer Meteorology*, **38**(4), 369-394.
6. Larsen, S.E., M. Courtney, F.A. Hansen, et al., 1992. *Turbulence intensity and power spectra 70m above the water surface of the Great Belt*, Risø-M-2898, Risø National Laboratory: Denmark. p. 50.
7. Charnock, H., 1955. Wind stress on a water surface. *Quarterly Journal of the Royal Meteorological Society*, **81**, 639.

8. Simpson, J.E., 1994. *Sea breeze and local winds* Cambridge: Cambridge University Press. p. 234.
9. Barthelmie, R.J., B. Grisogono, and S.C. Pryor, 1996. Observations and simulations of diurnal cycles of near-surface wind speeds over land and sea. *Journal of Geophysical Research (Atmospheres)*, 101(D16), 21,327-21,337.
10. Van Wijk, A.J.M., A.C.M. Beljaars, A.A.M. Holtslag, et al., 1990. Evaluation of stability corrections in wind speed profiles over the North Sea. *Journal of Wind Engineering and Industrial Aerodynamics*, **33**, 551-566.
11. Coelingh, J.P., A.J.M. van Wijk, and A.A.M. Holtslag, 1998. Analysis of wind speed observations over the North Sea coast. *Journal of Wind Engineering and Industrial Aerodynamics,* **73**, 125-144.
12. Bergstrom, H., P.-E. Johansson, and A.-S. Smedman, 1988. A study of wind speed modification and internal boundary-layer heights in a coastal region. *Boundary-Layer Meteorology*, **42**(4), 313-335.
13. Smedman, A.-S., H. Bergstrom, and B. Grisogono, 1997. Evolution of stable internal boundary layers over a cold sea. *Journal of Geophysical Research*, **102**(Cl), 1091-1099.
14. Kallstrand, B., 1998. Low level jets in a marine boundary layer during spring. *Contributions to Atmospheric Physics*, **71**(3), 359-373.
15. Sandstrom, S., 1997. Simulations of the climatological wind field in the Baltic Sea area using a mesoscale higher order closure model. *Journal of Applied Meteorology*, **36**, 1341-1552.
16. Bergstrom, H., 1990. *A study of the wind climate over the sea southeast of Blekings*, 1990, Meteorological Institute Uppsala University: Uppsala, Sweden.
17. Smedman, A.S., 1991. Occurrence of roll circulations in a shallow boundary layer. *Boundary-Layer Meteorology*, **57**(4), 343-358.
18. Coelingh, J.P., A.J.M. van Wijk, and A.A.M. Holtslag, 1996. Analysis of wind speed observations over the North Sea. *Journal of Wind Engineering and Industrial Aerodynamics*, **61**, 51-65.
19. Krohn, S., 1998. Offshore wind energy: full speed ahead. CADDET, **December** 16-18.
20. Barthelmie, R.J. 1999. *Monitoring offshore wind and turbulence characteristics in Denmark. In Proceedings of the 21st British Wind Energy Association Conference*. Cambridge: pp. 311-321.
21. Barthelmie, R.J., Nieben, N.M., B. Lange, et al. 1999. *Offshore wind resources at Danish measurement sites.* in *Proceedings of the 1999 European Wind Energy Conference and Exhibition.* Nice, March 1999: pp. 1101-1104. James and James Scientific Publishers.
22. Højstrup, J., B. Lange, R.J. Barthelmie, et al., 1997. *Offshore wind resources at selected Danish sites*, Risø-I-1156(EN), Riso National Laboratory: Roskilde, Denmark.
23. Mortensen, N.G., L. Landberg, I. Troen, *et al.*, 1995. *Wind Analysis and Application Program (WASP) Utility Programs,* Riso-I-666 (EN), Risø National Laboratory: Roskilde, Denmark.
24. Mortensen, N.G., L. Landberg, I. Troen, et al., 1993. *Wind Analysis and*

Application Program (WASP), Risø-I-666, Riso National Laboratory, Roskilde, Denmark.
25. Barthelmie, R.J., M. Courtney, B. Lange, et al., 1998. *Offshore wind resources at Danish measurement sites*, Risø-I-1339, Risø National Laboratory: Roskilde, Denmark.
26. Barthelmie, R.J., B. Lange, and M. Nielsen, 1999. *Wind resources at Rødsand and Omø Stålgrunde,* Riso-I-1456, Riso National Laboratory: Roskilde.
27. Johnson, H.K., 1998. On modelling wind-waves in shallow and fetch limited area using the method of Holthijsen, Booji and Herbes. *Journal of Coastal Research,* **14**(3), 917-932.
28. Højstrup, J. 1999. *Vertical extrapolation of offshore wind profiles in Proceedings of the 1999 European Wind Energy Conference and Exhibition.*Nice, March 1999.
29. Barthelmie, R.J., 1999. The effects of atmospheric stability on coastal wind climates. *Meteorological Applications*, **6**(1), 39-48.
30. Beljaars, A.C.M., A.A.M. Holtslag, and R.M. van Westrhenen, 1989. *Description of a software library for the calculation of surface fluxes*, KNMI: De Bilt, Netherlands.
31. Pal Arya, S., 1988. *Introduction to micrometeorology*. San Diego, California: Academic Press. p. 307.
32. Kallstrand, B. 1996. *Do simple models give a correct description of the wind conditions in a coastal area?* in *European Union Wind Energy Conference*. Goteborg: H.S. Stephens and Associates, Bedford, UK.
33. Troen, I. and E.L. Petersen, 1989. *European Wind Atlas Roskilde*, Denmark: Risø National Laboratory, p. 656.
34. Barthelmie, R.J., B. Lange, A.M.S. Sempreviva, *et al.* 1999. *Application of WAsP to offshore wind power prediction.* in *Proceedings of the 21st British Wind Energy Association Conference*. Cambridge, September 1999: in press.
35. Barthelmie, R.J., M.S. Courtney, J. Højstrup, *et al.*, 1996. Meteorological aspects of offshore wind energy - observations from the Vindeby wind farm. *Journal of Wind Engineering and Industrial Aerodynamics*, **62**(2-3), 191-211.
36. Mortensen, N.G. 1994. *Wind measurements for wind energy applications - a review. in Proceedings of the 16th British Wind Energy Association Conference.*
37. Barthelmie, R.J. and J.P. Palutikof, 1996. Predicting offshore wind speeds in coastal regions suitable for wind energy. *Journal of Wind Engineering and Industrial Aerodynamics*, **62**(2-3), 213-236.
38. Barthelmie, R.J., 1999. *The wind resource at Middelgrunden*, Risø-I-3334, Risø National Laboratory: Roskilde, Denmark.

Renewable Energy: Trends and Prospects. Edited by S.K. Majumdar, E.W. Miller and A.I. Panah

Chapter Eleven

INTEGRATION OF PHOTOVOLTAIC CELLS INTO THE GLOBAL ENERGY SYSTEM*

BENT SØRENSEN

Energy and Environment Group, Institute 2
Roskilde University
Bld. 02, Universitetsvej 1, DK-4000
Roskilde, Denmark

INTRODUCTION

Photovoltaic cells, also called solar cells, are devices that convert light into an electric current. Their present use is in a number of niche areas, from powering space vehicles and remote installations to powering small-scale consumer electronics. The cost of solar cells has dropped consistently over the last four decades, but is still very high compared to other bulk-power devices such as wind turbines (leaving apart the question of fossil or nuclear power plants, the cost of which include poorly known environmental externality costs). Therefore, solar cells compete in terrestrial uses only for rural electrification and telecommunication installations away from existing grids, and for special application, where ease of maintenance, dependability and the absence of emissions and noise are considered important. In addition, there has been, and is currently being implemented, several demonstration programmes of building integrated or centralised photovoltaic power, which signals the belief that increased volume of production, along with technological improvements, will bring the cost further down and eventually allow solar power to enter the area of mainstream applications.

---

* Part of this work was performed for the Danish Energy Department Energy Research Programme under contract EFP 1753/96-0002 and has been previously published in Bent Sørensen: Renewable Energy, 2nd Edition (Academic Press). Reproduction of this publication was granted by the author Dr. Bent Sørensen.

The present chapter is aimed at describing the options for integration of photovoltaics into future energy systems. First, the current and emerging solar cell technologies of interest to scenario builders are briefly reviewed. The purpose is to demonstrate that there are still many quite different solar technologies competing, and there is scope for more than marginal improvements of performance and cost, which is a natural state of affairs for a technology that still needs to lower cost by a factor of 5-10. Second, the solar resource is assessed, as well as the potential power production on a geographical basis, taking into account constraints of land use and availability of surfaces suitable for building integration. Third, I present scenarios for integrating photovoltaic technologies together with other renewable energy technologies into a viable overall system capable of providing the power needed by future societies, with consideration of temporal and spatial requirements in different areas of the world. Finally, some concluding remarks are offered, dealing with the conditions needed and the decisions that have to be made in order to make such systems become realities.

PHOTOVOLTAIC TECHNOLOGIES

Below, the main types of currently considered solar cells are reviewed, with comments on module and system layouts. The basic physics of solar cells is described in [1].

Monocrystalline cells

Mono-crystalline cells are cells constructed from single crystals, usually in the form of ingots sliced into a number of cells. Refinements over the years have brought the cell efficiency for mono-crystalline silicon cells from a few percent up to about 25%. The light capture is improved through trapping structures that minimise reflection in directions not benefiting the collection area, and by back-side designs reflecting light rays back into the active areas. The doping degree is altered near electrodes, and a thin oxide layer further helps to prevent electrons from reaching the surface rather than the electrode (this process being termed ìpassivationî). Further, top electrodes may be buried in order not to produce shadowing effects for the incoming light [2]. Simulation of the light trapping and electron transport processes in 1, 2 or 3 dimensions has helped in selecting the best geometry and degree of doping [3,4].

Multicrystalline cells

A newer technology for producing solar cells uses multicrystalline (sometimes referred to as 'polycrystalline') materials, instead of the single-crystal materials. Multicrystalline materials consist of small domains or grains of crystalline material, randomly oriented relative to each other. The crystal grains in multicrystalline materials sustain conductivity in the same way as single crystals do, but the transport of electrons across grain boundaries induces losses, reduces conductivity and thus makes the cells less efficient. On the other hand they can be produced by simpler methods than those needed for mono crystals, e.g. by evaporating suitable coatings onto a substrate. This field is in rapid development, as it is becoming possible to deposit only a few atomic layers onto a substrate, and with suitable techniques (such as using magnetic fields to align grains) it may soon be possible in this way to form near-monocrystalline layers without having to grow crystals.

Initially, it was believed that the additional losses at grain boundaries would necessarily make the efficiency of multicrystalline cells substantially lower than what is obtained by crystalline materials. However, the difference has recently narrowed, as a result of better understanding of the options for optimising performance of complex cell structures. One problem has been the damage inflicted upon multicrystalline cells by attempting to copy to them some of the efficiency-improving techniques that have worked well for monocrystalline cells (surface texturing, rear passivation by oxide layers). Yet, etching of inverted pyramids on the surface of multicrystalline cells has improved efficiency considerably and less damaging honeycomb texture patterns have brought the efficiency up to 20% [5]. This is a trend likely to complete the long-predicted change from expensive ingot-grown monocrystalline cell materials to deposition techniques for multicrystalline materials on suitable backings, much more suited for mass production and price reduction efforts. The advantages of thin-film multicrystalline solar cells over monocrystalline ones would seem to more than compensate for the remaining 5% efficiency difference.

Stacked cells

It is possible, instead of basing a solar cell on just a single p-n junction, to stack several identical or different cells on top of each other. The use of different cell materials aims at capturing a wider range of frequencies than possible with a single junction. In this case, materials of different band gaps will be stacked. In the case of stacked identical cells, the aim is to be able to use low-quality material (such as the previously mentioned thinly sprayed crystalline cells in contrast to ingot-grown ones) and still get an acceptable overall efficiency by stacking several layers of low individual efficiency [6]. A broad maximum of efficiency over 15% has been achieved for a stacked cell structure with about six layers, using vapour deposition to deposit each layer and laser grooving to make room for electrodes. Module assembly is eased by using a design allowing automatic series connection. The dependence on the number of layers is weak for the optimum bulk doping concentration of about 10^{18} cm^{-3} and a total thickness of around 10 μm.

Amorphous cells

Amorphous semiconductor materials exhibit properties of significant interest for solar cell applications. While elemental amorphous silicon has a fairly uniform energy distribution of electron levels, composite materials have been constructed which exhibit a pronounced energy gap, i.e. an interval of energy essentially without any states, as in a crystal. Spear and Le Comber [7] first produced an amorphous material, which was later proved to be a silicon-hydrogen alloy. Doping (introduction of boron or phosphorus atoms) is possible, so that p- and n-type amorphous semiconductors can readily be made, and a certain amount of ìengineeringî of materials to exactly the desired properties with regard to gap structure, doping efficiency, conductivity, temperature sensitivity and structural stability (lifetime) can be performed.

Shortly after the discovery of amorphous solar cells, Japanese scientists succeeded in creating designs of such cells that were suited for industrial production [8] and soon found a market in powering calculators and similar small-scale devices where the cell cost was relatively unimportant. Band gaps in the range from 1.0 to 3.6 eV can be engineered with different silicon alloys (SiGe, Si, SiC), and such cells may be stacked to obtain a broader frequency acceptance. However, the simplest version has just one type of material: an intrinsic layer of an a-Si:H compound is the main area of light absorption, and adjacent p- and n-type layers ensure the transport to the electrodes, of which the front one is made of a transparent material. The whole structure is less than 1 μm thick and is deposited onto a plastic backing material.

Maximum efficiencies of around 13% have been demonstrated, but one problem has persisted: because the structure of the material is without order, bombardment with light quanta may push atoms around, and the material degrades with time. Current practice is to degrade commercial cells before they leave the factory, thereby decreasing the efficiency by some 20%, but in return obtaining reasonable stability over a 10-year period under average solar radiation conditions [9]. Several layers of p-, i- and n-layers may be stacked, and the highest efficiency is obtained by replacing the amorphous n-layers by a multicrystalline pure Si, or silicon, compound layer.

Other materials and other thin-film cells

Use of materials from the chemical groups III and V, such as GaAs, CdS and CdTe, instead of silicon allow better engineering of band gaps in crystalline solar cells to suit particular purposes and bring forward new properties suitable for certain tasks (notably space applications and use in concentrating collectors). Important considerations in selecting materials include temperature dependence where crystalline silicon cell efficiency drops rather fast with the increasing temperature likely to prevail (despite possible active cooling) for cells operating at high levels of solar radiation (the efficiency drop is about 0.4% per °C for silicon [1]).

The GaAs bandgap of 1.43 eV is well suited for the solar spectrum, and, with a tandem cell of GaAs plus $GaInP_2$, an efficiency of over 30% has been reached [10]. At the moment, these cells are expensive and are mainly used in space. However, thin film versions may be developed as in the Si case, as they already have for CIS cells (copper-indium-diselenide). The highest efficiency obtained so far is about 17% for a Cu (In,Ga) Se_2 structure [11].

Organic solar cells

Suitable organic materials can trap sunlight and convert radiation into other forms of energy. Various ways have been attempted to copy this process. Calvin [12] considered a double membrane that would separate the ionised reactants of a photo-excitation process,

$$A + \text{light quantum} \rightarrow A^* ; \qquad A^* + B \rightarrow A + B^- .$$

In addition, a transport system is needed to get the ions to an electrode. No practical version of this idea has been produced. The same is the case for a concept aimed at both removing CO_2 from the atmosphere and at the same time producing methanol [13]. The absorption of solar radiation is used to fix atmospheric carbon dioxide to a ruthenium complex [Ru], which is then heated with water steam,

$$[Ru]CO_2 + 2H_2O + 713 \text{ kJ mol}^{-1} \rightarrow [Ru] + CH_3OH + \frac{1}{2}O_2$$

One scheme that has been realised is the attachment of a ruthenium complex as a dye to TiO_2, which may then transport electrons formed by photo-excitation in the Ru-complex to an electrode. The process is similar to the dye-excitation processes used in conventional photographic prints. The Ru-complex is restored by a redox process in an electrolyte joining the other electrode, with use of a platinum catalyst [14]. Because a monolayer of a dye such as [Ru] = cis-(NCS)2bis(4,4-dicarboxy-2,2-bipyridine)-Ru(II) will absorb less than 1% of in-coming solar radiation, it is proposed to attach the dye to a three-dimensional conglomerate of small (nanocrystalline) balls of TiO_2, in order to increase the overall cell efficiency to about 10%. It should still be possible for the TiO_2 balls to transfer the electron to the electrode through a series of collisions, as laboratory experiments have actually confirmed.

Module construction

Individual cells based on monocrystalline materials have a linear size of typically 1 to 10 centimetres (limited by techniques for growing ingots of monocrystalline material). Multicrystalline cells formed by deposition of silicon onto a backing (usually glass) may have larger cell size, and for amorphous cells, there are essentially no limitations. Amorphous cells have been produced on rolls of flexible plastic backing materials, with widths of 1-2 m and of any length. The same may also become possible for other thin film types, such as spray-deposited multicrystalline materials.

It is customary to assemble cells into modules by a mixture of parallel and series connections so that the resulting voltages become suitable for standard electricity handling equipment, such as inverters transforming the DC currents into AC currents of grid specifications and quality. Alternatively, microprocessor inverters may be integrated into each module or even into each cell, in order to minimise transport losses. In recent years, specific inverters optimised for solar cell applications have been produced, with an inverter efficiency increase from 90% to some 98% [15].

The technology is then characterised by two main solar radiation conversion efficiencies: the efficiency of each cell and the efficiency of the module sold to the customer. The latter is currently about 5% lower than the former, notably because of phase mismatch between the individual cell current components. However, this does not need to be so, and the difference between the two efficiencies is expected to diminish in the future.

Optical subsystem and concentrators

Optical manipulation of incoming solar radiation is in use for both concentrating and non-concentrating solar cells. In the latter case, it serves the purpose of reducing the reflection on the surface to below a few per cent over the wavelength interval of interest for direct and scattered solar radiation [5]. The collection efficiency exceeds 90% for wavelengths between 0.4 and 1.05 μm, and exceeds 95% in the interval 0.65 to 1.03 μm.

For a large-factor concentration of light onto a photovoltaic cell much smaller than the aperture of the concentrator, the principles used for thermal systems apply unchanged. Most concentrators are designed to focus the light onto a very small area, and thus tracking the Sun (in two directions) is essential, with the implications that scattered radiation largely cannot be used and that the expense and maintenance of non-stationary components have to be accepted.

One may think that abandoning very high concentrations would allow the construction of concentrators capable of accepting light from most directions (including scattered light). However, this is not so easy. Devices that accept all angles have a concentration factor of unity (no concentration), and even if the acceptance angular interval is diminished to say 0-60°, which would be suitable because light from directions with a larger angle is anyway reduced by the incident cosine factor, only a very small concentration can be obtained (examples such as the design by Trombe are discussed in [16]).

Use of residual energy

Given that practical efficiencies of photovoltaic devices are in the range of 10-30%, it is natural to think of putting the remaining solar energy to work. This has first been achieved for amorphous cells, which have been integrated into window panes, such that at least a part of the energy not absorbed is passed through to the room behind the pane. Of course, conductors and other non-transparent features reduce the transmittance somewhat. The same should be possible for other thin-film photovoltaic materials, including the emerging multicrystalline silicon cells.

Another possibility, particularly for photovoltaic panels not serving as windows, is to convert the solar energy not giving rise to a current into heat (as some of it in actuality is already). One may think of cooling the modules of cells by running pipes through their bottom side, carrying the heat away by some fluid. Again, the energy available for this purpose is the incoming radiation energy minus the reflected and the converted part. Reflection from the back side of the semiconductor material, aimed at increasing the path length of the photons in the material, could be chosen to optimise the value of the combined heat and power production, rather than only the power production. For concentrator cells, active cooling is needed in any case, because of the temperature dependence of the characteristics of the photovoltaic process [17]. The maximum electrical efficiency of a photovoltaic device implied by semiconductor physics is about 40% (unless many different materials are used), and the possible associated heat gain is thus of a similar magnitude.

SOLAR RESOURCE AVAILABILITY

Recent work [18,19] employs a geographical information system (GIS) to map solar resources on the basis of satellite data (radiation at top of atmosphere, albedo, downward radiation at surface) and to match it with demand modelling on a habitat basis (population density, energy demand intensity). PV potential use is based on estimates of practical areas for collection use (building roof areas, suitably inclined and oriented surfaces) combined with land use data (important for central receiver fields). Local measurement data have been converted to the GIS grid employed. This is a novel approach compared to previous assessment, planning and scenario work in the photovoltaic field, which has traditionally been country based.

The solar radiation model is obtained by using data for solar radiation incident on a horizontal plane. One set of data is based on an analysis [20] of satellite measurements of radiation, albedo (reflectance) cloud cover and attenuation of radiation in the air. Another study collects many types of weather and climate data [21] and uses balance equations and a global circulation model (horizontal resolution about 210 km and 28 vertical levels) to improve the consistency of data [22]. As no ground-based observations are used for solar radiation, this is entirely based on top-of-the atmosphere fluxes and various absorption, reflection and attenuation processes plus the balance requirements at the Earth's surface. The two sets of data give similar results for solar radiation. The balance equations (difference between upwards and downwards short- and long-wavelength radiation and heat fluxes) do suggest a too high albedo over oceans, but the land data that we are using appear to be reliable. Figure 1 shows the radiation on a horizontal plane at the Earth's surface for selected months of 1997 [21].

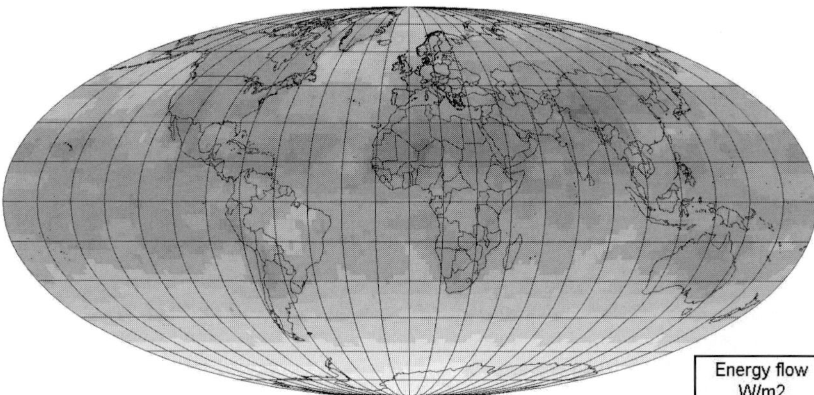

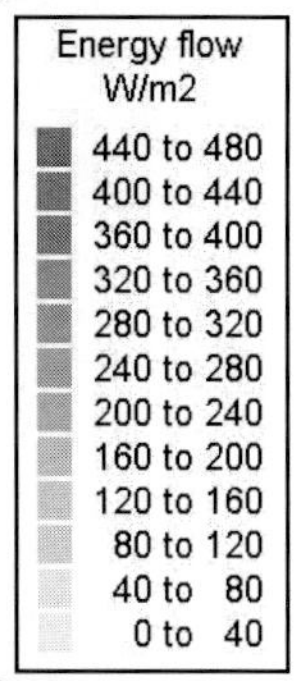

FIGURE 1. (a): April 1997 average solar radiation on horizontal surface (above, downward energy flux at Earth's surface, W/m^2).

FIGURE 1. (b): July 1997 average solar radiation on horizontal surface (below, downward energy flux at Earth's surface, W/m^2) Based on [1,18] using data from [21].

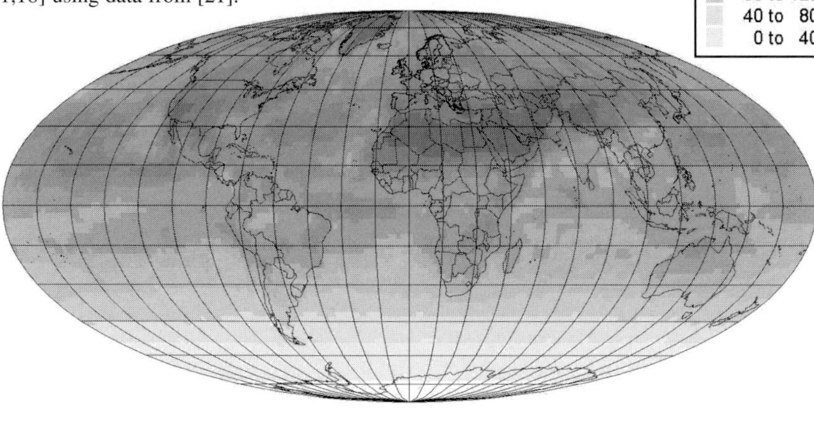

In order to calculate the solar radiation incident on inclined surfaces, such as the ones characterising most solar panel installations, one would ideally need hourly data for direct and scattered radiation (or equivalently for direct and total global radiation). One would then assume that the scattered radiation is uniform to calculate its value for differently inclined surfaces. For the direct part, ray calculations with the appropriate angles between the direction to the sun and the normal to the collector surface have to be performed hourly. Such calculations have been performed and compared with measurements on inclined surfaces at several major solar installations and some weather stations [1]. Here, global data are needed and only monthly average radiation data are readily available (at least with data sizes suitable for the present type of investigation), so an approximate relation has to be invoked.

The aim is to estimate the radiation on a surface tilted towards either North or South by an angle approximately equal to the latitude (as this gives the optimum performance) from the horizontal surface solar radiation data available. No great accuracy is aimed at, because the actual solar panel installations in the 2050 scenario will be of two types:

(a): Building integrated panels that will anyway be influenced by the structures at hand: some solar panels will be mounted on vertical facades, others on roofs being either flat or tilted, typically by angles 30°, 45° or 60°. In all cases the orientation may not be precisely towards South or North, although we estimate the resource as comprising only those buildings facing approximately the correct direction and not being exposed to strong shadowing effects from other structures. The penalty of incorrect orientation and tilting angle is often modest and mostly influences the distribution of power on seasons [1].

(b): Centralised solar farms will generally be oriented in an optimum way, using total production maximising for panels not tracking the sun. However, the majority of locations suited for central photovoltaic installations will be the desert areas of the Sahara, the Arabian Peninsula, the Gobi desert and Australian inland locations. As these are all fairly close to the Equator, there is modest seasonal variations in solar radiation, and the horizontal surface data are often quite representative.

As for the many unknown factors regarding the precise inclination of actual installations, the following simple approximation was adopted [18,19] (originating from an analysis of the Danish latitude 56°N data by [1]): The radiation on a latitude-inclined surface in January and July are noted to be very nearly equal to the horizontal radiation in October and April, whereas the horizontal surface data for January and July are lower and higher than the inclined surface measurements, respectively. It is therefore decided to use the October and April horizontal data as a proxy for January and July inclined surface radiation, and construct the April and October inclined surface values as simple averages of the January and July adopted values. Furthermore, this procedure, which works well for the Danish latitude, will be less inaccurate for low latitudes, because of the relative independence of seasons mentioned above, and we simply use it for all locations.

The horizontal surface data shown in Figure 1 have been checked against predictions of the European general circulation model HADCM2-SUL [23], and the 1997 monthly solar radiation values were found to be very similar.

In order to derive the actual energy extracted from solar cell panels, a fixed conversion efficiency of 15% is assumed, which is a fairly conservative estimate for year 2050 technology, considering that the current efficiency of the best mono-crystalline cells is above 20% and that of amorphous cells near 10%, with multicrystalline cells falling in between. The 2050 technology is likely to be thin-film technology, but not necessarily amorphous, as new techniques allow crystalline or multicrystalline material to be deposited on substrates without the complicated process of ingot growth and cutting. The efficiency of such cells is currently fairly low, but as they are much less expensive than crystalline cells, they may be stacked in several layers and thereby still reach the efficiency envisaged here [6].

Finally, the availability of sites is estimated, for mounting solar cell panels either in a decentralised fashion or centrally in major energy parks: The decentralized potential is based on availability of suitably inclined, shadow-free surfaces. It is assumed that the area of suitably oriented surfaces that may be used for building-integrated PV energy collection is 1% of the urban horizontal land area plus 0.01% of the cropland area. The latter reflects the density of farmhouses in relation to agricultural plot sizes and is based on European estimates, roughly assuming that 25% of rural buildings may install solar panels. The potential PV production is thus taken as 15% of radiation times the fraction of the above two area types and the stated percentage for each. A final factor of 0.75 is applied in order to account for transmission and storage cycle losses, assuming about 5% transmission losses and that roughly half the energy goes through a storage facility of 60% round-trip efficiency (e.g. reversible fuel cells).

TABLE 1.

Regional averages and sums of solar energy-supply related quantities [18].

Region (see below)	1	2	3	4	5	6	**Total**	Unit
Region area	20.1	15.4	28.3	26.3	20.1	30.9	**141**	M km2
Urban area fraction	0.0063	0.0133	0.0096	0.0178	0.0290	0.0122	**0.0147**	
Marginal land fraction	0.1155	0.2166	0.1239	0.0999	0.3038	0.4058	**0.2109**	
Decentralized PV potential, January	38.9	40.5	71.3	146	164	109	**570**	GW
Decentralized PV potential, July	60.2	66.6	118	141	217	122	**724**	GW
Decentralized PV potential, ann. Av.	49.5	53.5	94.5	143	190	115	**647**	GW
Centralized PV potential, January	2220	7310	6680	4500	7910	18200	**46900**	GW
Centralized PV potential, July	3870	5990	9460	3980	11500	22700	**57500**	GW
Centralized PV potential, annual av.	3040	6650	8070	4240	9690	20500	**52200**	GW

Region 1: United States, Canada. **Region 2:** Western Europe, Japan, Australia. **Region 3:** Eastern Europe, Former Soviet, Middle East. **Region 4:** Latin America, SE Asian high-growth countries. **Region 5:** China, rest of Asia. **Region 6:** Africa.

The flow of energy reduced by all these factors represents the energy delivered to the end-use customers in the form of electricity. It is shown in Figure 2 for each season, with regional sums given in Table 1.

For centralized PV, the estimated potential is based upon 1% of all rangeland plus 5% of all marginal land (deserts and scrubland), again times 15% of the incoming radiation and times 0.75 to account for transmission and storage losses. As shown in Figure 3 and Table 1, this is a huge amount of energy, and it shows that such centralized PV installations can theoretically cover many times the

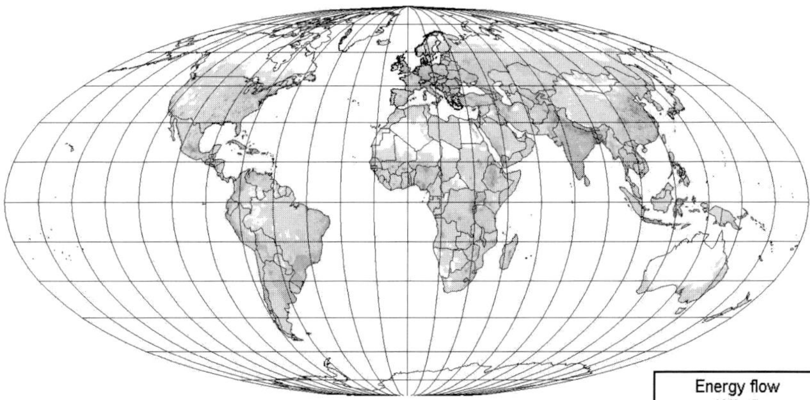

FIGURE 2a. July 1997 average solar radiation on horizontal surface (below, downward energy flux at Earth's surface, W/m^2) Based on [1,18] using data from [21].

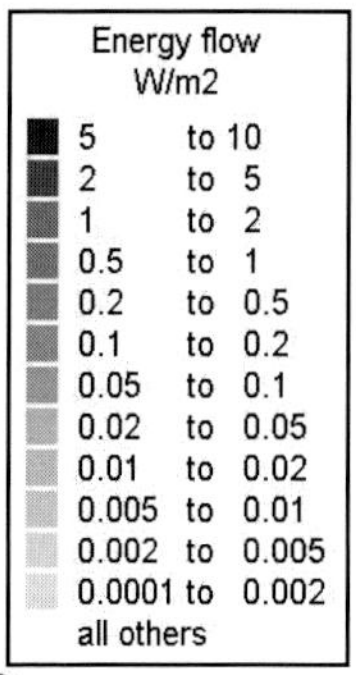

FIGURE 2b. April and October potential power production from building-integrated solar cells, with transmission and storage cycle losses subtracted (plain average of January and July values; W/m^2, below).

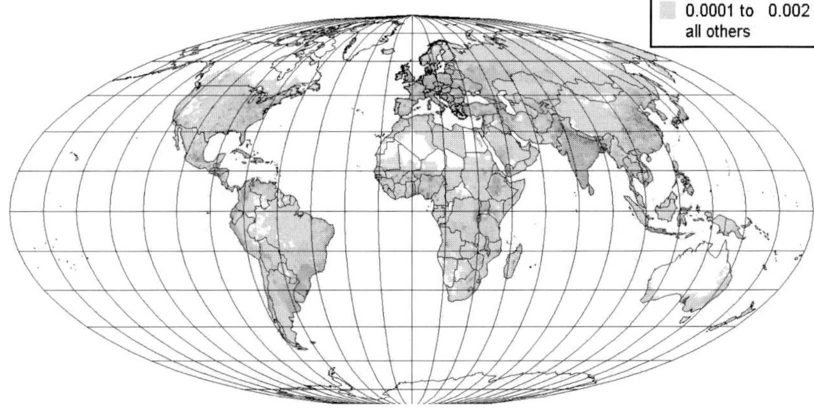

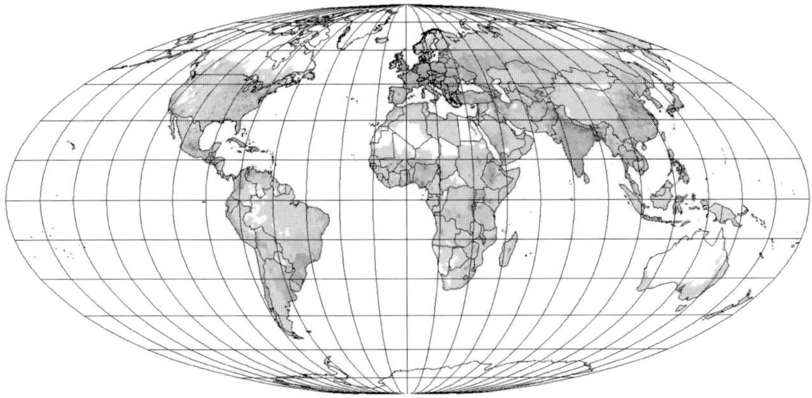

FIGURE 2c. July potential power production from building-integrated solar cells, with transmission and storage cycle losses subtracted (W/m^2, above).Based on [18].

demand envisaged over the next centuries. Setting aside 1% of rangeland would imply an insignificant reduction in other uses of the land, even if this would be a rather small number of large plants. The same is true for desert and marginal land, where only the 5% most suited need to be invoked, which in reality might be 10% for the entire installation (including frames and access roads), of which the solar panels is only part. Because of the huge area of the Sahara desert this would suffice to supply more than the entire world need for energy. This would require the availability of intercontinental transmission, which may be quite realistic in the future, by use of superconducting trunk lines [24].

INTEGRATION OF PHOTOVOLTAICS INTO ENERGY SYSTEMS

In a project performed for the Danish Energy Agency [18], four global energy supply scenarios with zero greenhouse gas emissions have been constructed, based upon a common energy supply scenario. The supply scenario is constructed by a bottom-up approach, considering basic and secondary needs, development of social organisation and activities, region by region, and finally considering population growth and settlement patterns to the scenario year, 2050. The efficiency of energy conversions is assumed by 2050 to average the best efficiency in or close to the market today. Efficiency improvement is here meant to include the introduction of new technology to perform a task in a way different from the one used earlier, as well as straight improvements in the energy efficiency of a given piece of technology.

The supply scenarios comprise use of fossil fuels without carbon dioxide emissions to the atmosphere, using nuclear conversion techniques without risk of catastrophic accidents or proliferation of nuclear material, and renewable energy sources either in a purely decentralized mode, or with inclusion of some wind-farms, central solar collector fields, and energy crops (although the need for such "centralized" facilities is far lower than the resources available). Here, only the renewable energy scenarios will be discussed. Photovoltaic devices deliver about 20% and 30% of the total supply assumed in the two scenarios.

Energy end use

Demand scenarios for the 21st century found in literature are based either on extrapolation ("business-as-usual" scenarios) or on technically feasible, normative assumptions about the development of societies. In a greenhouse warming mitigation context, the interesting demand scenarios are those which aim at reducing emissions at a lower cost than that of supply-side measures (fuel shifts or transition to energy sources not emitting greenhouse gases). Studies have identified a number of measures not undertaken although they have no significant cost [25,26]. The reason is inertia or opposition to ìreductionsî in energy use is seen as negative in the context of economic growth. As a result, measures at the supply side have been financed, that entail a higher cost per energy unit than that of suitable demand-side measures. One aim of greenhouse policies could be to change this attitude, by legislative means (such as building codes, standards for appliances, cars etc.) or by taxation (differential tax on cars and other equipment according to energy efficiency). Both types of policy means are in use in a few countries.

As an example of a demand scenario placing emphasis on demand-side measures, a bottom-up analysis based upon a vision of future global societies with high levels of prosperity will be discussed [18]. It will be underlying the global supply scenarios to be further discussed in section 4.2. The assumption is that by the mid 21st century, the average technology in use will equal the best current technology, with respect to energy efficiency. This is compounded with increasing population (using middle scenario of UN forecasts [27]), increasing urbanisation (according to UN [28], and increased per capita activity level by an average factor 2.7 for energy use). The GNP activity growth factor will be larger due to the de-coupling of economic and energy growth, and the distribution between regions will not be even (because a larger growth rate is assumed for the presently poor regions).

Figure 4 shows the total energy delivered to the end-users in the 2050 scenario, including energy for space conditioning, process heat, stationary mechanical energy, electric energy, energy for transportation and energy in food for all sectors of society. The average and totals for different regions are shown in Table 2, for each major energy type. The average energy demand is 0.9 W/cap. or three times the amount made useful at the end-user today. The energy made useful at the end-user today is only about 12% of the primary energy, and the challenge is to increase this fraction in the future [18].

In terms of full satisfaction of all primary and secondary human goals, the

TABLE 2.
Energy delivered to end-user in 2050 scenario 9 from [18,19]).
The six regions were defined in the footnote to Table 1.

Regions: /Energy quality:	1.	2.	3.	4.	5.	6.	**Average**
Food based	30 45 17	30 45 24	30 45 47	25 37 52	25 37 148	20 25 51	**23%** **36 w/cap.** **339 GW**
Food based on grain & vegetables	70 119 45	70 119 63	70 119 124	75 128 177	75 128 506	80 114 232	**77 %** **36 w/cap.** **1148 GW**
Gross transportation energy	359 136	299 158	140 146	201 277	99 392	30 61	**125w/cap.** **1170 GW**
Heat pump input for low-T heat and cooling	103 39	110 58	87 90	43 60	80 318	22 45	**65 w/cap.** **610 GW**
Environmental heat	240 91	256 135	203 210	100 140	186 741	51 105	**151 w/cap.** **1421 GW**
Direct electric and all other energy	420 153	424 224	245 255	288 398	283 1116	47 96	**240 w/cap.** **2242 GW**
Total delivered energy*	**1272** **482**	**1252** **661**	**838** **871**	**800** **1104**	**814** **3225**	**290** **591**	**742w/cap.** **6934 GW**
Population 2050	379	528	1040	1380	3960	2040	**9340 millions**

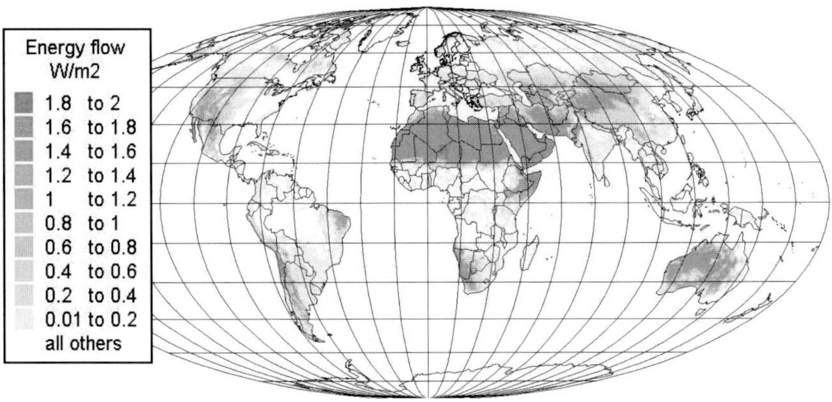

FIGURE 3. Potential annual average power production from centralized photovoltaic plants on marginal land, with transmission and storage cycle losses subtracted. Seasonal variations are modest (W/m²; scale is linear in contrast to that of Fig. 2). Based on [18].

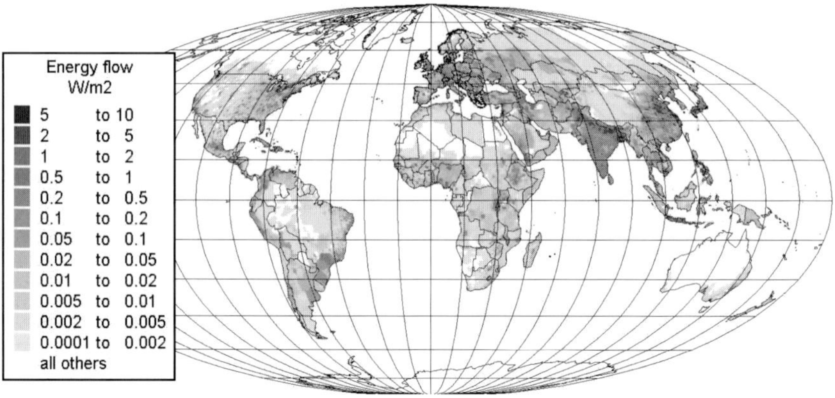

FIGURE 4. Total energy directly delivered to consumer in 2050 scenario (including environmental heat and food, transportation and electricity etc. columns of Table 2). The scale of average energy flow in watts per square metre of land area is given to the right (from [18]).

demand scenario assumes that for regions 1 and 2 of Table 2, there is nearly 100% goal satisfaction, for regions 3-5 some 2/3 of full goal satisfaction, and for region 6 a satisfaction level of 1/5. These levels are significantly higher than the ones characterising regions 3-6 at present, as the detailed analysis of Sørensen [18] shows.

Global scenarios

A number of issues speak against merely optimising energy supply systems on a national or regional basis. These have to do with the cost and supply security implications of creating, for some countries, a dependence on resources that have to be imported from far away, but also with the desirability of preserving levels of international trade to which the economy of the currently exporting countries have become dependent.

In this chapter, different scenarios for future energy supply systems addressing the greenhouse warming issue are analysed with respect to their requirements for energy transmission and energy trade, with emphasis on whether the problem is local, regional or global.

The renewable energy resources are fairly evenly accessible, although there are distinct variations with latitude (solar radiation, cf. section 3) and absence of obstacles (wind power). For biomass resources, limiting factors include solar radiation, nutrients, water and soil quality. Adding the other renewable resources to those of solar radiation discussed in section 3, one obtains a total estimate of practical exploitable resources as shown in Figure 5.

The global scenarios for the mid-21st century discussed in [18] includes two renewable energy scenarios, which may be characterised as follows:

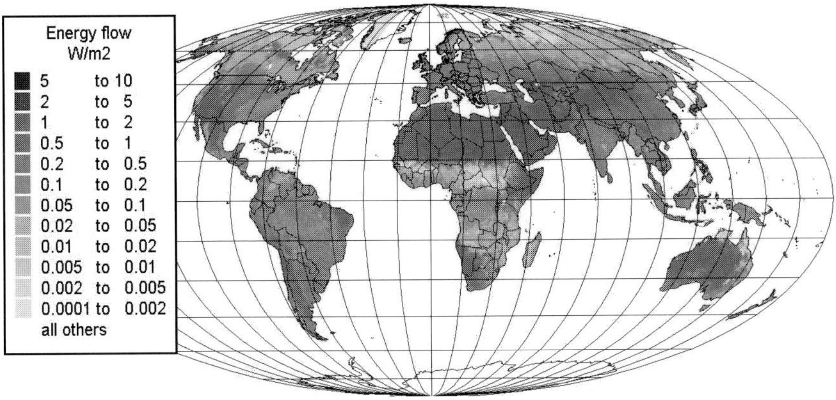

FIGURE 5. Total energy directly delivered to consumer in 2050 scenario (including environmental heat and food, transportation and electricity etc. columns of Table 2). The scale of average energy flow in watts per square metre of land area is given to the right (from [18]).

(A). The decentralized renewable energy scenario, based upon building-integrated solar systems and dispersed installations for utilising wind and biomass energy, the latter being based on integrated production of food, energy and bio-feedstock for industry.

(B) The centralized renewable energy scenario, which places additional solar collectors or wind turbines in areas of non-arable land, or off-shore in large farms. The scenario includes a cautious use of biomass plantations placed on land where competition with food production is considered minimal.

Figures 6 and 7 show the distribution of mismatch between supply and demand for the two scenarios. For both renewable energy scenarios, there are more surplus areas than deficit areas, and even in densely populated areas such as India or China, there are areas of surplus. The largest deficits are of course found in highly urban areas, where high-rise buildings make the surfaces suitable for solar collectors small compared with the demand, and where wind power and biomass production is not possible. In most cases, this implies a need for only local transport or transmission of energy, from cultivated country areas or marginal land, to the cities or particularly population- or energy-intensive regions, usually of modest dimensions.

Upon closer inspection of the forms of energy required, it is seen that even for the projected population increase, average food supply is adequate in all parts of the world, leading only to the well-known need for land to city transfer, also for countries such as India and China. This is due to the scenario assumptions on improved agricultural techniques, consistent with assumptions of technological and economic growth in all parts of the world, and despite the slight decrease in yields implied by an assumed high proportion of ecologically grown food.

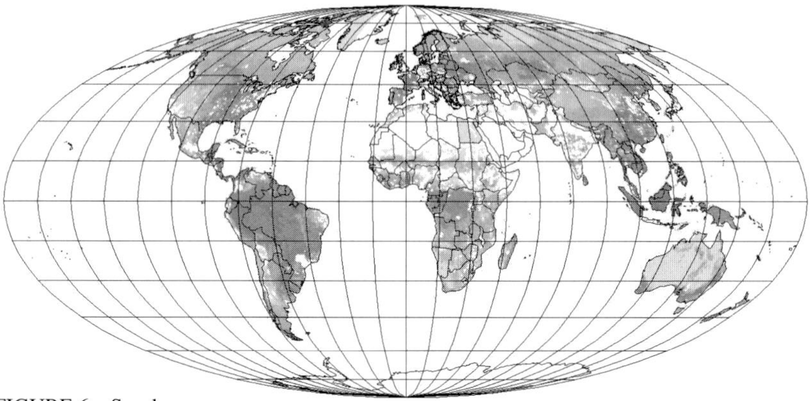

FIGURE 6a. Surpluses.

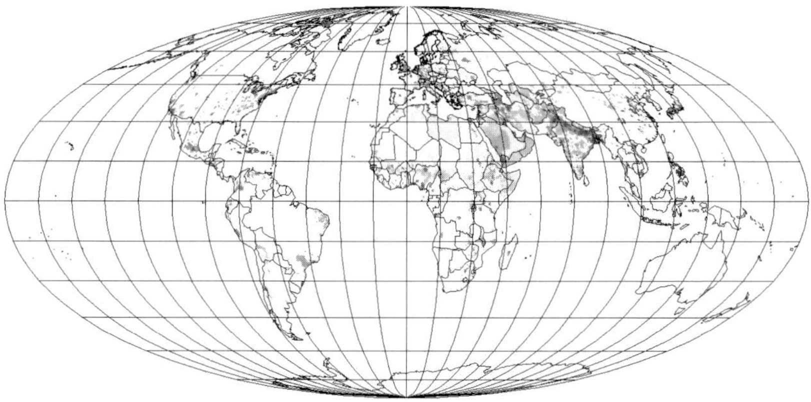

FIGURE 6b. Deficits. Difference between supply and demand, for the decentralized renewable energy 2050 scenario, using only building-integrated solar cells and farm-attached wind turbines, together with pesticide-free agriculture and bio-energy production only from agricultural and forestry residues, plus existing hydro power. The delivered energy supply comprises food, electricity and biofuels for stationary and transportation uses. Scale in W/m_2 is given to the left (from [18]).

For (bio-)fuels to be used in the transportation sector, there is a significant deficit in most of Europe, the Middle East and India, to be matched with surpluses in China, South-East Asia, Northern countries of South America and Northern countries of Asia, Europe and North America. This implies that biofuels have to be traded internationally, which is feasible as they can be transported much like oil.

For electricity, there are strong deficits in urban areas and in much of Central Europe, India and Eastern China, whereas surpluses occur in the rest of the world and for the decentralized renewable energy scenario particularly in South America. This makes the required trade in the decentralized scenario nearly

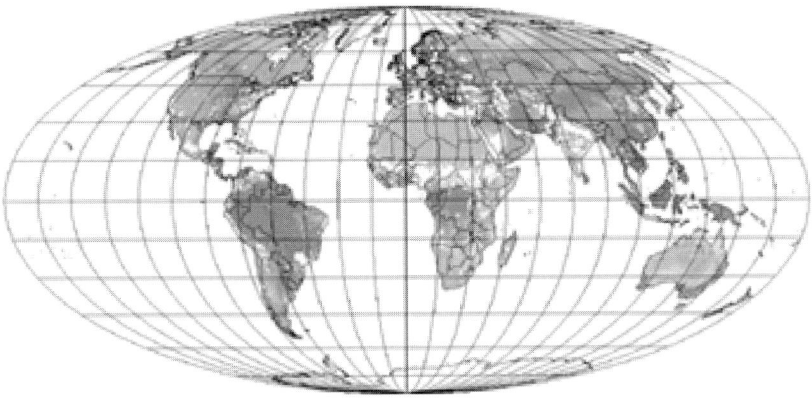

FIGURE 7a. Surpluses.

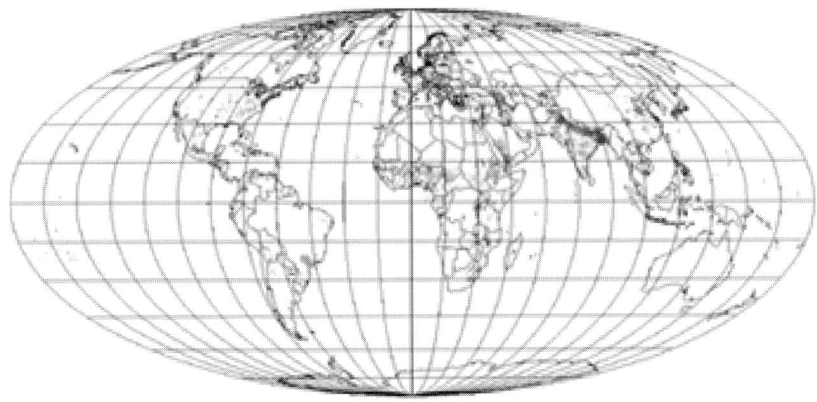

FIGURE 7b. Deficits. Difference between supply and demand, for the centralised renewable energy 2050 scenario, using building-integrated and centralized solar cells on marginal land, wind turbines near farms and in parks on marginal land or off-shore, together with pesticide-free agriculture and bio-energy production from residues and a limited number of energy-crops or energy-forests, plus hydro power in place or under construction. The delivered energy supply comprises food, electricity and biofuels for stationary and transportation uses. Scale in W/m_2 is given to the left (from [18]).

impossible, unless some intercontinental electricity transmission technologies emerge. Transformation of surplus electricity to portable fuels would solve the problem, but there are barely enough decentralized resources to allow for the losses. By contrast, the centralized renewable energy scenario has additional resources from agricultural and marginal land, making trade in the form of biofuels or electricity for more modest transmission distances capable of graciously solving the problem.

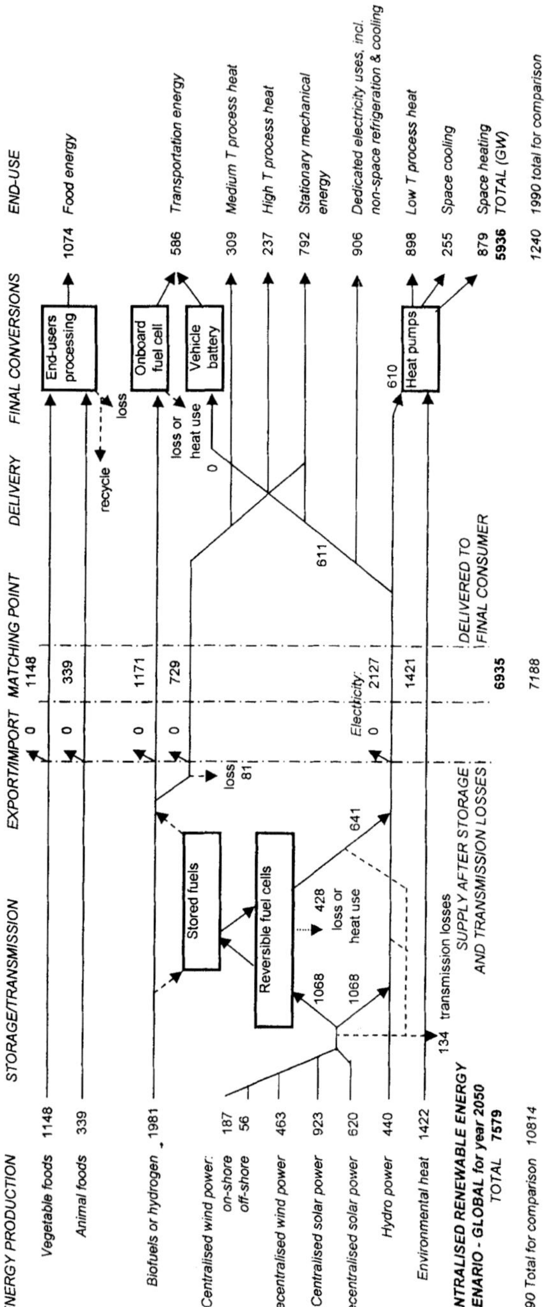

FIGURE 8. Overview of centralized 2050 renewable energy scenario (energy flows in GW or GWy/y). From [18].

TABLE 3.

Assumed energy supply (after storage conversion cycles and transmission but before interregional imports/exports) in centralised 2050 scenario, and corresponding supply-demand balances. The term "other energy" comprises all energy except energy in food and energy used for transportation. Source: [18]. Region

Region (cf. Table 1)	1	2	3	4	5	6	**Total**	**Unit**
Total food balance (as Table 10)	151	144	107	283	-23	-167	493	GW
Total biofuels used	250	190	300	640	327	192	1899	GW
Of which decentralized biofuels	250	166	236	640	295	192	1779	GW
Balance: total biofuels minus use for transportation	136	158	146	277	392	61	1170	GW
Hydro power	80	70	50	120	110	10	440	GW
Decentralized solar power	27	50	67	109	189	23	465	GW
Centralized solar power	40	20	114	19	400	100	693	GW
Decentralized wind power	20	37	80	100	100	10	347	GW
On-shore wind parks	12	10	20	38	46	0	126	GW
Off-shore wind power	0	40	0	0	13	0	53	GW
Balance: other energy, ann.av.	100	-63	139	288	-660	217	22	GW
Balance: other energy, Jan.	74	-120	169	306	-890	214	-250	GW
Balance: other energy, April	87	-51	72	243	-710	200	-160	GW
Balance: other energy, July	129	-17	180	352	-410	233	461	GW
Balance: other energy, Oct.	110	-64	129	252	-610	220	35	GW

Table 3 shows the regional totals of energy balances, with an indication of seasonal changes. It is seen that trade and transmission requirements vary from season to season, with quite large differences involved. The overall layout of the energy system is illustrated in Figure 8, for the centralized scenario, which has the highest share of photovoltaic, due to a modest addition of desert-located central solar power plants.

DISCUSSION AND CONCLUSIONS

For the scenarios based entirely on renewable energy flows of fairly low energy density, it is found that the decentralized scenario works well in some regions but on a global average basis (with the restrictions posed by renewable resources available locally and consistent with the decentralisation paradigm), it barely matches the demand of the 2050 population with the assumed massively improved living standards. This implies that the scenario requires import of energy by countries such as India, and as surpluses exist mainly in South America and is dominantly in the form of electricity, the transfer will be difficult, and seemingly in contradiction with the "local self-sufficiency" idea behind the decentralized scenario.

By contrast, the scenario adding a certain amount of centrally produced renewable energy exhibits supply in generous excess of demand, and trade of energy between regions will allow the system to be very robust against changes in assumptions on demand development and area use. For example, desert regions in North Africa and the Middle East can export photovoltaic electric power to Europe and thereby create a basis for continued economic development without resort to oil.

While the positive economic implications of adopting the energy demand scenario with high emphasis on energy efficiency is evident, the economic aspects of the supply scenarios cannot be stated with certainty. The renewable energy technologies comprise hydro and wind power, which are largely economic today. Biofuel technologies are currently about two times more expensive than the current sources, and photovoltaic technologies are currently more than ten times more expensive than conventional coal-fired power supply. Even considering externality costs, it is therefore clear that a cost reduction is needed, and projections suggest that such a cost reduction for photovoltaic power is indeed possible and forthcoming with continued market development support.

The more detailed discussion in [18] implies that the paradigm of extreme decentralisation (local or even individual control over energy supply) is not stable (e.g. towards variability between years of renewable resources) and that it requires international trade and transmission of energy of a size difficult to reconcile with the local self-sufficiency idea. The conclusion of this is that only the centralized renewable energy scenario offers a feasibly sustainable energy supply system for the long-term future, and that it is indeed very likely to become feasible in the near future, adding a modest amount of centralized wind and photovoltaic energy (wind-farms on-shore and off-shore, solar cell farms), plus a modest amount of biomass crops grown for biofuels, to the decentralized renewable energy systems. It is therefore imperative that policy measures (such as liberalisation of electricity trade) do not obstruct this development.

REFERENCES

[1] Sørensen, B. 2000. Renewable Energy. Academic Press, London, 2nd edition, ca. 900 pp.

[2] Green, M. 1992. Crystalline and polycrystalline silicon solar cells. pp. 337-360 in Renewable energy sources for fuels and electricity (Johansson et al., eds.), Island Press, Washington.

[3] Basore, P. 1991. PC1D v.3 Manual and User Guide. Report 0516/rev. UC-274. Sandia National Laboratory, Albuquerque.

[4] Sørensen, B. 1994. Model optimization of photovoltaic cells. Solar Energy Materials and Solar Cells, 34, 133-140.

[5] Zhao, J., Wang, A. and Green, M. 1998. 19.8% efficient multicrystalline silicon solar cells with "honeycomb" textured front surface, pp. 1681-1684 in Proc. 2nd World Conf. on PV Solar Energy Conversion (Vienna), JRC

EUR 18656 EN, Ispra.

[6] Wenham, S., Green, M., Edminston, S., Campbell, P., Koschier, L., Honsberg, C., Sproul, A., Thorpe, D., Shi, Z. and Heiser, G. 1995. Limits to efficiency if silicon multilayer thin film solar cells, pp. 1234-1241 in 1994 First World Conf. on Photovoltaic Energy Conversion (Kona), IEEE.

[7] Spear, W. and Le Comber, P. 1975. Solid State Commun. 17, 1193-1196.

[8] Hamakawa, Y., Tawada, Y. and Okamoto, H. 1981. Int. J. Solar Energy, 1, 125.

[9] Sakai, H. 1993. Status of amorphous silicon solar cell technology in Japan, pp. 169-172 in Technical Digest of 7th Int. PV Science and Engineering Conf. Nagoya Institute of Technology.

[10] Deb, S. 1998. Recent developments in high efficiency photovoltaic cells. Renewable Energy, 15, 467-472.

[11] Tuttle, J., Ward, J., Duda, A., Berens, A., Contreras, M., Ramanathan, K., Tennant, A., Keane, J., Cole, E., Emery, K. and Noufi, R. 1996. The performance of Cu(In,Ga)Se2-based solar cells in conventional and concentrator applications. In Proc. Material Research Society Symposium, 426, 143.

[12] Calvin, M. 1974. Science, 184, 375 ñ 381.

[13] Jensen, J. and Sørensen, B. 1984. Fundamentals of energy storage, Wiley, New York, 345 pp.

[14] O'Regan, B. and Grätzel, M. 1991. Nature, 353, 737.

[15] IEA 1999. Photovoltaic power systems in selected IEA member countries. 3rd survey report of the Power Systems Programme Task 1, International Energy Agency, Paris.

[16] Meinel, A. and Meinel, M. 1976. Applied Solar Energy. Addison-Wesley, Reading

[17] Wysocki, J. and Rappaport, P. 1960. J. Appl. Phys. 31, 571ñ578.

[18] Sørensen, B., 1999. Long-term scenarios for global energy demand and supply. IMFUFA Texts 359. Roskilde University, 212 pp.

[19] Sørensen, B. and Meibom, P., 1999. A global renewable energy scenario. International Journal of Global Energy Issues (forthcoming 1999).

[20] Pinker, R. and Laszlo, I., 1992. Modelling surface solar irradiance for satellite applications on a global scale. J. Applied Meteorology 31: 194-211.

[21] NCEP/NCAR, 1998. 40-year reanalysis project. Bulletin of the American Meteorological Society: data available at the website http://ingrid.ldgo.columbia.edu/sources in the catalogue of NOAA NCEP-NCAR CDAS-1 data sets.

[22] Kalney, E. et al., 1996. The NCEP/NCAR 40-year reanalysis project. Bulletin of the American Meteorological Society: available at the website quoted in [3].

[23] Mitchell, J. and Johns, T., 1997. On modification of global warming by sulphate aerosols, Journal of Climate, 10: 245-267. Model output (12 months by 240 years) available at IPCC Data Distribution (1998)

website HYPERLINK "http://www.dkrz.de/ipcc/ddc/html/gfdlrun2.html" http://www.dkrz.de/ipcc/ddc/html/gfdlrun2.html

[24] Nielsen, S. and Sørensen, B., 1999. Interregional power transmission. International Journal of Global Energy Issues (forthcoming).

[25] Sørensen, B. 1982. Energy choices: Optimal path between efficiency and cost. pp. 279-286 in Energy, resources and environment (S. Yuan, ed.), Pergamon Press, New York.

[26] Sørensen, B. 1991. Energy conservation and efficiency measures in other countries. Australian Department of the arts, sport, the environment, tourism and territories, Greenhouse Studies No. 8. Commonwealth of Australia, Canberra.

[27] UN 1996. Populations 1989, 2015, 2050. United Nations Population Division and UNDP, Washington.

[28] UN 1997. Urban and rural population estimates and projections as revised in 1994. United Nations Population Division and UNDP, Washington.

Renewable Energy: Trends and Prospects. Edited by S.K. Majumdar, E.W. Miller and A.I. Panah

Chapter Twelve

SOLAR PHOTOVOLTAIC ELECTRICITY: STATUS AND PROSPECTS FOR COST REDUCTION

A. K. BARUA AND DEBAJYOTI DAS

54 4 Energy Research Unit
Indian Association for the Cultivation of Science
Jadavpur, Calcutta-700 032, India

INTRODUCTION

It is well known that amongst the new and renewable energy sources, solar energy has the highest potential. It is abundant, everlasting and clean. In fact if only 0.1 % of the total solar radiation incident on the earth' s surface can be utilized, all of the present energy requirement of the world can be met. There are two main routes for the utilization of solar energy, viz.: (1) solar thermal and (2) solar photovoltaic (SPV). In the solar thermal route the solar radiation is absorbed in the form of heat energy and also utilized in a similar form. In the SPV route the solar radiation is directly converted into electricity through a device which is known as a solar cell. Since energy is obtained directly in the form of electricity it can be put to diverse use.

At present, with the available technologies SPV electricity is 5-10 times more expensive than conventional grid electricity which has, in general, a number of direct and indirect subsidies. In spite of the present cost of SPV electricity large scale investments are being for its development due to the following reasons, viz. (I) control of CO_2 emission generated from the fossil fuel based electricity by replacing it with SPV electricity and (2) supply of high quality energy to unelectrified rural and remote areas in developing countries which generally have good solar insolation.

In spite of its very high potential, a very large scale use of SPV will be possible if a number of techno-economical problems can be overcome. At present only about 275 MW/Year of PV modules are annually produced in the world, and it is increasing by 10-15% every year at the present level of technology. This very low production of SPV modules is mainly due to its high cost. Therefore a breakthrough in technology which will make SPV electricity economically viable in relation to more conventional routes of generation of electricity is necessary.

In this article we shall deal with the present status of SPV technologies and also the prospects of its cost reduction. Apart from cost of SPV electricity one has to consider the sustainability of SPV electricity for control of CO_2 emission and as a real source of energy as solar cells require a considerable amount of energy at the time of its fabrication.

BASICS OF SOLAR PHOTOVOLTAICS

A solar cell is basically a n/p junction semiconductor device and a typical structure is shown in Figure 1. When a n-type semiconductor is brought into contact with a p-type semiconductor, the Fermi level of the two types of semiconductors try to come to the same level. The Fermi level of n-type semiconductor, which is towards the conduction band, comes down and that of the p-type goes up. Thus, there is band bending at the n/p junction. If the semiconductor material has proper band gap (e.g., Si) electrons and holes are generated when solar radiation is incident on the junction. The electrons move down the gradient of the built-in potential and the holes tend to move up the gradient. The electrons and holes are driven in opposite directions by the electric field and are collected. Thus, in a solar cell one gets both voltage and current

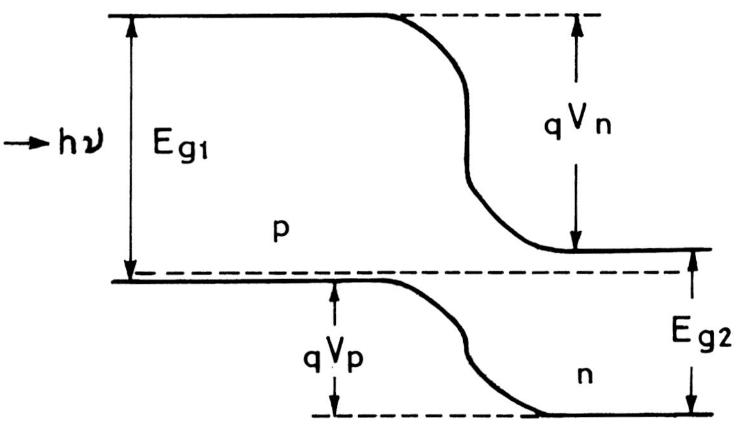

FIGURE. 1. (a) Energy band model for a p/n junction.

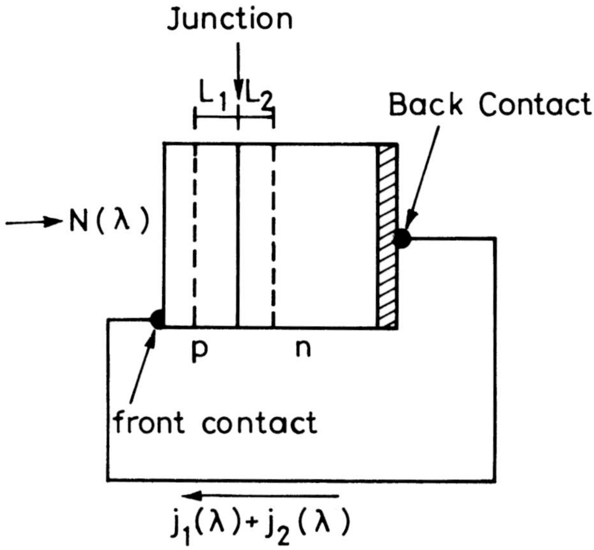

FIGURE. 1. (b) Solar Cell Geometry.

without using an external field.

A solar module consists of a number of solar cells connected in series to achieve desired voltage and current outputs. A number of modules can be connected to build up a SPV array to generate electric power of high capacity.

The efficiency of a solar cell is defined as the percentage of the incident solar radiation, which is converted into electrical power. This can be expressed as follows

$$\eta = (V_{oc}\, I_{sc} / P_{inc}) \times F.F \times 100 \qquad (1)$$

In Eqn. (1) η is the conversion efficiency, P_{ins} is the energy of incident radiation, and V_{oc} and I_{sc} are the open-circuit voltage and the short circuit current of the solar cell. A solar cell is a diode and F.F is the Fill Factor, which determines the quality of the device. The maximum theoretical value of F.F is one which corresponds to a rectangular shape of the I-V curve (current vs. voltage) of the cell.

The other important characterization of the solar cell is to the measure of its spectral response as a function of the wavelength.

A SPV system has two components, viz.: (1) Solar Module and (2) Balance of Systems (BOS). In determining the cost of SPV electricity the most important factor is the cost of solar modules including its lifetime. The BOS comprise the storage system, electronic component, frame etc.

SUSTAINABILITY OF SPV ELECTRICITY

The most important issues with the future of SPV electricity are whether it is sustainable in terms of control of CO_2 emission and is a real source of power vis-a-vis the energy that is required for the fabrication of solar cells.

The amount of CO_2 saving by replacing fossil fuels with SPV depends on the regional fossil fuel mix and the available solar insolation. It varies from 270 gm of CO_2/kWh to >1050 gm/kWh. Other pollutants as NO_x, SO_2 and particulates are avoided. With the average value of 662 gm of CO_2/kWh and an average of 5.5 hr of sunlight per day, 1 kWh SPV panel could save 1330 kg of CO_2 annually. CO_2 pay back time (the time required for the PV panel to save the amount of CO_2 generated for its fabrication by using fossil based fuels) also depends on the total energy mix and the solar panel efficiency. Assuming the energy cost of the single crystal silicon panel at 600 kWh/m^2 and average of 602 gm of CO_2/Wh, the manufacture of lm^2 panel produces 400 kg of CO_2 At 10% module efficiency and solar insolation of lkW/m^2 it takes 10 m^2 of PV panel to produce lKW. Manufacturing of 10 m^2 solar panel produces 4000 kg of CO_2 which is paid back in 3 years [1]. The lifetime of a single crystal solar cell is greater than 20 years. Thus, even at the present level of technology SPV electricity is sustainable in terms of CO_2 emission. By similar calculation, it may be shown that the energy pay back period of single crystal silicon SPV panel is 6-8 years. Thus, SPV even at its present level of technology is sustainable on the two major issues. If properly developed and applied it can be similar to the breeder reactor concept of nuclear energy.

With improvement of existing technologies and introduction of new technologies, about which we shall discuss later in some detail, the energy cost of fabrication of PV modules will come down very significantly. This in turn will reduce CO_2 and energy pay back times.

APPLICATIONS OF SPV ELECTRICITY

The present and future applications of SPV electricity are the following:

1. Rural applications
2. Urban applications
3. Industrial application
4. Specialized applications

We shall deal with the various applications listed above briefly:

1. Rural applications

Even at the present level of cost of SPV electricity it is economically viable in areas 5-10 km away from the grid line and remote areas including islands which are not yet electrified. The major applications in unelectrified rural areas are:

a. Domestic and community lighting
b. Source of power for pumps for irrigation and drinking water
c. Medical applications

Solar lanterns and domestic lighting systems are being used already in reasonable scale and becoming quite popular. In addition, experiments have started in having centralized SPV power stations to supply grid quality power to villages. SPV electricity can be used fruitfully for the running of pumps for irrigation and drinking water. The special advantage is that in this case storage of electricity is not necessary, which at present is a significant factor contributing to the cost of SPV electricity.

One major medical application in rural areas is to supply power for running the refrigerators required for the preservation of vaccines, medicines, etc.

2. Urban applications

A major thrust in the area of urban application of SPV has been putting SPV panels on rooftops. Thus each building can be self sufficient in energy. This is also connected to the conventional electricity grid where any excess power generated through SPV is fed into the grid and vice-versa. A beginning has been made in integrating SPV modules in building architecture. These types of applications have the advantage that no extra space is occupied by SPV modules.

To encourage this activity, most countries are providing substantial subsidies. If proved successful, this will be a major market for SPV in the near future, as, if technologically successful, cost may not be a barrier in advanced countries.

3. Industrial applications

Although not yet used significantly, industries can use SPV electricity together with the conventional grid electricity for reducing the peak load. The technoeconomics of this application has not yet been worked out fully.

4. Specialized applications

There are many specialized applications of SPV electricity e.g., source of power in spacecrafts, off-shore oil exploration platforms, rural telephone exchanges, railway signaling etc.

Thus, possible applications of SPV even at the present level of technology are many and varied. But there are quite a few obstacles, which are not directly technology related.

The final goal of SPV is to set up multi-megawatt power plants.

STATUS OF SOLAR PHOTOVOLTAIC TECHNOLOGIES

In spite of the sustainability of SPV electricity in terms of control of CO_2 emission and as a real source of energy, it is necessary to bring down drastically the cost of solar modules so that SPV is really viable in the next decade. While determining the appropriate technologies suitable for large-scale production of SPV modules for terrestrial use, the following factors are important:

a. Availability of materials
b. Production technology suitable for mass production
c. Low energy processing for materials and fabrication technology
d. Environmental friendliness of process and product.

We shall divide the SPV technologies into two classes, viz.: (A) silicon based technologies and (B) non-silicon based technologies.

A. Silicon based solar cell technologies:

As a SPV material, silicon is most important due to the following reasons: viz., (i) it is one of the most abundant materials available on earth so that very large production of SPV modules is possible using this material and (ii) it is cheap and free from environmental hazards in terms of process and product.

At present the following silicon based solar cell technologies are in different stages of development and commercialization:

1. Single crystal and Multicrystalline silicon solar cell
2. Silicon ribbon solar cell
3. Amorphous silicon thin film solar cell
4. Silicon film solar cell
5. Microcrystalline silicon film based solar cell
6. Combination of amorphous with other types of silicon based solar cells.

1. *Single Crystal and Multicrystalline Silicon Based Solar Cells:*

Crystalline silicon (both single and multicrystalline) occupy at present more than 80% of the total world production of SPV modules [2]. The single-crystal silicon solar cells have conversion efficiency of 12 to 16% at the production level. At the laboratory level the efficiency achieved is 24.7% [3]. These types of cells have excellent stability and reliability. Properly fabricated modules are expected to have life times of more than 20 years. This technology, although already proven and being used for power applications, has the following basic problems in lowering the cost from the present level of US$ 4.00/pW:

a. High requirement of material (wafer thickness ~400 μm)
b. High purity of the material resulting in high material cost
c. High energy processing
d. Comparatively small area cells which have to be connected to fabricate modules.

Since the material cost is high, it is not possible to make a cost breakthrough by increasing volume of production. It is estimated that by increasing the volume of production by a factor of ten the cost may be halved. However, there are other problems, which are discussed below.

A large amount of highly purified feedstock of silicon is needed. At the present level of technology for production of 1 kWp of modules, 20 kg of silicon is needed. Thus the feed stock problem is a major one for this technology. Until 1995, mainly inexpensive silicon wafers, which are rejects from the microelectronics industry, were being used. With the increase in volume of production there is a shortfall in supply of wafers from the source. New industries for the production of solar grade silicon will be necessary for increasing production.

Another serious problem is that wafers are cut from ingots with a mechanical saw so that there is a considerable loss and decrease in yield. The energy payback period for this type of solar cell is also high.

In a multicrystalline silicon wafer unlike a single crystal wafer there are a large number of smaller crystals having different orientations. The square shaped multicrystalline wafers can be cut from cast polysilicon ingots and are cheap to assemble. Multicrystalline wafers are produced from silicon of lower purity than those used for single crystal silicon and are therefore cheaper. But the conversion efficiency is 2 to 4% lower than that for single crystal silicon solar cells so that cost reduction potential of this technology is not much.

2. *Silicon Ribbon Technology*

One way of reducing the cost of crystalline silicon solar cells is to reduce its thickness from the present level of ~400 μm. The ways to do this are to produce a much thinner polycrystalline silicon layer (ribbon) or deposit such a layer on a substrate which is not a silicon wafer. The polycrystalline layer is directly produced from the liquid phase, which is called the liquid phase epitaxy (LPE). For obtaining liquid phase, silicon has to be heated to its melting point of ~1400°C. Thus, only costly substrates are being used at such a high temperature which stands in the way of cost reduction. In addition, the energy pay back period is quite high. Such high temperatures also create some additional problems described below. Because of a high thermal gradient (~500°C/cm) present at the solid/liquid interface, there may be thermal and mechanical stress on the film. These types of solar modules are not yet commercially produced and because of the high temperature of process and relatively large thickness of the wafer (50-100 μm) and expected lower yield, the available technology may not lead to significant cost reduction.

Considerable R and D is going on for the development of silicon ribbon technology at a lower temperature and lower thickness.

3. *Amorphous Silicon Thin Film Solar Cell*

A Thin film solar cell (TFS) has very high potential for cost reduction of PV modules due to the following reasons:

a. Low material cost

The thickness of the cell may be 0.5 to 3 micron depending on the technology used. This has to be compared with the thickness of ~400 μm for crystalline silicon solar cells.

b. Low energy cost for processing

In general TFS cells are produced at 200-300°C compared to >1200°C for crystalline silicon solar cells.

c. Capability of production of large area modules in a single step:

These factors help in the reduction of production cost. A large number of cells can be integrated in series and parallel by a laser scribing method.

Since the material cost is low in principle, very large scale cost reduction is possible by increasing the volume of production.

Amongst the various types of TFS cells, only amorphous silicon (a-Si) solar cells are being produced commercially. Thus, at present the world production of PV modules comprise only crystalline silicon and amorphous silicon modules. Therefore, we shall consider the status of a-Si technology in some detail.

Generally, the active material of a-Si solar cell is a thin film of a-Si:H produced from the RF glow discharge of silane gas (SiH_4). When SiH_4 is decomposed by this method (plasma processing), the a-Si film deposited on the substrate has bonded hydrogen which improves the quality of the film and makes it suitable for use in solar cells. a-Si:H can also be doped p- and n-type by adding suitable dopant gases with silane. It is also possible to vary the band gap of this material by alloying it with suitable elements. Undoped a-Si:H has band gap ~1.75 eV and responds to the visible part of the solar spectrum. By alloying with C or O, the band gap can be widened to >2.0 eV so that it responds to the shorter wavelength part of the solar spectrum. By alloying with an element like Ge the band gap of a-SiGe:H can be lowered to ~1.5 eV. This material responds to the longer wavelength part of solar radiation (900-1000 nm).

a-Si solar cells have p-i-n structure whereas crystalline silicon solar cells have p/n structures. This is due to the fact that a-Si:H has much higher defect density than crystalline silicon and its doped version has even a much higher defect density. Thus, it is necessary to have p-i-n structure with very thin p and n layers which produce the voltage when light is incident. The i-a-Si:H layer produces the current.

There are two types of a-Si solar cells, viz.: (1) single junction and (2) multi-junction. A single junction has the typical structure given in Fig. 2 (a). In a multijunction structure one cell is deposited on the other and at present double and triple junction cells are being produced. Typical structure of double and triple junction a-Si solar cells are shown in Fig. 2 (b) and 2 (c).

If multigap materials are used in the fabrication of multijunction a-Si solar cells then the width of the spectral response of the cell can be increased so that the cell current is also increased.

The a-Si:H material has a basic disadvantage (Staebler-Wronski effect) of degradation of optoelectronic properties due to light soaking, which stabilizes after some time of exposure of light. For single junction a-Si solar cell, the light induced degradation before stabilization is ~25 % and for multijunction cells it is <15%.

a-Si solar cells can be fabricated on different types of substrates, e.g., glass, stainless steel, flexible substrates etc.

At the production level Si:H based layers are deposited sequentially by the plasma CVD method (usually 13.56 MHz RF is used). The substrate can move from one chamber to the other under vacuum. It is also possible to produce it in a roll-to-roll machine when a stainless steel like substrate is used. The production technology of a-Si solar cells is now quite matured although it is expected to be improved further.

At the R and D level for single a-Si solar cells of small area, the highest

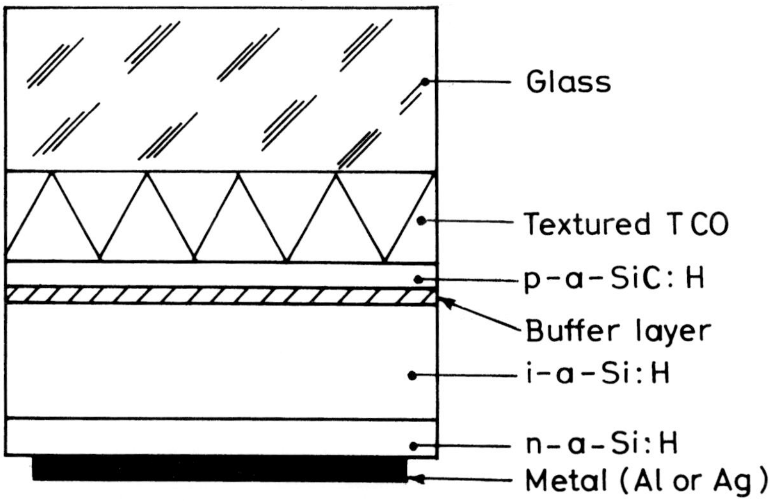

FIGURE 2. Schematic diagram of a (a) single junction.

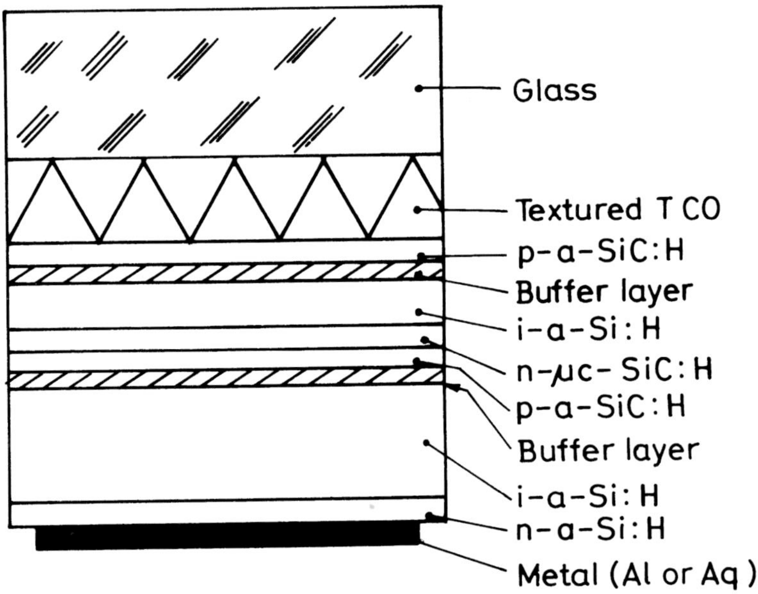

FIGURE 2. Schematic diagram of a (b) double junction.

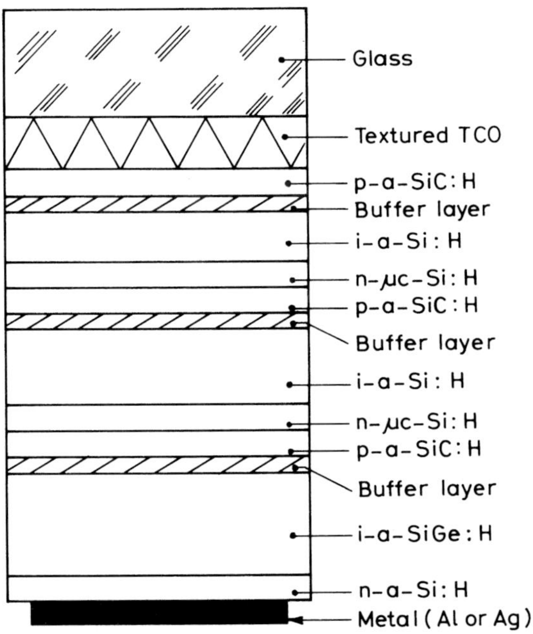

FIGURE 2. Schematic diagram of a (c) triple junction amorphous silicon solar cell.

efficiency (initial) achieved is 12-13% and light induced degradation before stabilization is 20-30%. At the production level till now, stabilized efficiency of large area modules is 4.5-5.5%. However, recently Kaneka Corporation, Japan has set up a 20 MW/year capacity plant which is expected to produce large area modules with stabilized efficiency ~8.0% [4].

For double and triple junction small area solar cells, the initial efficiencies are in the range 12-13% but the effect of light induced degradation is <15%. At the R and D level, large area modules of minimum size (40 cm x 120 cm) have been produced with stabilized efficiency 9-10%. At the production level, the stabilized efficiency of the modules is 6-8%. Solarex, USA has set up a double junction a-Si plant of capacity 10 MW/year. USSC, USA and Canon, Japan have set up a production facility for triple junction modules having capacity 5 MW/year.

At present a-Si modules are being mostly used for small power applications. Multijunction modules with higher stabilized efficiency have proved their use for power applications. This type of module can be very easily integrated in a building structure so that no additional space is needed. It is expected that in the next five years there will be a significant improvement in the stabilized efficiency. With the setting up of larger size plants (30-100 MW/year) the cost of a-Si modules will come down in the range US$ 1 .0-2.0/pW from the present cost of US$ 2.5-4. l/pW.

4. *Silicon Film Solar Cell*

A silicon film solar cell is made of a thin polycrystalline silicon layer of thickness 1-3 μm. The challenge of this technology is that the polycrystallization of silicon has to be made at comparatively low temperatures < 550°C so that cheap substrates like glass can be used. Considerable R and D is going on to achieve this. One route is the chemical route and the other route is the PECVD method. In the latter method, amorphous or microcrystalline silicon film is deposited by the RF or VHF PECVD method and, by heating, a polycrystalline layer is produced. In principle it is possible to produce polycrystalline silicon film by the laser ablation method.

By using the PECVD method, conversion efficiency of ~9% has been achieved. This route has the problem of a low deposition rate which stands in the way of successful commercial production.

This technology is receiving considerable attention at the R and D level to overcome the problems before it can be commercialized.

5. *Microcrystalline Silicon Based Solar Cell:*

It has been mentioned above that a-Si solar cells show significant light induced degradation before stabilization. This problem is probably inherent to the material itself and most probably cannot be fully eliminated, although it can be minimized. Recently it has been shown by a group at the University of Neuchatel [5] that it is possible to produce microcrystalline (μc) Si:H film, which under certain conditions shows significant photosensitivity and also responds to a larger part of the solar spectrum than what is done by a-Si:H film. In this type of material, the very high frequency (VHF) plasma CVD method is used that enhances the rate of deposition as well as the grain size of the micro-crystallites. Conversion efficiency of 8.5% has been achieved by using all the layers as μclayers with cell thickness of 2.7 μm [5]. With usual RFPECVD method (13.56 MHz) and a deposition temperature 500°C a cell efficiency of ~10.1% has been achieved at thickness ~2 μm. These cells, unlike a-Si solar cells, are stable with respect to light induced degradation.

However, this technique of making solar cells has the following problems viz.: (1) low deposition rate, (2) stringent requirement of crystal structure for optimum results. These problems may stand in the way of commercialization of this technology.

6. *Combination of a-Si with other types of Silicon Based Solar Cells*

It has been discussed earlier that a-Si thin film solar cells have a proven production technology and by varying band gap of the active i-layer it can be made to respond to different parts of the solar spectrum. In addition, if the i-layer is sufficiently thin there is no significant light induced degradation as the charge carriers generated by the incident light are collected by the built-in-field of the solar cell before they can recombine. Due to these reasons the following types of combinations are being tried at the R and D level.

i) a-Si/single crystal silicon
ii) a-Si/polycrystalline silicon
iii) a-Si/microcrystalline silicon

The bottom cell of the three structures mentioned above respond better in the larger wavelength part of the solar radiation. Therefore, a tandem cell with the top cell a thin p-i-n structure a-Si solar cell, makes the combination more responsive to the shorter wavelength part of the solar radiation without losing stability.

Sanyo Electric, Japan has developed an a-Si/single crystal silicon tandem cell quite well with efficiency ~18%. Plans have been made to commercialize this type of solar cell. Kaneka Corporation of Japan has recently claimed [6] a major breakthrough in the fabrication of a-Si/polycrystalline silicon structure solar cell. The polycrystalline silicon bottom cell has also been prepared by plasma CVD method. For 1 sft module, stabilized efficiency of 11.3% has been achieved. The company expects to achieve similar types of efficiency for larger area cells and commercialize it in the year 2003.

B. Non-Silicon Based Solar Cell Technologies

The most important non-silicon based technologies are the following:

a) CIS (CIGS)/CdS Thin Film Cell
b) CdTe/CdS Thin Film Cell

We shall describe briefly the status and prospects of each of these technologies:

a) Copper Indium Diselenide Based Solar Cell

Copper indium diselenide (CIS) and copper indium gallium diselenide (CIGS) are direct band gap polycrystalline semiconductors with a very high optical absorption coefficient. The CIS/CIGS layer is p-type and a layer of CdS serves as the n-layer. Thus this is a heterojunction based solar cell. There are a number of methods for the deposition of these layers which are simpler than those for a-Si solar cells.

For small area (0.449 cm^2) highest efficiency, ~ 18.8% has been achieved [7], which is excellent for a thin film cell of thickness ~3 μm. The band gap of CIS is in the range 1.1-1.2 eV which is good for the absorption of solar radiation. The band gap of CIS can be increased by the introduction of gallium. The n-CdS functions as the window layer. Like a-Si solar cell, large area integrated modules can be fabricated by using laser scribing technique. For modules of area ~ I sft the efficiency, is 9-12%, which is much lower than that for small area cells, has been achieved.

The CIS cell is stable with respect to light induced degradation. However, they show some instability in hot and humid environment. In principle for large volume of production ~100 MW/year the cost of CIS module will be much lower than that of crystalline silicon cells.

However, in spite of its achievements, these types of modules are yet to be commercially produced. Many attempts have been made in the past and attempts are also being made at present to commercialize this technology. Industry like Siemens Solar is interested in this technology. The problem has been failure to produce high efficiency large area modules on a repetitive basis. Due to the problems associated with production this technology has remained at the pilot plant level for a long time.

There are few other problems associated with this technology. A major one is the availability of materials if the annual production exceeds 100 MW/year. Materials like indium, cadmium are much less abundant than silicon. The other problem is the toxicity of the materials and the product. Cd, Se, etc. are highly poisonous.

b) CdTe Solar Cell

Similar to CIS, CdTe is a direct band gap material which fully absorbs solar radiation at a thickness ~1 μm. The band gap of 1.45 eV is excellent for single junction cells. The n-CdS layer is used as the window layer and integrated modules can be fabricated in one step. Since CdTe is a binary compound, it is easier to deposit than CIS or CIGS. A typical CdTe solar cell is deposited on a

Transparent Conducting Oxide coated glass substrate like the a-Si solar cell.

For CdTe solar cells (1 cm^2 area) efficiency, ~16% has been achieved at the laboratory level [8]. Two industries are building up plants with production capacity ~10 MW/year. Many other industries are trying to commercialize this technology. The efficiency expected is 8-9% at the commercial production level. However, the problems seem to be the same as those for CIS modules.

C. Other Types of Solar Cell Technologies:

There are few other types of solar cell technologies which are being tried but are at present purely at the R and D level. A few of them are described below.

(a) Dye sensitized PV Solar Cells

A new type of solar cell has been fabricated by Gratzel [9] which has been drawing considerable attention. In this cell a rough TiO_2 layer is covered by a light absorbing dye on a redox couple in a suitable electrolyte. Charge transport is realized by the conduction band of the TiO_2 on one side and an ion flux within the electrolyte on the other side. With this type of cell efficiency, ~10% has been achieved for small area cells.

However, these cells are not stable as yet. The use of a liquid electrolyte may be the main cause of this instability. It will be difficult to overcome this problem.

(b) Carbon Based Solar Cells

Apart from silicon, Carbon is another material which is cheap and abundant. Attempts are being made to fabricate cells with this material. It is, however, in a very preliminary stage.

TECHNO-ECONOMIC SCENARIO OF PV-TECHNOLOGY

For the developing countries, off-grid use of SPV is a major application. For developed countries, grid connected systems are important. Table 1 shows the world PV shipment region-wise for the period 1992 to 1998. During 1997 and 1998 the growth has been a spectacular 72%, whereas prior to 1997 the growth rate was on an average 12-15 %. The major boost has been due to a heavily subsidized grid connected residential program in advanced countries like Japan [10].

Among the rest of the world, Indian PV market is 7-8 MW/year. Table 2 shows the world PV market sector wise.

It may be seen from Table 2 that apart from grid connected systems the growth in other sectors has remained in the range 12-15%. However, in developing countries like India, off-grid applications in rural areas is receiving

TABLE 1.
World PV module shipments in MW

Region	1992	1993	1994	1995	1996	1997	1998
US	18.1	22.44	25.64	34.75	38.85	51.0	53.7
Japan	18.8	16.7	16.5	16.4	21.2	35.0	49.0
Europe	16.4	16.55	21.7	20.1	18.8	30.4	31.0
Rest of world	4.6	4.4	5.6	6.35	9.75	9.4	18.7
Total	**57.9**	**60.09**	**69.44**	**77.60**	**88.60**	**125.8**	**152.4**

TABLE 2.
World module shipments by market sector

Market Sector	1990	1993	1996	1997	1998
Consumer Products	16	18	22	26	30
US off-grid residential	3	5	8	9	10
World off-grid rural	6	8	15	19	24
Communications/signal	14	18	23	28	31
PV/diesel commercial	7	10	12	16	20
Grid-connected	1	2	7	27	36
Central power	1	2	2	2	2
Total	**48**	**63**	**89**	**127**	**153**
Average price ($/W)	4.5	4	4	4.2	4
Low price ($/W)	4	3.5	3	2.75	2.6

increasing attention, and this sector is likely to grow rapidly in the near future in all developing countries.

Thus, even at the present level of technology, PV module production is likely to rise much higher than 15%. However, it is expected that in the next decade the price of the PV module will be below $ 2.00 /pW, and this will greatly boost the PV market. The goal of $1.00/pW is likely to be achieved in the next 10-15 years with the maturity of one or more Thin Film Technologies, of which a-Si is the forerunner at present.

The costs of manufacturing of different types of PV modules (1997) are given in Table 3 [10]. The cost includes recovery of capital cost at 20% per year.

This is based on 1997 figures. Since then there has been significant improvement in a-Si technology in terms of plant capacity and conversion efficiency so that the present figure is ~$ 2.0/pW.

It is projected that during the next decade the production capacity will rise to 3-5 GW/year and cost will be $1.5-2.0/pW. At this price the market will expand even more rapidly in the future.

TABLE 3.
Cost of manufacturing of different types of PV modules (1997)

Type of module	**Cost of manufacturing/pW (US $)**
Single crystal silicon	4.0
Multicrystalline silicon	4.0
Amorphous silicon	2.5

CONCLUSIONS

From what has been described and discussed above, it is clear that in the next century solar photovoltaics will be a sustainable source of clean energy. With progress in technology already made and large scale investments in R and D by Government and Industries it is very likely that the cost goal of $ 1.00/pW (selling price) will be reached within next 10-15 years. Apart from awareness and incentive, a global policy has to be introduced for the promotion of Renewable Energy Sources including Solar Photovoltaics.

With an increase in production volume, material will be a major factor choosing technology. In that context, silicon based technologies are very much ahead of others.

REFERENCES

[1] J. A. Turner; Science 285 (1999) p.687.

[2] G. Girout-Matelowski, U. Theden and A. W. Blakers; Proc. 2nd World Conference on Photovoltaic Energy Conversion, Vienna, July (1998) p. 3403.

[3] M. A. Green, J. Zhao, A. Wary and S. R. Wenham; Tech. Digest, Int. PVSEC- 1 l, Sapporo, Japan, Sept (1999) p.21.

[4] Y. Tawada and H. Yamagishi; Tech. Digest. Int. PVSEC-11, Sapporo, Japan, Sept (1999) p.25.

[5] J. Meier et. al.; Mater. Ser. Symp. Proc. 507 (1998) 139.

[6] K. Yamamoto, M. Yoshomi, Y. Tawada, Y. Okamoto and A. Nakajima; Tech. Digest, Int. PVSEC-11, Sapporo, Japan, September (1999) p.225.

[7] T. M. Friedalmeier and H. W. Schock; Proc. 2nd World Conf., Vienna, July (1998) p. 1117.

[8] H. Ohyama et. al.; Proc. 26th EEE Photovoltaic Specialist Conference, (1997) p. 988.

[9] B. O'Regan and M. Gratzel; Nature 353 (1991) 737.

[10] P. Maycock; Renewable Energy World, July (1999).

Renewable Energy: Trends and Prospects. Edited by S.K. Majumdar, E.W. Miller and A.I. Panah

Chapter Thirteen

APPLICATIONS OF SOLAR THERMAL ENERGY

JONATHAN J. MILES[1]
and
CHARLES E. ISKANDER YOUSIF[2]

College of Integrated Science and Technology
[1]James Madison University
Harrisonburg, VA 22807
and
[2] Institute for Energy Technology
University of Malta, Roman Port Street,
Marsaxlokk, ZTN09, Malta

INTRODUCTION

The Sun's energy reaches the surface of the Earth in abundance, delivering a peak spectral solar flux of approximately 1000 W/m^2 at sea level on a clear day. This virtually limitless source of energy can be absorbed and used for a multitude of purposes including desalination of water, space heating, cooling by absorption refrigeration, and cooking. But perhaps the most common application is the heating of hot water for domestic and commercial needs. Solar collectors designed for this purpose are available and utilized throughout the world, and implementation of this technology is cost-effective in most populated regions of the world.

The Sun's energy arrives at the Earth in the form of electromagnetic energy, a form of energy that when absorbed by materials is converted into thermal energy or heat. Thus it is sometimes stated that heat is the most basic form of the Sun's energy. Approximately 25% of the total energy we use, for all applications, goes to heating in residences and commercial buildings.[1] Water heating accounts for about 20% of the total energy consumed in typical homes in the United States, as well as in many developing countries[2], and natural gas and electricity are the most common sources of energy to heat water, generally to a temperature between 50 and 65°C (120 to 150°F). Homes equipped with electric water heaters, however, can sometimes dedicate as much as one-half of electrical energy consumption to water heating.[3]

There are numerous technologies that exploit a variety of mechanisms to convert solar energy into other forms. Photovoltaic cells convert short-wave solar energy directly into electrical energy. Wind turbines convert the mechanical energy of the wind into electricity. However, these two examples, like most others, involve multiple conversions of energy (i.e. solar to electric to some other form such as thermal). Alternatively, solar heating of hot water represents a direct effect of the Sun's power, and avoids the losses inherent in all energy conversions.

This chapter describes the basic features of the modern solar water heater (SWH, which may also represent the term *solar water heating*), and describes some of the recent trends and projects in both residential and commercial settings. The authors also present a treatment of the economics of SWH, and describe some recent successes regarding utilization of SWH on a large scale. Finally, elements that could enhance increased adoption of SWH in the future are discussed.

SOLAR WATER HEATER TECHNOLOGIES

The first commercial solar water heater was built by Clarence Kemp in 1891. By 1897, thirty percent of the houses in Kemp's home town of Pasadena, California were equipped with solar water heaters.[4] Modern flat-plate collectors provide an economically viable means for heating water for residential and commercial applications. These collectors employ the greenhouse effect to trap solar radiant energy and transfer this energy to a fluid, typically a mixture of water and an anti-freezing agent.

The configuration of components of a typical roof-mounted, flat-plate collector is shown in Figure 1.[1] Incident solar radiation passes through the glass plates and is absorbed by the black, selective-coated metal surface. The selective coating ensures maximum absorption of solar radiation and minimizes radiative losses. The glass cover traps solar energy and reduces convective heat losses. Energy absorbed by the metal plate is conducted through the tubes to the working fluid. For best results, the tubing is soldered to the metal plate. Although this process is labor-intensive, it maximizes surface contact, and therefore heat transfer, between the metal sheet and the tubes, thus optimizing the overall performance of the collector. The insulation minimizes heat losses to the surroundings, and the working fluid flowing through the tubes is heated and transported to the load.

A typical SWH installation in a modern household utilizes one or more panels (termed collectively as the *collector* in this chapter) similar in construction to that shown in Figure 1. The collector is mounted on a south-facing slope of the roof of a house, and is connected in such a way that solar-derived hot water is delivered directly to the existing conventional water heater inside the building. The overall system is typically controlled so that the conventional energy source (i.e. resistive heating element or gas burner) does not activate unless solar heating is insufficient to meet the load demand.

Figure 2 shows one approach to tie a solar panel into the existing hot-water system of a modern house. Note that the working fluid of the SWH system does not mix with the household water. This feature is critical to maintaining the quality of potable water and to avoid freezing. The working fluid is typically an anti-freeze or water-alcohol mixture.

The SWH system shown in Figure 2 is an active system, so called because an external power source is required to drive the pump that moves the working fluid through the collector and other components. With the exception of the collector, the components that constitute an active SWH system are typically installed indoors. The anti-freeze solution is circulated when the temperature of the fluid in the collector is greater than that inside the water heater. This promotes maximum energy transport to the water heater, and minimizes losses to the environment during cold and cloudy conditions.

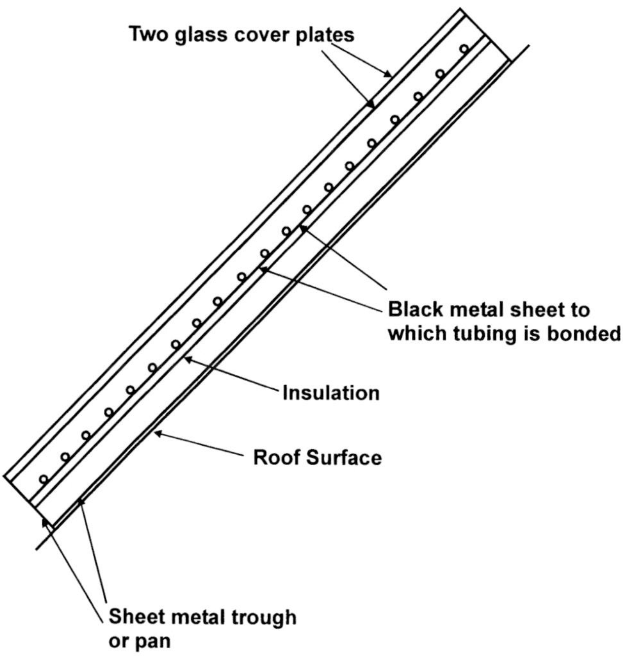

FIGURE 1. Solar panel for heating water.

An alternative SWH technology is a simple thermosiphon solar water heater as pictured in Figure 3. Circulation is induced by density variations between the liquid in the collector and that within the storage tank. This cost-effective technology is used in thousands of homes in Australia, Japan, and North Africa. It is also common in the Mediterranean region and the sub-tropical islands of the

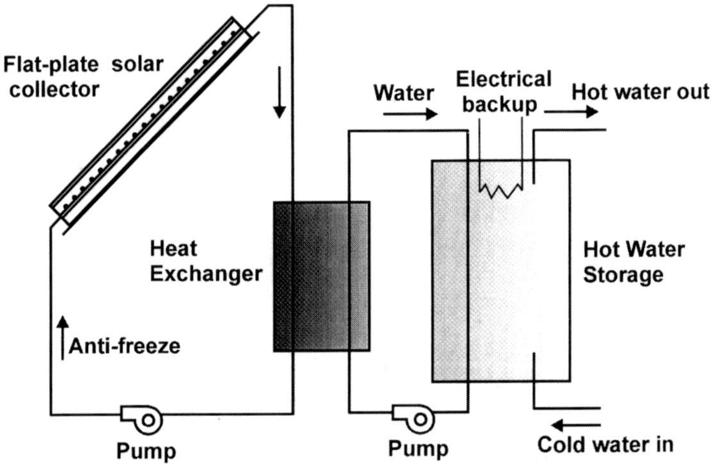

Figure 2. Circulating-liquid solar collector system.

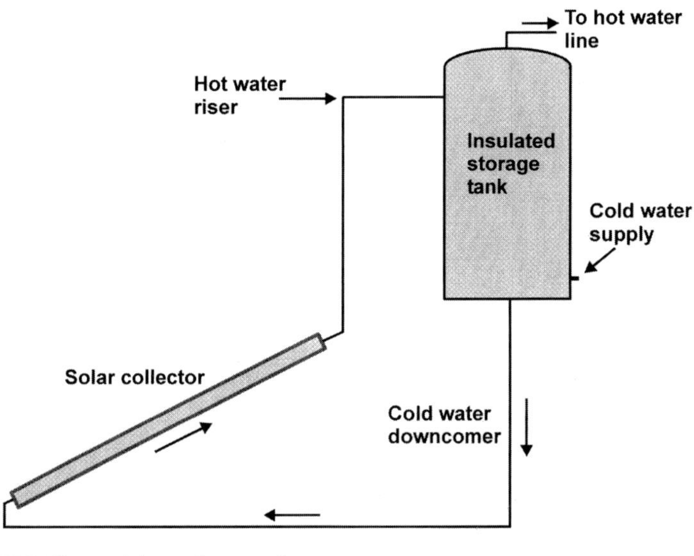

FIGURE 3. Thermosiphon solar water heater.

Caribbean. Under favorable climatic conditions of abundant sunlight and moderate winter conditions, a thermosiphon system can produce 110 to 190 liters (30 to 50 gallons) of hot water at temperatures up to 70°C (160°F) in summer and 50° (120°F) in the winter. An auxiliary electric heater placed inside the building would typically be used to boost temperatures during unfavorable weather conditions.[5]

A computer modeling and simulation of a proposed thermosiphon system was performed at the Institute for Energy Technology in Malta in June 1999.[6] The simulation utilized a radiometric and weather data set representing average meteorological conditions for the Maltese archipelago. A conservative set of assumptions was applied to this study, and simulations were performed with both copper and EPDM tubes. The results of this analysis predict an energy cost savings of 45% for EPDM and 86% for copper, and a simple payback period on the order of seven years for either material. A more realistic set of assumptions would reduce the predicted payback to fewer than five years.

Collector technologies more advanced than flat plate are also available for SWH. Evacuated tube collectors that offer higher performance than conventional flat-plate technology are manufactured by Nippon Electric Glass (NEG) of Japan and Thermomax® of the United Kingdom. The efficiency of all solar collectors decreases as the temperature of the fluid flowing through increases. However, the evacuated tube reduces conductive and convective losses due to vacuum separation between absorber and glass. This results in a greater overall efficiency and a higher maximum fluid temperature.

FIGURE 4. NEG Sun Family® RK-20 evacuated tube solar water heater.

The Sun Family® RK-20 SWH system incorporates eight vacuum-insulated, passive solar collectors that also provide hot water storage. It is pictured in Figure 4.

According to NEG, the RK-20 system provides hot water that consistently exceeds the outlet temperature of a comparably sized, conventional flat-plate system, under a wide range of ambient conditions. This conclusion is drawn from measurements obtained in Nagoya, Japan with the unit facing directly south at a 30-degree angle.

In the NEG evacuated tube, water flows through the core of the tube and absorbs thermal energy directly. Alternatively, the Thermomax® evacuated tube encloses a heat pipe within which is sealed a working fluid. The natural convection of this fluid is driven by the energy absorbed, and consequently delivers thermal energy to the water flowing across a manifold in thermal contact with the tip (condenser) of the tube. An array of these tubes linked by a common manifold would constitute the collector of a complete solar water heating system. This technology is illustrated in Figure 5.

The Thermomax® product has been tested independently and, like the NEG product, out-performs a conventional flat-plate collector. However, the higher cost of these vacuum-based technologies would eliminate them from consideration in systems where cost is a principal factor. The expense associated with these advanced designs reflects their performance. But for applications such as absorption cooling that require optimal efficiency and higher temperatures than

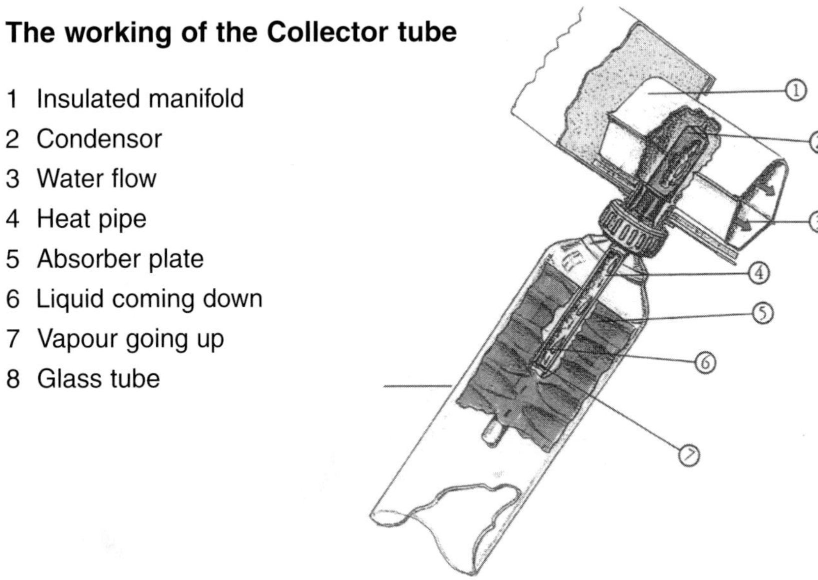

FIGURE 5. Thermomax® evacuated heat-pipe solar collector.

flat-plate technology can provide, or for applications in which cost is secondary, the evacuated-tube technology provides an attractive alternative.

RECENT TRENDS IN SOLAR WATER HEATING

Low-cost SWH in Florida residences

People with low income are especially sensitive to energy costs, since their utility bills generally represent a larger proportion of total expenditures than for their middle-income counterparts. The United States Department of Energy (DOE) operates a weatherization assistance program to keep the homes of low-income residents comfortable and affordable in the winter months.

Naturally, weatherproofing measures do not make as much economic sense in warmer climates. Therefore, the DOE joined with the Florida Department of Community Affairs (DCA) and the Florida Solar Energy Center (FSEC) in 1994 to reduce monthly energy bills for low-income residents in Florida by providing solar water heating systems.[7] The program has three main objectives: to reduce energy use and power bills for low-income residents; to evaluate the feasibility of SWH systems as a weatherization program measure; and to provide a niche market for the solar industry.

The program provides "bare bones" SWH systems at a cost of approximately $1500 each, about $1000 less than that for a conventional system. The low-cost system uses a 24-ft^2 flat-plate collector (rather than 40 ft^2 as with a conventional system) and the existing 50-gallon electric water heater (rather than an 80-gallon solar storage tank). The overall solar fraction is around 50%, compared to 70% for a conventional system. The use of standard electric water heaters also enables low-income clients to afford replacements since they are available at a much lower cost than larger solar storage tanks.

Most systems installed are integral collector storage (ICS) or direct-pumped systems, with either one using the flat-plate collector. The ICS system, essentially a pre-heater that feeds solar-heated water to the conventional electric water heater during household draws, was chosen for its simplicity and freeze tolerance in warm climates. The pumped system is used primarily in south Florida where freeze conditions are very rare. Both systems utilize a backup electric water heater during unfavorable weather conditions. In support of this program, the State of Florida certified solar contractors to install the systems. FSEC coordinated installation efforts, developed a detailed solar training program for agency staff, and developed training materials and inspection checklists.

Since the program began, over 800 systems have been installed in Florida at an average system cost of $1555. Monitoring has revealed that the average water heating energy consumption has dropped by more than half, from 3100 kWh per system per year prior to installation to 1500 kWh afterward. The cost for water heating for an average family has subsequently dropped from $250 prior to $120 afterward, despite a slight increase in hot water consumption following installation of the systems.

FESC conducted surveys of all clients who had received SWH systems. Over 35% responded, with a majority reporting a high level of satisfaction with their system. The survey also revealed that system owners needed further information and education regarding the operation and maintenance of SWH systems.

Solar Thermal Applications and Guaranteed Results in Europe

Solar water heating for domestic use is the most common application of solar thermal technology across Europe. The installed collector area in the European Union (EU) today is over 8 million square meters, up from 5.5 million square meters five years ago. This adoption results annually in a reduction in emissions of 1.4 million tonnes of CO_2, and the displacement of 450 thousand tonnes of oil.[8]

Combined water and space heating to support a conventional heating system is common in areas of Europe that endure a long heating season, countries such as Austria, Germany, France, Switzerland, and the Netherlands. District heating plants in Denmark, Sweden, and Germany have been augmented with solar collectors since 1979. There are even a number of industrial heating applications in Greece.

After experiencing a setback following the latter part of the 1970s and early 1980s, the market for solar thermal applications is again on the rise. Solar systems manufactured and installed in the U.S. and abroad during the 1970s often fell short of predicted performance as a result of faulty materials, errors in systems design, and lack of maintenance. Because solar systems frequently provide only marginal economic benefits, a minor breakdown can undermine completely the economic practicality of an installation.

In an effort to restore confidence among potential residential and commercial clients, the industry has developed measures to ensure that installations will perform satisfactorily. These steps include the development of improved SWH designs; simple and accurate methods for sizing and estimating the annual productivity of SWH systems; and the development of affordable and reliable means for monitoring system performance and operation and alerting the customer to operating anomalies.

An example of a project that applies these measures yet takes the additional step of offering a "guarantee of performance" is the 1988 installation at Castres Hospital in southern France. The implementation of this project included the preparation of a "contract" for guaranteed solar results that would compensate the client if the guaranteed results were not attained, would be consistent with the building codes, and would be legally effective.[9] This solar demonstration project has provided hot water reliably to the hospital for over ten years. There are now over 60 large solar thermal systems, with guaranteed results, operating in France, Spain, Germany, and several other countries, including the recently-opened, luxurious Novotel Fleur d'Epée in Guadeloupe. The SWH system in this hotel features a 96-m^2 collector array and an estimated productivity of 886 kWh/m^2 per year.

The concept of Guaranteed Solar Results (Garantie de résultats solaires or GRS), as applied to domestic hot water production, dictates that a solar-thermal system is expected to supply a minimum quantity of energy each year given a fixed rate of hot-water load consumption. If the energy supplied fails to meet the energy guarantee, compensation is awarded commensurate with the difference between the guaranteed solar energy supply and the energy actually available for heating water during the contract.

The GRS contract involves a party known as the guarantor, the party who is responsible for offering the guarantee and providing any compensation that may be due. The guarantor might be a solar manufacturer or installer, a consultant, or a developer, any party with the ability to fulfill its contractual financial obligations. The guarantor must evaluate the predicted solar production of a SWH system, and would generally guarantee solar energy supply in the amount of 80 to 90% of the estimated annual figures in order to avoid the problem of exceptional climatic variation.[9] The Novotel Fleur d'Epée system provides a guaranteed supply limited to 80% of the calculated estimate of performance.

The GRS contract offers several essential advantages to the consumer. Since the customer is guaranteed a certain annual supply of energy, the investment payback time is also guaranteed, and in the absence of financial risk there is less difficulty in acquiring financing. Also, a guaranteed system is monitored permanently, so any breakdown is quickly identified and repaired. The Guaranteed Solar Results concept, with third-party investment, could play a major role in developing the SWH market worldwide.

ECONOMICS OF SOLAR WATER HEATERS

The cost of solar water heaters varies significantly from one country to the next, and this is one factor that determines the economic feasibility of SWH. A solar water heating system may not supply all of the hot water that a modern, western household requires. It can, however, supply enough hot water to provide significant savings resulting in reduced energy bills. Also, it is the only viable means to heat water in some rural communities around the world. In this section, the life-cycle costing (LCC) method is presented to evaluate the feasibility of using a solar water heater at home in place of an electric water heater.

Life-cycle costing takes into account the total costs involved in purchasing and maintaining a SWH system, and discounts them to a common point in time, usually taken as the present. In practice, one can determine the cost of energy with this system, and compare it to that of using a more conventional means to produce that energy. The advantage of using the life-cycle costing method over a simple costing method is that it takes into consideration the initial costs as well as all other future costs, and discounts them to their present value. As a result, the real payback period can be determined.

The life-cycle costing, *LCC*, is expressed by Equation 1

$$LCC = \frac{C + M - S}{E} \quad (1)$$

in which *C, M,* and *S* represent respectively the total initial capital costs, maintenance and replacement costs anticipated for the whole life of the system, and the salvage or scrap value at the end of system life, all measured in dollars. The value *E* represents the energy produced by the system in kWh. These quantities must be converted to their present values by using the appropriate discount factors that are found in tables of standard textbooks[10] or by calculation using the equations presented below.[11]

For a fixed amount of money to be paid in *n* years time, the present value interest factor, $PVIF_{k,n}$, is given by

$$PVIF_{k,n} = \frac{1}{(1 + k)^n} \quad (2)$$

in which *k* represents the net discount factor. In light of inflation, a unit of money paid now is worth less in the future; but if this unit were invested, it would earn interest over the assigned period. The difference between these two factors gives rise to the net discount factor, *k*, that is used in this analysis and defined by

$$1 + k = \frac{1 + m}{1 + i} \quad (3)$$

where *m* is the highest interest that could be acquired by investing the money elsewhere and *i* represents the average inflation rate over the period. For an annual, fixed recurrent expenditure, the present value interest factor annuity, $PVIFA_{k,n}$, is calculated by Equation 4.

$$PVIFA_{k,n} = \sum_{t=1}^{n} \frac{1}{(1 + k)^t} = \frac{1}{k}\left[1 - \left(\frac{1}{(1 + k)^n}\right)\right] \quad (4)$$

Equations 2 and 4 can be applied to discount a financial sum to its present value by multiplying that sum by the appropriate factor. Other quantities pertinent to carrying out a life-cycle costing are the costs of daytime electricity and the solar water heater, and the lifetime of the solar water heater.

Finally, it is necessary to calculate the amount of energy that a solar water heater will save. This can be estimated in a simplified fashion by determining the product of the efficiency of the system and the total insolation. This calculation is performed by Equation 5

$$E = \eta A I \tag{5}$$

in which E is the energy saved by the solar water heater in kWh, η represents overall efficiency of the solar water heating system, A is the area of the solar panel in m^2, and I represents total solar radiation incident on the panel measured in $kWh/m^2/day$.

Example: A solar water heater is to replace an electric heater in a house. Find the life-cycle costing per unit of energy output when utilizing a thermo-siphon solar water heater costing $1,200. The system is operating at an average efficiency of 50%, and is subjected to an average insolation of 5.4 $kWh/m^2/day$ on the plane of the panel. Assume the information given below:

- Area of solar panel: 3 m^2
- Day-time cost of electricity $0.08/kWh
- Inflation rate: 3%/year
- Interest rate: 8%/year

Solution: Given the following assumptions:

- Lifetime of solar water heater: 8 years
- Installation costs: $200 in Year 1
- Operation and maintenance costs: 2% of initial cost, annually
- Salvage value: 5% of initial cost in Year 8
- Interest and inflation rates remain constant during the lifetime of system
- Cost of system and installation (assumed to be paid in full in the first year) equals $1,400

Application of Equations 3 and 4 provides the net discount factor, k, equal to 4.85% and the present value interest factor annuity, *PVIFA*, of 6.5 for this example. These values are required to find the total discounted recurrent maintenance expenditure over the lifetime of the system.

To calculate the salvage value, the appropriate discount is determined using Equations 2 and 4 respectively, with n equal to 8 years. The salvage and maintenance values, S and M, are $41.08 and $156 respectively. The total discounted costs incurred are $C + M - S = \$1514.92$, and the annualized costs are this value divided by the present value interest factor annuity of 6.5, or $233/year. The annual energy output can now be determined by multiplying the system efficiency by the incident solar radiation and the area of the solar panel for 365 days of the year to produce 2957 kWh/year. Finally, the life-cycle costing (LCC) per unit of energy is determined and compared to the day-time cost of electricity. For this particular example, LCC/kWh equals $233/2957 kWh or 7.9

cents/kWh. In comparison, this cost is slightly lower than that of electric hot water heating during the day, suggesting that the investment in a solar water heater is worthy of consideration from an economic point of view.

This analysis can be modeled readily in any spreadsheet program, providing a straightforward means to compare different types of solar water heaters operating under varying conditions. It should be noted that in carrying out a life-cycle cost analysis, neither revenues nor profits are considered.

Another factor commonly used to evaluate the economic viability of a system is the payback period. This is defined as the year during which the cumulative profits or savings surpass the costs. Considering the example presented, the discounted savings occur because the solar water heater replaces an electric heater that would otherwise have consumed electricity at the daytime rate of \$0.08/kWh. Using Equation 2, the annual savings for year n, is calculated as

$$\boldsymbol{Savings = 2957 \times 0.08 \times PVIF_{0.0485,n}}$$

with the results of calculation of n shown in Table 1. The example presented demonstrates a payback occurring during the 7th year.

Successful Applications of SWH on a Large Scale

Solar water heating has been introduced, with varying degrees of success, in many countries including Australia, Cyprus, Greece, Israel, and the United States. In some instances, the widespread application of this technology occurred in response to the oil crisis of the 1970s. However, the incentives offered by governments to promote the use of solar energy applications could not sustain a continued development of this industry after the end of the crisis. Nonetheless, solar applications in some countries, such as Cyprus and Israel, have evolved from a cottage industry in the 1960s to major production enterprises in the 1980s and beyond.

TABLE 1.
Discounted annual costs and savings.

Year	Capital	Maintenance	Salvage	Savings	Net Profit
0	-1400	0		236.52	-1163.48
1		-22.89		225.58	-960.79
2		-21.83		215.14	-767.48
3		-20.82		205.19	-583.20
4		-19.86		195.70	-407.26
5		-18.94		186.65	-239.55
6		-18.06		178.02	-79.60
7		-17.23		169.78	**72.95**
8		-16.43	41.08	161.93	259.53

The estimated cumulative area of solar collectors in the principal markets around the world in 1994 is shown in Figure 6 and illustrates that the Mediterranean region is the leader in this respect. This trend is expected to continue into the future.[12] The popularity of solar water heating in this region is due not only to a generous solar resource, but may be attributed also to the development of societies that accept solar energy readily as a viable source of sustainable energy at an affordable and reasonable price.

The island nation of Cyprus, with a population of 754,000, has become one of the largest and most concentrated users of solar water heaters per capita, and this trend can be expected to continue into the future. The reasons behind this achievement are neither political nor purely economic; Cyprus's story is that of a nation without an oil resource, which has therefore chosen to utilize its solar potential for the benefit of its people. Implementation of solar thermal energy applications in Cyprus has grown consistently; and as the market has grown, costs have dropped, leading to increased adoption of SWH. The present total area of solar collectors per capita is close to 1 square meter (m^2).

Most of the solar systems installed in Cyprus today are manufactured locally, as the industry has matured significantly since its beginning in 1960. Figure 7 shows that the installed capacity is about 0.85 m^2 per capita. Almost 90% of residential buildings and 50% of hotels have solar water heaters, most of which incorporate an open- or closed-loop thermosiphon design.

In Israel as well, solar water heating has been applied successfully on a large scale. Installation of SWH systems has been mandatory in all new buildings since 1980, thus helping Israel to reduce its national energy consumption by about 4 percent.[12] The technology has so far penetrated up to 85% of the residential sector.

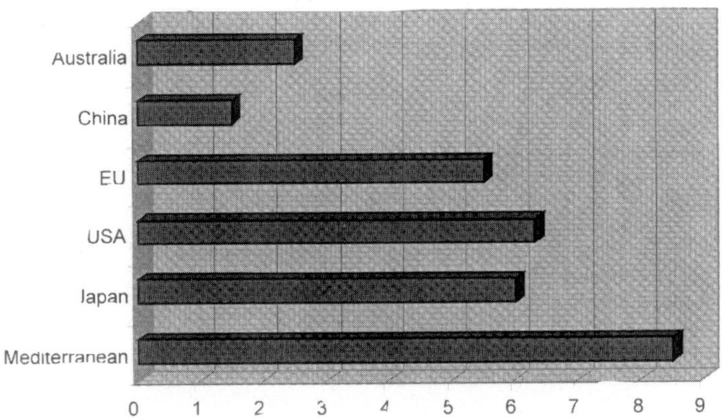

FIGURE 6. Estimated cumulative area of solar collectors in the principal world markets.

From these two examples, it is clear that political decisions and economic incentives can help promote the use of renewable sources of energy for our everyday needs. But such measures cannot guarantee positive results unless the mentality of the public changes to embrace these new technologies. There is still much progress to be made. Solar water heating has yet to invade the large market for space heating in these countries.

The Future of Solar Water Heating

The demands and customs of every country around the world are so unique that one cannot offer a single set of guidelines that might accelerate the implementation of solar water heating universally. However, the one element that is critical to the success of SWH in any region is education. Without an effective educational campaign that infuses the public with a general awareness and understanding of SWH, no guarantees can be offered for the consistent and long-term growth of the market for SWH technologies.

The building of highly visible demonstration projects in public buildings, such as schools and hospitals, can help to promote confidence in solar technologies. But it is very important that installations be designed and implemented to the highest standards in order to avoid unreliable systems that fall short of their predicted performance. Appropriate training of technicians and engineers is especially important to support a growing solar infrastructure.

If one considers the established markets of Greece, Germany and Austria, Israel, or the developing markets in northern Europe, it is clear that a combination of factors has contributed to their successes. Public awareness campaigns running in tandem with government incentives, and in some cases special agreements between governments and major utilities, have played an important supporting role in the consistent growth of SWH adoption and market penetration.

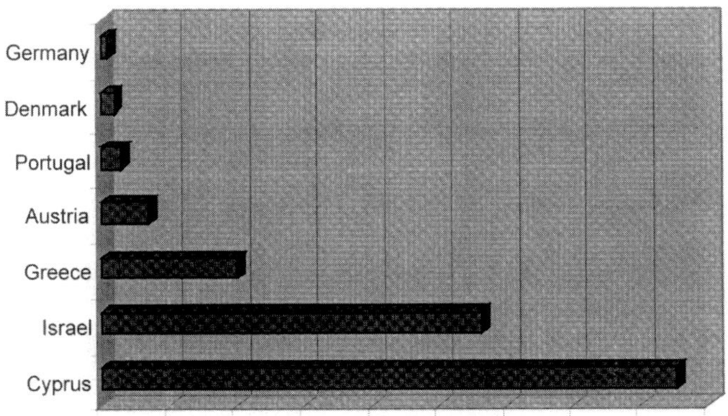

FIGURE 7. Total area per capita of glazed solar collectors.

According to the European Solar Industry Federation, the most important aspect of national and EU policies for promoting solar thermal is that they be focused on the support of market penetration. They advise the following actions:

- Subsidy programs for solar thermal systems for a limited time
- Stimulation of demand through public awareness campaigns and demonstration projects
- Training programs for designers and installers
- Development of standards and codes of practice
- Expansion of procedures for guaranteeing solar output
- Technology transfer across borders
- Building regulations governing energy consumption in new houses

An additional measure that might be considered is the gradual phase-out of subsidies for fossil-based electricity, subsidies that could be diverted to projects involving renewable energy systems. New incentives could also be created such as tax relief for new projects that integrate solar systems into new and renovated buildings.

There is no intended or suggested hierarchy regarding the implementation of any of the measures described above. Any strategy designed to support the growth and implementation of SWH depends solely on the resources and intentions of the country and its people; the measures described here suggest only a set of guidelines that in specific instances have proved effective. But given the vast solar resource available in most populated regions, and the maturity of SWH technologies available today, there lies a cautious optimism among those who support solar water heating that the technology will continue to improve, and implementation will become increasingly widespread.

REFERENCES

[1] Rosenberg, P. 1993. The Alternative Energy Handbook. The Fairmont Press, Prentice Hall, Lilbyrn, Georgia, p. 19.

[2] Reddy, A. K. N. and J. Goldemberg. 1990. Energy for the Developing World. *Scientific American*. 263/3: 111-118.

[3] Kraushaar, J. J. and R. A. Ristinen. 1993. Energy and Problems of a Technical Society. John Wiley & Sons, Inc., New York, p. 252.

[4] Center for Renewable Energy and Sustainable Technoogy (CREST) web page: http://solstice.crest.org/renewables/re-kiosk/solar/solar-thermal/history/history-2.shtml

[5] Avallone, E. A. and T. B. Baumeister. 1987. Markís Standard Handbook for Mechanical Engineers. McGraw-Hill, New York, p. 9-164.

[6] Flora, A. F., D. E. Brewer and G. Q. de Windt. 1999. Computer Modeling and Simulation of Solar Water Heating Systems with TRNSYS. Institute for

Energy Technology at Marsaxlokk, University of Malta, Malta.

[7] Harrison, J., S. Long and K. Sheinkopf. 1999. 'I'm Getting Free Hot Water.' Renewable Energy World. 2/1: 19-23.

[8] Berkmann, R. 1999. Promoting Solar Thermal: Comparison and Evaluation of Experiences. Renewable Energy World. 2/4: 93-95.

[9] Roditi, D. 1999. No Risks, No Worries: Guaranteeing Solar Results. Renewable Energy World. 2/2: 35-41.

[10] Brealey R. A. and S. C. Myers. 1988. Principles of Corporate Finance. McGraw Hill.

[11] Larson, R. W. et al. December, 1992. Economics of Solar Energy Technologies. *American Solar Energy Society (ASES).*

[12] Roditi, D. February, 1996. Sun in Action. *European Commission Altener Programme.*

Renewable Energy: Trends and Prospects. Edited by S.K. Majumdar, E.W. Miller and A.I. Panah

Chapter Fourteen

GROUND-SOURCE HEAT PUMPS

GARY PHETTEPLACE

US Army Cold Regions Research and Engineering Laboratory
72 Lyme Road
Hanover, NH 03755-1290

INTRODUCTION

Heat normally flows from a warmer medium to a colder one. This basic physical law cannot be reversed without the addition of energy. A heat pump is a device that does so by essentially "pumping" heat up the temperature scale transferring it from a cold material to a warmer one by adding energy, usually in the form of electricity. A heat pump functions by using a refrigerant cycle similar to the household refrigerator. In the heating mode a heat pump removes the heat from a low temperature source, such as the ground or air, and supplies that heat to a higher temperature sink, such as the heated interior of a building. In the cooling mode the process is reversed and the heat is extracted from the cooler inside air and rejected to the warmer outdoor air or other heat sink. For space conditioning of buildings, heat pumps that remove heat from outdoor air in the heating mode and reject it to outdoor air in the cooling mode are common. These are normally called air-source or air-to-air heat pumps. Air-source heat pumps have the disadvantage that the greatest requirement for building heating or cooling is necessarily coincident with the times when the outdoor air is least effective as a heat source or sink. For this reason, in cold climates air-source heat pumps are essentially unfeasible.

The efficiency of any heat pump is inversely proportional to the temperature difference between the conditioned space and the heat source (heating mode) or heat sink (cooling mode), as can be easily shown by a simple thermodynamic analysis (1). For this reason, air-source heat pumps are less efficient and have a lower heating capacity in the heating mode at low outdoor air temperatures. Conversely, air-source heat pumps are also less efficient and have a lower cooling capacity in the cooling mode at high outdoor air temperatures. Ground-source heat pumps (GSHP), however, are not impacted directly by outdoor air temperatures. GSHP use the ground, ground water, or surface water as a heat source or sink. Figure 1 shows some soil temperatures recorded from heat pump demonstration projects conducted at Ft. Polk, LA (2). It is clear from Figure 2 that the soil temperature does not vary significantly over the annual cycle below a depth of about 2 meters (6.5 feet). Although outdoor air temperatures in this climate may range from wintertime lows of 0°C (32°F) (3) or lower to summertime highs around 33°C (95°F) (3), the soil temperature at depths greater than 2 m never falls below about 18°C (64°F) or rises above approximately 25°C (77°F), averaging around 20°C (68°F).

Several different terms have been used to describe GSHP systems. They include: Geothermal Heat Pumps, Earth Source Heat Pumps, Geo-source Heat Pumps, and Geo-exchange systems among others. All systems which embody the ground-source concepts described below have been lumped into the general category of GSHP here. Primarily for marketing reasons, other names, such as those above, will be found in use.

Ground-Source Heat Pump Types

There are three basic types of GSHP: ground-coupled heat pumps (GCHP), ground water heat pumps (GWHP), and surface water heat pumps (SWHP). Each of these is described in more detail below.

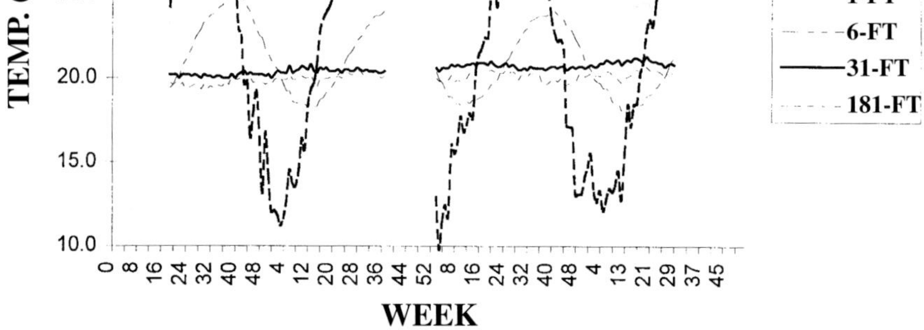

FIGURE 1. Ft. Polk, LA soil temperatures.

Ground-coupled heat pumps (GCHP) use buried closed piping loops which exchange heat with the ground to "couple" the heat pump systems with the ground. This process of heat exchange is accomplished by circulating a fluid, usually water or a water based antifreeze solution, in buried pipe loops. The buried pipe loops consist of either vertical (Figure 2) or horizontal (Figure 3) arrays of buried special purpose "plastic" pipes that form heat exchangers with the ground. Vertical heat exchangers are fabricated simply by drilling boreholes into the ground and inserting a "u-tube" into the borehole. This type of drilling often does not require drilling equipment or techniques as elaborate as those used for normal water well drilling. In addition to low land area requirements, vertical ground coupling has

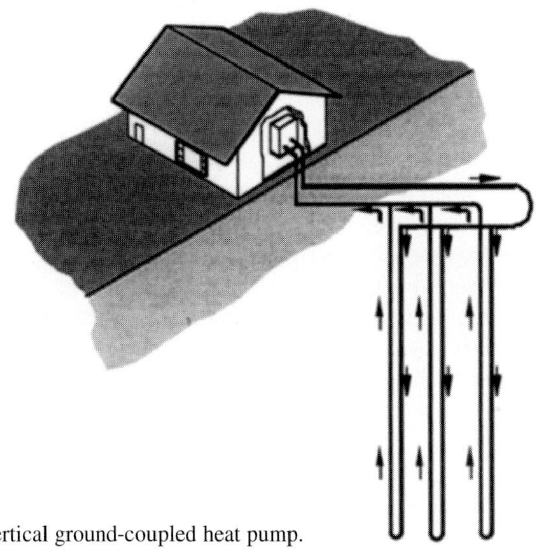

FIGURE 2. Vertical ground-coupled heat pump.

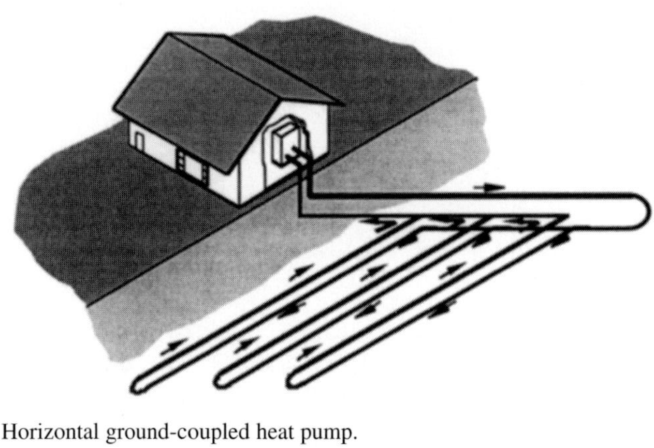

FIGURE 3. Horizontal ground-coupled heat pump.

several other advantages: stable deep soil temperatures with greater potential for heat exchange with ground water, and adaptability to most sites. Among vertical ground-coupling's disadvantages are potentially higher cost, problems in some geological formations, and the need for an experienced driller/installer.

Horizontal heat exchangers may be installed in trenches excavated by trenching machines, backhoes or excavators. Piping may be placed in the trenches either singly or in multiple-pipe arrangements. The primary advantage of horizontal systems is lower cost. This results primarily from fewer requirements for special skills and equipment combined with less uncertainty in subsurface site conditions. The disadvantages of horizontal ground-coupling are its high land area requirernents, its limited potential for heat exchange with the groundwater, and the wider temperature swings of the soil at the typical burial depths.

An alternate method of installing a horizontal heat exchanger is the "slinky" method (Figure 4). When using the slinky method a wide pit is excavated with a bulldozer, excavator, backhoe, or loader. The coils of plastic piping, rather than being uncoiled, are spread out in a spiral pattern resembling a deformed slinky toy. Usually, a fixture is used to obtain uniform coil spacing before the coils are tied to one another to maintain the appropriate spacing. The material excavated is then carefully pushed or placed back over the piping coils. It is also possible to use the slinky method with the coil placed vertically in trenches. Obtaining adequate compaction of the backfill can be difficult for the vertical slinky configuration. Slinky systems have the same advantages as conventional horizontal systems but require less land area and are adaptable to a wider range of construction equipment. In general, the most favorable means of heat exchanger construction will depend on local soil conditions and the type of construction equipment available.

Ground-water heat pump (GWHP) systems are the oldest form of GSHP. These systems extract water from the ground, exchange heat with this water, and

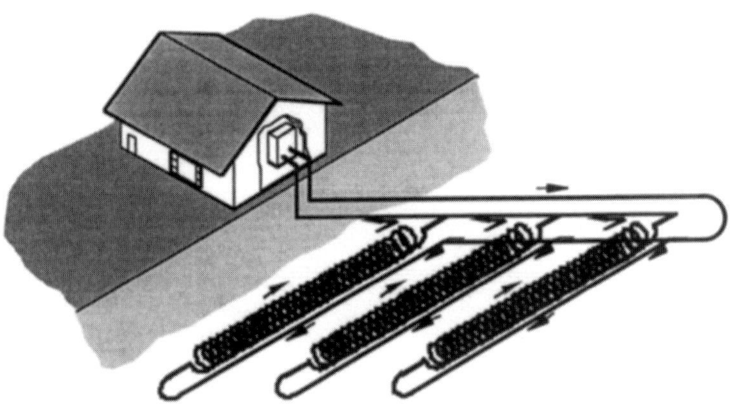

FIGURE 4. Horizontal slinky-coupled heat pump.

then return it to the ground (Figure 5) or dispose of it at the surface (Figure 6) where permitted. These systems have the lowest installed cost in most cases, especially in larger applications. However, their use is limited by the availability of ground water. For larger applications water quality is not as much of an issue as one might imagine, as heat exchangers are used to isolate the heat pumps from

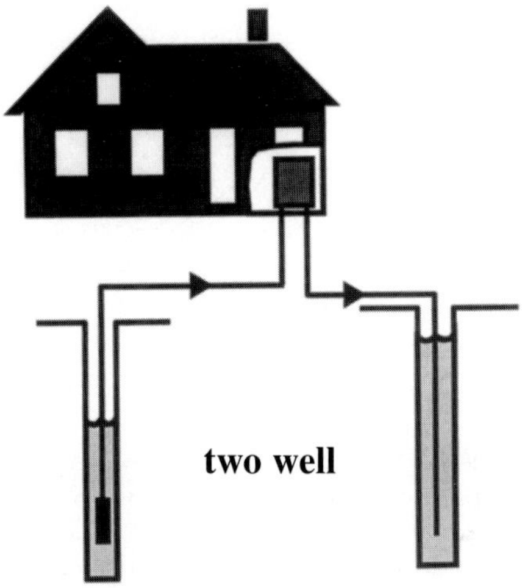

FIGURE 5. Ground-water heat pump with reinjection.

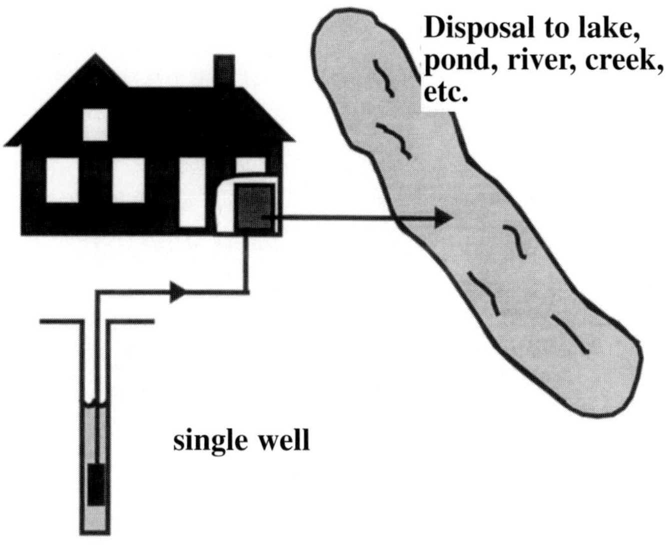

FIGURE 6. Ground-water heat pump with surface disposal.

the ground water. By isolating the heat pumps from the ground water it becomes possible to provide a central heat exchanger for an entire building, normally of the plate and frame type, that may be easily cleaned as necessary. Avoiding contact between the ground water and the atmosphere (i.e. Oxygen) is paramount to eliminating problems with GWHP systems. Failure to recognize this fact and provide an isolating heat exchanger lead to the premature failure of many early GWHP systems.

Surface water heat pump (SWHP) systems extract or discharge heat to surface water bodies. In some instances this may be done directly by piping water from the water body to and from the heat pump (Figure 7), while in other instances it is done by "coupling", again using plastic pipes, submersed in the water body (Figure 8).

A number of variations of these basic types have been proposed or have seen limited use. One example is the so called "standing column well". In this approach ground-water is taken from and returned to the same well with a fraction being extracted and not returned to the well. The extracted water is usually consumed for domestic purposes. Net extraction of water from the well forces new water to flow in, thus allowing "thermal regeneration" of the well.

In commercial buildings air-conditioning loads almost always exceed heating loads, often even for climates in the Northern United States. This presents a special set of problems for ground coupled systems because the heat rejection to the earth far exceeds the extraction. One method that has been used to solve this problem has been the addition of a cooling tower to the system to provide supplemental heat rejection capacity. These systems are often called Hybrid GCHP systems.

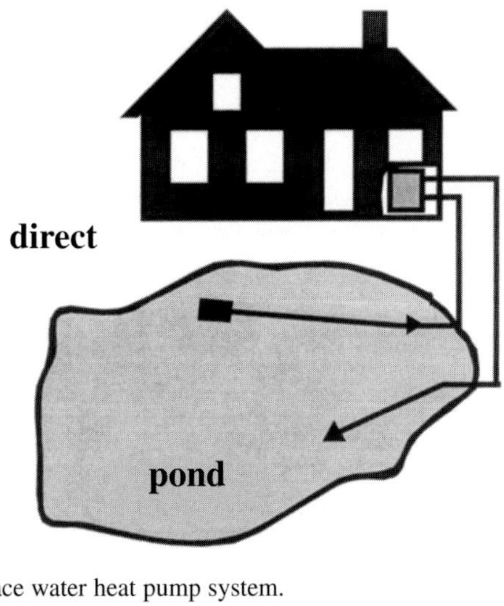

FIGURE 7. Direct surface water heat pump system.

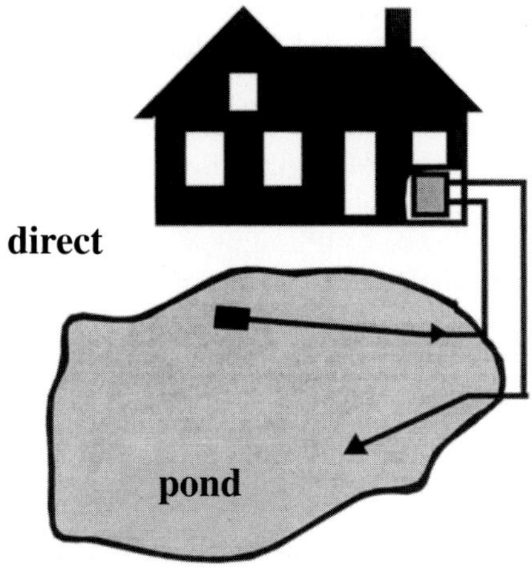

FIGURE 8. Indirect surface water heat pump system.

Another type of ground-source heat pump system is the direct expansion GCHP. The direct expansion GCHP system circulates refrigerant directly to piping buried in the soil rather than using water or an antifreeze solution as an intermediate heat transfer fluid. The direct expansion GCHP has the potential for higher efficiency than a GSHP system using water or an antifreeze solution as an intermediate heat transfer fluid. However, it lacks the prominent advantage of having all the refrigerant contained in the factory fabricated unit like the GSHP. To date, applications of direct expansion GCHP have been very limited and the environmental concerns of such a system have not been fully investigated.

Regardless of the type of ground-source heat pump system used, the heat pump itself will be a single package unit. Larger buildings use multiple heat pump units, one for each zone whose temperature is to be controlled separately. A number of manufacturers in the United States produce these units. They are usually an "extended range" version of a normal water loop heat pump, see (4) for a description of water loop heat pump systems. The extended range allows them to operate over a broader range of entering water temperatures (EWT), as may be encountered in GSHP systems.

The classic parameter which has been used to describe the performance of a heat pump is the coefficient of performance (1). A simple definition of the coefficient of performance (COP) adequate for our purposes is:

$$\text{COP} = \frac{\text{Useful Thermal Effect}}{\text{Purchased Energy Input Rate}}$$

The COP is a dimensionless parameter on which upper bounds can be found using the laws of thermodynamics. It is common in the US air conditioning industry to report the cooling performance in terms of what is called the "energy efficiency ratio", or EER. The EER is not a dimensionless parameter, since the energy input rate is in Watts and the output thermal effect is in Btu/hr. The EER can be readily determined by multiplying the COP by 3.413. While COP's will vary dependant on the quality of the heat pump unit and its operating conditions, in the heating mode COP's in the range of 3 to 4.5 can be expected for GSHP. In the cooling mode performance will generally be lower, *net* COP's of 2 to 3.5 can be expected when the heat load within the conditioned space that is generated by the fans, compressors, and parasitics is subtracted from the gross cooling effect.

Where a demand for hot water heating exists, many GSHP manufacturers have heat pump units which incorporate desuperheaters. A desuperheater recovers heat from the high pressure side of the refrigeration cycle before the refrigerant is routed to the condenser. In the cooling mode this is waste heat that would have been otherwise rejected to the ground. In the heating mode a portion of the heat pump's output is diverted for hot water heating. Since the heat pump normally operates at a coefficient of performance (COP) in the heating mode of about 3 or more, this heat pump generated hot water heating is done at a cost of about 1/3 or less of the cost of direct electric hot water heating. In both the heating and cooling mode, a small, typically 50 watt, circulating pump is needed to circulate the hot water between the existing hot water heater storage tank and the heat pump unit. This is the only cost for operating the desuperheater and thus it is possible to meet some of the hot water demands at a very low cost in the air conditioning mode and at costs that are often very favorable in the heating mode. The circulating pump and controls for the desuperheater are normally built into the heat pump unit.

The only other major component in a GSHP system is the circulating pump or pumps. For residential scale applications this pump is modest in size, typically several hundred watts. For larger scale systems multiple heat pumps are often combined on one or more common circulating loops to the ground-source. In this case circulating pumps are larger and often they are equipped with variable speed drives that adjust pump capacity to meet the demand of only the heat pump units in operation, thus increasing overall system efficiency.

Prospects for GSHP Technology

In the United States, GSHP have found extensive application in the residential marketplace. Successful GSHP applications can be found in many of the different climatological regions of North America. However, the most significant development for residential applications has been in the southern and south central regions of the United States. Widespread use in Canada attests to their economic viability in heating dominated climates with low electrical costs. In contrast, few installations have been made in the northeastern United States where electric costs tend to be higher than average. Development in the western United States has been sporadic to date. Application is primarily dependent on the economics. In turn, the economics of GSHP installations are dependent on a number of factors, the primary ones being: building type, electrical energy costs, geological conditions, ground and surface water availability and quality, and the cost of competing technologies. A life cycle cost (LCC) analysis should be conducted for any application where it's economic feasibility is uncertain.

While most development of GSHP has historically been in residential applications, in the past several years they have been receiving increased acceptance for commercial scale applications. While in residential applications, economics often will not be favorable for GSHP technologies, for commercial scale applications such is not the case. GSHP can successfully compete on a first cost basis with all but the least expensive technologies and when operating costs are considered in a LCC analysis, GSHP will often be the most cost effective alternative.

Because the temperature of the heat sinks/sources used by GSHP are much more stable than the air temperature, GSHP are able to operate at much higher efficiencies during much of the year when compared to their air-source counterparts. This often results in reduced electrical energy usage, and in addition, reduction in peak electrical demand as well. This is particularly true in the heating mode where air-source heat pumps usually require electric resistance heat for supplemental heating at low outdoor air temperatures. Occasionally, in colder climates it may be cost effective to provide supplemental electric resistance heating on a GSHP system as well. However, the amount needed and its usage will usually be modest compared to that required for an air-source heat pump in the same climate.

The potential of GSHP for reduction in electrical demand has caused many of the electric utilities in the United States to embrace GSHP technology. While deregulation of the electric industry in the United States has caused electric utilities to curtail most demand side management programs, such as those that would have given rebates or other incentives for GSHP, some utilities do still offer services such as design assistance for GSHP installations. Electric utility companies, along with others in the GSHP industry provide extensive marketing and some technical outreach activities via the Geothermal Heat Pump Consortium (www.ghpc.org). This organization has existed until present using member support as well as support for the US Department of Energy (DOE). DOE support for this organization has recently been discontinued, but current plans are to continue with member support and seek additional sources of support.

A recent study (5) of operating buildings has shown that those equipped with commercial GSHP systems have 37% lower energy consumption and 20% lower energy costs than those with conventional systems. These results are from a limited sample of building, some with GSHP systems of less than optimal design, but are felt to be reasonably representative by the author. Another study (6) of schools in Lincoln, NB found that GSHP equipped schools used 26% less energy than their similarly situated counterparts. Schools, incidentally, have been a very successfully application of GSHP technology, consult the Geothermal Heat Pump Consortium's web site (www.ghpc.org) for an impressive list of application in various regions of the United States. For residential scale systems, measured results (7) from the world's largest GSHP project to date at Ft. Polk, LA found energy saving of 33% and reductions in peak summertime electrical demand of 43%.

As well as Energy saving, GSHP also offer the potential for reduced maintenance when compared to their air-source counterparts, and most other heating and/or cooling system alternatives, due to the absence of any exposed outdoor equipment and inherent simplicity. The more stable operating temperature/pressure of the compressor in a GSHP also promises extended compressor life when compared to air-source heat pumps. For example, (8) estimates the life expectancy of a commercial water-source heat pump to be over 25% greater (19 years) than its commercial air-source counterpart (15 years). Another major source of problems with air source "split system" heat pumps is leaks in the refrigerant piping connecting the indoor and outdoor units. GSHP do not have any external refrigerant piping, it's all contained within the indoor unit. Because all components of the refrigerant cycle of the GSHP system are contained within the single indoor unit, the refrigeration system is sealed at the factory and will seldom require service, much like a conventional household refrigerator. The amount of refrigerant is also reduced in GSHP systems for this same reason.

Environmentally, GSHP represent a superior alternative in many aspects. Lower refrigerant charges and reduced leakage when compared to air-source systems are two advantages they share over these systems. When compared to fossil fuel burning alternatives, source energy emissions are almost always less because of the improved environmental control possible on large electric power plants when compared to distributed small scale fuel burning equipment. The reduced consumption of energy by the GHSP, since much of the energy consumed comes from the ground-source, also results in greatly reduced emissions when compared to most alternatives.

The limitations on GSHP systems are site specific. At some sites soil conditions may be unfavorable for heat transfer causing pipe loop lengths for groundcoupled heat pumps to be excessive. In areas where land is limited, horizontal systems are probably not feasible, as is the case for most commercial scale systems. Where drilling conditions are unfavorable, such as bedrock, vertical ground-coupling may be cost prohibitive. In areas where ground water is readily available and is of satisfactory quality it may be possible to use it directly with a ground water heat pump. In such cases, particularly in larger applications, it is almost always more economical to use the groundwater directly in a groundwater heat pump system. In all cases, one must ensure that the application is consistent with the prevailing environmental regulations.

Issues

A few issues regarding the application of GSHP technology are not yet fully resolved. The basic concept of a GSHP is not new, yet only in the past few years have they become popular. While the design and installation infrastructure is fully in place in some parts of the United States, in many areas the necessary infrastructure is underdeveloped. Thus, it may be difficult to find qualified designers and installers in a particular area. For larger projects involving many individual GSHP systems or large commercial installations, it's quite feasible and often advisable to seek experienced designers and installers from outside the local area when necessary. Many installers of GSHP have shown themselves to be competitive even at locations significantly distant from their normal operating area. Design guidance has been evolving very rapidly in this field. The most recent comprehensive design guidance can be found in (9) and (10). The selection of competent designers and installers cannot be overemphasized.

Another one of the major unresolved issues is the environmental regulation of GSHP. While GSHP systems offer tremendous environmental benefits they also have potential for ground water contamination if not properly installed. In GCHP systems that use antifreeze mixtures as circulating fluids, the antifreezes used are those normally considered to be less harmful to the environment. ASHRAE funded a study (1 1) of antifreeze compounds used in the industry that provides a very good summary of the materials used and the issues surrounding their use.

Vertical GCHP have an additional environmental issue regarding the vertical boreholes. In order to prevent surface water from contaminating ground water aquifers, or cross contamination of aquifers, it is necessary to backfill portions of the borehole with an impermeable material normally called "grout". Initially the industry and some state regulators, endorsed complete grouting of the borehole as a conservative practice. However, in was soon recognized that the low thermal conductivity of bentonite based grouts had a very significant adverse thermal impact on heat transfer from the ground loops. Standard bentonite based grouts, the most popular grouting materials, typically have thermal conductivity values of 0.73 W/m-°C (0.42 Btu/hr-ft°F) according to (9). With few exceptions, the thermal conductivity of the native rock and soil formations are much higher than standard bentonite grout thermal conductivity (9). Thus, grouting around the vertical u-tube heat exchanger with standard bentonite based grouts is equivalent to insulating the heat exchanger from it's heat source/sink. To solve this problem, grouts with enhanced thermal conductivity have been developed. The two principal thermally enhanced grouts are cementitious mixtures and bentonite/quartzite sand mixtures, the later being commercially available as a premixed product.

Some recent experimental work (12) has confirmed the negative impact of grout on borehole heat transfer. Under heat rejection loading, average water temperature was nearly 6°C (11°F) higher for a 16.5-cm (6.5-in.) diameter borehole backfilled with standard bentonite grout when compared to a 12.1-cm (4.75-in.)

diameter borehole backfilled with thermally enhanced bentonite grout. Using fine sand as backfill in a 16.5-cm (6.5-in.) diameter borehole lowered the average water temperature over 8°C (14°F) when compared with the same diameter bore backfilled with standard bentonite grout. For a typical system (12) with a 16.5-cm (6.5-in.) diameter borehole, the use of standard bentonite grout would increase the required bore length by 49% over fine sand backfill in the same borehole. By using thermally enhanced grout in a smaller 12.1 -cm (4.75-in.) borehole, the bore length is increased by only 10% over fine sand backfill in the larger 16.5-cm- (6.5-in.-) diameter borehole. Thus, the results of this study (12) suggest three steps that may be taken to reduce the impact of grout on system performance. First, reduce the amount of grout used to the bare minimum. Sand or cuttings may be used where allowed, but take care to ensure that the entire interstitial space between the piping and the borehole diameter is filled. When grout must be used, use thermally enhanced grout. Finally, reduce the borehole diameter as much as possible to mitigate the effects of whatever grout or backfill is used.

The overall outlook for expanded application of GSHP technology is very favorable. The technology itself is well established, yet as noted earlier the infrastructure remains underdeveloped in some areas of the United States, and the world at large. There will be continued refinements to the application of these systems as the technology is not fully mature at this point. However, it is the author's opinion that one should not expect major breakthroughs that will significantly change the current configurations of systems and application of the technology. Growth can be expected to be significant in the commercial sector where GSHP systems offer all their inherent advantages discussed above while remaining cost competitive with the conventional alternatives. In residential applications growth will be slower due to the initial cost premium for GSHP when compared to the alternatives.

Sources of Additional Information

Aside from the references cited here, there are a number of other sources of information available to those interested in GSHP systems. The previously mentioned Geothermal Heat Pump Consortium (GHPC), located in Washington, DC is one such source. They may be contacted at (888) ALL-4GEO or visit their Website at http://www.ghpc.org. The GHPC is a public/private venture funded by DOE, electric utilities, and manufacturers of heat pumps and allied equipment used in the industry. Their primary focus is marketing of the technology.

Another source of information is the International Ground Source Heat Pump Association (IGSHPA) located at Oklahoma State University. They can be reached at (800) 626-4747 or via their Website at http://www.IGSHPA.okstate.edu. IGSHPA provides various training opportunities, primarily to systems installers, as well as design software and a number of publications.

The Geo-Heat Center, funded by DOE and located at the Oregon Institute of

Technology, also provides information and assistance. They can be reached at (541) 885-1750 or via their Website at http://www.oit.edu/~geoheat. The University of Alabama and the Geo-Heat Center have recently initiated the publication of a quarterly newsletter (*Outside the Loop*) with a focus on information of interest to engineers and contractors, this publication is available via the referenced website.

REFERENCES

1. Reynolds, William C. and Herry C. Perkins. 1977. Engineering Thermodynamics, Pp 287-289, McGraw-Hill Book Co., New York, NY.
2. Phetteplace, G., H. Ueda, and D. Carbee. April 1992. Performance of Ground-Coupled Heat Pumps in Military Family Housing Units, Proceedings of the 1992 ASME International Solar Energy Conference, Maui, Hawaii, 4-8 April, 1992, pages 377-383.
3. Department of the Army. July 1978. Engineering Weather Data, Technical Manual TM 5-785.
4. ASHRAE. 1996. 1996 ASHRAE Handbook, Systems and Equipment. Pp. 8.15-8.19. American Society of Heating, Refrigerating, and Air-Conditioning Engineers (ASHRAE), Atlanta, GA.
5. ASHRAE. 1998. Operating Experiences with Commercial Ground-Source Heat Pump Systems. Pp. viii. American Society of Heating, Refrigerating, and Air-Conditioning Engineers (ASHRAE), Atlanta, GA.
6. Shonder, J.A., M.A. Martin, T.R. Sharp, D.J. Durfee, and P.J. Hughes. 1999. Benchmark for Performance: Geothermal Applications in Lincoln Public Schools. Proceeding of the American Society of Heating, Refrigerating, and AirConditioning Engineers (ASHRAE), V. 105, Pt. 2, ASHRAE, Atlanta, GA.
7. Oak Ridge National Laboratory. Undated. Big Savings from the World's Largest Installation of Geothermal Heat Pumps. Pp 1. Oak Ridge National Laboratory, Oak Ridge, TN.
8. ASHRAE. 1999. 1999 ASHRAE Handbook, HVAC Applications. Pp. 35.3. American Society of Heating, Refrigerating, and Air-Conditioning Engineers (ASHRAE), Atlanta, GA.
9. ASHRAE. 1997. Ground Source Heat Purnps - Design of Geothermal Systems for Commercial and Institutional Buildings. American Society of Heating, Refrigerating, and Air Conditioning Engineers (ASHRAE), Atlanta, Ga.
10. ASHRAE. 1999. 1999 ASHRAE Handbook, HVAC Applications. Pp. 31.1431.23. American Society of Heating, Refrigerating, and Air-Conditioning Engineers (ASHRAE), Atlanta, GA.
11. Heinonen, E.W., R.E. Tapscott, M.W. Wildin, and A.N. Beall. 1997. Assessment of Antifreeze Solutions for Ground-Source Heat Pump Systems. Report 908RP, American Society of Heating, Refrigerating, and AirConditioning Engineers (ASHRAE), Atlanta, Ga.

12. Spilker, E. H. 1998. Ground-Coupled Heat Pump Loop Design Using Thermal Conductivity Testing and the Effect of Different Backfill Materials on Vertical Bore Length, Proceedings of the American Society of Heating, Refrigerating, and Air-Conditioning Engineers (ASHRAE), Atlanta, GA.

Renewable Energy: Trends and Prospects. Edited by S.K. Majumdar, E.W. Miller and A.I. Panah

Chapter Fifteen

GEOTHERMAL ENERGY RESOURCES

PHILLIP M. WRIGHT[1]
and
ASSAD I. PANAH[2]

[1]Idaho National Engineering and Environmental Laboratory
P.O. Box 1625
2525 Fremont Avenue
Idaho Falls, ID 83415-2214
and
[2]Department of Geology and Environmental Science
University of Pittsburgh at Bradford
300 Campus Drive
Bradford, Pennsylvania 16701-2898

INTRODUCTION

World population is expected to double to 10 billion people by the year 2050. In the same time frame, world economic output is expected to increase by a factor of three to five and a 1.5- to 3 fold increase in energy use is projected (Nakicenovic et al., 1998). The long-predicted decline in availability of fossil fuels has not materialized, and expectations now are that economic development over the next century will not be constrained by their availability. However, the capacity of the earth to absorb higher levels of pollution will become an ever more limiting factor in the use of fossil fuels. As pollution becomes more of a problem, energy quality considerations will become increasingly important to people – environmental cleanliness and efficiency. Because of their environmental and other advantages, renewable energy resources will supply a growing portion of our energy use.

Geothermal energy, the natural heat of the earth, is a global resource that can be used partially to replace conventional fossil and fissile fuel sources. The enormous potential of this clean, reliable energy source is just beginning to be developed with the construction of geothermally powered electrical generating stations and with increased use of geothermal heat in district heating systems, crop-drying plants, and greenhouses throughout the world.

In this chapter, we will present a discussion of the nature of geothermal energy resources, how they form and are classified, their renewability, environmental aspects of geothermal energy utilization, and the status of geothermal technology. A companion chapter in this volume (*Geotherrnal Energy for Heating and Electrical Generation*) discusses the use of geothermal resources for electrical power generation and a myriad of direct uses of geothermal heat.

NATURE OF GEOTHERMAL RESOURCES

Geothermal energy is heat from inside the earth. Since temperature universally increases with increasing depth in the earth, heat flows continuously outward to the surface and is lost to space by radiation in the infrared band. Highly viscous or partially molten rock at temperatures between 650°C and 1,200 °C exists everywhere beneath the earth's surface at depths of 80 km to 100 km, and the temperature at the earth's center, nearly 6,400 km deep, is estimated to be 4,000 °C to 6,000 °C. The measured mean value of outward surface heat flow is 82 milliwatts/m^2, and since the surface area of the earth is 5.1 x 10^{14} m^2, the continual rate of heat loss to space is about 42 billion thermal kilowatts. This enormous heat flux is the driving force behind such large-scale geologic processes as plate tectonics, volcanism, mountain building, and earthquakes. However, since the normal heat flux reaches the surface at ambient temperature, it is too low in grade to be captured and used. Therefore, we must seek places where the earth's heat is concentrated at shallow depths (0.3 km to ~5 km) in order to encounter temperatures high enough to be useful. Fortunately, geologic processes do concentrate earth's heat.

Igneous Processes Associated with Geothermal Systems

Magma (molten rock) bodies are generated in the earth at depths of 20 km to 100 km or more, and move upward by buoyancy along lines of weakness in the earth's crust. Geologic processes involving molten rock are known as igneous processes. A volcanic eruption occurs if the magma reaches the surface, but not all magma bodies reach the surface. Many of them cool and solidify at depths of 2 to 10 km, with the cooling process taking a few tens of thousands to a few million years, depending on the size of the magma body, whether or not cooling is aided by convection of heated ground waters, and other factors. In areas where volcanic activity and subsurface magmatic intrusion have taken place during the past million or so years, the host rock and the intrusive body may contain very large quantities of heat. It is these concentrations of the earth's heat that we seek in exploring for geothermal resources. We also look for hot springs whose waters have come to the surface after circulation to depths of a few kilometers, where they are heated in the normal temperature gradient.

In a revolution of geologic thinking that began in the 1960s and continues today, it was discovered that the earth's lithosphere (the solid upper layer, which includes the crust) is divided into about 12 large, rigid plates. These plates move relative to one another in response to the slow convective motion of the rock of the earth's mantle, the layer lying below the lithoshpere. Mantle convection is driven by the flow of heat out of the earth's core toward the surface. Where mantle rock rises up, the crust splits apart along long, globe-encircling fissures that occur mostly in the ocean basins. Magma enters the widening fissures to form new oceanic crust. The volcanism associated with this injection of magma occurs mostly in the deep ocean basins, away from our ready view. Submersibles have photographed this submarine volcanism and the accompanying "ocean smokers", jets of high-temperature, mineral-laden water that results from hydrothermal convection powered by the intense heat in the newly emplaced, porous volcanic rocks.

Since the earth is not expanding, we can reason that if new crust is created at spreading centers, older crust must be destroyed at another location. This, indeed, proves to be the case. In areas where the rigid plates, moving away from spreading centers, meet an obstacle such as a continent, the oceanic plate is usually thrust under the continental plate in a process known as "subduction." The downgoing, subducted plate descends into regions of increasing temperature. Finally, at a depth of ~80 km or more, the upper parts of the subducted plate begin to melt, and the resulting magma bodies rise upward by buoyancy through the mantle and overlying crust, some reaching levels of 1 to 6 km below the surface. These magmas bring up enormous quantities of heat. Some of them breach the surface, resulting in continental volcanism. The processes discussed above are illustrated in Figure 1.

Most of the earth's magmatic and volcanic activity occurs along the principle plate-tectonic features, the mid-ocean ridges where crustal plates spread apart, and the subduction zones where one crustal plate slides slowly beneath another crustal plate. Both the mid-ocean spreading centers and the associated subduction zones form globe-encircling systems, in places broken and offset by transform faults. The so-called "Pacific Ring of Fire" that surrounds the Pacific ocean is formed of volcanoes resulting from intrusion of magma above subduction zones. The Andes mountains in South America, the volcanoes of Central America and Mexico, the Cascades mountains in the western U.S. and Canada, the Aleutian Island chain, the volcanoes on Kamchatka, and those in Japan, the Philippines, Indonesia, and elsewhere in Asia, as well as the volcanic islands in the South Pacific and New Zealand, are all part of this ring of fire. The volcanoes of east Africa and the Mediterranean also result from subduction processes.

Other volcanic areas, such as Iceland, sit astride midocean ridges, where strings of mostly submarine volcanoes create new sea-bed floor, occasionally extending above the ocean surface.

At certain locations, single, hot upwelling plumes of plastically flowing rock occur in the earth's mantle, and the heat they bring up causes volcanoes within crustal plates. A trail of dead and dying volcanoes will mark the movement of the crustal plate across the hot spot. Such a plume is believed to have caused the Hawaiian Island chain. Active volcanism is currently taking place on the Big Island and in a newly forming subset volcano southeast of the Big Island. Along the island chain to Maui, Oahu, and Kauai the volcanic rocks get progressively older, those in Kauai being typically ~2 million years old. This chain actually continues northwest across the Pacific as underwater volcanic mountains to Midway Island, where it makes a bend and then continues in a fairly straight line to the Kamchatka Peninsula. Such has been the movement of the Pacific plate across the mantle hot spot during the last ~20 million years.

Volcanic and intrusive rocks associated with magma coming directly from the mantle are generally basaltic in mineral composition. This is the case on Hawaii and Iceland, to name only two examples. Basaltic rocks form black lavas and are relatively low in silica, relatively high in iron and manganese, and melt at temperatures ~1,200 °C. Due to this high melting temperature and the low silica

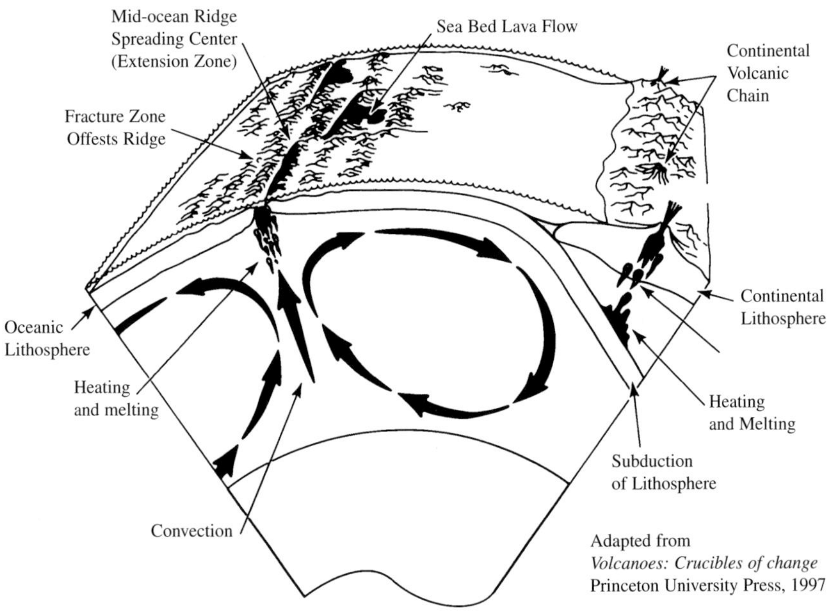

FIGURE 1. Earth Dynamics – Plate Tectonic Processes.

content, basaltic magmas have low viscosity, and can move up narrow channels such as faults, to spew out on the surface. As a result of their fluidity, eruptions of basaltic magma are usually relatively quiescent. By contrast, magmas that are generated in subduction zones and rise through the continental crust, are lighter in color, higher in silica content and lower in iron and manganese. They typically melt at temperatures ~650 °C. Due to their lower melting temperature and high silica content, these so-called "andesitic magmas" are highly viscous and are not able to flow in narrow channels. They move upward only in large bodies. The presence of an andesitic igneous rock can, in general, be taken as evidence of a large intrusive body in the subsurface. Such large intrusions bring up enormous quantities of heat. It is the andesitic volcanoes that display the explosive eruptions as seen in Mt. St. Helens in Washington, and Mt. Pinatubo in the Philippines, to name two recent examples.

Geothermal resources are associated with all types of intrusive and volcanic activity, but the largest resources are often associated with andesites. The basaltic volcanism in Iceland has created hydrothermal systems that are used to provide hot water to heat about 87 % of the homes and buildings, and also to generate electrical power. Hydrothermal systems associated with andesitic volcanoes in the Philippines furnish hot water and steam to generate 26 % of that country's electricity. Clearly, geothermal resources can make a significant contribution to a nation's energy requirements.

Geothermal Resource Classification

There are several systems used to classify geothermal resources based on their origin, temperature, and chemistry. One system based on origin is the following:

- **Magmatic resources** are found in areas of active geologic processes such as plate tectonic spreading centers, subduction zones, and transform faulting, and in areas of mantle hot spots (Hawaii, Yellowstone). They consist of molten magma and hot rocks that have crystallized from magma in the last ~1 million years. Such rocks may produce (1) high-temperature hydrothermal convection systems, and (2) high-temperature rock whose natural permeability is too low to allow production of geothermal fluids at rates sufficient to operate a commercial power plant. This later category contains the majority of the heat as a rule, but can not be exploited economically using today's technology.

- **Crustal rocks** (both continental and oceanic) are hot at sufficient depth everywhere on earth, since temperature increases with depth. The geothermal gradient (rate of increase in temperature with depth) normally varies from ~10 °C/km to more than 50 °C/km, but within some geothermal fields it may exceed ten times these values. In much of the world, the average gradient is 30 °C/km to 40 °C/km, and temperatures of 150 °C, suitable for electric-power generation, are found at depths of ~3.5 km, even outside of areas of volcanic activity. In regions where there has been geologic stability for long periods, such as the flat, granitic shield areas found on most continents, the average gradient is in the range 10 °C/km to 20 °C/km, and temperatures of 150 °C would normally be reached at depths of 6 km or more - deeper than economic drilling depth given today's level of drilling technology and current energy prices. Thermal energy in deep crustal rocks forms an enormous resource that awaits use in the future.

- **Thermal aquifers** are aquifers filled with warm and hot water. These aquifers are a common and widespread occurrence throughout the world. Examples include the Madison and Dakota aquifers in the north-central U. S., the aquifers in the Paris basin in France, widespread aquifers in the Beijing and Tianjin area of China, and the extensive aquifers in Poland, Hungary, and much of the rest of the Middle East. In some cases, a hot spring might be caused by water circulating deeply along a fault and returning to the surface. Thermal aquifer -resources could be used on a much, much broader scale than they are at as present to provide heating for homes and buildings, industrial process heat, aquaculture, greenhouse heating, and a wide variety of other uses. Most thermal aquifers are too low in temperature to contemplate generating electricity, but there are a few exceptions.

- Many petroleum basins worldwide contain horizons in which the fluid is **geopressured** - it bears part of the weight of the rock column (lithostatic load) and is, thus, substantially above hydrostatic pressure. The water in these horizons generally contains dissolved methane and is often hot from deep burial depth. Thus, geopressured geothermal resources contain three types of energy; (1) thermal energy, (2) mechanical energy from their high pressure, and (3) natural gas. In the Gulf Coast states of Texas and Louisiana, geopressured geothermal resources having temperatures in the range ~100 °C to ~130 °C are common at depths below ~4 km. Generation of electricity has been demonstrated from these resources using a small plant, while simultaneously recovering natural gas, but the demonstration project did not indicate economic feasibility at today's energy prices.

In addition to the above genetic classification, geothermal resources are also divided by their temperature:

- High-temperature resources: T > 150 °C
- Moderate-temperature resources: 90 °C < T <= 150 °C
- Low-temperature resources: T <= 90 °C.

Temperatures as high as ~360 °C have been observed in hydrothermal resources in the Salton Sea geothermal field in the Imperial Valley of Southern

California and in the geothermal field at Puna, Hawaii, among several other locations. The Japanese have measured a temperature >500 °C in impermeable granitic rock (age < 1 million years) in a test hole 3.7 km deep in the Kakkonda area, located in the northern part of Honshu (Ikeuchi, et al., 1996). From these maximum temperatures, there is a continuum of geothermal resource temperatures down to ground-water temperature. As one might expect, many more low-temperature resources are observed than high-temperature resources. In fact, the number of lower temperature resources increases exponentially with decreasing temperature. Similar distributions are found elsewhere in nature. For example, if the grade of copper ore in a deposit is plotted against the number of tons having that grade, the amount of copper ore increases exponentially with decreasing grade down to the background content of average rocks.

Most geothermal electrical generating stations today are built on resources that produce fluids in the high-temperature range, 150 °C or higher. Binary-cycle power plants (see Geothermal Energy for Heating and Electrical Generation, this volume) operate on lower temperatures, in the range 100 °C to 200 °C. Direct uses of geothermal heat for heating buildings, greenhouses, aquaculture, bathing, and many other uses generally operate on resources in the low- and moderate-temperature range.

There are other classifications of geothermal resources as well. Hydrothermal resources are those in which hot water or steam occurs that can be brought to the surface in wells. **Hydrothermal** resources are found associated with magmatic

TABLE 1.
Estimate of World Geothermal Energy Resource Base[1]

GEOLOGIC REGIME	U.S. RESOURCES Joules bbl oil equivalent	WORLD CONTINENTAL RESOURCES[1] Joules bbl oil equivalent	TECHNOLOGY NEEDED FOR DEVELOPMENT
Magmatic Systems (surface to 10 km)	1×10^{24} J $160,000 \times 10^9$ bbl	15×10^{24} J $2,400,000 \times 10^9$ bbl	Hydrothermal (part) EGS[2] (part)
Crustal Heat[3] (3 km to 10 km)	14×10^{24} J $2,300,000 \times 10^9$ bbl	490×10^{24} J $79,000,000 \times 10^9$ bbl	EGS
Thermal Aquifers	55×10^{18} J 9×10^9 bbl	810×10^{18} J 130×10^9 bbl 1	Hydrothermal
Geopressured Basins (surface to 6.9 km)	0.17×10^{24} J $28,000 \times 10^9$ bbl	2.5×10^{24} J $410,000 \times 10^9$ bbl	Oil Field plus Hydrothermal
Total Oil Reserves[4] (for comparison)	890×10^9 bbl	$5,300 \times 10^9$ bbl	

[1] Excluding Antarctica. Only a rough estimate is possible.
[2] EGS = Engineered or Enhanced Geothermal Systems (aka Hot Dry Rock).
[3] Excluding heat in magmatic systems, thermal aquifers, and geopressured basins.
[4] Includes crude oil, heavy oil, tar sands, and oil shale (various sources).

systems, thermal aquifers, and in geopressured basins. This is currently the most important geothermal resource type, and is discussed in some detail below. There are also **hot dry rock** and **hot wet rock** resources which contain an insufficient quantity of water and are too low in producibility to be economically usable.

CONCEPTUAL MODELS OF HYDROTHERMAL RESERVOIRS

Convective hydrothermal resources are geothermal resources in which the earth's heat is carried upward by convective circulation of naturally occurring hot water or steam. Underlying some high-temperature convective hydrothermal resources is presumably an intrusion of still-molten or recently solidified rock. Other convective resources result from the circulation of water down fractures to depths where the rock temperature is elevated, even in the absence of an intrusion, with heating and buoyant transport of the water to the surface.

Because of geological and geochemical diversity among hydrothermal resources, each resource is unique. We devise conceptual models to help understand them and to guide exploration and development activities. Actual resources in nature are always much more complex than our conceptual models of them. The models presented below will help illustrate a few of the general types into which actual hydrothermal systems can be classified. For a more complete discussion of the nature of geothermal systems, see Wright and Culver (1998).

Water-Dominated Systems

Figure 2 illustrates a high-temperature, hot-water dominated geothermal system. Models for such systems have been discussed in papers by White et al. (1971), Mahon et al. (1980), Henley and Ellis (1983), and Norton (1984), among others. The heat source is typically molten or recently solidified rock lying at a depth of perhaps 3 to 10 km. Groundwater circulates downward in open fractures and removes heat from these deep, hot rocks as it rises buoyantly and is replaced by cool recharge water moving in from the sides. Water is an ideal heat-transfer medium because it has a high specific heat (amount of heat absorbed per degree increase in temperature) and a high heat of vaporization (amount of heat needed to turn liquid water at the boiling point into steam at the same temperature). Rapid convection produces uniform temperatures over large volumes of the reservoir. There is typically an upflow zone at the center of each convection cell, an outflow zone or plume of heated water moving laterally away from the center of the system, and a recharge zone where water moves downward. The upflow zone may contain local areas of boiling and a two-phase region, containing both water and steam may exist, but the pressure is controlled by the water phase. Escape of hot fluids is often minimized by a near-surface sealed zone or cap-rock formed by precipitation of minerals in fractures and pore spaces. Surface manifestations include hot springs, fumaroles, geysers, travertine or tufa

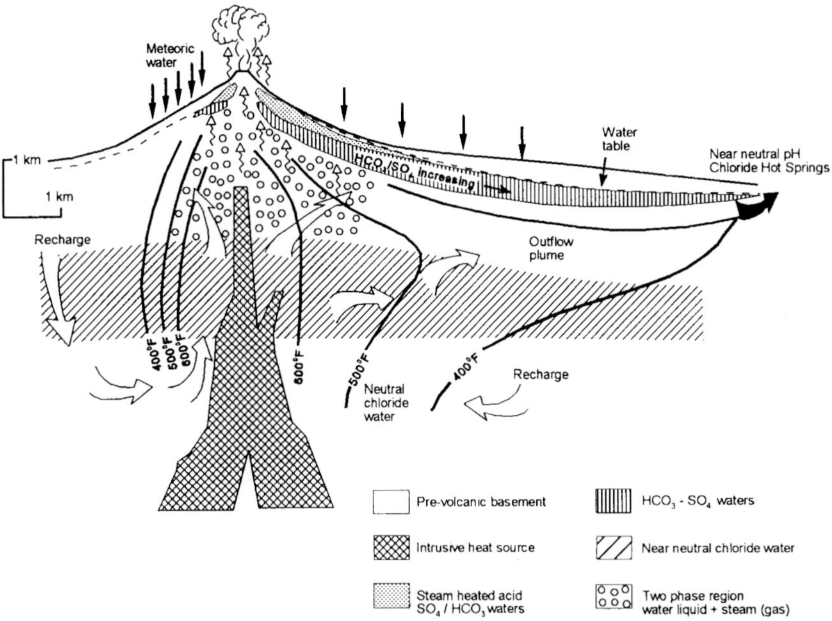

FIGURE 2. Conceptual model of a hydrothermal system in volcanic terrane.

deposits, chemically altered rocks, or alternatively, no surface manifestation at all. If there are no surface manifestations, discovery is difficult and requires sophisticated geology, geophysics, geochemistry and hydrology.

Chemical isotopic studies of hydrothermal fluids show that the bulk of the water and steam is derived from meteoric water (rain or snow), with the exception of those few systems where the fluids are derived from seawater or connate brines (Craig, 1963). Only a small percentage of the water comes from the intrusive rocks at depth. As the fluids move through the reservoir rocks, the compositions of both the fluids and the rocks are modified by the dissolution of primary, preexisting minerals and the precipitation of secondary minerals. The entire hydrothermal convection system (rocks and fluids) is, in fact, a large-scale chemical reactor with interactions that are not completely understood today. The deep convecting waters are generally enriched in sodium chloride and depleted in magnesium. Salinities of high-temperature geothermal fluids range from less than 10,000 ppm total dissolved solids in some volcanic systems to over 250,000 ppm total dissolved solids in basin environments such as the Salton Sea, California (Helgeson, 1968; Ellis and Mahon, 1977).

The pressure and temperature in most high-temperature hydrothermal convection systems lie near the curve of boiling point versus depth for saline water (Figure 3), and sporadic, local boiling occurs in many systems. Because

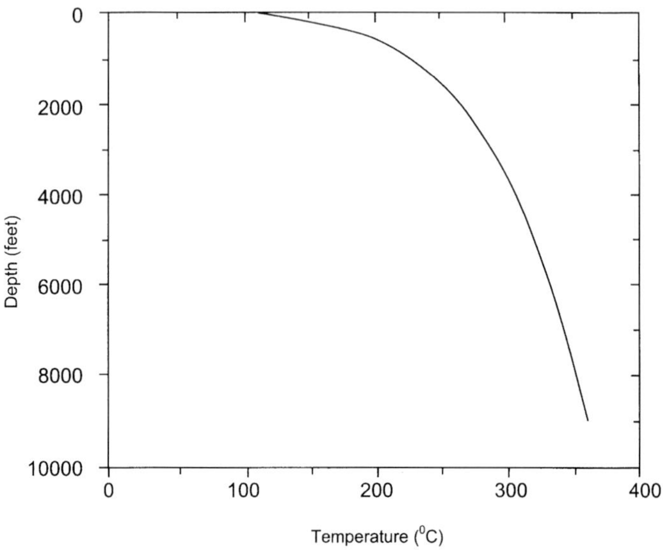

FIGURE 3. Boiling point versus depth in water-dominated geothermal systems.

boiling concentrates acidic gases (carbon dioxide and hydrogen sulfide) in the steam, the oxygenated meteoric water overlying a boiling reservoir is heated and acidified, often becoming bicarbonate/sulfate in nature. The resulting acidic waters interact with the near-surface rocks to form hydrothermal minerals, most prominantly clay minerals, that can be used to help locate zones of subsurface boiling.

Vapor-Dominated Systems

Water-dominated hydrothermal systems are by far the most common kind, but in a few systems, there is steam in the ground. Figure 4 shows a conceptual model of a hydrothermal system where steam pervades the rock and is the pressure-controlling fluid phase, a so-called "vapor-dominated" geothermal system (White et al., 1971). Convection of deep water brings heat upward to a level where boiling can take place. Boiling removes the latent heat of vaporization, thereby cooling the rock and water at the boiling water table, and allowing more heat to rise from depth. Steam generated by boiling moves upward through fractures and may be superheated by the hot surrounding rock. At the top and sides of the system, heat is lost from the vapor and condensation results, with the condensed water moving downward to be vaporized again. If an open fracture penetrates to the surface, steam may vent or may heat shallow ground water to boiling. Within the vapor-filled part of the reservoir, temperature is nearly uniform due to rapid upward steam flux. Pressure within the reservoir is controlled by the vapor phase and, therefore, increases slowly with depth. Because the rocks surrounding the steam reservoir contain ground water under hydrostatic pressure, a large horizontal pressure difference exists between the steam in the reservoir and the water in adjacent rocks, with the steam reservoir being underpressured with respect to hydrostatic pressure. We postulate that the permeability at the boundaries of the reservoir must be low, otherwise water in the surrounding rocks would inundate the steam reservoir. Low peripheral permability could be a result of pre-existing geological features such as impermeable beds or faults, or it could have formed by deposition of minerals in the fractures and pores at the sides of the resevior, making a sealed zone.

The formation of a vapor-dominated system appears to require the venting of steam at a rate in excess of water recharge to prevent flooding of the reservoir (White et al., 1971). In fact, vapor-dominated systems are believed to form from preexisting water-dominated systems by boiling down. Thermal evolution studies have shown this to be the case at The Geysers geothermal area, about 80 miles north of San Francisco, California. A hydrothermal system that is basically water dominated can have one or more natural zones that are two phase (containing both liquid and vapor) or are vapor dominated, and local steam zones can even result from the production of water from a well if local water recharge is insufficient to keep pace with water production.

At The Geysers, steam is produced from depths of 1 km to 3 km and is used to run turbine engines which turn electrical generators. The Geysers is the largest geothermal electrical producing area in the world, sustaining more than 1,200 megawatts of power generation. Other producing vapor-dominated resources occur at Larderello and Monte Amiata, Italy, and at Matsukawa, Japan.

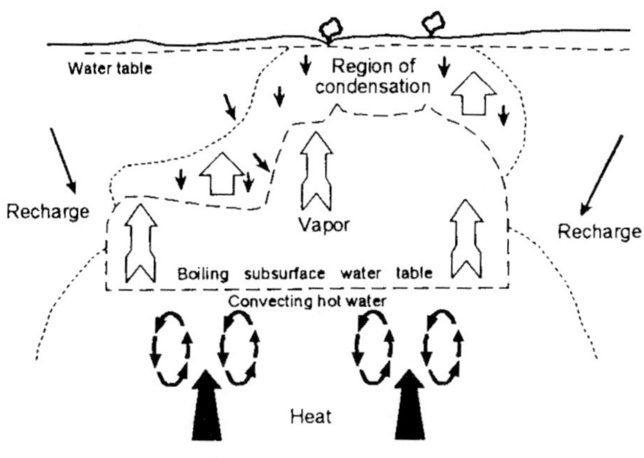

FIGURE 4. Conceptual model of a vapor-dominated resource.

The Hydrothermal Reservoir

At this point, it is desirable to discuss the term reservoir. The reservoir is the volume containing hydrothermal fluids at a useful temperature. The porosity of the reservoir rocks determines the total amount of fluid available, whereas the permeability determines the rate at which fluid can be extracted. However, one must *not* envisage a large bathtub of hot water that can be tapped at any handy location. Open spaces filled with water or steam in the interconnected faults, fractures, and pores forming the reservoir usually constitute only 2 % to 5 % of the rock volume. Both porosity and permeability vary over wide ranges at different points in the reservoir. A typical well will encounter tight rocks over much of its length, with steam or hot water inflow mainly along a few open fractures or over a restricted stratigraphic interval. Apertures of producing fractures are sometimes as small as a few millimeters, but in other cases they reach 1 centimeter or more. Areas where different fracture or fault sets intersect, or where fractures intersect permeable rock units may be especially favorable for the production of large volumes of fluid. The longevity of a well depends upon how completely the producing zones are connected to the local and reservoir-wide network of porosity. If this inter-zone permeability is poor, the local open spaces are drained quickly and fluid production drops. However, if the well intersects a thorough-going geologic structure, such as a major fault or fracture, the local producing volume around the well is recharged continuously, and fluid production can be maintained for many years.

Most of the thermal energy in hydrothermal reservoirs actually resides in the rock, not in the water or steam. In liquid-dominated reservoirs, about 85 % of the total heat in the reservoir area resides in the rock, and about 15 % resides in the water. In vapor-dominated reservoirs, only about 5 % of the thermal energy resides in the steam.

Virtually all of the geothermal electrical-generation industry's exploration effort is presently directed at locating vapor- or water-dominated hydrothermal systems having temperatures above 300°F (150 °C). Some of the highest grade resources are capable of commercial- electrical power generation today, and the majority of the growth in geothermal energy production is expected to come from hydrothermal resources until well into the present century. There is also great potential worldwide for development of direct-heat uses at or near the sites where low- and moderate-temperature hydrothermal resources occur. Of course, geothermal heat pumps (see Geothermal Energy for Heating and Electrical Generation, this volume) can be used virtually anywhere, for their operation does not depend on having enhanced temperature such as is found in the types of geothermal areas we have been discussing.

Low- and Moderate-temperature Geothermal Resources

Geothermal resources suitable for direct-heat application can be of any usable temperature, but resources whose temperature is above 300 °F (150 °C) are usually used to produce electricity. The outflow areas of high-temperature hydrothermal convection systems often produce water of low and intermediate temperature. However, there are also a number of other models of hydrothermal systems used for direct-heat applications, and a representative few will be discussed in this section.

Low- and moderate-temperature resources commonly occur in areas of active geologic faulting, folding, and areas where the regional heat flow is above normal, such as in much of the western U. S. There are literally tens of thousands of such occurrences in the world. Some sample models of convective-hydrothermal resources are shown in Figure 5, taken from Sorey et al. (1983). Figure 5(a) shows a resource located along a fault zone. The water sinks in one part of the fault zone to a depth where it is sufficiently heated that it can rise along another part of the same fault zone. Recharge for the thermal springs could also come from the rock in either wall of the fault. This simple model is believed to represent the basic mechanism for many fault-related hydrothermal resources. In detail, of course, each resource is individual - the model in each of the figures is meant to be illustrative only of general mechanisms. Drilling to tap the resource would have to intersect the fault at depth, and a well may produce more water and/or higher temperature water than the surface springs discharge, since there is often near-surface dilution of thermal waters by shallow aquifers.

Figure 5(b) represents a possible mechanism for certain convective-hydrothermal resources in areas of fault-block valleys separated by up-faulted mountain ranges. Water sinks along a fault zone on one side of the valley, moves laterally in an aquifer at depth where it is heated, and discharges on the other side of the valley by movement up a second fault zone. Studies of geochemistry have demonstrated this as a viable mechanism in the Jordan Valley in Utah. The aquifer at depth may or may not be close enough to the surface to be economically tapped by wells. Figure 5(c) shows a situation where water infiltrates into a permeable bed in a mountain block and sinks deeply enough to be heated. Folding of the rock strata brings the permeable aquifer near the surface at another location, and the thermal fluid is channeled upward, where it can be reached by wells and perhaps from thermal springs. There is no need for the permeable horizon to outcrop in order to have thermal springs- thermal fluid may migrate up fracture zones from the aquifer depth. Alternatively, there may be no surface expression of the underlying thermal aquifer at all. Warm wells may be the only clue to such an occurrence.

THERMAL ENERGY IN GEOTHERMAL RESOURCE AREAS

All geothermal resources producing electrical power today are of the hydrothermal reservoir type. High-temperature hydrothermal reservoirs are formed in volumes of rock surrounding and including the hot intrusive body

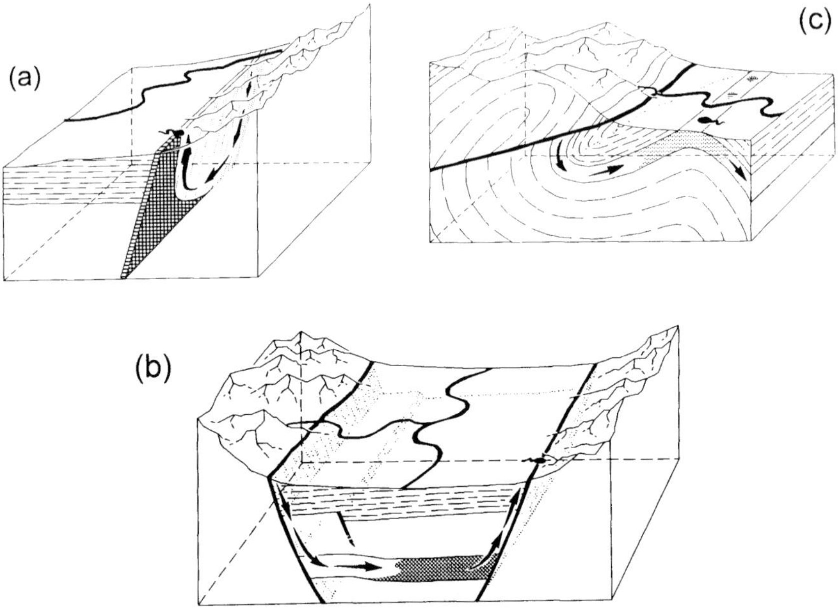

FIGURE 5. Conceptual models for low- and moderate- temperature hydrothermal resources: (a) Fault Plane Model, (b) Deep-Basin Model, (c) Permeable Bed in Folded Strata Model.

volumes of rock that develop fractures through one or more geologic processes. However, it is important to understand that most of the volume of hot rock in volcanic/intrusive systems does not develop a dense enough interconnected network of natural fractures to host a hydrothermal convection system. Therefore, most hydrothermal reservoirs are surrounded on one or more sides and below by volumes of hot rock in which porosity and permeability are quite limited. Of course, the total thermal energy in a geothermal system includes not only the energy in the hydrothermal reservoir, but also the energy in surrounding and underlying rock.

Geothermal reservoirs and their host surroundings are enormous thermal resources by any standards. An example is furnished by the steam-dominated hydrothermal reservoir and its hot host rock at The Geysers field. The area of the high-permeability reservoir, now being used for electricity generation, is roughly 75 km^2, its thickness is about 1,500 m, and its average temperature is ~235 °C. The amount of heat contained in this reservoir at temperatures above 100 °C is about 4.1 X 10^{19} Joules. Permeability decreases downward in The Geysers field, but temperature increases. The 1,500-m thick block immediately below the permeable reservoir is at an average temperature of ~340 °C, and has a thermal energy content of 7.4 X 10^{19} J. Thus, the total heat content at temperatures above 100 °C in the immediate reservoir area in The Geysers vapor-dominated reservoir is about 12 X 10^{19} J. The combustion of roughly 20 billion barrels of oil would be required to warm these rocks to their present temperatures if heating started at 100 °C. To provide a comparison, we note that a 20 billion-barrel oil field would be among the top five oil fields ever discovered!

Figures 5a, 5b, and 5c give a rough estimate of the thermal energy just in the immediate area of The Geysers reservoir proper. The reservoir itself is imbedded in a thermal anomaly extending northeast of the steam reservoir and having an area of ~750 km^2 (Walters and Combs, 1989), whose three-dimensional temperature distribution is poorly known due to lack of exploration drilling. Clearly, the area has many times more heat than given by the figures above for the reservoir proper. We must conclude that geothemmal resource areas can contain an enormous amount of thermal energy.

RENEWABILITY OF GEOTHERMAL ENERGY

The use of geothermal resources has been known and documented since the dawn of mankind. Bathing and cooking are activities carried out worldwide by civilizations both ancient and modem. Electricity was first generated from the vapor-dominated geothermal reservoir at Larderello in Tuscany, Italy in 1904. Although the Larderello power plants were destroyed during the Second World War, they were quickly rebuilt after war's end. Today this historic field produces more than 500 megawatts of electricity, enough for the home needs of perhaps half a million people. At Wairakei, New Zealand, electrical power has been generated from a water-dominated geothermal resource since 1958. Today the field produces 164 megawatts of power. At The Geysers field, power generation began in 1960, and continues today. It can readily be shown that only ~2 % of the heat has been mined since electricity production began in 1960. Projections indicate that each of these fields will generate power at commercial rates for many more decades, perhaps centuries.

From the geological viewpoint, we know that natural geothermal systems have occurred throughout geologic time. We see evidence of this in fossil hydrothermal systems, whose ages are tens to hundreds of millions of years old, and are now exposed at the surface by erosion. Some of these fossil geothermal systems are ore bodies that are being mined today! Age dating, using naturally radioactive minerals deposited by the hydrothermal systems, tells us that a typical lifetime for geothermal activity around magmatic intrusive centers ranges from a minimum of perhaps 5,000 years to more than 1,000,000 years as a magma body cools underground. Computer simulations of the cooling of intrusions also give ages in this range. Geothermal resources are, thus, renewable by any reasonable definition of the word.

We further observe that the natural rate of recharge of geothermal reservoirs from deep heat sources varies from a few to more than 1,000 thermal megawatts. The natural recharge rate represents the minimum rate at which geothermal systems could, in principal, be produced for tens of thousands of years. In commercial development, however, geothermal systems are typically produced at rates that exceed their natural recharge. This is done for economic reasons. Thus, high-temperature geothermal systems have a finite lifetime during which they will produce economic electrical power, although this lifetime may perhaps be supplemented by a further period during which they can be used for direct-heat applications that require lower temperatures. The exact lifetime of any given geothermal system is difficult to determine early in its history, but projections into the future become more reliable as a production history is established.

WORLDWIDE GEOTHERMAL RESOURCE BASE

Geothermal resources are to be found in essentially every country. Even Antarctica has hot springs! However, the occurrence of higher temperature resources at economic drilling depths is generally restricted to regions having volcanic activity within the last few million years (Figure 6). Lower-temperature resources are found in volcanic areas, but they are also found in areas of thick sedimentary basins and areas where geologic structure allows water to circulate to depths of several kilometers and return to the surface.

The total worldwide geothermal resource base represents a truly enormous amount of energy. Only an insignificant fraction of this energy is now being used for mankind's benefit. Geothermal energy's real promise is based on the enormous size of the resource base, the ready availability of technology to exploit some of the resources, and relatively low risk that research will lead to enabling technology for use of a much larger portion of the resource base. Table 1 presents a rough estamate of the worldwide geothermal energy resource base, including energy in magmatic systems, crustal heat, thermal aquifers, and geopressured basins. The thermal energy is given in units of Joules and in barrels of oil equivalent, i.e. the amount of oil that would need to be burned to produce the same amount of thermal energy. For comparisons, the total oil reserves are also shown in the bottom row. The implications of this table are truly impressive. The worldwide geothermal energy resource base in approximately 15,000 times greater than the thermal energy in all of the world's oil! Think what recovery and use of even 1% of this energy would do for humanity and the environment. In fact, this is not the end of the story. The total geothermal energy resource base is much larger than that contained in all of the fossil (oil, coal, natural gas) and uranium (nuclear power) worldwide.

ENVIRONMENTAL ASPECTS OF GEOTHERMAL ENERGY UTILIZATION

Human activities throughout the world release large quantities of damaging

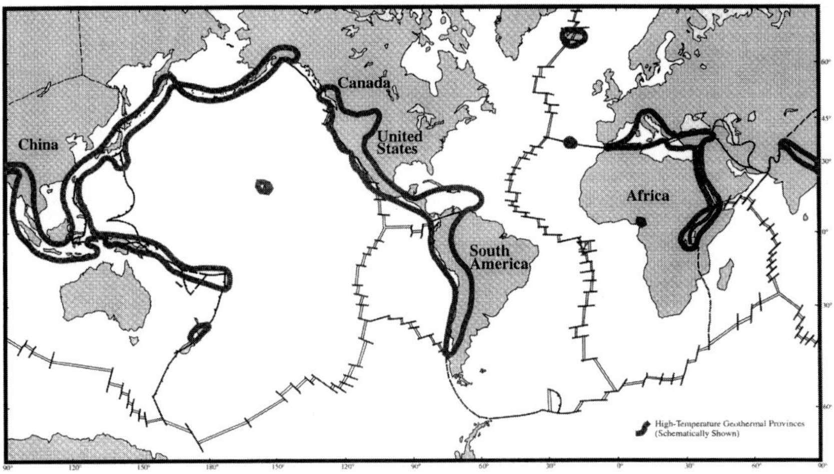

FIGURE 6. World high-temperature geothermal provinces.

pollution into the environment. A major contributor to atmospheric and groundwater pollution is the extraction and consumption of fossil and nuclear fuels. In the United States, for example, electric utilities account for (1) 70 % of sulfur dioxide emissions (the main cause of acid precipitation), (2) 33 % of nitrogen-oxide emissions (which combine photochemically with hydrocarbon vapors to form ground-level ozone, a gas which harms crops, animals, and humans), (3) 20 % of the releases of gases linked to the atmospheric greenhouse effect (mostly carbon dioxide), and (4) 50 % of all nuclear waste, whose long-term storage is a major problem with no permanent solution to date.

Compatible with sustainable development anywhere in the world - regardless of stage of development, population, or level of sophistication - geothermal energy resources are an attractive option to partially replace fossil and fissile fuels. Its history has proven that geothermal energy utilization has relatively minor adverse environmental impacts, while at the same time having significant environmental advantages compared with other energy sources. For example, the use of geothermal energy reduces emissions of greenhouse and acid-rain gases and requires much less land use than the fossil or fissile fuels. Geothermal energy utilization not only helps solve pollution problems, it also reduces dependency on imported energy sources.

In order to avoid adverse environmental impacts from geothermal energy utilization, careful investigation is necessary prior to development. Early study and consistent application of careful management practices are essential in protecting the environment. The entire process, from exploration, to drilling of wells, to design and construction of the utilization system, and then to monitoring of operations, must be well planned. A detailed Environmental Impact Assessment

(EIA) or other such study is usually undertaken prior to the issuance of permits to proceed with development. Regulations to ensure safe operations are in place at the federal, regional and local levels in essentially every country using geothermal energy. Some of these regulations deal with protection of ground water supplies, others deal with minimizing levels of atmospheric emissions, while still others deal with construction activities at the site. These precautions have been found to be sufficient at hundreds of sites around the world.

Atmospheric Emissions

Eighty-four percent of the total U. S. energy consumption comes from fossil fuels (DOE/EIA, 1997). Combustion of these fossil fuels is believed to be a major source of the observed carbon dioxide buildup in the atmosphere over the past century - an accumulation which has accelerated during the past fifty years. Carbon dioxide is a greenhouse gas whose effects keep the earth's surface temperature at a level conducive to life as we know it. However, adding more carbon dioxide, as we have been doing, is believed to cause global warming and could have dire consequences.

Geothermal electrical generating plants have marked advantages over fossil-fuel powered generating plants in terms of atmospheric emissions of harmful gasses (Figure 7). A geothermal power plant emits <5 % of the carbon dioxide, <1% of the sulfur dioxide, and <1 % of the nitrogen oxides produced from a coal power plant of equal size (Houk, 1966). Even natural gas, the cleanest burning of fossil fuels, emits significant quantities of carbon dioxide.

Geothermal energy use in the United States and in the world has a significant effect in helping to avoid air pollution, but if resource development could be accelerated, this positive effect would be much greater. Using data from Huttrer (2000) on worldwide electrical generation, and from Lund and Boyd (2000) on worldwide direct uses of geothermal energy, one can make very rough calculations of the amount of atmospheric pollution avoided compared with furnishing the same amount of energy using coal.

According to Wright (1995), in any given year in the United States, electricity produced from geothermal resources eliminates the emission of 22 million tons of carbon dioxide, 200 thousand tons of sulfur dioxide, 80 thousand tons of nitrogen oxide, and 110 thousand tons of particulates compared with the production of the same amount of electricity from conventional coal-powered plants.

An important consideration usually overlooked is the fact that the small quantities of gases emitted from geothermal power plants are naturally occurring and result from geologic processes. Because the Earth's crust is porous and permeable, these gases would eventually find their way to the surface anyway, even in the absence of beneficial utilization of the geothermal resource. Therefore, in reality, geothermal projects actually have zero net emissions of gases to the atmosphere, although they do concentrate and speed the natural emissions process somewhat.

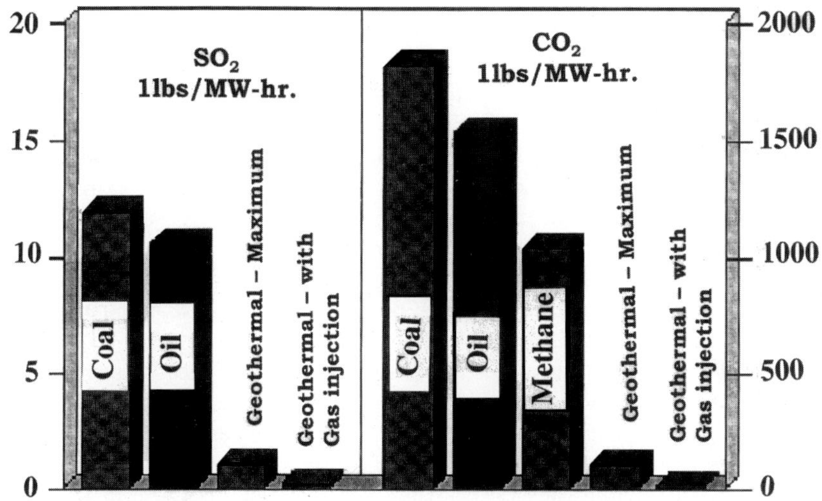

FIGURE 7. Emissions of atmospheric pollutants from fossil-fuel and geothermal power plants.

In geothermal power projects, it is sometimes feasible to inject all gases along with the spent geothermal fluid by pumping it down injection wells and into the same geologic horizons from which they originated. In these cases, atmospheric emissions are reduced to zero.

Hydrogen sulfide (so-called "rotten-egg gas") occurs naturally in solution in many geothermal fluids. Since this gas is harmful in large enough concentrations, and since it produces an odor nuisance if not mitigated, it must be removed from vapors exiting the power plant. Several methods are available to remove hydrogen sulfide from geothermal plants. Perhaps the most common are the Stretford redox process and incineration. In the Stretford process, elemental sulfur, in the amount of ~1.5 kg per megawatt-hour of electricity generated, is produced through reaction using a vanadium catalyst (Ecoview Environmental Consultants, 1974). This sulfur can be sold for use in the chemical industry to produce sulfuric acid, thus enhancing the economics of geothermal power plant operation. The Geysers field, in Califomia, is a major facility generating some 1,200 megawatts of electrical power. Each of the power plants in this large field is equipped with hydrogen sulfide abatement equipment. This abatement is so successful that the two counties where the power plants are located, Lake and Sonoma counties, have received awards for environmental cleanliness. According to the current EPA standards, hydrogen sulfide and carbon dioxide emitted from California geothermal power plants are not harmful in any way to public health.

Land Use

Energy resources that require mining and/or subsequent treatment such as refining (including coal, oil, natural gas and uranium fuels) usually require large land areas for extraction, refining, transportation and waste disposal, in addition to dedicated space for the plant. Unlike utilization of fossil fuels, geothermal energy utilization encompasses the whole fuel cycle, from resource extraction to final use or transmission of product, at one location. This is because geothermal energy must be utilized directly or converted to electricity in the near proximity of the resource-hot water and can not economically be transported to another location for use. As a result geothermal power projects have significantly lower total land requirements than other energy sources.

It is estimated that for production of one megawatt of electric power, geothermal plants use less than one-eighth of the land used by a typical coal mine and power plant system (Sheiler, 1976). In addition, land utilized for geothermal exploitation can also be simultaneously used for other purposes, such as agricultural activities or cattle grazing. For example, the Imperial Valley of southern California is one of the most productive agricultural areas in the United States. This area is also home to some 15 individual geothermal power plants, currently producing more than 400 megawatts of electrical power.

Water Issues

Geothermal fluids usually contain dissolved salts. Since solubility of most salts increases with increasing temperature, high-temperature geothermal fluids are sometimes highly saline, containing many dissolved species, some of which are potential environmental contaminants. Among the gasses are hydrogen sulfide (discussed above) and occasionally radon, while sometimes among the dissolved solids are boron, mercury and arsenic salts (MacCabe, et. al., 1993). Since these species can be harmful even in low concentrations, reliable handling methods are required in order to avoid contamination. However, technology for the safe, non-polluting use of geothermal waters has been well developed and tested. Production and injection wells are lined with steel casing and cement to isolate the geothermal fluids from the environment. Injection of spent geothermal fluids back into the deep reservoir not only disposes of the water in the geologic horizon from which it came, it also helps to maintain reservoir pressure and increase the longevity of the resource. Another method of dealing with hazardous components in geothermal fluids is use of a heat exchanger. The geothermal fluid enters one side of the heat exchanger, and gives up its heat to a clean water on the other side of the exchanger. The heated, clean water is then used as a source of heat, say in a greenhouse. Geothermal fluid and the clean water never come into contact. The geothermal fluid from the exit of the heat exchanger is injected back into the subsurface horizons from which it came.

Another water issue concerns its consumption in some geothermal applications. The cooling towers of geothermal power plants are of one of two types, evaporative cooling or air cooling. The evaporative type of coolers are more efficient but consume water, which is evaporated into the atmosphere. Air coolers use no water, but are bulkier and less efficient than water coolers. Where water consumption is a consideration, air coolers can be used. The usual application of air coolers is with binary-type power plants (See Geothermal Energy for Heating and Electrical Generation, this volume).

Other Potential Impacts

Although geothermal energy is environmentally safe, its potential adverse impacts must be taken into consideration and any needed mitigation measures instituted. In addition to those considerations discussed above, there are several others that are less common and usually not a significant problem. These include induced seismicity, land subsidence, waste disposal, disruption of local ecosystems, noise pollution, and aesthetic issues.

In a few high-temperature geothermal fields, seismicity is known to be associated with the production of geothermal fluids. This induced seismicity is of a low magnitude, the vast majority of the seismic events being well below the level of felt shaking. Geothermal production from low- and rnoderate-temperature geothermal systems, such as those used for greenhouse heating or fish farming, is not generally associated with induced seismicity. To the authors' knowledge, induced seismicity is not believed to be a limiting problem at any geothermal field.

Land subsidence has been observed in some – certainly not all – geothermal fields, attributed to removal of subsurface fluids. This is a problem only in the large geothermal fields used for electrical power production, where large volumes of fluids may be withdrawn from the subsurface. Extreme displacements of several feet to several tens of feet have occurred. However, such subsidence has been observed mainly in fields where spent geothermal fluids are not injected back into the reservoir. Where subsidence is an important issue, it has been mitigated successfully by injection of 85 % to 100 % of the fluids removed. In some areas, where the geothermal field is remote from other development, a reasonable amount of subsidence poses no particular problem. In no case has subsidence been on the broad scale usually found above underground coal mines, either in terms of amount of sinking or of the area affected.

A few geothermal fields produce waste products, such as precipitated silica, that may contain contaminants such as boron which must receive proper disposal. Local regulations usually specify what must be done to prevent contamination of the environment. Waste disposal has not proved to be a problem at any operating geothermal field.

In the process of drilling geothermal wells and building a geothermal installation, local road building, grading and leveling are required. Care is taken to ensure that local plants and animals are not harmed and that new hazards are not created. As one measure to reduce surface impact, several wells may be drilled from the same well pad but in different directions to access different parts of the reservoir. Drilling and construction sites are restored as nearly to their previous condition as possible. Since many geothermal fields are in mountainous areas, measures must be taken to ensure that erosion or landslide hazards are not created, and that nearby streams are protected from pollution. These are, of course, normal precautions during any construction project. When construction is complete, the affected areas are usually restored as prescribed by ordinance. To cite one example of such reclamation, during geothermal exploration at Las Tres Virgenes, in Baja, Mexico, more than 4,000 plants, including many endemic species found in the roads and drilling pads, were transplanted to help restore the area with about half of these plants surviving the process (Geothermal Energy Association, 1997).

In some cases, noise pollution may be voiced as a concern. Geothermal electrical generating plants are usually quiet except, of course, inside the plant. Other types of geothermal utilization, such as greenhouse heating or crop drying, usually pose no noise problems. More noise is usually associated with the drilling and flow testing of geothermal wells, but this problem is fugitive, and is gone when the work ceases. Noise pollution has not proved to be a limiting problem in geothermal development.

NIMBY

Concerns are almost always raised by local people, often supported by regional or national advocacy groups, whenever a construction project of any type is proposed. Geothermal projects are no exception. Although, for geothermal projects, the concerns are usually couched in terms of one or more of the supposed hazards discussed above, the real, often thinly disguised, underlying reason for these concerns is the NIMBY (Not in My Back Yard) issue. People are happy to be able to turn on the light switch, or eat a greenhouse-grown tomato in the winter, or gas up their car, but the same people do not want a power plant, greenhouse, or filling station located near them. Generally, if legal and regulatory processes are scrupulously carried out, the real need, or lack thereof, for a particular project will become apparent, and the will of the people will be respected. As population continues to increase, however, the NIMBY effect will lead to greater conflict between project developers and the populace.

STATUS OF GEOTHERMAL TECHNOLOGY

One primary barrier to more rapid development of geothermal resources worldwide continues to be inadequate technology. In the United States, costs range from 4 to 7 cents per kilowatt-hour for new power developed at highgrade geothermal resources, whereas new generation capacity from natural gas produces power at 2 to 4 cents per kilowatt-hour. The key to solving our technical problems is continued research and development. With a modest level of R&D, we can realize the full potential of geothermal energy. An estimated $200 million annually are spent worldwide on developing new and improved geothermal technology. The U.S. Department of Energy budgets some $30 million of this amount, while the government of Japan spends considerably more.

Although use of geothermal fluids has a long history, reaching back to the first recordings of civilization, modern uses on a wide scale are fairly recent. As a result of this, and the fact that the geothermal industry worldwide is very small compared to the mining or petroleum industries, geothermal technology has not at all reached a mature stage. Most of the exploration techniques used to find geothermal reservoirs has been adapted from mineral exploration, while most of the well drilling and production technology has been adapted from the petroleum industry. However, little of this borrowed technology is optimum in the geothermal environment, leaving room for significant improvement. The technical areas where the most severe barriers to further economic devolpment of geothermal energy currently exist, and where we would recommend continued R&D, are in drilling, exploration and reservoir technology, and energy conversion.

Drilling. Drilling is one of the most expensive activities in any geothermal development project. It is necessary to drill production wells to bring geothermal fluids to the surface, and to drill injection wells to return the spent fluids to the reservoir. Because of the high temperatures and corrosive nature of geothermal fluids, geothermal drilling is significantly more difficult and expensive than conventional oil and gas drilling. Each well in a high-temperature geothermal field costs $1 million to $3 million, and an average field consists of 10 to 100 or more wells. Drilling costs account for one-third to one-half of the total cost of a geothermal project. Improvement in existing drilling technologies and development of new, advanced drilling technologies will significantly lower the cost of geothermal energy utilization, enabling expanded use of clean, reliable geothermal energy. We believe that drilling costs can be lowered by 10 % to 20 % in the next decade through incremental improvements to existing technology, and by 30 % to 50 % in the long term through development of advanced technologies.

Exploration and Reservoir Technology. In exploration, finding concentrations of geothermal heat is actually not a major problem. The major challenge facing developers of geothermal projects is how to remotely detect permeable zones deep in the subsurface - zones that will produce large quantities of geothermal fluids - so that drill holes can be sited and steered to intersect them. No two geothermal reservoirs are alike, and their permeabilities (their ability to transmit geothermal fluids to the well bore) vary widely in all three dimensions over short distances. Present exploration techniques are not specific enough and result in too many dry wells, driving up development costs. Further, inadequate knowledge of the physical and chemical properties of the subsurface makes it difficult to mine the heat in the most efficient way and ensure the sustainability of geothermal reservoirs. Better geological, geochemical, and geophysical techniques, as well as improved computer methods for modeling heat-extraction strategies from geothermal reservoirs, are needed. We believe that geothermal development costs can be lowered by 10 % to 20 % with improvements in earth science techniques.

Energy Conversion. The efficiency in converting geothermal steam into electricity in the power plant directly affects the cost of power generation. During the past decade, the efficiency of dry- and flash-steam geothermal power plants was improved by 25 %. Power plants installed at The Geysers geothermal field in California during the 1960s required 20 lbs of steam to produce 1 kilowatt-hour of electricity. The newest plants at The Geysers, installed in the mid and late 1980s, require only 14.5 lbs of steam to produce that same kilowatt-hour of electricity. We believe that geothermal power-plant efficiency can be improved by an additional 10 % to 20 % over the next decade or two with a modest investment in R&D.

REFERENCES

Barbor, D.C., et al., 1992. A geothermal development assessment of likely carbon dioxide emissions and proposed methods for reduction. Barnett, P.R. (KRTA Ltd., Auchland (New Zealand)); pp. 605-610 of Transactions, Volume 16, 20th anniversary, Geothermal Resources Council, Davis, CA. 680p. (CONF-921083—: Geothermal Resources Council (GRC) annual meeting, San Diego, CA, 4-7 Oct.

Blaydes, P.E., 1994. Environmental advantages of binary power plants can enhance development opportunities. pp. 121-125. Restructuring the geothermal industry. Transactions, Volume 18. Geothermal Resources Council, Davis, CA, 655p. (CONF-941043—: 1994 Annual meeting of the Geothermal Resources Council: Restructuring the Geothermal Industry, Salt Lake City, UT, 2-5 Oct.

Craig, H., 1963. The isotopic geochemistry of water and carbon in geothermal areas: in Nuclear geology of geothermal areas, Spoleto: Pisa, Consiglio Nazionale delle Recerche, Laboratorie de Geologia Nucleare.

Duffield, W. A.., Sass, J. H., and Sorey, M.L., 1994. Tapping the Earth's Natural Heat, U.S. Geological Survey circular; 1125, 63 pp.

Ellis, A. J., and Mahon, W. A. J., 1977. Chemistry and geothermal systems: Academic Press, New York, 392 p.

DOE/EIA, 1997. Renewable Energy Annual for 1996: U.S. Department of Energy, Energy Information Administration, DOE/EIA00603(96), 192 p.

Geothermal Energy Association, 1997. 1025 Thomas Jefferson Street, N.W., Suite 109, Washington, DC, 20001, U.S.A.

Goddard, W. B., and Goddard, C. B., 1990. Energy fuel sources and their contribution to recent global air pollution trends. Geothermal Resources

Council Transactions, Vol. 14-Part I, pp. 643-649.

Grande, M. G., 1995. The economic and environmentally sound handling of geotherrnal sulfur byproducts: A case history. (Northern California Power Agency, Middletown, CA). pp. 445-450 of Accomplishments of the past and challenges of the future. Transactions, Volume 19. Geothermal Resources Council, Davis, CA . 604p. (CONF-951037—: Annual Meeting of the Geothermal Resources Council, Reno, NV, 8-11 Oct.

Hare, R., et al., 1995. Emission factors of geothermal power plants in California. (California Energy Commission, Sacramento, CA). pp. 147-151 of Accomplishments of the past and challenges of the future. Transactions, Volume 19 Geothermal Resources Council, Davis, CA. 604p. (CONF-951037—: Annual Meeting of the Geothermal Resources Council, Reno, NV, 8-11 Oct).

Helgeson, H. C., 1968. Geologic and thermodynamic characteristics of the Salton Sea geothermal system: American Journal of Science, v. 266, p. 129-166.

Henley, R. W., and Ellis, A. J., 1983. Geothermal systems ancient and modern, a geochemical review: Earth Science Review, v. 19, p. 1-50.

Houk, J.E., 1996. Air quality assessments in support of the Environmental Imapct Statement (EIS), Oregon Air Contaminant Discharge Permit (ACDP) and Oregon Energy Facility Siting Council certificate (EFSC) for the Newberry Geothermal Pilot Project. (AGI Technologies, Portland, OR. McClain, D.W. pp. 19-27 of Geothermal Resources Council, Davis, CA, 886p. (CONF-960913—: Annual Meeting of the Geothermal Resources Council, Portland, OR, 29 Sept-Oct).

Huttrer, G. W., 2000, The status of world geottermal power generation, 1995 1999: Proceedings of the World Geothermal Congress 2000, Japan International Geothermal Association, Pisa, Italy.

Ikeuchi, K., Komatsu, R., Doi, No., Sakagawa, Y., Sasaki, M., Kamenosono, H., and Uchida, T., 1996. Botton of hydrothermal convection found by temperature measurements above 500 °C and fluid inclusion study of WD-1 in Kakkonda geothermal field, Japan: Geothermal Resources Council Transactions, 20, 609616.

Lund, J. W. and T. L. Boyd, 2000. Geothermal Direct Use in the United States, Update: 1995-1999, Proceedings of the World Geothermal Congress 2000, Japan International Geothermal Association, Pisa, Italy.

MacCabe, P.J., Gautier, D.L., Lewan, M.D., and Turner, Christine, 1993, The Future of Energy: U.S. Geological Survey Circular 1115, 58p.

Mahon, W. A. J., Klyen, L. E., and Rhode, M., 1980. Neutral sodium/ bicarbonate/sulfate hot waters in geothermal systems: Chinetsu (J. of the Japan Geothermal Energy Assn.), v. 17, 11-24.

Nakicenovic, N., Grubler, A., and McDonald, A., 1998. Global Energy Perspectives: Cambridge University Press, 299p.

Norton, D. L., 1984. Theory of hydrothermal systems: Annual Reviews of Earth and Planetary Science, v. 12, 155-177.

Salinas, F., Naranjo, A., 1997. Vegetation ecological restoration during geothermic exploratory perforation: A case study in Mexico. Ortega-Rubio, A. (and others); pp. 141-144 of Meeting the challenge of increased competition. Geothermal Resources Council, Davis, CA. 665p. (CONF-971048—: Geothermal Resources Council (GRC) Annual Meeting, San Francisco, CA, 12 Oct.

Sheiler, L., 1976. Geothermal Effluents, their toxicity and Prioritization. In proceedings of the first workshop on sampling geothermal effluents. Las Vegas: U.S. Environmental Protection Agency, Environmental Monitoring and Support Laboratory.

Sorey, M. L., Nathenson, M. and Smith, C., 1983, Methods for assessing low-temperature geothermal resources: in Assessment of Low-Temperature Geothermal Resources of the United States — 1982, U.S. Geological Survey Circular 892, M. J. Reed, Editor, 73 p.

U.S. Department of the Interior, 1973. Draft Environmental Impact Statement on the Federal Geothermal Development Program, Volumes 1-111, and Washington, DC: Government Printing Office.

Walters, M., and Combs, J., 1989, Heat flow regime in The Geysers Clear Lake area of Northern California: Geothermal Resources Council Transactions, v13, 491-502.

Western Energy Resources and the Environment, 1977. Geothermal Planning by Resource Planning Associates, Inc. for Environmental Protection Agency, Office of Research and Development. Office of Energy, Materials and Industry.

White, D. E., Muffler, L. J. P., and Truesdell, A. H., 1971, Vapor-dominated hydrothermal systems compared with hot-water systems: Economic Geology, v. 66, p. 75-97.

Wright, P. M., 1995, Geothermal Energy – Clean, Sustainable Energy for the Benefit of Mankind and the Environment: Brochure published by the Earth Sciences and Resources Institute, University of Utah, 8 p.

Wright, P. M., and Culver, Gene, 1998, Nature of Geothermal Resources: in Lund, J. W., ed., Oregon Institute of Technology, 27-69.

Renewable Energy: Trends and Prospects. Edited by S.K. Majumdar, E.W. Miller and A.I. Panah

Chapter Sixteen

GEOTHERMAL ENERGY FOR HEATING AND ELECTRICAL GENERATION

PHILLIP M. WRIGHT[1], JOHN W. LUND[2]
and
ASSAD I. PANAH[3]

[1]Scientific Fellow
Director, Subsurface Science Initiative
Idaho National Engineering and Environmental Laboratory
P.O. Box 1625
2525 Fremont Avenue
Idaho Falls, ID 83415-2214
[2]Geo-Heat Center
Oregon Institute of Technology
2301 Campus Drive
Klamath Falls, Oregon 97601
[3]Department of Geology and Environmental Science
University of Pittsburgh at Bradford
300 Campus Drive
Bradford, Pennsylvania 16701-2898

As the world's population continues to increase, infrastructure expands proportionately to meet people's needs. With expanding population and infrastructure come increased needs for heat and electric power. Because of pollution generated by the use of fossil fuels, and because the world's supply of fossil fuels is finite, alternative energy resources must be developed to meet future needs. To make the right choices among energy resource alternatives, we must evaluate each option on the basis of cost, reliability, environmental impact and economic impact, including jobs lost or gained. Since so much of today's energy supply comes from fossil fuels, the challenge of the new century will be to extract and consume fossil fuels with minimal environmental damage and/or to provide energy with fewer drawbacks and more convenience. Science, industry, and government bear the brunt of the responsibility for meeting this challenge.

Among the possible new energy resources, geothermal energy is being actively promoted in the U. S. and in many other countries. Geothermal energy is highly versatile, its reserves are virtually infinite, it is environmentally safe and can be used for the generation of electricity, as well as a host of applications where heat is needed. Furthermore, geothermal energy can be used over a wide range in temperature and volume, whereas for conventional energy resources to be profitably exploited, size and grade are more important. Expanded use of geothermal energy helps reduce a country's dependency on imported energy resources, thereby increasing self sufficiency and economic security. Currently, half of the oil consumed in the United States is imported, constituting the largest portion of our trade deficit. Based on these and other advantages of geothermal energy, we can predict with confidence that use of geothermal energy will continue to grow in the United States and worldwide.

DEVELOPING GEOTHERMAL ENERGY RESOURCES

The nature and origin of geothermal energy resources are reviewed elsewhere in this volume (Chapter 14). Of the several types of geothermal resources discussed there, hydrothermal resources, those consisting of naturally occurring thermal waters, are the only type in commercial use today. Hydrothermal energy is used to generate electricity and to provide heat for a wide variety of uses, including district heating, greenhouse heating, food processing, aquaculture, and bathing (the so-called "direct-heat applications"). For hydrothermal resources higher in temperature than ~150 °C, electricity generation is the option usually pursued. Electricity can be readily sold as a commodity, which usually provides maximum value from the resource. For resources having temperatures below ~150 °C, some form of direct use of the heat is generally undertaken. However, resources as low in temperature as 100 °C are used to generate electricity profitably under special circumstances.

The discussion below of the geothermal development process is general in that it applies equally well to projects for electricity generation or for direct use of the geothermal heat. After presenting these considerations, we will examine power generation and direct applications in some detail.

Geothermal Exploration

Exploration for geothermal resources is similar to exploration for petroleum or mineral deposits, making use of techniques from the earth-science fields of geology, geochemistry, geophysics and hydrology (Wright, 1998a). Such obvious surface features as geysers and hot springs alert the explorationist to the presence of an underlying resource. However, direct surface manifestations of a subsurface geothermal reservoir may be subtle or even absent, since some hydrothermal systems discharge through porous rocks underground while others are capped by an impermeable layer which impedes upward movement of the thermal fluid. Even in the absence of surface geothermal features, indications of prospective subsurface conditions may be recognized by the experienced geologist. Volcanic rocks, especially rhyolite, of recent geologic age (less than ~1 million years old), deposit such as sinter or tufa from past hot-spring activity. Chemical alteration of the minerals in surface rocks, when recognized as having resulted from their interaction with hydrothermal solutions arising from depth, and geophysical anomalies can all help pinpoint a prospect.

The prime objective of exploration work is first to find a prospective area, and then to site wells that test for a geothermal resource at depth (Wright, 1998a). Because drilling is expensive, it must be minimized as much as possible. Design and execution of an exploration program in the most efficient way possible, usually calls for a strategy to be developed based on existing knowledge of the regional and local geology. To be successful, a test well must intersect one or more permeable horizons that carry hydrothermal fluids in the subsurface. Although, fractures, faults and contacts between different rock types are common in nature, these and other potentially permeable features may or may not be continuous or extensive enough to be connected with the source of thermal fluids. Because there is no known way to detect, from the surface, the particular permeable zones, at depths of hundreds or thousands of feet that are interconnected with the source of geothermal fluids, exploration techniques are mostly indirect and provide only circumstantial evidence of the existence and location of a subsurface geothermal reservoir. Drilling exploration wells is the only way to provide definitive proof of the existence of a geothermal reservoir in the subsurface.

A successful wildcat test well is one which intersects thermal fluids (water or steam) that can be produced at the surface over an extended period of time. Exploration wells are tested by flowing them in order to decide whether to proceed further. During a flow test, the well is usually produced in a sequence of several different flow rates while the production pressure is continuously monitored. Measurements are also made of well-head temperature, flowing temperature at depth (sometimes), fluid chemistry and dissolved gas content. Tests may call for flowing the well continuously for durations ranging from several days to several months. The longer the test, the more confidence the engineer will have in predicting future production characteristics of the well and its ultimate useful lifetime. However, well testing is expensive and may involve fluid disposal problems, so from an economic viewpoint they are kept as short as possible consistent with developing the data needed. Prior to, during, and after flow tests, pressures or water levels are measured in any nearby wells to help determine the extent of interference among wells, i.e. the extent of hydrologic connectivity. From observation of pressure decline with increased flow rate and also with time, the reservoir engineer can make predictions of long-term production characteristics, needed for further project design and economic analyses.

With a successful flow test, the development program then moves on to drilling step-out wells to help determine the limits of the geothermal reservoir at depth, and then to the drilling and flow testing of production and injection wells. Eventually the stage is reached where enough is known about the temperature, pressure, flow rate and chemistry of the produced fluids to specify the design of the utilization system - the fluid-gathering pipes, pumping systems, electrical generating plant or direct-use application plant, and fluid disposal system. Small geothermal systems may support only one or two wells used for a direct-heat application such as heating greenhouses or drying crops. Larger geothermal systems may support many production wells, and be suitable for use in generating electrical power.

The chemistry of the geothermal fluids produced from the subsurface reservoir is an important concern. Most geothermal fluids having temperatures below ~100 °C have low concentrations of dissolved salts, and cause little or no environmental concern. Some low-temperature geothermal waters are even potable. However, since the solubility of most salts increases with increasing temperature, and since hot water interacts with reservoir rocks to leach mineral constituents much more effectively than does cool water, hotter geothermal resources can be laden with dissolved constituents. Dissolved constituents are usually dominated by sodium chloride, often with calcium bicarbonate or sulfate species. Silica is also present in increasing amounts with increasing temperature. Total dissolved solids (TDS) are usually in the range 5,000 parts per million by weight (ppmw) to more than 300,000 ppmw, compared with sea water at ~35,000 ppmw.

Some constituents may pose environmental problems if not handled properly. For this reason, and to help keep the reservoir pressurized, spent fluids from almost all power plants and many direct-heat uses are injected back into deep horizons near or within the reservoir. In general, one tries to produce fluids from zones of hot fluid upflow and inject spent fluids into zones of natural recharge. With proper design of the production/injection system, reservoir pressures are maintained while heat is mined from the rocks in an optimum fashion. Care is taken in all wells to prevent the geothermal fluids from entering near-surface aquifers by using steel well casing, securely cemented in place. Local, state and federal regulations stipulate how this is to be done, ensuring safe development of geothermal resources.

Risks in Geothermal Development

The chief risks in geothermal energy development are concerned with the reservoir (Wright, 1998a). With modern technology, there is little risk that the electrical generation plant or the direct-use equipment will not perform properly, perhaps after a little fine tuning.

The reservoir risk is that the pressure or temperature will decline at rates exceeding engineering predictions. This risk is highest in the first years of operation and decreases with time as the long-term production characteristics of the hydrothermal system become better known. It can be mitigated to some extent by long-duration flow tests during the exploration phase, which reveal the maximum production rates at acceptable pressure and temperature decline rates. Another risk mitigation method is to begin with a small installation on a geothermal field and gather reservoir-production data at the same time that electricity generation or greenhouse heating is going on. With a few year's operation, confidence in the production characteristics (temperature, pressure, chemistry) and reservoir longevity can be obtained and a larger plant installed.

Risks are also mitigated by maintaining and updating the geological and engineering models of the reservoir and its environs. Each well adds to the geological knowledge of the area, and with enough drilling, a reasonably detailed picture of the geology in the subsurface begins to emerge. The distribution of various rock types and the existence and configuration of faults and fractures are some of the parameters obtained by examination of rock chips from drilling and by interpretation of geophysical logs taken by instruments lowered into the wells. These are used to build up a three-dimensional picture of the subsurface. The reservoir engineer constructs a computer simulation - a reservoir model - incorporating information from the geologic model, from the production and injection histories of wells, and from measurements of the so-called "native state" of the reservoir, which includes temperature and water-level distribution in the subsurface before production began. All pertinent data are incorporated in these models, and their complexity can become surprising in larger, well-known fields where a great deal of drilling and well testing have resulted in a large data base. Of course, the subsurface *is* complex in most geothermal areas, and even our best models form only a crude approximation of what is going on beneath our feet.

Monitoring of Production and Injection

Monitoring the production characteristics of the reservoir is an important facet of geothermal operation - for generation of electricity or direct-use of the heat. Keeping good records of production temperature, pressure and chemistry of selected wells will generate data that can be used in the predictive, computer-based reservoir models discussed above to decrease risk of surprise. Monitoring is done by measurements down the well, at the well head, at the plant, and sometimes in special monitor wells drilled for the purpose, or in converted former production or injection wells.

During a geothermal reservoir's lifetime, an individual production well may decline in temperature or pressure or an injection well's ability to imbibe fluids may decline due to changes in the reservoir adjacent to the well, or due to deposition of scale in the well casing. Treatments are available to mitigate these effects, and are usually collectively called "work-over" of the well. They may entail mechanical or chemical removal of scale from the well casing, or treatment of the reservoir by injection of acid solutions to rejuvenate the permeability of the rocks.

GENERATION OF ELECTRICITY FROM HYDROTHERMAL RESOURCES

There are two basic types of geothermal power plant - the steam plant and the binary-cycle plant. As its name implies, the steam plant operates by using the expansion and cooling of steam, as it traverses a turbine engine, to turn an electrical generator. In the binary-cycle plant, a low boiling-point working fluid is used in a closed loop, and the heat energy is furnished by a heat exchanger which uses the geothermal fluid as its source. The working fluid is vaporized in a primary heat exchanger, the vapor expands through a turbine engine and is then condensed in a secondary heat exchanger, to return to the primary heat exchanger and repeat the cycle. There is no consumption of the working fluid, and the working fluid never contacts the geothermal fluid.

In some geothermal resources, steam is obtained directly from the well. So-called "vapor-dominated" hydrothermal systems contain steam as the pressure controlling fluid phase in cracks and pores in the rock. For generation of electricity from vapor-dominated geothermal systems, any sand debris and water droplets entrained in the steam are removed from the flow produced from the well, and the steam goes directly to the turbine. Vapor-dominated systems are quite rare in nature, however. It is much more common to find liquid water as the pressure-controlling reservoir fluid phase in the reservoir - the so-called "liquid-dominated" hydrothermal systems.

To provide steam from liquid water, special equipment is used to allow very rapid but controlled boiling (flashing) of the geothermal water after it arrives at the surface. The turbine-generator equipment itself is essentially identical to that used with vapor dominated systems. In the discussion below, we will concentrate on flash-steam and binary plants.

Flash-Steam Power Plants

As its name implies, the flash-steam plant uses steam to power a turbine engine, which turns the generator (Wright, 1998b). The steam is obtained from high-pressure, high-temperature geothermal water (or sometimes a mixture of water and steam) arriving at the surface through production wells. Of course, the thermal water arrives at the surface under high pressure and at high temperature. By lowering the pressure in a vessel known as a "flash tank" or a "separator", some fraction of the geothermal water - typically about 30 percent for waters ~250 °C in temperature - flashes to steam. The steam is separated from the remaining water, and the steam goes directly to the turbine engine (Figure 1). Steam moving through the turbine impinges on its blades, causing the rotor to turn. As the steam flows through the turbine, and its energy is converted to rotary motion, the temperature and pressure of the steam drop. Typical geothermal turbines rotate at 2,400 rpm, and are usually coupled directly to the electrical generator. The output of the generator goes to a switch yard for boosting to transmission voltages, and then is synchronized with and enters the electrical transmission grid.

Dual-flash plants have two stages of flashing, the first being a high-pressure flash and the second being a flash in a separate vessel at a lower pressure. Steam from the high-pressure flash is introduced into the front end of the turbine, and steam from the low-pressure flash is introduced at a mid-point, after the high-pressure steam has lost some of its energy. Dual-flash installations have higher overall efficiency of conversion of thermal energy into electricity than single-flash installations, but are more expensive to build and operate. Today, both types of plants are in use.

After exiting the turbine, the steam is condensed by rapid cooling in order to decrease back pressure on the turbine, or equivalently, to increase the temperature drop across the turbine, and thereby increase its efficiency. Pressures at the turbine exhaust are usually sub-atmospheric. Non-condensible gases (NCGs), occur naturally in the geothermal reservoir, dissolved in the geothermal water. NCGs separate and move with the steam during flashing, with only a small fraction remaining in the residual, unboiled water. These NCGs, predominately carbon dioxide, but also often small quantities of hydrogen, hydrogen sulfide,

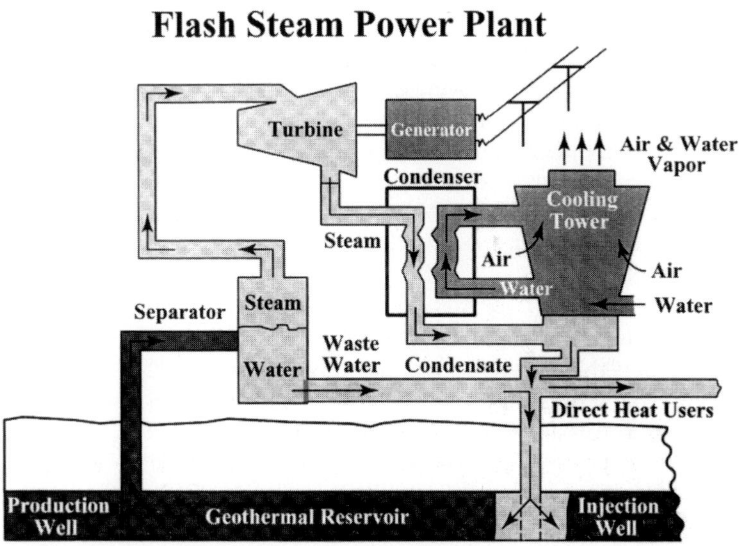

FIGURE 1. Geothermal Energy for Heating and Electronic Generation.

methane and other gases, are collected and compressed to atmospheric pressure for scrubbing to remove environmental contaminants before release from the plant. Alternatively, the NCGs may be compressed to even higher pressure and injected back into the reservoir horizons along with residual, cooled water from the flash process and the steam condensate.

Cooling towers provide the heat rejection to cool the turbine exhaust. Some cooling towers use evaporation of water while others simply draw air through heat exchangers. Evaporative coolers are more efficient, but consume water by evaporation into the atmosphere. The lower efficiency of air-cooling may be compensated by providing a larger cooling-tower facility. The efficiency of air coolers is significantly affected by daily and annual variations in ambient air temperature, while the efficiency of evaporative-coolers depends on the relative humidity of the intake air. Air coolers have an obvious advantage in areas where water is scarce or expensive, but water coolers are usually the preferred option because of their greater efficiency in cooling for an installation of given size and cost.

Binary-Cycle Power Plants

Binary-cycle power plants can be used where the temperature of the resource is too low for efficient generation of steam by flashing. Generally, the flash-steam cycle will be chosen if the production temperature is ~200 °C or more, while the binary cycle will be chosen for production temperatures in the range 100 °C to 200 °C. In the binary plant, the geothermal fluid is circulated through one side of a heat exchanger. A working fluid is vaporized on the other side of the heat exchanger in a separate, closed-loop system. The vapor from the working fluid runs a turbine engine and is then condensed to liquid again before re-entering the heat exchanger (Figure 2). Choice of binary working fluids is based on their temperatures of vaporization and abilities to carry large amounts of heat. An isobutane/isopentane mixture is commonly used. Some organic working fluids engender a potential fire hazard, but the danger has been reduced to essentially zero by using modern engineering design methods. Binary plants operating on an ammonia-water cycle are somewhat experimental and show considerable promise. To date, there are no ammonia-cycle geothermal power plants operating commercially, although a small one is under construction in Iceland. Studies have shown that ammonia-water cycle plants should be several percent more efficient in conversion of thermal energy to electrical power than plants operating on organic fluids. Important research remains to be done in improving the conversion efficiency of binary-cycle plants.

Binary plants are often built in modular form, with individual modules being less than one to a few megawatts in size. This arrangement has several advantages. First, a power facility can start at small size until confidence in the reservoir is built up through production from a few wells. Later, additional modules can be added. Another advantage is that individual modules can be taken off line for repair or maintenance without affecting overall plant output unduly.

Combined-Cycle Power Plants

Some recent power-plant installations, such as that at Puna on the Big Island of Hawaii, that at Rotokawa in New Zealand, and the recently commissioned plant in Ethiopia, combine flash and binary cycles. The geofluid is first flashed, and the steam used to turn a turbine in the usual way. The fluid remaining from the flash and the vapor exiting the steam turbine are then used to drive a binary plant. This method improves the extraction of heat from the geofluid before it is returned to the subsurface.

The Contribution of Geothermal Electricity Generation

Most of the large, high-temperature geothermal fields worldwide support single flash or dual flash geothermal power plants, or combined-cycle power plants. Binary plants are being successfully used-on many lower-temperature reservoirs. Table 1 shows a sampling of geothermal power plants, along with their locations and some of their characteristics. Note that power is being successfully and economically generated by single-flash, double-flash, binary and combined-cycle plants ranging in size from 2 megawatts to 120 megawatts and using fluid ranging in temperature from 106 °C to > 300 °C.

Table 2 gives the current installed geothermal electrical generating capacity in

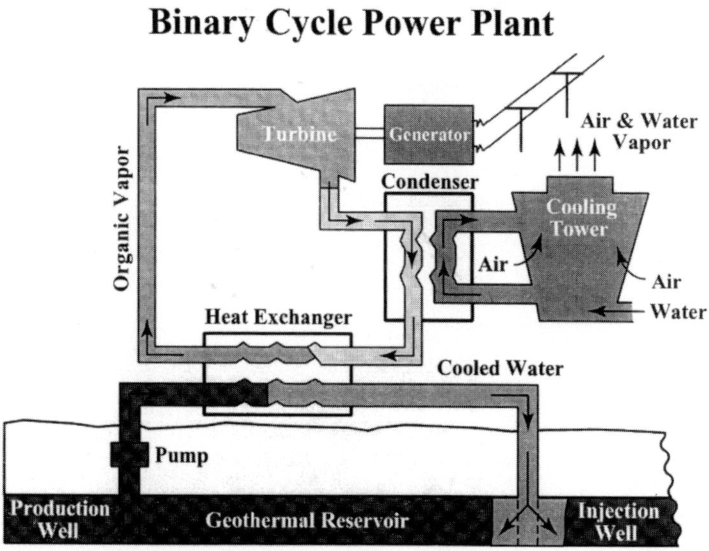

FIGURE 2. Geothermal Binary Turbine Generator.

TABLE 1.
Characteristics of Selected Geothermal Power Plants.
(From information on Geothermal Resources Council web site <www.geothermal.org>)

Plant Name	Location	Date Operational	Cycle Type	Gross Output (megawatts)	Fluid Temp (°C)
Geysers 18	Geysers Field, CA	2/83	Direct Steam	1 19	170
West Ford Flat	Geysers Field, CA	12/88	Direct Steam	32	238
Dixie Valley	Churchill Co., NV	7/88	Double Flash	67	249
Blundell	Beaver Co., UT	7/84	Single Flash	25	195
Heber	Imperial Val., CA	1/85	Double Flash	52	188
Salton Sea	Imperial Val, CA	2/89	Double Flash	50	316
Ormesa 2	Imperial Val., CA	1/88	Binary	24	160
Amedee	Lassen Co., CA	1/88	Binary	2	106
PGV1	Puna, Hawaii	4/93	Single Flash/Binary	25	316
Ahuachapan I	E1 Salvador	7/75	Single Flash	30	230
Olkaria	Kenya	6/81	Single Flash	48	299
Cerro Prieto 3-2	Mexico	1/86	Double Flash	110	337
Larderello 3	Italy	4/50	Double Flash	98	193
Valle Secolo	Italy	1/92	Double Flash	120	200
Yangbajain	Tibet	9/77	Double Flash	25	156

TABLE 2.
World Geothermal Electrical Production (Megawatts, MWe).
(After Huttrer, 2000)

United States	2,228	Kenya	45
Philippines	1,909	Guatemala	33
Italy	785	China	29
Mexico	755	Russia	23
Indonesia	590	Turkey	20
Japan	547	Portugal (Azores)	16
New Zealand	437	Ethiopia	9
Iceland	170	France (Guadeloupe)	4
El Salvador	161	Australia	0.2
Costa Rica	143	Thailand	0.3
Nicaragua	70	TOTAL	7,975

the world. We note that the current (beginning of 2000) installed capacity is 7,975 megawatts, an increase of ~6,700 megawatts over the installed geothermal generating capacity of 1,300 megawatts in 1975. An estimated 49,000 gigawatt-hours of electrical energy were generated from geothermal resources worldwide in 1999. For comparison, the International Energy Agency's web site (www.iea.org) quotes a figure for worldwide electricity consumption in 1996 at

13,700,000 gigawatt-hours. In other words, geothermal energy accounts for less than 0.4% of the world's electricity. While this figure is small, it must be understood that the potential for electricity production from geothermal energy is a much, much larger figure. A recent study by the U. S. Geothermal Energy Association (Gawell et al., 1999) estimates that eventually geothermal power could serve the needs of 865 million people worldwide. Development of new technology and improvement of market incentives will both be necessary to reach this goal, but the good news is that the worldwide geothermal resource is more than adequate to support this level of development. We see geothermal energy being one component in a mix of the renewable energy technologies and resources that will form the future energy resource.

Recovery of Minerals and Metals

Minerals deposited from geothermal fluids have been used by humans since time immemorial. Sulphur mined for gunpowder, borax for soap and salt for table consumption, are only three examples. However, until the present time it has not been economically attractive to recover metals from geothermal waters. However, there has been the recovery of various borate compounds at Lardarello, Italy using geothermal water since the late 1700's, and diatomaceous earth in Iceland since 1967. Some highly-saline geothermal waters contain significant amounts of recoverable salts, including valuable metals. The Salton Sea geothermal field in Southern California produces waters containing ~300,000 ppm dissolved salts, and these salts carry significant concentrations of valuable metals. During 2000, a plant will be commissioned by the field operator, CalEnergy Operating Corp., to recover zinc from geothermal fluids presently being brought to the surface to run the existing power plants (Clutter, 2000). Of course, the concentration of zinc in the geothermal fluids is very low, in the parts per million range, but the power plants require large volumes of water - 20 million pounds per hour in this case - so the amount of metals moving through the system and back into the ground through injection wells is significant. The zinc-recovery process will include ion exchange of zinc onto an organic resin, solvent extraction from the resin, and finally electrowinning to produce about 30,000 metric tons annually of 99.99 % pure zinc. Other contained metals of interest to CalEnergy are manganese, lithium, boron, and precious metals. Income from the metal recovery at the Salton Sea field will augment income from electricity sales and allow expansion of generation from the field at a time when geothermal power generation has difficulty competing with electricity generation using natural gas.

Few geothermal fields have fluids of high-enough salinity to carry significant quantities of most metals, and such metals' recovery systems are not generally feasible using today's technology. However, most high-temperature geothermal fluids contain dissolved silica, and geothermal field operators are now considering ways to recover this silica and produce a product high enough in grade to be marketable. Initial studies are very encouraging.

DIRECT USES OF GEOTHERMAL RESOURCES

Direct, or non-electric, utilization of geothermal energy refers to the immediate use of the heat energy rather than to its conversion to some other form such as electricity. The primary forms of direct-use include swimming, bathing and balneology (therapeutic use), space heating and cooling (including district heating), agriculture (mainly greenhouse heating and some animal husbandry), aquaculture (mainly fish pond and raceway heating), industrial processes, and heat pumps (for both heating and cooling). In general, the geothermal fluid temperatures required for direct use are lower than those needed for economic electric power generation.

Most direct applications use geothermal fluids in the low- to moderate-temperature range, between 50 °C and 150 °C. In general, wells needed to produce the reservoir can be drilled using conventional water-well drilling equipment. Throughout the world low-temperature geothermal systems are much more widespread than high-temperature systems, and are, thus, more likely to be located near potential users. In the U.S., for example, of the 1,350 known or identified geothermal systems, 5 % are above 150°C, and 85 % are below 90°C (Muffler, 1979). In fact, almost every country in the world has some low- temperature geothermal resources, while fewer have accessible high-temperature resources.

Traditionally, direct-use of geothermal energy has been on a small scale, mostly by individuals. More recent developments involve large-scale projects, such as district heating (Iceland, France, the U.S.), greenhouse complexes (Hungary, Russia, the U.S.), or major industrial use (New Zealand and the U.S.). Heat exchangers are becoming more efficient and better adapted to geothermal projects, allowing use of lower-temperature water and highly saline fluids. Heat pumps utilizing very low-temperature fluids have extended geothermal development into traditionally non-geothermal countries such as Switzerland and Sweden, as well as areas of the mid-western and eastern U. S. Most equipment used in these projects is of standard, off-the-shelf design and needs only slight modifications to handle geothermal fluids (Gudmundsson and Lund, 1985; Geo-Heat Center Quarterly Bulletin, 1997).

Worldwide, the installed capacity of direct geothermal utilization is almost 15,200 MWt (thermal megawatts) and the energy use is about 191,700 TJ/yr (53,200 GWh/yr) distributed among 58 countries (Lund and Freeston, 2001). This amount of energy is equivalent to that in 13.5 million tonnes of fuel oil per year (TOE). The distribution of energy use with type of application is shown in Figure 3. The installed direct-use geothermal capacity in the U.S. is 3,766 MWt and the annual energy use is 20,300 TJ (5,640 GWh), equivalent to the energy in 0.70 million TOE (Lund and Boyd, 2000). Internationally, the largest uses are for bathing, swimming and balneology (42%), and for space heating (23%) (2/3 of which is due to district heating); whereas, in the U.S., the largest use is geothermal heat pumps (59 %). In comparison, Iceland's largest geothermal energy use

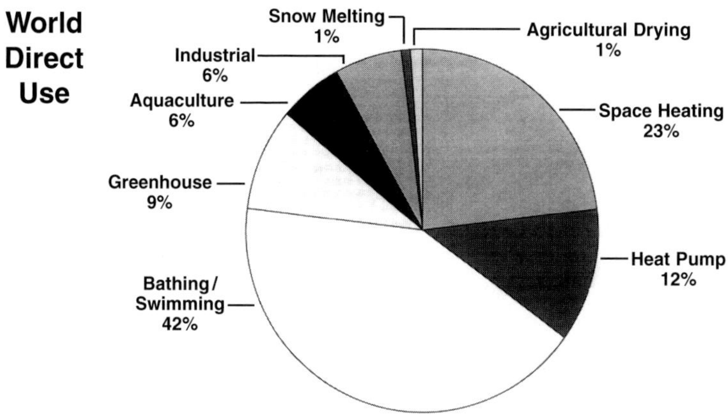

FIGURE 3a. Distribution of geothermal energy use in the world and the U.S.

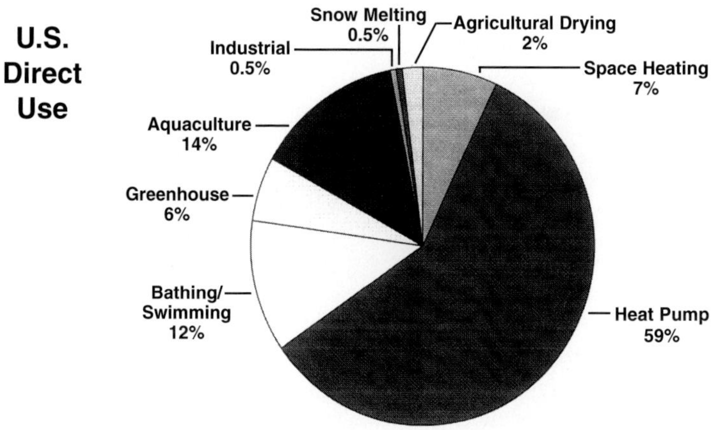

FIGURE 3b. Distribution of geothermal energy use in the world and the U.S.

is 77 % for space heating (4,330 GWh/yr) primarily with district heating systems (Lund and Freeston, 2001).

Data on direct-use applications are available for approximately 58 countries; however, the quality of the data is highly variable. The data are based on Lund and Freeston (2001) from country update papers prepared for the World Geothermal Congress 2000 (WGC2000) in Japan. Table 3 shows the installed capacity and total production by continent (Lund, 2000). As can be seen from this table, a large percentage of the geothermal direct utilization (35%) occurs in Europe, with only 14 % in the Americas. The opposite is true for geothermal electric power generation - 47 % occurs in the Americas, and only 12% in Europe. Table 4 lists the 10 top countries in direct utilization (Lund and

TABLE 3.
Direct Use of Geothermal Energy 2000
(Lund, 2000; Lund and Freeston, 2001)

	Installed Capacity (MWt)	Total (GWh/a)	Production %
European, Union	1,849	4,761	
Europe, other	3,865	14,143	
Europe, total	5,714	18,904	35
N-America	4,308	7,013	
C&S America	47	253	
America, total	4,335	7,226	14
Asia	4,623	24,505	46
Oceania	342	2,065	4
Africa	125	503	1
World Total	15,159	53,243	

TABLE 4.
Top Ten Countries in Direct Utilization
(Lund and Freeston, 2001)

	Installed (MWt)	Production (GWh/a)
China	2,282	10,531
Japan	1,167	7,662
USA	3,766	5,639
Iceland	1,469	5,603
Turkey	820	4,377
New Zealand	308	1,967
Georgia	250	1,752
Russia	308	1,707
France	326	1,360
Sweden	377	1,147

Note: The above tables include geothermal heat pumps (heating mode only).

Freeston, 2001). Based on country update papers for WGC2000, these numbers increased by over 35% for many countries since 1995. Table 5 presents a breakdown of the types of direct-use in the world and the top four countries. Of interest is the emphasis on bathing/swimming/balneology in Japan (80%) and China (62%), spaceheating (mainly in the form of district heating) in Iceland (77 %), and geothermal heat pumps in the United States (59%). Aquacultural use of heat is also prominent in China and the U.S.

TABLE 5.
Types of Direct Use in the World and the Top Four Countries (in %)(Lund, 2000, Lund and Freeston, 2001)

	World	China	Japan	USA	Iceland
District Heating	15	9	<1	4	77
Space Heating	8	0	14	3	0
Heat Pumps (heating and cooling)	12	0	<1	59	<1
Bathing/Swimming/Balneology	42	62	80	12	5
Greenhouse Heating	9	5	2	6	3
Aquacultural Pond/Raceway heating	6	12	2	14	3
Industrial	5	4	<1	<1	8
Snow Melting	<1	0	1	<1	2
Agricultural Srying	<1	<1	0	2	0
Others	2	8	<1	0	0
	100	100	100	100	100

In terms of market penetration, the leading user of geothermal energy is Iceland, where more than 86 % of the population enjoys geothermal heat in their homes from 26 municipal district heating services, and 50 % of the country's total energy use is supplied by direct heat and electrical energy derived from geothermal resources (Ragnarsson 2000).

The Lindal Diagram

The Lindal diagram, named for Baldur Lindal, the Icelandic engineer who fist proposed it (Gudmundsson et al., 1985), indicates the temperature range suitable for various direct use activities (Figure 4). Typically, agricultural and aquacultural uses require the lowest temperatures, with values from 25 °C to 90 °C. In some resources, dissolved salts include hazardous species (arsenic, boron) that could pose a problem for plants and animals. To mitigate this potential problem, heat exchangers are often used, and the hazardous geothermal water is injected back into the subsurface geothermal reservoir. Space heating requires temperatures in the range 50 °C to 100 °C, with 40 °C being useful in some marginal cases. Ground-source heat pumps extend the range down to 4 °C. Cooling and industrial processing normally require temperatures over 100 °C.

Equipment

Standard equipment is used in most direct-use projects, provided allowances are made for the nature of geothermal water and steam. Temperature is an important consideration, as is water quality. The primary components of most low-temperature direct-use systems are downhole and circulation pumps, transmission and distribution pipelines, peaking or back-up plants, and various forms of heat extraction equipment (Figure 5). Fluid disposal is sometimes on the surface, but more often in the subsurface by injection. Of course, all fluid

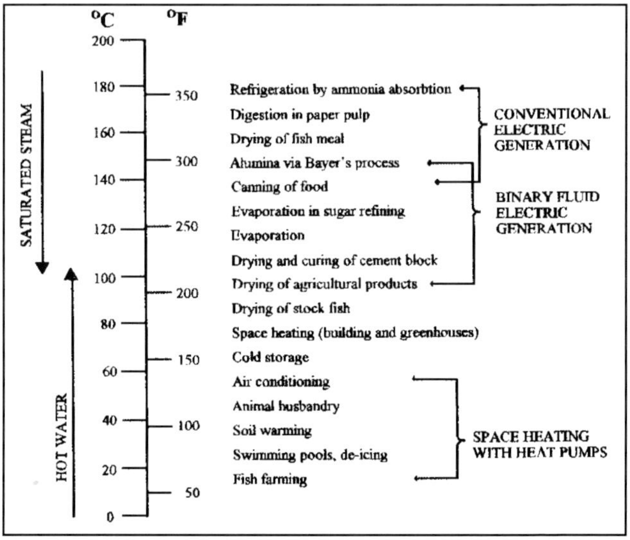

FIGURE 4. Lindal Diagram.

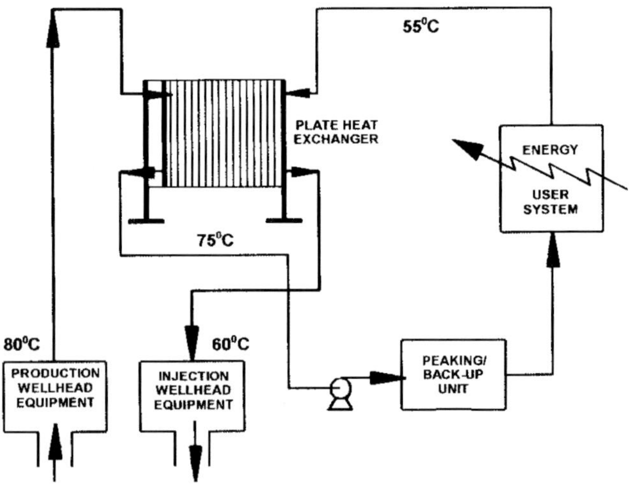

FIGURE 5. Geothermal direct-utilization system using a heat exchanger.

disposal must meet project permitting requirements to ensure safety and preclude environmental damage.

A peaking system may be necessary to meet maximum load. Peaking can be provided by increasing the water temperature or by providing tank storage of extra thermal water. Both options mean that fewer producing wells are needed, resulting in a cost savings. When the geothermal water temperature is merely warm (below 50°C), heat pumps can often be used to boost water temperature at a low cost. Detailed discussion of equipment design and use can be found in Lund, *et al.* (1998).

Corrosion of equipment and precipitation of scales in wells and surface equipment are a potential problem, depending on the chemistry of the geothermal fluid. In almost all instances, such problems can be handled by proper engineering design of the system. One common problem concerns dissolved oxygen, which is absent in most geothermal waters, except those that may have been mixed in the subsurface with near-surface, meteoric water. Dissolution of oxygen in geothermal water can alter the chemistry remarkably, and lead to both corrosion and scaling problems that would not otherwise occur. Care is usually taken to prevent atmospheric oxygen from entering direct-use systems by proper design of pumps, piping, and storage tanks. The isolation of geothermal water by installing a heat exchanger may also solve this and similar problems. If a heat exchanger is needed, a clean secondary fluid can be circulated through the consumer side of the system, with the geothermal water from the primary side of the heat exchanger being injected into the ground after use.

Space Conditioning

Space conditioning includes both heating and cooling. Space heating with geothermal energy has widespread application, especially on an individual basis. Buildings heated from individual wells are popular in Klamath Falls, Oregon; Reno, Nevada; and Taupo and Rotorua, New Zealand. Absorption chilling for air conditioning with geothermal energy has not been much used because of the high temperature requirements of chillers and the low efficiency of the cycle. Geothermal heat pumps (groundwater and ground-coupled) have become popular in the U. S., Canada, Switzerland, and Sweden, used for both heating and cooling.

An example of space heating and cooling with low- to moderate-temperature geothermal energy is the Oregon Institute of Technology in Klamath Falls, Oregon. Here, eleven buildings with approximately 62,000 m^2 of floor space are heated with water from three wells. Average production temperature is 89 °C. Up to 62 L/s (930 gpm) of fluid can be provided to the campus from these wells, with the average heat utilization rate >0.53 MWt and the peak utilization at 5.6 MWt. In addition, a 541 kW (154 tons) chiller, requiring up to 38L/s of geothermal fluid, produces 23 L/s of chilled fluid at 7 °C to meet the campus cooling base load.

District Heating

District heating is the heating *en masse* of buildings connected to a common system. The heat source is usually in a central location, and supplies hot water or steam through a network of pipes to individual dwellings or blocks of buildings. The heat is used variously for space heating and cooling, domestic water heating and industrial process heat. In geothermal district heating, a geothermal well field is, of course, the primary source of heat. However, depending on the resource temperature, the system may be a hybrid, including the occasional use of fossil fuel or heat pumps to meet peak loads.

Geothermal district heating systems are in operation in at least 12 countries, including Iceland, France, Poland, Hungary, Turkey, Japan and the U. S. The Warm Springs Avenue project in Boise, Idaho, dates back to 1892 and originally heated more than 400 homes. It is the earliest such project in the U. S. The Reykjavik, Iceland, district heating system (Figure 6) is probably the most famous in the world (Frimannsson 1991, Lund 1996, and Ragnarsson, 2000). This system supplies heat for a population of around 160,000 people. The installed capacity of 830 MWt is designed to meet the heating load at temperatures above -10°C. During the occasional colder periods, the increased load requirement is met by large storage tanks and an oil-fired booster station.

In Paris, France, production wells in sedimentary basins provide heat to more than 200,000 housing units from 41 projects (LaPlaige, et al., 2000). These wells provide from 40 °C to 100 °C water from depths of 1,500 to 2,000 m. In the Paris

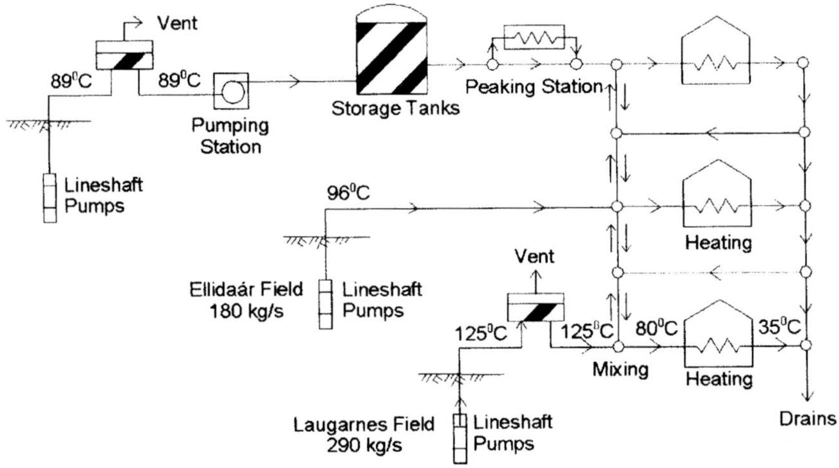

FIGURE 6. Reykjavik district heating system (prior to the Nesjavellir connnection).

basin, a doublet system (one production well and one injection well) provides 70 °C water, with the peak load met by heat pumps and conventional fossil fuel burners.

Agribusiness Applications

Geothermal energy is particularly attractive in agriculture and aquaculture applications due to the fact that their temperature requirements are at the lower end of the geothermal temperature range, where there is an abundance of resources. Use of waste heat or the cascading of geothermal energy uses also has good potential in agribusiness uses. Agribusiness applications that can be considered include greenhouse heating, aquaculture and animal husbandry, soil warming and irrigation, mushroom culture, and bio-gas generation.

Numerous commercially marketable crops are being raised in geothermally heated greenhouses in Hungary, Russia, New Zealand, Japan, Iceland, China and the U.S. These include vegetables, such as cucumbers and tomatoes, potted and bedded flowers, house plants, tree seedlings, and cacti. Using geothermal energy for heating reduces operating costs, which can account for 35% of the product cost, and allows operation in colder climates where commercial-scale greenhouses heated with fossil fuels would not normally be economical. Because the geothermal heat is so inexpensive, once installed, greenhouses can be ventilated economically with large volumes of outside air to help keep down plant disease.

The use of geothermal energy for raising catfish, shrimp, tilapia, eels, and tropical fish has produced crops faster than by conventional heating or solar heating. Using geothermal heat allows better control of pond temperatures, thus optimizing growth.

Industrial Applications

Although the Lindal diagram shows many potential industrial and process applications of geothermal energy, the world's uses are relatively few. The oldest industrial use is at Larderello, Italy, where boric acid and other borate compounds have been extracted from geothermal brines since 1790. Today, the two largest industrial uses are a diatomaceous-earth drying plant in northern Iceland and a pulp, paper and wood processing plant at Kawerau, New Zealand. Notable U.S. examples are two onion dehydration plants in northern Nevada (Lund, 1995), and a sewage digestion facility in San Bernardino, California. Alcohol fuel production has been attempted in the U.S. using geothermal heat; however, the economics were marginal and this application has not been successful.

Cascaded Uses

By cascading of geothermal applications we mean extracting heat from geothermal water in two or more applications in sequence, those first in the sequence requiring higher-temperature water than those later in the sequence. For example, water used to heat a greenhouse complex could subsequently be used for aquaculture prior to being disposed. In principal, cascading has not been practiced at more than a few sites, but there is much interest in the concept. One example of cascading is at the Utah State prison, where geothermal heat has been used for space heating, with the water then being used to raise tropical fish as a business within the prison, and the effluent then going to a hog farm.

A seemingly obvious application of cascading might be the use of waste heat from a geothermal power plant, or for that matter from any large power plant, fossil-fired or nuclear, since they all have heat that must be rejected. However, there are practical and legal difficulties to this scheme, and it is not usually done.

Heat Pumps

No technology for home heating and air conditioning is more efficient than the geothermal heat pump (GHP). GHPs reduce electricity use 30% to 60% compared with traditional electrical heating and cooling systems. The net savings in energy consumption mean fewer adverse environmental impacts from power plants. Domestic hot water can be produced essentially free by the GHP during the air-conditioning season and for small cost in winter. Maintenance costs for GHP systems have proven to be very low. Best of all, the customer enjoys lower utility bills.

GHP systems can be used for a variety of applications, including the heating and cooling of homes, schools and other buildings, heating of water, and commercial and industrial heating and refrigeration applications. Used for years in Europe, geothermal heat pumps are making a big impact on energy efficiency in the U.S. and Canada.

The heat pump itself operates on the same principle as the home refrigerator, which is actually a one-way heat pump. The GHP, however, can move heat in either direction. In the winter, heat is removed from the earth and delivered into the home or building (heating mode). In the summer, heat is removed from the home or building and delivered for storage into the earth (air-conditioning mode). Because electricity or natural gas, which powers the GHP motor, is used only to move heat, not to produce it, the GHP will deliver 3 to 4 times more energy than it consumes.

The GHP unit sits inside the home or building, at the site normally used for a gas furnace (Figure 7). In a typical installation, a loop of plastic pipe is placed in a vertical hole bored several hundred feet deep, and the hole is then backfilled. An environmentally safe antifreeze solution is circulated through the loop and through the heat pump for removing heat from or transferring heat to the ground. There is no consumptive use of groundwater, nor is there any contact between the solution in the plastic pipe and the earth or ground water. In other applications, ground water from a well may be used as the source or sink for heat. Several

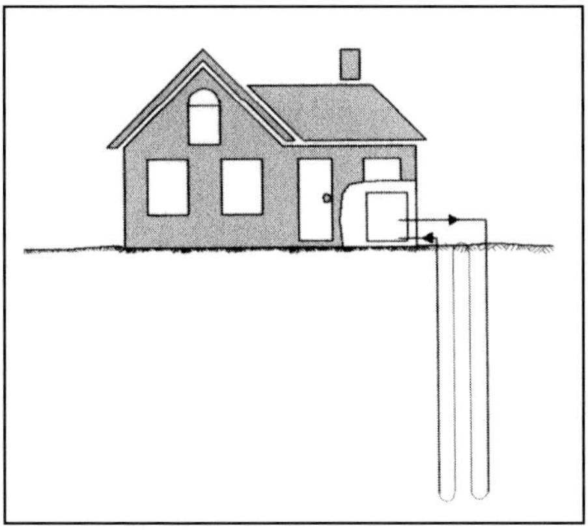

FIGURE 7. Typical ground-source heat pump installation.

alternative groundloop installations have also been successfully used. Loops of plastic pipe can be placed in a horizontal trench, either stretched out or in the form of coils (the so-called "slinky" configuration). Placement at the bottom of a nearby pond is also possible if the pond does not freeze completely in winter. Heat pumps are available with heating capacities of less than 3 kW to over 1,500 kW.

It is estimated that almost 120,000 groundwater systems, and more than 100,000 closed-loop vertical and 200,000 closed-loop horizontal systems are already in use in the U. S. In Europe, an estimated 120,000 units have been installed (Rybach and Sanner, 1999).

ECONOMICS OF GEOTHERMAL ENERGY USE

The economics of geothermal development are affected by many variables. These include, but are certainly not limited to: quality of the resource (temperature and fluid chemistry); subsurface permeability, i.e. rates of production from wells; drilling depth to the resource; access to the field site for operations such as drilling and construction; costs of environmental control such as hydrogen sulfide scrubbing; and comparative costs of competing energy sources.

Regarding resource quality, it is generally true that the higher the temperature, the higher the resource value since more energy can be extracted from a given quantity of thermal water. The lower the salinity and the hydrogen sulfide content, the easier the resource will be to develop. Due to the high cost of geothermal drilling, shallow reservoirs are less expensive to develop than deep ones. Production rates from geothermal wells have a wide range. Some so-called "super wells" have been drilled that produce enough thermal fluid from one well to generate 50 megawatts of electrical power. Flow rates less than ~250,000 pounds per hour from high temperature resources are usually considered to be non-commercial. Of course, direct-use applications may require much lower production rates than this figure.

Since geothermal resources are spawned by geologic activity such as volcanism and faulting, many geothermal resources are located in mountainous regions where access might be difficult. Forests and jungles compound access problems, increasing drilling and construction costs. Skilled workers are in short supply in many developing countries, another important cost element. Other developing countries have inadequate laws and regulations to encourage investment, and highly bureaucratic federal and local governments. With all of the above considerations, we must conclude that the economics of geothermal energy development vary widely from resource to resource and country to country, and with topographical aspect, geographic location, and the socioeconomic structure of the country. It is, thus, very difficult to give more than a range of possible costs for geothermal energy development and utilization. In this section, we will discuss in a general fashion some of the factors that affect the cost of geothermal electrical power or thermal water for direct uses.

In terms of electrical power generation, a number of assertions that are more or less true at each project site can be made. First, the cost of installing the well field, including exploration, drilling, and well testing is usually in the range 35 % to 50 % of the total project costs. As a rough guide, the cost of the power plant equipment and construction is in the range $800 to $1,200 per installed kilowatt of generation capacity, and total project costs are in the range $2,000 to $3,000 per installed kilowatt. Thus, a 55 megawatt plant on a new geothermal field in the United States with easy access and reasonable drilling costs will represent an investment of $100 million to $150 million. Such a plant, again on a new field, will generate electrical power at a cost in the range 5 cents to 7 cents per kilowatt hour. These costs may be compared with today's generation costs, using a combined-cycle combustion turbine operating on natural gas, of 2 cents to 3 cents per kilowatt hour (kwh). It is easy to see that geothermal energy is not very competitive.

However, after the costs of the plant and well field are paid off, geothermal power generation is quite inexpensive. More will be said about this later.

Of course, the costs discussed above are for locations having reasonable access to the field, good service and construction infrastructure (ports, rail lines, roads), and governments that have created a proper investment environment. Such conditions do not apply in many developing countries, and geothermal power generation costs will be higher.

Costs of direct use of geothermal resources are extremely variable. Recent international data (Freeston, 1996) gives US$ 270/kW of installed capacity for all projects reported, with a range from US$ 40 to US$ 1880/kW. In the U. S., the annual operation and maintenance cost is estimated at 5% of the installed cost.

Competitors to Geothermal Energy

From several viewpoints, the chief competitor to increased geothermal energy utilization is the fossil fuels. Fossil fuels are the cheapest and most plentiful energy resource in most locations in the world today. Fossil fuels are portable, and form a worldwide market commodity, unlike geothermal energy. In addition, the fossil fuel industry is enormous in size and is well entrenched. It has a vast, worldwide network of infrastructure for production, distribution and consumption of its products, and fossil fuels themselves are plentiful. Most people studying energy availability now believe that fossil fuels will continue to be plentiful for at least a century into the future (Nakićenović, et al., 1998). Much remains to be discovered. For example, the vast stores of methane hydrate on the ocean floors, just now being discovered and understood, will no doubt contribute to our use of natural gas in years to come.

Further challenges for geothermal energy will derive from the on-going deregulation of the energy sector and privatization of national energy organizations. Geothermal energy will be forced to compete head-on in energy markets where cheaper fuels are dominant, and could easily be placed in a difficult situation. There is some up-side to this situation, however- in the United States, geothermal operators have been competing very successfully in the spot energy market, earning a premium for their electrical power during periods of high demand.

In terms of renewable electrical power generation, wind power is a competitor in some markets, but the advent and growth of wind generation is not expected to affect the market for geothermal power significantly. In some countries having hydropower resources as well as geothermal resources, such as the Philippines, the two options do compete at some level. Of course, solar electrical power generation is not and will not be a serious competitor of geothermal energy for years to come.

It is unfortunate that some of the major benefits of using geothermal energy are not recognized in the energy marketplace today (Wright, 2000). Two of these are the environmental benefits of having clean, non-polluting energy available, and the low life-cycle costs of geothermal energy utilization.

Uncompensated Environmental Advantages

Today in most countries, there is little or no penalty for emissions to the atmosphere of polluting gasses - gasses that cause global warming, acid precipitation, and ground-level ozone. Without penalties reflected in the energy marketplace, pollution will continue and even grow, along with growing damage to the environment, ecosystems, and human health. What can be done about world markets that do not value the environment? This is a knotty problem. The World Bank, the United Nations, and the governments of some far-seeing countries have been exploring economic systems that value the environment through strategies for sustainable development (Serageldin and Steer, 1944). At the present time, however, there is little political will in most industrialized nations, especially in the world's largest economy, the United States, to implement dramatic changes.

As global warming proceeds, it will cost billions of dollars to combat the flooding, hurricane damage, poor crop and timber harvests, and population resettlement. However, the advent of the Kyoto Protocol has spawned serious attempts by governments of some countries to reduce carbon dioxide emissions in an attempt to halt potentially catastrophic global warming. Also, there appears to be momentum developing in industrialized countries toward a green energy market. A certain fraction of energy users appear to be willing to pay a somewhat higher price to have clean energy available to them. Such a green energy market is helping sustain geothermal production in a deregulated electrical market in the United States today.

Other market-effecting methods of spurring the utilization of renewable energy include the efforts by some governments to mandate a certain percentage of a country' s energy supply to come from renewable sources. This so-called "renewable portfolio standard" will likely be adopted in more countries to help the fledgling renewable industries reach and exceed a critical, self-sustaining size, while at the same time decreasing use of fossil fuels and their emissions of atmospheric pollutants. Another incentive being used in some countries is the production tax credit. In general, these incentives are good in that they help achieve the goal of accelerated geothermal energy development.

Life-Cycle Costs

A second aspect of geothermal energy utilization that is not recognized adequately in today's energy marketplace is its comparatively low life-cycle cost (Wright, 2000). Although the initial costs of geothermal installations are high, the life-cycle costs of geothermal energy utilization are reasonably low. A significant portion of the start-up costs are concerned with securing the fuel supply, that is, performing exploration, and then discovery, production and injection drilling. However, when costs of installation of a geothermal well field and the utilization plant (electrical or direct-heat) have been recovered, the operation and maintenance costs are very competitive in energy markets. In the United States, geothermal power plant operations and maintenance costs are in the range 1.5 to 3.5 cents per kilowatt hour, comparable with natural gas generation costs. Since we can expect geothermal power plants to have a 30-year life or more, and geothermal reservoirs to produce for decades or even centuries, the average cost of geothermal energy over this lifetime is quite low.

However, today's energy markets take a comparatively short-term view and low life-cycle costs confer little advantage compared with the problem of high startup costs. With the market economy becoming a more dominant force in the world economy, the advantage of low life-cycle costs for geothermal energy is being lost, at least to some degree. The intervention of governments would appear to be needed in the matter, so that the low life-cycle costs of geothermal energy are captured. Such intervention should, of course, result naturally from a country's or even a company's long-term planning, and need not be overly intrusive. It could take the form of an incentive consciously to support development of geothermal resources in order to obtain their long-term benefits of low cost and environmental friendliness.

OUTLOOK FOR THE FUTURE

There appears to be a very large potential for the development of geothermal electrical power generation and direct-use across the world. Only a small fraction of available resources are being used today. Financial constraints in many countries and the low price of competing energy sources offers competition and has delayed market penetration of this clean, reliable energy source. Given the right environment, and as gas and oil supplies dwindle, the use of geothermal energy will provide a competitive, viable and economic alternative source of renewable energy.

Development of new and improved technology will continue to make a larger fraction of the geothermal resource base useable. The conventional energy sources have had a long developmental history, with a great deal of research support and subsidy from world governments. The use of geothermal energy is quite recent by comparison, except, of course, for the ancient uses of bathing and cooking. As the fledgling geothermal industry matures worldwide, we can reasonably expect applications to expand. And if global warming poses the threat that some people, including an apparent majority of scientists, think it does, then we will see a jump in impetus to develop this abundant, clean energy resource.

REFERENCES

Clutter, T. J. 2000, Mining economic Benefits from Geothermal Brine: Geothermal Resources Council Bulletin, 29, 1, 31-2.

Freeston, D. H., 1996. Direct Uses of Geothermal Energy 1995, *Geothermics*, 25(2): 189-214.

Frimannsson, H., 1991. Hitaveita Reykjavikur After 60 Years of Operation – Development and Benefits, *Geo-Heat Center Quarterly Bulletin*, Vol. 13(4): 1-7.

Gawell, Karl, Reed, M.J., and Wright, P.M., 1999, Preliminary Report – Geothermal Energy Potential for Clean Power from the Earth: Geothermal Energy Association, 7 April, 13 p.

Geo-Heat Center, 1997. *Quarterly Bulletin*, 19(1), Geothermal Direct-Use Equipment: 38 p.

Gudmundsson, J.S.; D.H. Freeston and P. J. Lienau, 1985. The Lindal Diagram, *Geothermal Resources Council Transaction*, 9(1): 15-19.

Gudmundsson, J.S. and J.W. Lund, 1985. Direct Uses of Earth Heat, *Energy Research*, 9: 345-375.

LaPlaige, P., F. Jaudin, A. Desplan, and J. Damange, 2000. The French Geothermal Experience – Review and Perspectives. *Proceeding of the World Geothermal Congress 2000*, International Geothermal Association, Pisa, Italy, pp. 283-295.

Lund, J.W., 1995. Onion Dehydration, *Geothermal Resources Council Transaction*, 19: 69-74.

Lund, J.W., 1996. Balneological Use of Thermal and Mineral Waters in the USA, *Geothermics*, 25(1): 103-148.

Lund, J.W. 2000. World Status of Geothermal Energy Use – Overview 1995-1999, *Proceedings World Geothermal Congress 2000*, International Geothermal Association, Pisa, Italy, pp. 4105-4111.

Lund, J.W. and T.L. Boyd, 2000. Geothermal Direct Use in the United States, Update: 1995-1999, *Proceedings of the World Geotherrnal Congress 2000*, Japan, International Geothermal Association, Pisa, Italy, pp 297-306.

Lund, J.W. and D.H. Freeston, 2001. World-wide Direst Uses of Geothermal Energy 2000, *Geothermics*, 30/1, Elsevier Science, Ltd., Oxford, Great Britain, pp. 29-68.

Lund, J.W.; Lienau, P. J. and B. C. Lunis (editors), 1998. *Geothermal Direct-Use Engineering and Design Guidebook*, Geo-Heat Center, Klamath Falls, OR: 470 p.

Muffler, L. P. J., (Ed.) 1979. Assessment of Geothermal Resources of the United States - 1978, *USGS Circular 790*, Arlington, VA.

Nakićenović, Nebojsa, Grübler, Arnulf, and McDonald, Alan, eds., 1998, Global Energy Perspectives: published for the International Institute for Applied Systems Analysis and the World Energy Council by Cambridge University Press, 299 p.

Ragnarsson, A., 2000. Iceland Country Update, *Proc. of the World Geothermal Congress 2000.* International Geothermal Association, Pisa, Italy. pp 363-375.

Rybach, L. and B. Sanner, 1999. Ground-Source Heat Pump Systems – The European Experience, *Proceedings of the 1999 Workshops, International Geothermal Days – Oregon 1999*, Geo-Heat Center, Klamath Falls, OR: 159-170.

Serageldin, I. and Steer, A., eds., 1994, Making Development Sustainable – From Concepts to Action: Environmentally Sustainable Development Occasional Paper No. 2, The World Bank, Washington, D.C., 40 p.

Wright, P.M., 1998a, Exploration for Direct Heat Resources: *in* Lund, J.W., ed., Geothermal Direct Use Engineering and Design Guidebook, 3rd Edition, GeoHeat Center, Oregon Institute of Technology, pp 71-113.

Wright, P.M., 1998b, The earth gives up its heat: Renewable Energy World, 1, 3, 20-25.

Wright, P.M., 2000, Geothermal energy in the new century: International Geothermal Association Newsletter, Quarterly No. 40, Pisa, Italy. Special Address delivered to the World Geothermal Congress 2000, 8 June 2000, Morioka, Japan.

Renewable Energy: Trends and Prospects. Edited by S.K. Majumdar, E.W. Miller and A.I. Panah

Chapter Seventeen

HYDROGEN POWER IN SPACE VEHICLES

BRYAN PALASZEWSKI

Leader, Advanced Fuels
NASA John H. Glenn Research Center
Lewis Field
MS 5-10
Cleveland, Ohio 44135

Over the last 5 decades, a wide variety of hydrogen-based propellants for space vehicles have been developed. This article will discuss some historical background for and the properties of these hydrogen-based fuels, as well as their applications for space missions. Mission analyses will be presented to show the future benefits of these fuels, which included gelled hydrogen, metallized gelled propellants, and atomic propellants. Also discussed are the technologies needed to create atomic boron, carbon and, hydrogen fuels in solid hydrogen.

NOMENCLATURE

Al	Aluminum
APS	Auxiliary Propulsion System
ESA	European Space Agency
GLOW	Gross Lift Off Weight
He	Helium
H_2	Hydrogen
I_{sp}	Specific Impulse (s)
LEO	Low Earth Orbit
LRB	Liquid Rocket Booster
MMH	Monomethyl hydrazine

NASA	National Aeronautics and Space Administration
NASDA	National Space Development Agency of Japan
NTO	Nitrogen tetroxide
O_2	Oxygen
RP-1	Rocket propellant 1 (highly refined kerosene)
SEI	Space Exploration Initiative
SSTO	Single Stage to Orbit
STS	Space Transportation System
STS-C	Space Transportation System-Cargo
wt%	Weight percent

INTRODUCTION

In the development of space exploration, hydrogen has been a critical fuel for numerous applications. Hydrogen was first used in space vehicles on the Centaur upper stage[1] in the early 1960's. The Centaur, aboard an Atlas rocket, was the first vehicle to deliver the Surveyor soft landing vehicles to the Moon and to deliver Mariner 9, the first orbital mapping mission to Mars.[2] The Centaur, in many improved forms, is still used to this day to expand our horizons in the Solar System and the Universe. After the development of the Centaur, NASA was convinced that hydrogen was an excellent candidate fuel to assist humans to reach the Moon in Project Apollo.[1] The Saturn I, IB, and Saturn V rockets all used high energy hydrogen stages to propel both robotic and human missions to the Moon and into near Earth space. The Former Soviet Union, and now Russia, had developed the Energia Booster, which used hydrogen for its upper stage.[3] This vehicle has been flown for test missions, and is a potential option for very heavy lift space missions. The Ariane series of rockets of the European Space Agency (ESA), and the Japanese H-2 rocket series (of the National Space Development Agency of Japan, NASDA) represent current vehicles with hydrogen fueled upper stages.[3] The NASA Space Shuttle is also one of the largest current users of hydrogen in a large External Tank that supplies oxygen and hydrogen to three Space Shuttle Main Engines.[3] These engines power the upper stage of the Shuttle, and are returned to Earth for reuse on successive flights.

WHY HYDROGEN?

Liquid hydrogen is one of the most powerful and energetic fuels. This fuel has been identified through analyses and testing to be the most attractive for space missions. Because of its high energy, hydrogen is able to reduce the launch mass of rocket vehicles, thereby reducing the overall size of the rocket and making space travel more practical. Early concepts for lunar transportation, including the Nova rocket, relied solely on liquid kerosene and solid rocket propellants.[4, 5, 6] Using these lower energy propellants required rockets that were more massive than the Saturn V used for Moon missions. Some engineers estimated that the Saturn V would be twice as tall if it were to exclusively use liquid kerosene propellants.

While liquid hydrogen can reduce the size of rockets, it has some important features, such a cryogenic storage, and high boil off rates. Liquid hydrogen's storage temperature is 20 degrees Kelvin, and this presents unique storage and handling requirements. However, over the many decades of using hydrogen, engineers and scientists have harnessed its power, and solved many of the problems that plagued early space flight missions. Materials issues, such as hydrogen leakage through metal skins, caused early Atlas-Centaur flights to fail far short of orbit.[2] Hydrogen leakage and control have been mastered since, and successful flights are achieved with a relatively high rate of success.

FUTURE HYDROGEN-BASED FUELS

A critical issue with the future use of hydrogen is the liquid fuel density. Liquid rocket propellants are often the largest volumes aboard a rocket vehicle. Their large volume must be contained by very large, bulky, heavy tanks that keep the liquids at the right pressure and temperature until they are used in a rocket engine. An increase in the density, or mass per volume, can reduce the mass of the tanks, structure, and other parts of the rocket that are dependent upon the volume of the propellants. Higher density propellants are therefore an excellent way to improve the performance of rocket vehicles because they will reduce the mass of the tanks and the associated components. Reducing the mass of the rocket will also lead to reducing the weight of fuel needed to move the rocket, so there is a powerful cascade effect that occurs when reducing the mass of rockets and the amount of fuel.

Reducing the weight of a rocket is also made possible by using a higher energy fuel, because less fuel is needed as its energy increases. This energy is related to the exhaust velocity or specific impulse (I_{sp}) of the rocket engine, essentially a miles-per gallon analogy for rockets. Typically, the higher the energy of the fuel, the higher the specific impulse. Adding energy to the fuel is accomplished by changing its chemistry, gelling the fuel, gelling and adding metal particles, or adding other higher energy compounds. One of the highest energy compounds is a fuel with boron, carbon, or hydrogen atoms. These atomic fuels represent the most futuristic of chemical propellants, but they have the greatest potential for extremely high specific impulse, hundreds of seconds greater than O_2/H_2 powered engines.[7]

Several ways of increasing the density of the propellants have been used in the past. One of these ways is to add small metal particles, or frozen liquid particles, and suspend them in the fuel or oxidizer.[8-23] A gelling agent is used to thicken the fuel and allow the suspension of the solid particles. The gelling agent may be frozen liquid particles, solid particles, or long-chained liquid polymers. Gelling the fuel, without adding metal particles, can also increase the density and change the energy of the fuel. The gelling agent can add energy and increase the fuel density, but the energy and density increases are much larger if metal additives are also used. Density of the fuel can also be increased by using solid hydrogen, a technique that is suggested to trap and store the very high energy boron, carbon, or hydrogen atoms.[7, 13, 24-26] The solid hydrogen particles would be stored in liquid helium, which would maintain their temperature at 3 or 4 degrees Kelvin.

In this article, the methods of gelling propellants, increasing energy with metal combustion, increasing fuel energy with stored atoms, and the overall vehicle and mission benefits will be outlined. Several propellant types are illustrated, and the benefits of gelled hydrogen (with no added metal), metallized gelled propellants, such as $O_2/H_2/Al$ and O_2/RP-1/Al, and atomic hydrogen are described.

GELLING PROPELLANTS

Gelling the propellant is a way of increasing the viscosity, or thickness, of a liquid. For gelled liquids, a shear thinning liquid is desired.[11] When such a liquid begins to flow, its viscosity drops and makes the liquid easier to flow. The thinner low viscosity liquid is easily pumped from the propellant tank to the rocket engine. When the liquid is not flowing, the thickened liquid helps to suspend the additives. Thus, the additives or metal particles are uniformly distributed in the fuel, and can remain there until needed by the engine.

METAL COMBUSTION

To create a metallized gelled propellant, fine particles of metal are added to a gelled fuel. Fine particles typically have better combustion properties. A size from 5 to 10 microns (or millionths of a meter) in diameter are used in metallized gelled fuels. In past research, many different metals have been investigated, but aluminum is an excellent overall choice.[11] It is easily produced and handled, and has very good combustion energy. Boron and carbon are also metal additive candidates that have good combustion properties. In most cases, there are improvements in rocket engine performance: 5 to 25 seconds increase in I_{sp}. Though the performance benefits seem small, these increases in specific impulse have a powerful effect on reducing the overall mass of rocket vehicles, which will be discussed in a later section.

Boron, carbon, and hydrogen atoms are also potential fuel additives.[7, 13, 24-26] The atoms would be stored in a solid cryogenic ice, made of hydrogen at 4 Kelvin (-452 degrees F), a very low temperature. In the more traditional liquid rocket propellants, only metal molecules are desirable. The metal atoms are very unstable at temperatures above 10 Kelvin (-441 degrees F) and will tend to recombine into molecules before their energy can be used in the rocket engine. These very advanced atomic propellants have not been demonstrated in large scale rockets, and are relatively far in the future, but their potential is very exciting, as their specific impulse can be greater than 750 seconds, more than 50% higher than traditional O_2/H_2 propellants.[7, 13]

WHY GELLED AND METALLIZED GELLED PROPELLANTS?

Gelled and metallized gelled rocket propellants have been considered for many different applications. Numerous studies have shown the potential benefits of gelled fuels and oxidizers.[8-12, 14-23] While gelled fuels are not in operational use, there are many technology programs underway to eliminate the unknowns with gelled propellants and the propulsion systems that will use them. Technology programs to prove the combustion performance of gelled propellants have been conducted most recently by the U.S. Army Missile Command, with their industry and university partners, for tactical missile applications.[23] The NASA Glenn Research Center (formerly NASA Lewis) and its partners have investigated O_2/H_2 /Al and O_2 /RP-1 /Al for NASA missions and conducted experimental programs to validate elements of the combustion and fuel technology. Gelled and metallized gelled hydrogen and RP-1 have been emphasized because hydrogen and RP-1 are typical propellants for NASA launch vehicles and upper stages. Derivatives of these propellants are therefore preferred to minimize the incremental risk for a newly introduced propulsion concept.

APPLICATIONS AND BENEFITS OF GELLED AND METALLIZED GELLED PROPELLANTS

Gelled and metallized gelled propellant applications would be for rocket powered launch vehicles, upper stages, interplanetary craft, rocket based combined cycle airbreathing vehicles, as well as vehicles using combination (rocket and airbreathing) propulsion. Airbreathing space vehicles will typically have a large hydrogen tank, making up the largest volume of the vehicle. Any method of reducing the volume of that tank will benefit these space vehicles.

There are five major benefits of these gelled and metallized gelled propellants: safety increases, boiloff reductions, density increases with the attendant area, and volume related mass reductions for related subsystems (thermal protection system, structure, insulation, etc.), slosh reductions, and I_{sp} increases (in some cases).

Safety can be significantly increased with gelled fuels. A higher viscosity reduces the spill radius of the gelled fuel and limits the potential damage and hazard from a fuel spill. Another important advantage is the potential for leak reduction or elimination. The leak paths from the feed systems would be minimized and the possible explosion potential would be reduced. Boiloff reduction is another feature of gelled hydrogen, and they are up to a factor of 2 to 3 over ungelled liquid hydrogen. This feature will assist in long term storage of hydrogen for upper stages that must be stored in orbit or if the stage must have a long coast time between firings. Also, lunar flight and interplanetary missions with large hydrogen fuel loads will derive a benefit.

Significant density increases are possible and have been demonstrated with gelled and metallized gelled propellants. Figure 1 depicts the density for gelled hydrogen with a range of weight percent (wt%) values of methane gellants. A 10% density increase is possible with a mixture of 90-wt% H_2 and 10-wt% ethane or methane additive. These gellants are introduced into the hydrogen as frozen particles that form a gel structure in the hydrogen. A material called a nanoparticulate can also be used for gelling, and it can reduce the amount of gellant needed to 7 to 8-wt%. Further theoretical studies in adding metal particles have shown that there are much greater increases in density (over 3 times the density of liquid hydrogen, with a 70-wt% aluminum loading), but

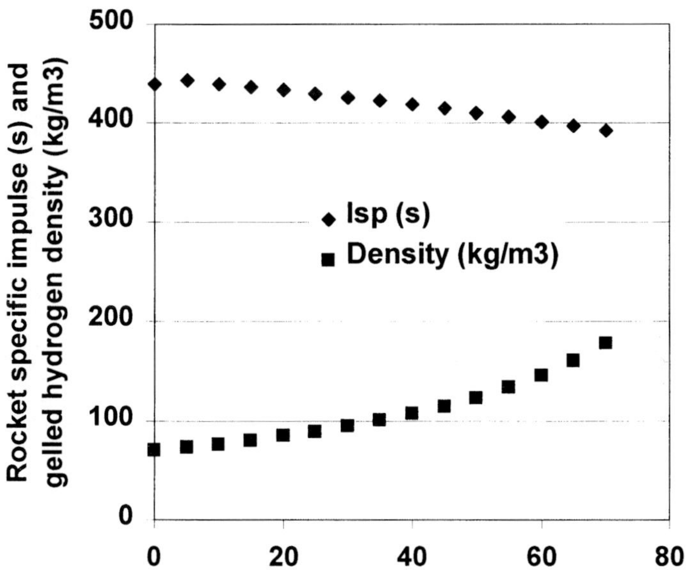

FIGURE 1. Gelled Hydrogen Rocket Engine Specific Impulse
(Oxygen as oxidizer, 2250 psi chamber pressure, expansion ratio = 40:1, Ref.12)

the complexities of metal combustion must be overcome to make this fuel fully successful.

Vehicle Performance Benefits:

Many different applications of gelled and metallized gelled propellants have been investigated. The vehicle types range from launch vehicles, which take payloads to Low Earth Orbit (LEO), upper stages which help payloads move from LEO to the outer planets, and missions to Mars.

Single Stage to Orbit (SSTO): Gelled hydrogen was investigated for SSTO polar missions.[12] The engine performance and gelled hydrogen fuel density are illustrated in Figure 1.[12] The peak I_{sp} is delivered at a 5-wt% gellant loading, but the most attractive and stable gel is created at a 10-wt% gellant level. At this gellant level, the I_{sp} of the gelled hydrogen engine is equivalent to the O_2/H_2 rocket engine (with no gelled fuel). This design point was selected for the analyses of gelled hydrogen SSTO rockets. Figure 2 shows the payload capacity of two SSTO rockets: one without gelled fuel and the other using gelled hydrogen.[12] The gelled hydrogen rocket is able to deliver nearly the same payload

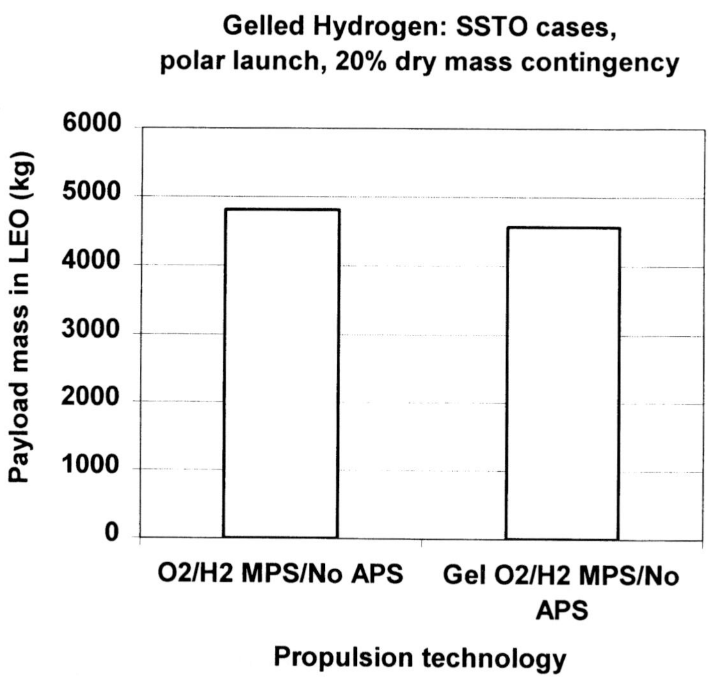

FIGURE 2. Gelled hydrogen propellants for SSTO flight (Ref. 12)

as the baseline O_2/H_2 vehicle. While the payload of the gelled hydrogen vehicle is somewhat lower than the baseline, the gelled hydrogen rocket offers the operational advantages of increased safety, reduced fuel boiloff, leakage and fuel slosh.

Mars Flights: Using metallized gelled propellants for human Mars missions, more payload can be delivered on every mission, and this increase is enabled with the higher specific impulse of the $O_2/H_2/Al$ propellants. Figure 3 illustrates the payload increases for the fast expedition flights, where a metal loading of 60-wt% Aluminum (with 5 seconds more I_{sp}) increased the payload by 22%.[8] The baseline payload for the O_2/H_2 propellants, without metal additives, is 25 metric tons (55,115 pounds). With a 70-wt% Aluminum loading, which improved the I_{sp} by 10 seconds, the payload increase is 33%, with over 33 metric tons (72,752 pounds) of payload delivered to Mars.[11] Using these higher density fuels, we are also able to deliver payloads to Mars at a higher rate, thus reducing the overall transportation costs. Figure 4 provides the comparison of STS-C launches needed to place Mars vehicles using oxygen/hydrogen and metallized gelled fuels in Earth orbit.[8] Large savings in the number of rocket launches are possible over many Mars missions. After 5 Mars flights, a total of 16 rocket launches are

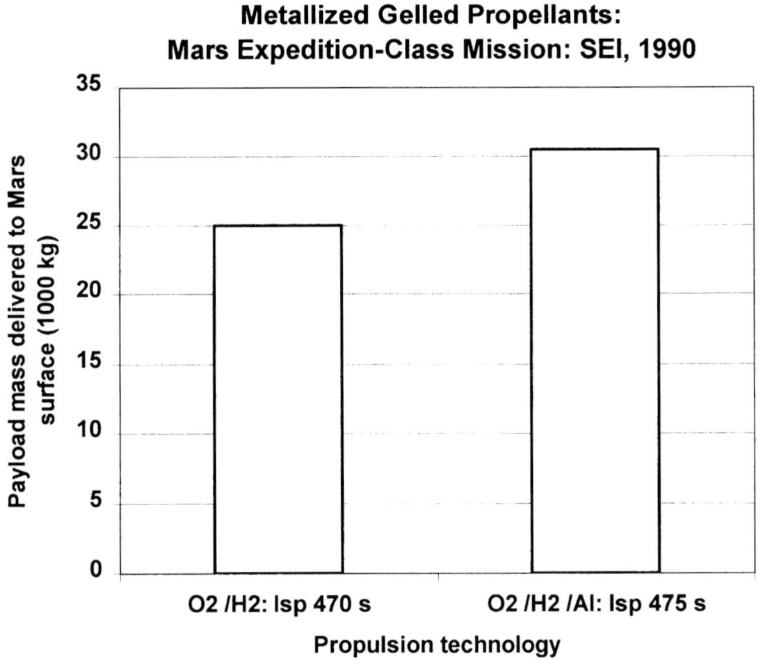

FIGURE 3. Metallized gelled propellants for Mars expedition-class mission (Ref. 8)

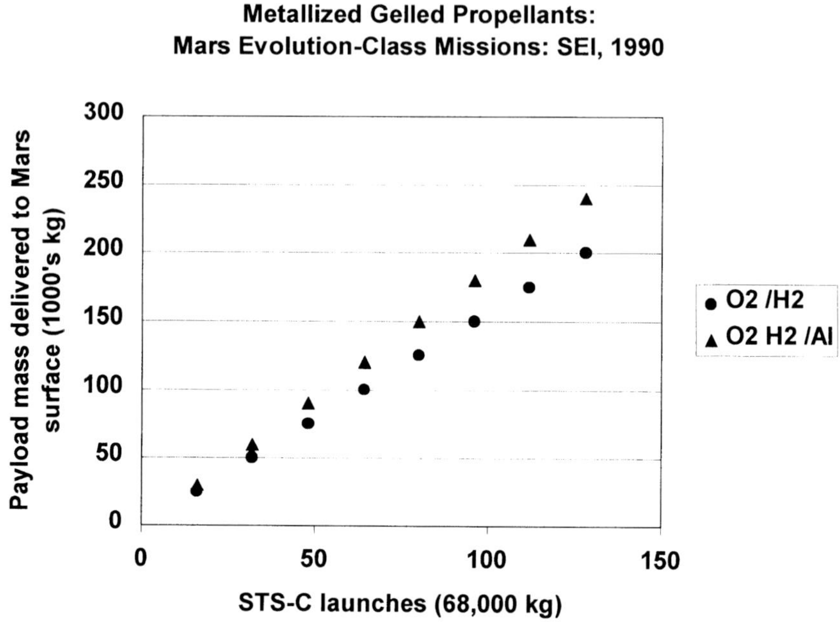

FIGURE 4. Metallized gelled propellants for Mars expedition-class mission (Ref. 8)

saved, reducing the cost of launches, but also saving many years in the on-orbit assembly time for a Mars exploration initiative.

Space Shuttle Boosters: One way of using higher density fuels is to replace the hydrogen-based fuel with a hydrocarbon, such as metallized gelled RP-1, using Al as the additive. Figure 5 illustrates the Liquid Rocket Booster (LRB) length using O_2/RP-1/Al propellants for the Space Shuttle.[10] In this example, the Solid Rocket Boosters (SRB) of the Shuttle have been replaced with LRBs. These analyses were conducted to find ways to improve Space Shuttle's payload performance to the Space Station. The performance can be increased several ways. Very significant booster length reductions are possible with these high density metallized gelled fuels. These length reductions can ease the ground handling of the boosters, and reduce the drag during ascent, thus improving the Shuttle's performance. Alternatively, the metallized gelled booster length can be allowed to grow to the Solid Rocket Booster length, and the payload performance of the Space Shuttle can be increased by 14% with 55-wt% RP-1/Al and 35% with 50-wt% MMH/Al.[10] Additional increases in payload performance were possible with small diameter increases in the metallized gelled LRBs. A 0.31 meter (1-foot) diameter increase (from 12 to 13 feet) would increase the Shuttle payload in LEO from 22,680 to 31,980 kg (50,000 to 70,500 pounds) using 55-wt% RP-1/Al.[10]

Upper Stages: Special, very fast, (high energy) missions can use metallized gelled fuels to the greatest advantage. With any fast mission, very high velocity changes are needed, and the increased specific impulse is required to make the upper stage size manageable and affordable. On high energy missions, metallized gelled propellants can sometimes double the payload of robot missions.[9] Studies of fast missions to Uranus showed that $O_2/H_2/Al$ could allow heavy payloads that are beyond the capacity of current chemical propellants.[9]

ATOMIC PROPELLANTS

New technologies in atom formulation and the manipulation of matter have led to the discovery and synthesis of new rocket propellants.[7, 13, 24-26] Solid cryogenic propellants storing atoms of boron, carbon, hydrogen, or other atomic fuels require a unique propulsion system design, in which the fuels are stored at liquid helium temperatures during ground handling and flight. In atomic fuels employing recombination, the individual atoms are stored in a solid cryogenic matrix. An example of the solid cryogen may be hydrogen. When warmed, the atoms would recombine, creating a very high specific impulse. Figure 6 depicts the I_{sp} for atomic hydrogen for a wide range of atom storage percentages and mixture ratios with oxygen as an oxidizer. Atomic hydrogen recombination can

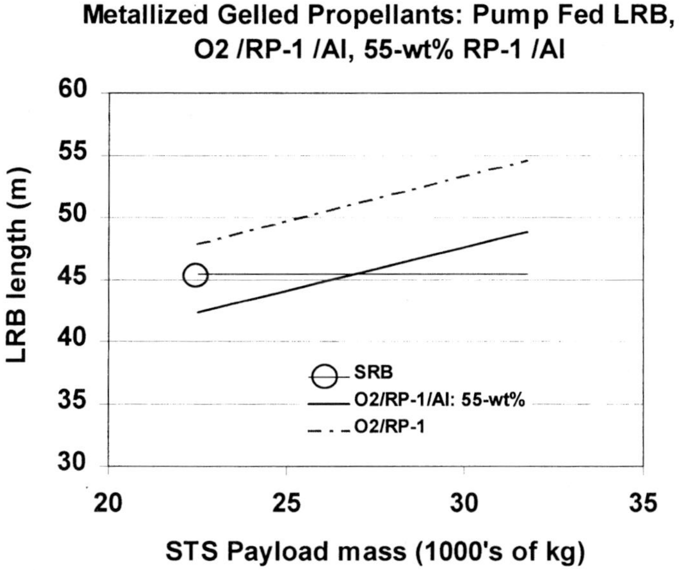

FIGURE 5. Metallized O_2/RP-1/AL LRB length versus STS payload mass (Ref.10)

theoretically deliver over 1500 seconds of specific impulse.[7, 13] These theoretical performance levels are for nearly pure atomic hydrogen, which is not likely to be isolated in a practical aerospace propulsion system. Potential performance benefits in the nearer term are more likely to be in the 500 to 750 second specific impulse range, which require much lower percentages of atomic hydrogen: 5 to 15-wt%.

The benefits of atomic hydrogen as a launch vehicle propellant for several vehicles were analyzed and are shown in Figure 7.[7, 13] An important figure of merit is the reduction of rocket launch mass, or Gross Lift Off Weight (GLOW). The baseline case for comparison is the National Launch System (NLS) using O_2/H_2 propellants at an I_{sp} of 430 s. The reductions in GLOW that are possible with atomic hydrogen are nearly 50% with a 750 second specific impulse. This I_{sp} performance level requires a 15-wt% of atomic hydrogen stored in solid H_2.

Formulations of monopropellants and bipropellants with high energy density materials are taking advantage of the extensive theoretical developments of the last 50 years and turning them into realistic propellants and additives. Using atomic propellants is more complex than using traditional propellants because of their unique chemistry. The high energy atoms must be formulated very meticulously because they do not occur in a stable form at high densities in nature. These formulations offer increased energy density, but they must be manufactured and stored in a storage medium. This storage technology is being refined and experimentally studied.[7, 24-26] If successful, very important advances

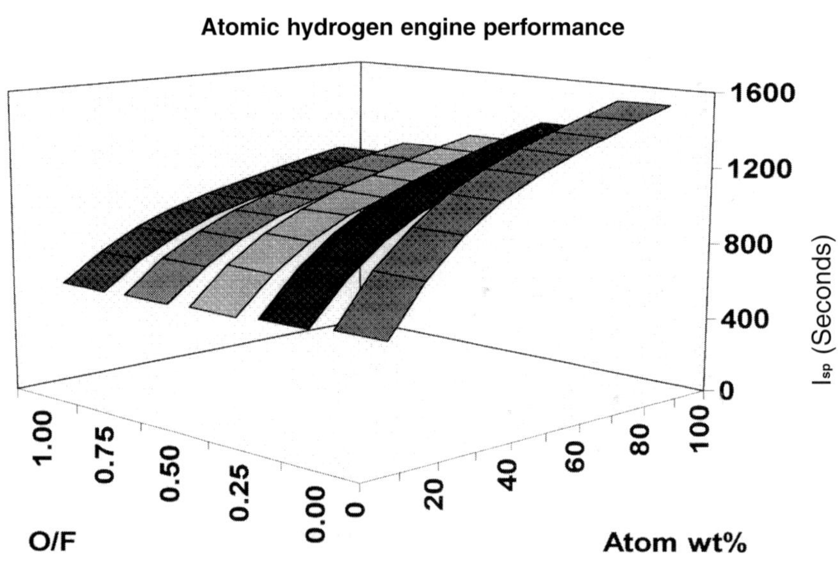

FIGURE 6. Atomic hydrogen engine performance (Ref. 7)

can be made in rocket vehicle design that may lead to low cost access to the Solar System.

CONCLUDING REMARKS

The costs of launching to Earth orbit are a major challenge to reducing in-space transportation costs. If the payload to orbit per flight were increased, the number of launches to effect missions is reduced. This reduction in launches and the attendant cost reductions are important for missions throughout the Solar System. The effective mission travel time is also reduced in that the reduced number of launches reduces the time for on-orbit assembly of large space vehicles. These time reduction analyses were conducted for human Mars missions, and the effects were staggering. The time to deliver payload to orbit, and ultimately to Mars, was reduced by many years using gelled metallized propellants, with increases in I_{sp} of 5 to 10 seconds. New hydrogen-based propellants can provide near term incremental benefits in increased I_{sp} that will save years of assembly costs for large space missions. Future far-term atomic propellants may deliver 100s of seconds of increased I_{sp} over O_2/H_2 propulsion. These performance increases can translate into more compact space vehicles, with GLOW values that are 50 to 80% less than that of current launch vehicle designs. These GLOW reductions can be translated into hundreds of percents (>300%) increases in payload mass delivered to orbit. These increases can allow one heavy lift launch vehicle to deliver a complete human interplanetary space vehicle to orbit, eliminating on-orbit assembly. When speed is essential, atomic propellants can ultimately open up the entire Solar System, from the ground up.

CONCLUSIONS

Higher density hydrogen-based propellants can reduce the size and weight of rockets, as well as make them cheaper to operate. The density of liquid fuels is typically relatively low, and methods of increasing density, such as gelling, using metal additives, or different chemistry, are good ways to improve the performance of rockets and airbreathing space vehicles.

Using high energy density propellants, the cost of space access can be reduced for future space vehicles. By increasing in payload mass per flight, fewer flights are needed and the overall cost of the space project is reduced. Alternatively, the vehicle designer can use advanced fuels to reduce the complexity of vehicle operations. The ways to increase vehicle payload performance with advanced fuels are reducing the Gross Lift Off Weight (GLOW), reducing the dry weight, and reducing vehicle size due to increased fuel density, increased specific impulse, or both. Using gelled or other more viscous fuels, accidental fuel releases are prevented, making propulsion systems safer.

The future also beckons with new atomic propellants born of the understanding of advanced materials and the most basic nature of matter.[7, 13, 24-26] Many rocket designers dream of harnessing the most powerful chemical bonds between individual atoms of boron, and carbon, or hydrogen. The atoms, once created, are arrested within a cryogenic solid and released as they enter the rocket engine. Though these propellants are currently difficult to fabricate in large quantities, there is hope that the power of molecular manipulation from microtechnology, and ultimately nanotechnology, will make these new, and in some cases, as yet unknown, propellants a shining reality.

REFERENCES

1) Sloop, J. L., *Liquid Hvdrogen as a Propulsion Fuel.* 1945-1959, NASA Special Publication (SP)-4404, 1978.
2) Richards, G. R. and Powell, J. W., "The Centaur Vehicle," *Journal of the British Interplanetary Society*, Mar. 1, 1989, V. 42: pp. 99-120.
3) Isakowitz, S. J., *International Reference Guide to Space Launch Svstems*, Washington: AIAA (American Institute of Aeronautics and Astronautics). 1991.
4) Ertel, I. D. and Morse, M. L., *The Apollo Spacecraft - A Chronology*, NASA SP- 4009, Volume I, 1969.
5) Benson, C. D. and Faherty, W. B., *Moonport: A History of Apollo Launch Facilities and Operations*, NASA SP-4204, 1978.
6) Kalitinsky, A., "Nova Launch Vehicle Design Studies," in *Exploration Of Mars*, Proceedings Of The American Astronautical Society Symposium On The Exploration Of Mars, Denver, Colo., Jun. 6, 7, 1963. /Advances In The Astronautical Sciences, Vol. 15, George W. Morgenthaler (Ed.), North Hollywood, Western Periodicals Co., 1963, pp. 107-131.
7) Palaszewski, B., "Launch Vehicle Performance for Bipropellant Propulsion using Atomic Propellants with Oxygen," AIAA 99-2837, June 1999.
8) Palaszewski, B., "Metallized Propellants for the Human Exploration of Mars," AIAA *Journal of Propulsion and Power*, V. 8, No. 6, Nov.-Dec. 1992, pp. 1192-1199.
9) Palaszewski, B., "Advanced Launch Vehicle Upper Stages Using Metallized Propellants," NASA TP-3191, NASA TM-103622, presented at the JANNAF Propulsion Meeting, Anaheim, CA, October 3-5, 1990.
10) Palaszewski, B. and Powell, R., "Launch Vehicle Propulsion Using Metallized Propellants," AIAA *Journal of Propulsion and Power*, V. 10, No. 6, Nov.-Dec. 1994, pp. 828-833.
11) Palaszewski, B. and Rapp, D., "Design Issues for Propulsion Systems Using Metallized Propellants," AIAA 91-3484, NASA TM-105190, September 1991.

12) Pelaccio, D., Palaszewski, B., and O'Leary, R., "Preliminary Assessment of Using Gel and Hybrid Propellant Propulsion for VTOL/SSTO Launch Systems," AIAA 97-3216, NASA TM 1998-206306, July 1997.
13) Palaszewski, B., "Atomic Hydrogen Propellants: Historical Perspectives and Future Possibilities," AIAA 93-0244, NASA TM-106053, January 1993.
14) Wong, W., Starkovich, J., Adams, S., and Palaszewski, B., "Cryogenic Gellant and Fuel Formulation for Metallized Gelled Propellants: Hydrocarbons and Hydrogen with Aluminum," AIAA 94-3175, June 1994.
15) Wiswell, R. and Huggins, M., "Launch Vehicle and Upper Stage Liquid Propulsion at the Astronautics Laboratory (AFSC) - A History Summary", AIAA 90-1839.
16) Van der Wall, E., "Carbon Compounds /Liquid Hydrogen Fuels", Aerojet Liquid Rocket Company Final Report, Technical Report FR02-W396, October 1970, Contract SNP-1.
17) Keller, C.W., "A Study of Hydrogen Slush and/or Hydrogen Gel Utilization Vol. 1: Saturn S-4C Manned Mars Flyby Vehicle Application Study", NASA CR98936, NAS8-20342, Lockheed Missiles and Space Co., K-11-67-1K-Vol. 1, October 31, 1968.
18) Keller, C.W., "A Study of Hydrogen Slush and/or Hydrogen Gel Utilization Vol. II: Test Program 41.5-inch Diameter Tank", NASA CR-98935, NAS8-20342, Lockheed Missiles and Space Co., K-11-67-lK-Vol. 2, October 31, 1968.
19) McKinney, C.D., and Tarpley, W., "Gelling of Liquid Hydrogen," Technidyne, Inc., Contract Number NAS3-4186, NASA CR-54967, RR 66-49, June 1, 1966.
20) Palaszewski, B., "Metallized Gelled Propellant Experiences and Lessons Learned: Oxygen /RP-1 /Aluminum Rocket Engine Testing," in CPIA Publication 627, 1995 Gel Propulsion Technology Symposium, Huntsville, AL, September 1995.
21) Palaszewski, B. and Zakany, J., "Metallized Gelled Propellants: Oxygen/RP-1/Aluminum Rocket Combustion Experiments," AIAA 95-2435, July 1995.
22) Munjal, N., Gupta, B., and Varma, M., "Preparative and Mechanistic Studies on Unsymmetrical Dimethyl Hydrazine-Red Fuming Nitric Acid Liquid Propellant Gels," in *Propellants, Explosives, and Pyrotechnics*, V. 10, No. 4, 1985, pp. 111.
23) Arndt, Hallit, Ramona E., Hawk, Clark W., Allan, Barry D., Logan, Norman, "Compatibility Of Gelled Inhibited Red Fuming Nitric Acid With Aluminum Alloys," AIAA Paper 95-2533, *Journal of Propulsion and Power*, V. 13, No. 4, July-Aug. 1997, pp. 481 -487.
24) Fajardo, Mario E., "Limitations On Stored Energy Densities In Systems Of Separated Ionic Species, " *Journal of Propulsion and Power*, V. 8, Jan.-Feb. 1992, pp. 30-36.
25) Fajardo, M., Carrick, P., and Kenney, J. III, "Matrix Isolation Spectroscopy Of

Metal Atoms Generated By Laser Ablation. I - The Li/Ar, Li/Kr, And Li/Xe Systems," *Journal Of Chemical Physics*, V. 94, May 1, 1991, p. 5812- 5825.

26) Brazier, C. and Carrick, P. "Observation Of Several New Electronic Transitions Of The B2 Molecule," *Journal Of Chemical Physics*, V. 96, No. 12, June 15, 1992, pp. 8684-8690.

Renewable Energy: Trends and Prospects. Edited by S.K. Majumdar, E.W. Miller and A.I. Panah

Chapter Eighteen

HYDROELECTRIC POWER PRODUCTION: REALIZING THE BENEFITS BY RESOLVING ENVIRONMENTAL ISSUES

GLENN F. CADA

Environmental Sciences Division
Oak Ridge National Laboratory
Oak Ridge, Tennessee 37831-6036

INTRODUCTION

Hydroelectric power is the most mature of the renewable energy technologies. The energy of flowing water has been used to perform work directly since ancient times, and hydropower turbines have generated electricity since the 19th century. Hydropower can be installed anywhere in the world with flowing water, whether on tiny headwater streams or major rivers, in tropical lowlands or high-altitude boreal climates. The reservoirs associated with hydroelectric facilities often serve other valuable purposes, such as flood control, irrigation, and drinking water supply.

Although the technology for generating electricity by hydropower turbines is well understood, the consequent environmental impacts have often been overlooked or unresolved. This chapter considers existing and potential future hydropower production, the benefits of this form of renewable energy, and the social and environmental impacts that must be mitigated in order to ensure that new projects are environmentally sound and contribute to sustainable development.

EXISTING AND FUTURE HYDROPOWER RESOURCES

Just as there is a wide variety of natural settings for hydropower sites, there is a variety of developmental schemes that have different benefits and impacts.[1] The most common distinction is between *large dams* and *small dams*. Alternatively, such dams are often referred to as high head and low head dams, respectively, to reflect the concept that hydraulic head (the height from which water drops before reaching the turbine) is a key aspect of electrical generation. As a general rule of thumb, the electrical generating capacity of large hydropower projects can range from about 25 MW to over 10,000 MW. Small hydropower projects are often considered to have a total capacity ranging from 1 MW to 25 MW. Smaller yet are the mini-hydropower projects (less than 1 MW) and microhydropower projects (less than 100kW). Mini- and micro hydropower projects are often located on very small streams to provide decentralized electrical power in remote locations. Many of the most controversial aspects of hydroelectric development, e.g., social impacts of human resettlement or inundation of terrestrial habitat, arise from the creation of large reservoirs associated with large dams. However, large dams are far less numerous than small dams. For example, the International Commission on Large Dams estimates that worldwide there are about 40,000 large dams (dams with a height of 15 meters or more, or shorter dams with high discharges) and 800,000 small dams.[2]

In terms of effects on the environment, the distinction between *storage projects* and *run-of-river projects* is often more useful. Storage projects usually have large dams and reservoirs so that water can be stored on a daily or seasonal basis and released as needed for power production or other uses. Thus, storage projects can cease generation and impound water flowing into the reservoir at times of the day when less electricity is needed, then commence generation when the power is most needed. In contrast, run-of-river projects generally have small reservoirs that do not have the capacity for storing large amounts of water; such projects are generally most useful for contributing to a relatively constant baseload of electricity. The daily or seasonal flow regulation associated with hydroelectric storage projects can be very valuable for matching generation to needs, but it can have greater environmental impacts than run-of-river operations.[3]

Finally, there is a distinction between *conventional hydropower projects* and *pumped storage projects*. Conventional projects are the most common, where water is impounded behind a dam and released downstream through the turbine. Pumped storage projects, on the other hand, utilize two reservoirs at different elevations. During hours when demand for electricity is low, water is pumped from the lower reservoir to the upper reservoir. Then, when electrical demand is high, the water is released from the upper reservoir through the pumps (which are now run in reverse as turbines) to augment power supplied by other sources. Because pumped storage projects consume more electrical power pumping water than they generate in the turbine cycle, they are a net consumer of electricity.

However, as with other storage projects, pumped storage can provide electricity quickly during the hours of the day when it is most needed.

Hydropower has always been an important source of electricity in the United States. Hydropower represented 15.6 percent of total electrical generation by utilities in 1975, but has since declined to less than 10 percent (Table 1). Even considering the reduction as a percent of total generation, hydropower continues to account for over 95 percent of the electricity generated by renewable energy sources. Hydropower produces around 5 percent of the total energy in the U.S., and around 50 percent of the energy produced by renewable sources. On a regional basis, hydroelectric power represents 14 percent of the net electrical power generation in the Rocky Mountain states and nearly 63 percent along the Pacific Coast. Hydropower is a relatively small component of the electrical generating capacity in the Midwestern U.S.

On a global scale, hydropower is a very important source of electricity, not only in terms of installed capacity, but also with regard to facilities under construction and undeveloped potential (Table 2). Although installed capacities (existing projects) are relatively uniformly distributed across the continents, construction of new hydropower plants and undeveloped potential are concentrated in Asia, Africa, and South America. North America and Europe are unlikely to experience substantial new developments of hydroelectric power projects.

TABLE 1.
The contribution of hydroelectric power and other renewables to electrical generation and energy production in the United States. Source: Reference 4.

Year	Total generation or production	Percent of total			Hydroelectric as a percent of total renewables
		Non-renewables	Hydroelectric	Other renewables	
		Net electrical generation by utilities (10^6 kWh)			
1975	1,917,649	84.2	15.6	0.2	98.9
1980	2,286,439	87.7	12.1	0.2	98.0
1985	2,469,841	88.2	11.4	0.4	96.3
1990	2,808,151	89.6	10.0	0.4	96.3
1995	2,994,529	90.0	9.8	0.2	97.9
2000	3,009,500	91.6	8.4	<0.1	99.1
		Energy production (Quadrillion Btu)			
1975	61.357	92.3	5.1	2.6	66.8
1980	67.241	91.8	4.3	3.9	52.8
1985	67.720	91.1	4.4	4.5	49.2
1990	70.835	91.4	4.3	4.3	49.6
1995	71.301	90.6	4.5	4.9	47.9
1998	71.902	90.9	3.9	5.2	43.3

Although the rate of new dam construction in the U.S. has slowed, there is still a significant capacity for new production by equipment upgrades (replacing old turbines with more efficient turbines) or retrofitting numerous existing flood control dams with hydroelectric turbines. At present, one fifth of the world's electricity is generated by hydropower, and an estimated 1,600 dams are being built in 42 countries.[6] Many of these dams also provide irrigation and drinking water for growing populations.

ADVANTAGES OF HYDROELECTRIC POWER

Hydroelectric power production is largely free of several major classes of environmental impacts associated with non-renewable energy sources.[7] Unlike fossil-fueled power plants, hydroelectric generation does not produce sulfur and nitrogen oxides that can cause acidic precipitation (Table 3). Although construction of the hydropower project may result in temporary increases in dust and equipment emissions, these impacts are at least two orders of magnitude less than the emissions from a coal-fired power plant.[8] Inundation of vegetation and peat by the reservoir can cause the release of greenhouse gases,[9] e.g., methane and carbon dioxide, but the significance of this phenomenon greatly depends on the age, size, and geographic location of the reservoir and the amount of vegetation and soil carbon flooded. The emissions of greenhouse gases from reservoirs decrease over time and, in many cases, may not be significantly larger than those associated with the natural (unimpounded) forests and wetlands and free flowing rivers. In any event, hydroelectric power provides an opportunity to reduce the production of greenhouse gases when compared to energy produced by the combustion of fossil fuels and biofuels.[3]

Unlike most other forms of energy production, hydroelectric power plants generate few solid wastes. Land may be required for the disposal of dredged material or waterborne debris that is collected on the trash racks, but the amounts are very small compared to that needed for the continuing disposal of coal ash,

TABLE 2

The percentage of total world hydroelectric generating capacity in different geographic regions. Source: Reference 5.

Region	Installed capacity (650,000 MW)	Under Construction (135,000 MW)	Undeveloped Potential (>1,500,000 MW)
Europe	25	17	5
North America	24	1	6
South America	16	18	25
Africa	3	2	17
China	8	39	21
Rest of Asia	24	23	26

and slag.[10] Similarly, hydropower production does not create hazardous or nuclear wastes that require safe, long-term storage facilities. Many other impacts of the overall fuel cycles of other energy sources are minor for hydroelectric power. These externalities include resource extraction (e.g., coal mining, oil drilling), preparation/beneficiation of the fuels, and transportation (e.g. oil spills or other accidents).

Hydroelectric projects have multiple use benefits that are not achievable by other forms of energy production. In addition to electricity, hydropower reservoirs can provide flood protection, a reliable water supply for drinking and irrigation, as well as reservoir-based recreation and fisheries. The water supply benefits of hydropower reservoirs is of increasing importance in areas of the world where populations are growing and precipitation is seasonal or unpredictable.[11] The development of reservoir fisheries and aquaculture can be an important source of food and income for local populations. If properly managed, the ability of storage projects to control stream flows can improve navigation and enhance the assimilation of downstream discharges of pollutants.

IMPACTS OF HYDROPOWER DEVELOPMENTS

Although hydroelectric power plants have many advantages over other energy sources, the potential adverse impacts are also well known.[9, 12-14] Most of the adverse impacts of dams are caused by habitat alterations. Reservoirs associated with large dams can inundate large amounts of terrestrial and river habitat and displace human populations (Table 3). Diversion of water out of the stream channel (or curtailing reservoir releases in order to store water for future electrical generation) can desiccate riparian (streamside) vegetation. Insufficient stream flow releases degrade habitat for fish and other aquatic organisms in the affected river reaches below the dam. Water in the reservoir is stagnant compared to a free-flowing river; consequently, water-borne sediments and nutrients can be trapped and result in the undesirable proliferation of algae and aquatic weeds (eutrophication). In some cases, water spilled from high dams may become supersaturated with nitrogen gas and cause gas-bubble disease in aquatic organisms inhabiting the tailwaters.

Hydropower projects can also affect aquatic organisms directly. The dam can block upstream movements of fish, which can have severe consequences for anadromous fish (e.g., salmon, steelhead, American shad), catadromous fish (e.g., eels), or riverine fish that make seasonal migrations to spawn (e.g., sturgeon and paddlefish). Downstream-moving fish may be drawn into the power plant intake flow (entrained) and pass through the turbine. Entrained fish are exposed to physical stresses (pressure changes, cavitation, shear, turbulence, strike) that may cause disorientation, physiological stress, injury, or mortality (Figure 1). Five percent or more of the turbine-passed fish may be killed in the best existing turbines, and mortality in some turbines may exceed 30 percent. A variety of mitigative measures (e.g., intake screens and spill flows) have been employed to reduce the numbers of fish that are entrained and killed by turbine passage, but such measures have had only mixed success.[15] Even effective, well-designed intake screening and bypass systems may protect only a portion of the fish entrained in intake flows; the remainder will pass through the turbines.

It is necessary to minimize these adverse environmental effects in order to maintain the renewable energy production potential and water supply benefits of hydropower. A recent survey of the hydropower industry and its regulators in the United States found that the most important environmental issues limiting hydropower production are obstruction of fish passage, water quality degradation, and adequacy of streamflow releases below the diversion dam. It is likely that these issues are also of considerable importance to hydroelectric power development in many other parts of the world.

The costs and benefits of numerous mitigative measures designed to deal with fish passage and water quality problems have been analyzed.[15,16] These environmental mitigation studies identified numerous measures for increasing

TABLE 3
Potential environmental benefits and consequences of hydroelectric power.

Benefits	Consequences
No emissions of sulfur and nitrogen oxides	Emissions of greenhouse gases from flooded vegetation (at some sites)
Few solid wastes	Inundation of wetlands and terrestrial vegetation
Few impacts from resource extraction, preparation, and transportation	Conversion of a free-flowing river to a reservoir
Flood control	Replacement of riverine aquatic communities with reservoir communities
Water supply for drinking, irrigation, and industry	Displacement of human and wildlife communities
Reservoir-based recreation	Alteration of river flow patterns below the dam
Reservoir-based fisheries	Desiccation of riparian (streamside) vegetation below the dam
Enhanced tailwater fisheries	Retention of sediments and nutrients in the reservoir
Improved navigation on inland waterways	Development of aquatic weeds and eutrophication
	Alteration of water quality and temperature
	Spread of water-borne diseases
	Interference with upstream and downstream fish passage

the levels of dissolved oxygen in hydropower discharges, aiding the upstream movements of fish blocked by dams, and preventing the passage through turbines of downstream-migrating fish. Although many of these measures are proven, uncertainty about the effectiveness of mitigative measures associated with downstream fish passage and protection remains. For example, screens that exclude fish from turbine intakes are expensive to install and operate, yet little is known about their benefits to fish populations.[15] In some cases, mortality among screened fish has been greater than mortality among turbine-passed fish.[17]

A number of organizations are actively involved in research to reduce the negative impacts of hydroelectric power, including government agencies, utilities, turbine manufacturers, and private/academic researchers. As an example of the wide-ranging efforts that are underway, the U.S. Department of Energy's (DOE) Hydropower Program has promoted research into environmentally sound hydroelectric projects since 1978. In 1994, DOE initiated the Advanced Hydropower Turbine System (AHTS) Program to develop "environmentally friendly" turbines, i.e. turbine systems in which environmental attributes such as fish passage survival and water quality improvements are emphasized. The AHTS Program was instituted to explore the possibility that advanced turbines

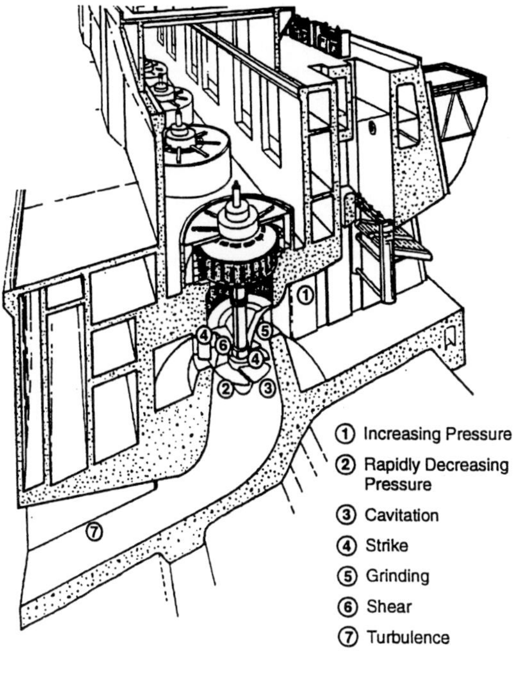

FIGURE 1. Locations within a turbine at which fish injury mechanisms are expected to be most severe.

could be used to (1) improve water quality by aerating hydropower discharges, and (2) eliminate the need for intake screens and other such measures by creating turbine passage conditions that are not damaging to fish. Such turbines could be employed either at new hydropower sites or to replace aging turbines at existing sites. To date, conceptual designs for turbines have been developed that are expected to reduce mortality among turbine-passed fish to 2 percent or less. Prototype and field tests of the advanced turbine designs are planned to verify these predictions.

Water quality problems in the discharges of hydroelectric power plants can result from the seasonal warming and thermal stratification of waters impounded by the dam. Decomposition of organic matter consumes dissolved oxygen (DO) in the bottom waters. Further decomposition, under chemically reducing conditions brought on by an absence of DO in the hypolimnion, may result in the buildup of toxic hydrogen sulfide and ammonia, and may mobilize iron, manganese, and some heavy metals from the reservoir sediments. Discharges from hydroelectric power plants that have low DO concentrations and elevated levels of contaminants may seriously affect other downstream water uses. However, not all reservoirs stratify or develop water quality problems. Stratification and low DO problems are influenced by such factors as temperature, reservoir surface area, depth, volume, flushing rate, and the degree of protection from wind. Reservoirs in most regions of the U.S. are more likely to experience low DO concentrations in summer than in winter, and such problems are more common for large than for small hydropower plants. The Southeastern U.S. and Ohio Valley regions are most likely to experience low DO episodes, probably because they have relatively warm water and deep reservoirs.[18, 19] Globally, tropical reservoirs with large amounts of vegetation are more likely to experience water quality problems than mountain reservoirs in cold climates.

Numerous techniques for increasing DO concentrations in reservoir releases have been explored (Table 4). These have been categorized as methods for

TABLE 4

Techniques for increasing dissolved oxygen concentrations in hydropower releases. Modified from references 16 and 19.

Reservoir Techniques	Powerhouse Techniques
Epilimnetic pumps Hypolimnetic air or oxygen diffusers Forebay destratification	Intake aeration Penstock air or oxygen diffusers Turbine aeration Draft tube venting
Tailwater Techniques	**Operational Techniques**
Submerged tailrace diffusers Surface aerators Side-stream aeration Aerating weirs	Sluice or spillway aeration Selective withdrawal Special turbine operations

increasing DO when the water is still in the reservoir (Reservoir Techniques), aerating the water as it passes through the intake, turbine, or draft tube (Powerhouse Techniques), aerating the water after it has left the draft tube (Tailwater Techniques), and assorted operational measures (Operational Techniques).[16, 19]

Aerating turbines are an attractive option for increasing DO concentrations in hydropower discharges because they can improve water quality while generating power with few additional costs and structures. Under a five-year program aimed at improving water quality in the Tennessee River basin, the Tennessee Valley Authority (TVA) actively developed many of the techniques listed in Table 4, including self-aerating (auto-venting) designs for Francis turbines. Results from one project were encouraging; initial tests showed that up to 5.5 mg/L of DO could be added to the turbine discharge (starting with an incoming DO concentration of zero). Compared to the original turbines, the advanced auto-venting turbines also provided overall efficiency and capacity improvements of 3.7 and 10 percent, respectively.[20]

Replacement of a free-flowing stream with a reservoir can encourage the spread of water borne diseases, most notably malaria, schistosomiasis, and Japanese "B" encephalitis.[3] These diseases are not a necessary consequence of hydropower development; however, the risks can be greatly reduced by effective public health and education programs. In fact, the ability of storage projects to reduce seasonal flooding may be a key weapon in the battle against diseases which rely on mosquitoes as vectors. Also, the potential for increased waterborne diseases is countered to some extent by the reduction in respiratory illnesses associated with atmospheric emissions from fossil-fueled power plants.

As noted earlier, many of the controversial impacts of hydroelectric power are mainly associated with large dams and/or storage projects. Large reservoirs will inundate large amounts of wetlands, floodplains, and associated terrestrial vegetation. This will displace human and wildlife populations, and the decomposition of the flooded vegetation may release greenhouse gases and degrade water quality. Large dams/reservoirs have the greatest potential to trap sediments and nutrients, support aquatic weeds, and interfere with the movements of fish. As a consequence of their ability to impound water, storage projects will alter flow patterns and thus aquatic and riparian habitats downstream from the diversion dam. Proper operation of hydropower projects must ensure that flow releases are optimized to maintain downstream habitats. Operational measures to accomplish these goals include minimum instream flow releases, moderate ramping rates (i.e., the rate at which flow releases are changed), pulse flows (to imitate brief, seasonal spates that maintain the proper sediment balance), and seasonal flow releases for upstream and downstream fish migration. These environmental mitigation measures are readily achievable by storage projects, but they can result in lost power production and revenues if they occur when demand for electricity is high.[16]

The impacts of small and medium dams are often much smaller and easier to mitigate. Small reservoirs will not flood significant amounts of human, wildlife, and riverine habitat, may not stratify or cause water quality changes, and can be readily equipped with fish passage measures. Run-of-river projects do not alter downstream flows as significantly as storage projects. Although the benefits of small hydropower are reduced as well, appropriately sited small projects can still provide valuable decentralized electrical power and localized water supply, fisheries, and recreation without the overwhelming impacts of large dams.

CONCLUSIONS

Hydropower has many advantages as a source of energy: it is renewable, the technology is well-understood, there is both a substantial installed base and a significant potential for growth, there are important multiple uses for the hydroelectric reservoir, and many environmental impacts (e.g., atmospheric emissions) are either small or nonexistent. However, hydroelectric power production is not without impacts. Large hydropower projects have caused major dislocations of human populations and have significantly reduced fish and wildlife populations, water quality, and habitats. Smaller projects can also cause impacts, but such impacts are often avoidable by proper siting and mitigation measures.

Recent developments in the design of advanced, environmentally friendly turbines indicate that there is a real potential for reducing some of the most common adverse impacts of hydropower.[21] For example, as a result of its success in developing dissolved oxygen enhancing techniques, TVA plans to install 26 auto-venting turbines at 13 hydroelectric projects that presently experience dissolved oxygen deficiencies in the tailwaters.[22] The efficiency improvements of these new turbines, combined with online performance monitoring systems and multi-unit optimization software, are expected to increase the electrical generation of the system while improving water quality in the basin.[20]

Conceptual designs for advanced turbines are being developed that have a potential to increase the survival of turbine-passed fish. It is expected that the environmental performance of hydroelectric turbines can be significantly improved through the use of advanced computerized modeling techniques and increased attention to biological performance criteria. The next step is to construct and test prototypes of these new environmentally friendly turbine designs. If the performance of the new turbines is proven, these advanced designs can be added to the suite of mitigative measures for hydroelectric projects to promote reliable, environmentally sound generation of renewable energy.

Mitigation measures to reduce environmental impacts can be expensive, both in terms of construction costs and the electrical power foregone due to operational constraints. It is usually preferable to invest effort in the proper siting of proposed hydropower projects in order to avoid environmental issues in the first place.[23] Because the social and environmental costs of large dams are coming under increasing scrutiny, future hydropower development is likely to focus on small- and medium-sized dams sited in areas that avoid environmental problems (and the need for mitigation) and provide decentralized electrical power and other services to local areas. As with any source of energy, to achieve sustainability the development must be balanced with public needs and public input, ideally through an effective Environmental Assessment process.

Through such assessments, the goals of the hydropower development project will be transparent to all and will include participation by all affected parties.[3] Given that human populations will increase and will have increasing needs for drinking water, irrigation water, and electricity, the multipurpose advantages of hydroelectric generation are likely to be emphasized in future development.

ACKNOWLEDGMENTS

Preparation of this manuscript was supported by the Office of Biopower and Hydropower Technologies, U.S. Department of Energy under contract DE-AC07-960R22464 with Lockheed Martin Energy Research Corporation. Mike Sale and Stan Hadley provided comments on the manuscript. This is ESD Publication No. 4897.

REFERENCES

1. Gulliver, John S. and Roger E. A. Arndt (Ed.). 1991. *Hydropower Engineering Handbook*. McGraw- Hill, Inc. New York, NY.
2. Oud, Engelbertus. 1997. Engineering and Economic Aspects of Planning, Design, Construction and Operation of Large Dam Projects. Pp. 17-39. In: T. Dorcey (Ed.) *Large Dams: Learning from the Past - Looking at the Future. Workshop Proceedings*. IUCN - The World Conservation Union and the World Bank Group, Washington, DC. pp. 145.
3. Goodland, Robert. 1997. Environmental Sustainability in the Hydro Industry-Disaggregating the Debate. Pp. 69-102. In: T. Dorcey (Ed.) *Large Dams: Learning from the Past - Looking at the Future. Workshop Proceedings*. IUCN The World Conservation Union and the World Bank Group, Washington, DC. pp. 145.
4. EIA (Energy Information Administration). 2001. *Annual Energy Review* 2000. U.S. Department of Energy, Washington, DC.
5. Armistead, Thomas F. 1998. Dam Conflicts Threaten Growth. *Engineering News Record*. 241(10): 30-34.
6. Houlder, Vanessa. 1999. Hydropower Threatened by Deluge of Objections. *The Financial Times*. November 2, 1999.
7. Cada, Glenn F. and James E. Francfort. 1995. Mitigation of Environmental Impacts at Hydroelectric Power Plants in the United States. Pp. 223-234. In: F. S. Sterrett (Ed.) *Alternative Fuels and the Environment*. Lewis Publishers,

Ann Arbor, MI, pp. 276.

8. ORNL and RFF (Oak Ridge National Laboratory and Resources for the Future). 1994b. *Estimating Externalities of the Hydro Fuel Cycles*. Report No. 6 on the External Costs and Benefits of Fuel Cycles: A Study by the U.S. Department of Energy and the Commission of the European Communities. U.S. Department of Energy, Washington, DC.
9. Rosenberg, D.M., F. Berkes, R.A. Bodaly, R.E. Hecky, C.A. Kelly, and J.W.M. Rudd. 1997. Large-Scale Impacts of Hydroelectric Development. *Environmental Review* 5: 27-54.
10. ORNL and RFF (Oak Ridge National Laboratory and Resources for the Future). 1994a. *Estimating Fuel Cycle Externalities: Analytical Methods and Issues*. Report No. 2 on the External Costs and Benefits of Fuel Cycles: A Study by the U.S. Department of Energy and the Commission of the European Communities. U.S. Department of Energy, Washington, DC.
11. Postel, Sandra L., Gretchen C. Daily, and Paul R. Ehrlich. 1996. Human Appropriation of Renewable Fresh Water. Science 271: 785-788.
12. Ackermann, William C., Gilbert F. White, E.B. Worthington, and J. Loreena Ivens. 1973. *Man-Made Lakes: Their Problems and Environmental Effects*. Geophysical Monograph 17. American Geophysical Union, Washington, DC. pp. 847.
13. Petts, Geoffrey E. 1984. *Impounded Rivers: Perspectives for Ecological Management*. John Wiley & Sons, New York, NY, pp. 326.
14. Mattice, Jack S. 1991. Ecological Effects of Hydropower Facilities. Chapter 8 In: J.S. Gulliver and R.E.A. Arndt (Ed.) *Hydropower Engineering Handbook*. McGraw-Hill, Inc., New York, NY.
15. Francfort, James E., Glenn F. Cada, Dennis D. Dauble, Richard T. Hunt, Donald W. Jones, Bennie N. Rinehart, Garold L. Sommers, and Ronald J. Costello 1994. *Environmental Mitigation at Hydroelectric Projects. Volume 2. Benefits and Costs of Fish Passage and Protection*. DOE/ID-10360(V2), U.S. Department of Energy, Idaho Operations Office, Idaho Falls, ID.
16. Sale, Michael J., Glenn F. Cada, Lisa H. Chang, Sigurd W. Christensen, Steven F. Railsback, James E. Francfort, Bennie N. Rinehart, and Garold L. Sommers. 1991. *Environmental Mitigation at Hydroelectric Projects. Volume 1. Current Practices for Instream Flow Needs, Dissolved Oxygen, and Fish Passage*. DOE/ID-10360, U.S. Department of Energy, Idaho Operations Office, Idaho Falls, ID.
17. Ferguson, John W. 1991. Relative Survival of Juvenile Chinook Salmon Through Bonneville Dam on the Columbia River. Pp. 308-317 In: D.D. Darling (Ed.) *Waterpower '91 Proceedings of the International Conference on Hydropower*. American Society of Civil Engineers, New York, NY, pp 2275.
18. Cada, Glenn F., K. Deva Kumar, Jean A. Solomon, and Stephen G. Hildebrand. 1983. An Analysis of Dissolved Oxygen Concentrations in Tail Waters of Hydroelectric Dams and the Implications for Small-Scale Hydropower Development. *Water Resources Research*, 19(4), 1043-1048.

19. Franke, Gary F., Donald R. Webb, Richard K. Fisher, Dilip Mathur, Paul N. Hopping, Patrick A. March, Michael R. Headrick, Istvan T. Lasczo, Yiannis Ventikos, and Fotis Sotiropoulos. 1997. *Development of Environmentally Advanced Hydropower Turbine System Design Concepts*. Voith Hydro, Inc. Report No. 2677-0141 for the U.S. Department of Energy, Idaho Operations Office, Idaho Falls, ID.
20. March, Patrick A. and Richard K. Fisher. 1999. It's Not Easy Being Green: Environmental Technologies Enhance Conventional Hydropower's Role in Sustainable Development. *Annual Review of Energy and the Environment*. 24 (November 1999).
21. Fisher, Richard K., Patrick A. March, Dilip Mathur, Fotis Sotiropoulos, and Gary Franke. 1998. Innovative Environmental Technologies Brighten Hydro's Future. *Proceedings of the XIX IAHR Symposium on Hydraulic Machinery and Cavitation*. Singapore: World Scientific Publishing Co. Pte. Ltd.
22. Hopping, Paul, Patrick March, Thomas Brice, and Joseph Cybularz. 1997. Update on Development of Auto- Venting Turbine Technology. Pp. 2020-2027 In: D.J. Mahoney (Ed.) *Waterpower '97 - Proceedings of the International Conference on Hydropower*. American Society of Civil Engineers, New York, NY, pp. 2267.
23. Cada, Glenn F. and Frank Zadroga. 1982. Small-Scale Hydroelectric Power for Developing Countries: Methodology of Site-Selection Based on Environmental Issues. *Environmental Conservation*. 9(4): 329-337.

Renewable Energy: Trends and Prospects. Edited by S.K. Majumdar, E.W. Miller and A.I. Panah

Chapter Nineteen

SUSTAINABLE WIND POWERED GENERATION

ALFRED B.O. SOBOYEJO

202 Aerospace Engineering
590 Wood Hedge Drive
The Ohio State University
Columbus, Ohio 43210

1.0 INTRODUCTION

Since the energy crisis of 1973, there has been considerable interdisciplinary research and development in order to have sustainable energy from other sources which are not linked with energy from petroleum. This has brought about the tremendous development of engineering systems for applications in sustainable wind-powered generation and utilization. This development has been witnessed in technologically advanced countries such as the United States of America and several European countries, where the abundant natural resources of the wind have been developed to provide viable alternative sustainable wind-powered energy for social and economic development.

It is also significant to remark that several countries with poor levels of technological and economical development, particularly in Africa, have experienced tremendous setbacks in economic development partly because the existing power generating systems in these countries have been working at poor levels of mechanical and electrical efficiencies. In these countries, particularly African countries with several years of military or non-democratic governments, engineering infrastructures for water supply, electrical and telecommunication systems in both urban and rural areas, cannot function properly because of the poor reliability of the existing power generating systems, which derive power largely from the natural resources of petroleum, gas, coal or hydroelectricity. In order to provide effective solutions to this problem, the use of alternative energy systems such as wind or solar powered systems have been recommended not only for these African countries but also for other countries, where fairly high and constant wind energy can be generated from the environment. In view of these problems, in both the developed and less developed countries, several countries in different parts of the world have shown deep interests in the generation of alternative wind-powered systems. Figures (1.1) and (1.2) are the diagramatic representations of two wind-powered systems. Figure (1.1) is a simple form of wind-powered high speed propeller for the generation of electrical power. Figure (1.2) is another simple form of wind-powered multi-blade turbine for pumping water.

The main objectives of this chapter are first to elucidate the applications of the basic engineering and scientific principles, which are necessary and crucial for the development of wind-powered systems for the development of sustainable structures and infrastructures, and for social and economic development of civil, mechanical, electrical and agricultural engineering systems, which can be used in both of the countries with high levels of economic development and others with existing relatively low levels of economic development. The second objective is to highlight the essential considerations of engineering design against the detrimental effects of wind forces on the structures and infrastructures, which have to be constructed as part of the economic development of an engineering system for sustainable wind powered generation.

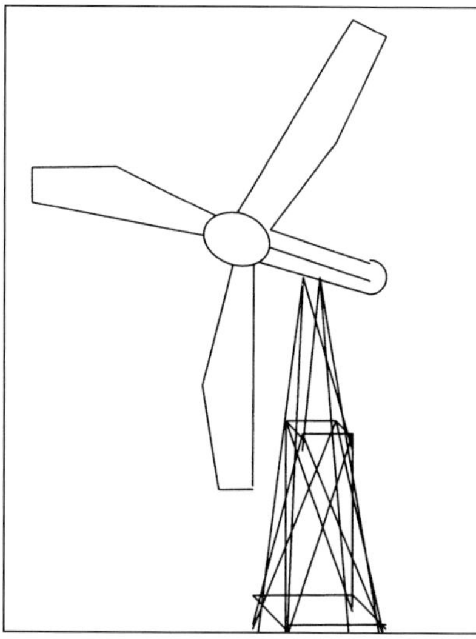

FIGURE 1.1 Wind-powered high-speed propeller for electrical power generation.

FIGURE 1.2 Wind-powered multi-blade turbine for pumping water.

In order to realize these objectives, the development in this chapter has been divided into the following sections:

(i) Overview of Wind Characteristics and Environmental Climatology.
(ii) Physical Properties of the Wind and the Wind Velocity Profile.
(iii) Wind Energy.
(iv) Statistical and Stochastic Process Models.
(v) Engineering Design of Structural Facilities Against Extreme Winds for Development of Wind-Powered System.
(vi) Conclusions and Outlook for Future.

2.0 OVERVIEW OF WIND CHARACTERISTICS AND ENVIRONMENTAL CLIMATOLOGY.

Wind is the physical measurement of air masses through the atmosphere, which is caused principally by the differential atmospheric heating over different areas of both the land and sea of the Earth. Because of the physical rotation of the Earth, the air flow near the Earth's surface is a complicated physical process and is not a simple process of air flow from high to low pressure zones but along the lines of equal pressures, which are known as isobars. It can also be shown from the basic principles of theoretical physics that the intensity of wind pressure is directly proportional to the magnitude of the local pressure gradient at any particular point of interest on the Earth's surface.

Many local climatological factors can affect the general wind flow, which can be quantified by the large-scale pressure field. There is a mixing of air by mechanical stirring caused by the friction between the ground and the air. The magnitude of environmental turbulence varies over a wide area of the Earth's surface; some of the major eddies may be several kilometers in magnitude, which can give rise to signals lasting several minutes. At the other end of the scale, small, though possibly severe, eddies may be due to the passage of the wind past a structure or any other minor obstruction which may have correspondingly small physical dimensions and duration measured in fractions of a second. Usually, the pattern is quite complex with the result that wind speeds can vary greatly from place to place and from moment to moment. The measurements of wind speeds will show dependence on the duration over which the sample is taken. Large averaging times will therefore allow the inclusions of large eddies; while a short averaging time may cover only a small superimposed eddy, which may ultimately have a higher wind speed.

Several countries have meteorological departments, which are responsible for the collection, quality-control, publication and storage of wind data, which are measured by anemometers on rigid masts or towers and if possible, are usually sited in open level country at a height of 10 meters above the ground level.

The above descriptions show that, compared to other sources of energy such as hydroelectricity, gas, or coal, wind energy is indeed a natural and environmental form of energy, which is a random variable, and it is a function of both time and space. Spatial variations are obviously dependent on local climatological and environmental factors.

In view of the variability of wind as a resource, the turning rotors of wind turbines do experience eddies of varying sizes. Furthermore, since the vertical wind profile of the complex wind field around the rotor exerts on the turning rotor considerable cyclic loading, fatigue stresses [2,3] due to wind shear [4] are developed in the turbine rotors. It is therefore important that engineering design of rotors of wind turbines for wind-powered generation should consider, among several essential factors, the design of such rotors against the detrimental effects of fatigue in the planning and development of projects for wind-powered generation. The probabilistic mechanics models of wind speed variability, with the applications of probabilistic or statistical methods from basic theoretical mechanics principles, have become important engineering tools which have been developed in recent times in order to ensure improved system reliability and efficiency of engineering designs of wind-powered systems [4].

2.1 MODELING WIND SPEED VARIATION

The annual mean wind speed can be used to estimate energy (kWh/m^2) or power per unit area (W/m^2), which indicates wind energy or wind power potential. The annual mean wind speed is important in the selection of a suitable site for a wind-powered system. The variation in the wind speed should also be estimated in order to find the total wind potential of the site. The longer the period of estimation of the wind speed data, the more accurate should be the wind power potential of the site.

Consider larminar airflow, which is perpendicular to the rotating blades and moving at a velocity V. The kinetic energy, E, of the flow is

$$E = \frac{1}{2} m V^2 \tag{2.1.1}$$

where $m = \rho AV$ is the mass of the air moved per unit time, ρ is the air density and A is the cross-sectional area of the blades swept through in one revolution. The kinetic energy expressed by equation (2.1.1) can be converted to power, P, since power is defined by kinetic energy per unit time. This gives the result

$$P = \frac{1}{2} \rho A V^3 \tag{2.1.2}$$

Since not all the energy in the flow can be extracted, the actual collectable energy, $\overline{P}$, from the wind machine is

$$\overline{P} = \eta P \qquad (2.1.3)$$

where η is the efficient of the machine, which has a maximum value of 0.593, which is known as the Betz Coefficient [5].

Assuming as air density $\rho = 1.22 kg/m^3$, the cross-sectional area, A, is expressed in m^2, and the air velocity, V, in metres per second, the wind power, P, can be expressed as

$$P = 0.612AV^3 \text{ (Watts)} \qquad (2.1.4)$$

The term (P/A) is known as the wind power potential. Figure (2.1.1) shows the annual average wind-power potential of the United States [5].

2.2 VERTICAL WIND PROFILE

In general, because of friction near the Earth's surface, wind speed decreases as the elevation above the ground level decreases. It is therefore expected that the

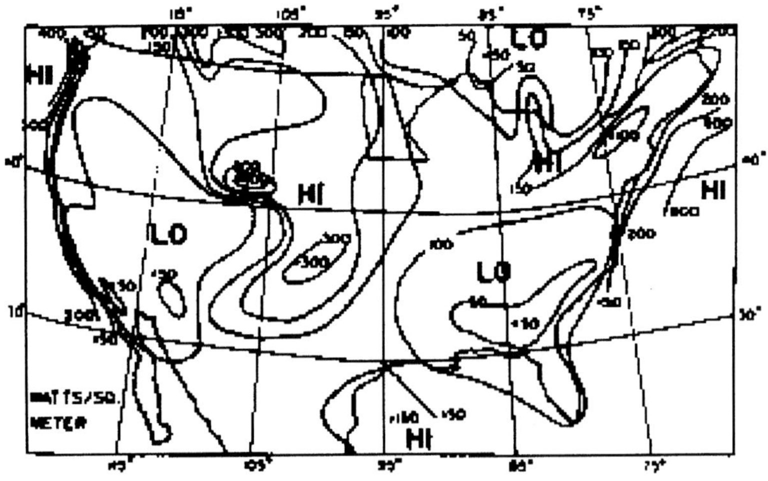

FIGURE 2.1.1 Annual average wind-power potential of the United States. [5]

horizontal component of the wind velocity increases with altitude above the Earth's surface. The variation of the horizontal component of the surface is known as the vertical wind profile or vertical shear.

The variation of wind shear with altitude imposes cyclic fatigue stresses on the rotor blades of turbines. In other words, at the top, the wind velocity is higher than at the bottom, see Figure (2.0.1). The resultant effect of the vertical variation in the wind shear profile is to increase the cyclic fatigue stresses at the top of the rotor blades [7,8,2,3]. Further details on mathematical modeling of wind speed as a function of height above the ground level are given in section (3.0).

2.3 CHANGE OF WIND DIRECTION

Changes in wind directions can occur because of seasonal changes due to the general circulation of winds and also because of diurnal and short-term variations in wind speeds and directions due to the Earth's rotation. In practice, wind direction changes are illustrated by a wind rose diagram, which is a circular graph which indicates the percentage of the winds from a particular direction. The mechanical design of reliable wind turbines should therefore ensure that there are devices that will make the turbines accept winds from any direction.

3.0 PHYSICAL PROPERTIES OF THE WIND AND THE WIND VELOCITY PROFILE

Experimental and physical evidences have shown that there is a non-linear variation of wind velocity profile with height above ground level. The wind, which moves freely and undisturbed at high elevations, H_G, and above, over the Earth's surfaces is the gradient wind with a wind velocity, which is known as the gradient wind velocity, V_G . This means that the gradient wind velocity, V_G, is unaffected by the frictional stresses near the Earth's surface, which will depend on the nature of the roughness of the terrain. The value of the gradient wind height, H_G, above the surface of the Earth depends on the type of terrain.

Near the surface of the ground, the wind is affected by frictional stresses from the roughness of the Earth surface and some of its kinetic energy is dissipated by wind turbulence or variations in wind velocities, which are more pronounced near the Earth's surface.

3.1 WIND VELOCITY PROFILE

Wind speed increases with height above the ground level. Davenport proposes a power law relationship of the form [9]

$$y(h) = \frac{V(h)}{V(H_G)} = \left[\frac{h}{H_G}\right]^{\gamma} \qquad 3.1.1$$

where $V(h)$ is the wind velocity at height h above the ground level; H_G is known as the gradient height, γ and HG are constants, which depend on environmental surface roughness of any of the three topographical zones, namely (i) urban, (ii) semi-urban and (iii) rural. The variable $y(h)$, is shown in equation (3.1.1). The values of these constants are as shown [9] in Table (3.1.1). In a recent study of wind speed profiles, Alfred B. O. Soboyejo proposed a new simplified linearized law of the form [10]:

$$Z(h) = \left[\frac{h}{y(h)}\right] = a_1 h + b_1 \qquad (3.1.2)$$

where a_1 and b_2 are constants for any particular zone. These values are shown in Table (3.1.2).

TABLE 3.1.1

Zone	The Exponent y	Gradient Height H_G (feet)	The Drag Coefficient, K
(i) Urban	0.40	1700	0.050
(ii) Semi-Urban	0.28	1300	0.015
(iii) Rural	0.16	900	0.005

TABLE 3.1.2

Zone	a_1	b_1
(i) Urban	0.9184	336.44
(ii) Semi-Urban	0.9433	165.92
(iii) Rural	0.9647	62.81

In equation (3.1.2), the height h above the ground level is in feet, $y(h)$ is dimensionless and it has a relationship, which is as shown in equation (3.1.1). Furthermore, the function below is as defined in the equation (3.1.2) above.

$$Z(h) = \left[\frac{h}{y(h)}\right]$$

4.0 WIND ENERGY

Wind energy is produced directly or indirectly from solar energy, which has produced nearly all the energy required by man. Winds with high speeds can lead to tornadoes, which have tremendous dynamic wind forces. In Europe, the Dutch windmills have produced wind energy for practical, mechanical, and electrical engineering applications in their rural areas. Furthermore, wind driven electric generators have been used widely in several parts of Europe, up till about forty years ago, when new sources of electrical power were produced more efficiently and at cheaper rates to consumers [5]. In several parts of rural Africa today, particularly in the northern areas of West Africa, there are simple village windmills, which are used to pump water from village wells [11].

4.1 WIND ENERGY PATTERN FACTOR (WEPF)

In view of the variability in wind speed, V, the effect of wind speed is an important consideration in the assessment of wind energy potential for any given location. The longer the averaging time in the collection of wind speed data for any given location, the smaller the calculated wind speed. Therefore, the wind power energy potential estimated from the yearly mean wind speed, $\overline{V}_{yr}$, would give a lower value than the value obtained from the hourly mean wind speed, $\overline{V}_{yr}$. The wind energy pattern factor (Q = WEPF) is therefore defined as follows:

$$Q = \frac{\frac{1}{n}\sum_{i=1}^{n}\overline{V}_{hr}^{3}}{\left(\overline{V}_{yr}\right)^{3}} > 1.0 \qquad 4.1.1$$

where $\overline{V}_{hr}$ and $\overline{V}_{yr}$ are the mean hourly wind speed and mean yearly wind speed respectively and n = 8760 (number of hours in a year). It is quite clear from equation (4.1.1) that (Q = WEPF) will always be greater than one or unity, since the numerator will always be greater than the denominator. The need to quantify the quantity (Q = WEPF) is important because Q is related to the cubic power of wind speed as shown in equation (4.1.1). This can be used for the engineering design specification of the wind-powered system, which is also a function of V^3, see equation (2.1.2) or (2.1.4).

4.2 WIND ENERGY

Since wind speed for any given location can vary in magnitude and direction, during any period when wind speed data are being collected wind energy potential, which can also be collected, therefore becomes a quantity, which can also vary. Therefore, the variability in the relationship between wind speed and time should be considered whenever the need arises to quantify the wind energy potential of any given site.

The total amount of energy, *E,* expressed in Watt-hours, collected from a wind-powered system is the integral or the summation of the output power, $\bar{P}$, expressed in Watt-hours, over a given period of time from say, t_1 hours to t_2 hours.

$$E = \int_{t_1}^{t_2} \bar{P}\,dt = \eta \int_{t_1}^{t_2} P\,dt \qquad 4.2.1$$

where the magnitude of the output power, $\bar{P}$, has been defined by equation (2.1.3).

For a good design of a wind-powered system for any site, it is therefore necessary to have long statistical records of the available wind speed data for the site. The longer the time of acquiring such data, the more accurate will be the estimation of the variability of the available wind potential for a given site. It is necessary to carry out such studies for different sites in order to determine the site with the greatest wind power potential for any given project. In several countries of the world, wind speed data, which can be used for the assessment of wind power potentials for different sites, can be obtained from the Department of the Meteorological Services.

5.0 STATISTICAL AND STOCHASTIC PROCESS MODELS

Probabilistic models of wind speed and wind energy can be developed in order to characterize wind speed and wind energy potential necessary for the specification of the size of wind turbines, operating characteristics, and the estimation of the annual energy output. The statistical concepts which can be used for such studies are hereby described briefly.

5.1 HISTOGRAMS, FREQUENCY AND DURATION MODELS

From the wind speed data of any site, histograms, frequency and duration models can be constructed in order to show the variability of the wind speed for the site. In general, such data can be presented as indicated below, assuming that wind speed magnitude, v, can be regarded as a random variable, V. The frequency, $p(v)$ of any wind speed is defined simply as

$$p(v) = \frac{N(v)}{N} \tag{5.1.1}$$

where *N(v)* is the recorded number of observations with magnitude *v*, and *N* is the total number of observations for all the available wind speed data for the given site.

The cumulative frequency *F(v = V)* of wind speed with magnitude *v* is given by

$$F(v = V) = \sum_{v<V} p(v) \tag{5.1.2}$$

The wind speed duration *R(v = V)* for the wind speed magnitude *v* = *V* is given by

$$R(v = V) = 1 - F(v = V) = \sum_{v>V} p(v) \tag{5.1.3}$$

For practical applications, which can be used for the plan and design of wind-powered systems for a given site, the wind speed data necessary for the applications of equation (5.1.1, 5.1.2 and 5.1.3) are normally collected at fixed time intervals throughout the year, which is assumed to have 8760 hours or (365 x 24 hours).

From the practical data of the histograms, frequency and duration curves for different locations for a given country or geographical zone, the wind power or wind energy potential resource maps can be prepared for the country. Such maps are used to find potential wind power generation sites, which can be used for the locations of suitable sites for wind-powered generation systems.

In view of the variability in the data of the wind speed, V, which can be observed at any given location and at specific time intervals, it is useful to characterize wind speed in terms of the mean, $\overline{V}$, and the standard deviation σ of the wind speed as follows:

$$\overline{V} = \frac{1}{N}\sum_{i=1}^{N} V_i \tag{5.1.4}$$

$$\sigma = \left[\frac{1}{N-1}\sum_{i=1}^{N}\left(V_i - \overline{V}\right)^2\right]^{\frac{1}{2}} \tag{5.1.5}$$

where *N* is the sample size. The quantity, σ^2, is known as the variance of the wind speed about the mean value, $\overline{V}$. For large sample size, *N*, wind speed data can be grouped into various intervals. This gives the diagrammatic representation known as the histogram, where u_i is the wind speed at the middle value of the interval, j is the number of intervals, m_i is the frequency or the number of observations in each interval.

For a discrete random variable, V, the wind speed, the probability for each interval is

$$p(V_i) = \frac{m_i}{N} \qquad 5.1.6$$

The total sum of all possible probabilities is unity.

$$\sum_{i=1}^{j} p(V_i) = 1 \qquad 5.1.7$$

The mean wind speed is therefore

$$\overline{V} = \frac{1}{N}\sum_{i=1}^{j} m_i V_i = \sum_{i=1}^{j} p(V_i)\cdot V_i \qquad 5.1.8$$

where $p(V_i) = \frac{m_i}{N}$ (Equation 5.1.6).

When the number of observations, N, is sufficiently large, then the wind speed V can be considered as a continuous random variable. In this case, the discrete probability, $p(V_i)$, becomes the probability density function, PDF, $f(V)$; and the total area under the probability density function is also unity

$$\int_0^{+\infty} f(V)dV = 1 \qquad 5.1.9$$

The lower limit of the integral of wind speed has to be zero because wind speed is positive and not negative.

The cumulative distribution function, $F(V)$ is given by

$$F(V) = \int_0^V f(V')dV \qquad 5.1.10$$

From equation (5.1.9) and (5.1.10) we have

$$f(V) = \frac{dF(V)}{dV} \qquad 5.1.11$$

The mean or the expected value, $\bar{V} = E(V)$ of wind speed V; and the variance, $\sigma^2 = Var(V)$ of the wind speed V are

$$\bar{V} = E(V) = \int_0^\infty V f(V) dV \qquad 5.1.12$$

$$\sigma^2 = Var(V) = \int_0^\infty [V - E(V)]^2 f(V) dV \qquad 5.1.13a$$

Equation (5.1.13a) can also be expressed as

$$\sigma^2 = Var(V) = E(V^2) - [E(V)]^2 \qquad 5.1.13b$$

Figure (5.1.1) shows the average wind speed for Canada [6].

5.2 PROBABILISTIC DISTRIBUTION FUNCTIONS FOR WIND SPEED

Depending on the size of statistical data, N for the assessments of the mean and standard deviation of wind speed and the proactical application, standard forms of probabilistic distributions are available for the characterization of wind speed variability.

The following are some of the important distributions, which can be used for the statistical modeling of wind speed variability

(i) Normal distribution or (Gaussian distribution)

(ii) Weibull distribution

(iii) Raleigh distribution

(iv) Extreme value distribution

The probability distribution functions, PDFs, for these distribution functions and other important statistical properties of these distributions are given below.

The Normal Distribution (when the mean $\bar{V}$ and standard deviation σ of wind speed are known). The PDF is

$$f(V) = \frac{1}{\sigma\sqrt{2\pi}} \exp\left[-\frac{1}{2}\left(\frac{V-\bar{V}}{\sigma}\right)^2\right]; \quad (-\infty < V < +\infty) \qquad 5.2.1$$

The Weibull Distribution.

The PDF is *(following page)*

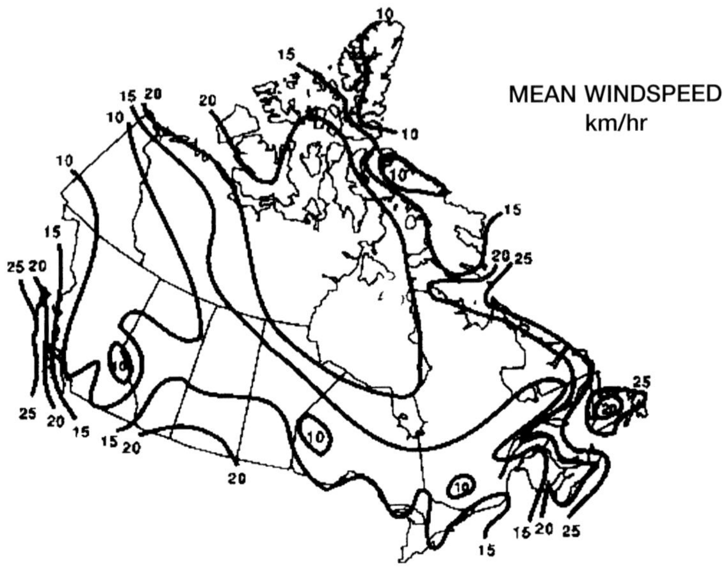

FIGURE 5.1.1. Annual average wind speed (km/hr) for Canada. [6]

$$f(V) = \frac{k}{(V_a - V_0)}\left[\frac{V - V_0}{V_a - V_0}\right]^{(k-1)} \exp\left[-\left(\frac{V - V_0}{V_a - V_0}\right)^k\right]; (k > 0, V > V_0, V_a > V_0) \quad 5.2.2a$$

where V_0 is the threshold or minimum value of V; V_a is the characteristic value of V, which occurs when the cumulative distribution function of V, $F(V=V_a) = (1-e^{-1}) = 0.632$; k is the Weibull's shape parameter. Therefore, the Weibull distribution is a distribution which normally has three parameters, namely V_0, V_a, and k. However, the parameters can be reduced to two, by assuming that $V_0 = 0$, and having k and making $V_a = C$. The PDF therefore becomes

$$f(V) = \frac{k}{C}\left(\frac{V}{C}\right)^{k-1} \exp\left[-\left(\frac{V}{C}\right)^k\right]; \quad (k > 0, V > 0, C > 0) \quad 5.2.2b$$

In this case, the probability density function *f(V)* is characterized by two parameters, one for shape (k) and the other for scale (C). The cumulative density function *F(V)* for the Weibull Distribution is *(following page)*

$$F(V)=\int_0^V f'(V)dV'=1-\exp\left[-\left(\frac{V-V_0}{V_a-V_0}\right)^k\right] \qquad 5.2.3a$$

$$\text{or} \quad F(V)=1-\exp\left[-\left(\frac{V}{C}\right)^k\right] \qquad 5.2.3b$$

Equation (5.2.3a) and (5.2.3b) are the cumulative distribution functions of the Weibull probability functions for the three-parameter and two-parameter Weibull distributions, respectively. The mean and standard deviation are

$$\overline{V}=E(V)=C\Gamma\left(1+\frac{1}{k}\right) \qquad 5.2.4$$

$$\sigma=\sqrt{Var(V)}=C\left[\Gamma\left(1+\frac{2}{k}\right)-\Gamma^2\left(1+\frac{1}{k}\right)\right] \qquad 5.2.5$$

where $\Gamma(\;)$ is the gamma function, the values of which can be found in standard mathematical tables.

The Weibull distribution can be used like any other distribution functions to quantify the wind energy pattern factor (WEPF = Q) which has been defined earlier in section (4.1) and as shown in equation (4.1.1). This can also be shown as follows:

$$Q=WEPF=\frac{\overline{V^3}}{\left(\overline{V}\right)^3}=\frac{E\left(V^3\right)}{\left[E(V)\right]^3} \qquad 5.2.6$$

Applying the Weibull distribution to equation (5.2.6)

$$E\left(V^3\right)=C^3\Gamma\left(1+\frac{3}{k}\right) \quad \text{and}$$

$$\left[E(V)\right]^3=C^3\Gamma\left(1+\frac{1}{k}\right)$$

Therefore wind energy pattern factor (Q = WEPF) is *(following page)*

$$Q = WEPF = \frac{\Gamma\left(1+\frac{3}{k}\right)}{\Gamma^{3}\left(1+\frac{1}{k}\right)} \qquad 5.2.7$$

Equation (5.2.7) shows that the wind energy pattern factor (WEPF) can be determined when the shape factor k of the Weibull distribution is known.

The Raleigh Distribution

The Raleigh probability density function, PDF, $f(V)$ is a special case of the Weibull probability distribution when the shape parameter $k = 2$. It is given by

$$f(V) = \frac{2}{C}\left(\frac{V}{C}\right)\exp\left[-\left(\frac{V}{C}\right)^{2}\right] \qquad 5.2.8$$

The cumulative distribution function CDF is also given by

$$F(V) = 1 - \exp\left[-\left(\frac{V}{C}\right)^{2}\right] \qquad 5.2.9$$

Since the mean value $\bar{V} = E(V) = C\Gamma\left(1+\frac{1}{2}\right)$ for $k = 2$, see equation (5.2.4) and making use of the following important properties of the gamma function

$$\Gamma\left(1+\frac{1}{2}\right) = \left[\frac{1}{2}\Gamma\left(\frac{1}{2}\right)\right] = \left[\frac{1}{2}\sqrt{\pi}\right] = \sqrt{\frac{\pi}{4}} \qquad 5.2.10$$

therefore, $\bar{V} = E(V) = C\sqrt{\frac{\pi}{4}}$; from which $C = \frac{2\bar{V}}{\sqrt{\pi}}$. This value can be substituted into equation (5.2.8 or 5.2.9), in order to model the Raleigh probability density function (PDF) and the cumulative density function (CDF) of wind speed.

Extreme Value Distributions

In order to estimate the peak gust speeds, which are greater than the average wind speeds recorded over long time intervals, the real practical data of peak wind speed at a given location do exhibit statistical scatters, the variability of which can be modeled by means of appropriate extreme value probability distribution. These extreme value probability distributions fall into two categories, namely the so-called Type I and Type II extreme value distributions of largest values of the maximum recorded annual wind speeds for a given site. [11,12,13]. The Type I extreme value distributions and its applications in wind engineering studies will be discussed in section (6.0).

5.3 STOCHASTIC PROCESS MODELS

Since wind speed is a statistical random variable, it is not possible to predict exactly the future magnitude and direction of wind speed at a given point in the Earth's environment. The best we can do is to find the probability that wind speed magnitude will be within certain limits. This variation is called a stochastic process or a random time dependent statistical process [14,15].

It is important to note that the value of a stochastic process at two times t_o and $(t_0 + \tau)$ with an interval τ are generally dependent. The dependence can be modeled in terms of the auto correlation function. This relates the value of wind speed $v(t)$ at time, t, to its value $v(t + \tau)$ at a later time $(t + \tau)$. This provides an indication of the degree of statistical correlation of the process with itself for various values of time intervals τ [6,15].

It is also possible to develop statistical procedures to simulate fairly horizontal wind speed by using Markov Chain [16], which is a finite state random process that develops over time in a probabilistic manner and where transitions from one state $(i\text{-}1)$ to another state, i, are made at discrete points in time. In general, a random process is a Markov process if the distribution of wind speed, $V(N)$, given the past records of the wind speed, can be shown to depend only on the previous samples $V(N)\text{-}1)$ and if $V(N)$ is at state i and $V(N)\text{-}1)$ is at state $(i\ \text{-}1)$.

6.0 ENGINEERING DESIGN OF STRUCTURAL FACILITIES AGAINST EXTREME WINDS FOR DEVELOPMENT OF WIND-POWERED SYSTEMS

In order to minimize the risks of failures of the structural facilities, which need to be designed as an integral part of the complete civil, structural, mechanical and electrical engineering system design, for wind-powered systems, against high wind speeds, it is necessary to consider the magnitudes of such extreme winds, which are likely to occur on the sites where such wind-powered systems are to be located. The methodology, which will be described here, can also be used for the design of other structural facilities, which have to be located in the same area [9 to 12; 17 to 22].

6.1 DESIGN CONSIDERATIONS

The equivalent static wind loadings on such structures can be evaluated using the guidelines from the relevant civil and structural engineering Codes of Practice [17,18]. Essentially, the Codes specify that the design wind speed, V_D, pressure q, and force F are given respectively by:

$$V_D = V_B S_1 \cdot S_2 \cdot S_3 \qquad 6.1.1$$

$$q = \frac{1}{2}\rho V_D^2 \qquad 6.1.2$$

$$F = qA = \frac{1}{2}\rho(V_B \cdot S_1 \cdot S_2 \cdot S_3)^2 \cdot A \qquad 6.1.3$$

where V_B is the basic wind speed, which is the maximum surface wind gust likely to be exceeded on the average only once in 50 years, at a 10 meter height above the ground level, in open country, S_1 = factor due to topographic influence, S_2 = factor due to surface roughness of the environment, gust duration and height of the structure above the ground level, and S_3 = factor to account for expected design life of the structure. Relevant values of the factors, S_1, S_2, and S_3 are tabulated in the Codes of Practice [18].

In view of the variability associated with the extreme wind speed data for any geographical location, the mathematical statistics of extreme value statistics of largest values can be used to find the probability distribution function of extreme wind speeds from the data of annual maxima of wind speeds recorded over several years in the particular location of interest. The basic wind speed, V_B, can be evaluated from the probability distribution function for the specific location which is the maximum surface wind speed likely to be exceeded on the average only once in 50 years, at 10 meters height above ground level, in open level country. This means that the probability of the wind speed V_B being exceeded will be (1/50) = 0.02.

Due to the variables in equations (6.1.1, 6.1.2, and 6.1.3) and in view of the climatological and environmental conditions associated with the high wind speeds in several locations, it is necessary to develop models for the quantification of such variability as follows:

$$\left(\frac{\Delta V_D}{V_D}\right)^2 = \left(\frac{\Delta V_B}{V_B}\right)^2 + \left(\frac{\Delta S_1}{S_1}\right)^2 + \left(\frac{\Delta S_2}{S_2}\right)^2 + \left(\frac{\Delta S_3}{S_3}\right)^2 \qquad 6.1.4$$

where ΔV_D, ΔV_B, ΔS_1, ΔS_2, and ΔS_3 are the possible variability which can be associated with the variables ΔV_D, ΔV_B, ΔS_1, ΔS_2, and ΔS_3, respectively.

Recommendations in this area based on field data gathered in West Africa during the past three decades have been given in another technical paper [11] in order to improve the reliability of the structural and mechanical design in meeting the unique environmental, climatological and other essential requirements of West Africa [11]. Similar useful data can be collected for other sites, anywhere in the world, where the need to plan and design wind-powered systems may be necessary.

6.2 STATISTICAL MODEL OF EXTREME WIND AND APPLICATIONS [10,11,13,21]

Using the Type I Gumbel distribution of extreme values in mathematical statistics, the maximum annual wind speed, y, for any geographical location can be shown to have a cumulative distribution function, *G(y)*, which is given by

$$G(y) = \exp\{-\alpha \exp(-\beta y)\}, \quad y \geq 0 \tag{6.2.1}$$

where α and β are constants for the region which can be determined from data. Equation (6.2.1) can be linearized to give

$$\ln\{-\ln G(y)\} = \ln \alpha - \beta y \tag{6.2.2}$$

When a regression analysis is carried out on equation (6.2.2) using data for any given region, the results can be used to evaluate constants α and β. The return period in years, for any annual maximum wind speed y is given by

$$T_y = \frac{\exp(\beta y)}{\alpha} \tag{6.2.3}$$

Moreover, the statistics of this problem indicate that for any set of data of annual maximum wind speeds, the mean and standard deviation of the data are respectively given by

$$\mu_y = \frac{\ln \alpha}{\beta} + \frac{\gamma}{\alpha} = \frac{\ln \alpha}{\beta} + \frac{0.577}{\alpha} \tag{6.2.4}$$

where γ = Euler's constant = 0.577

$$\sigma_y^2 = \frac{\pi^2}{6\alpha^2} = \frac{1.645}{\alpha^2}; \qquad \sigma_y = \frac{1.282}{\alpha} \qquad 6.2.5$$

Using the results of equation (6.2.4) and (6.2.5) from the knowledge of the mean and standard deviation of the data of extreme wind for any location, the values of the constants α and β can be evaluated in order to make probabilistic predictions about the magnitude of any extreme wind speed relevant to any desired level of return period from equation (6.2.3).

Furthermore, equations (6.2.1) to (6.2.5) can also be used to generate a map of extreme wind distributions in any given country, state, or geographical zone for any study period. By computing from the extreme wind data for every possible location in the country, the magnitudes of the extreme winds for any duration can be obtained. Contours of equal magnitudes of wind speeds for all the locations can be prepared in a map, for the same

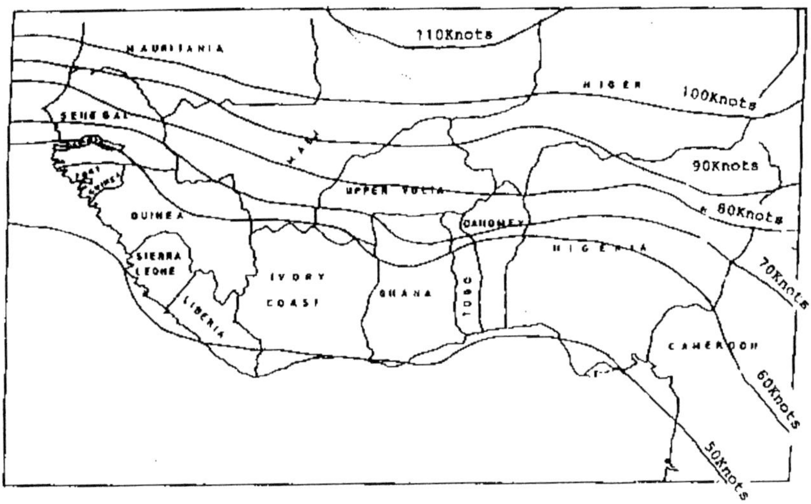

FIGURE 6.2.1 Annual extreme winds for West Africa [21] (30ft above ground) 50 year mean recurrence interval.

level of the desired return period. Such a map is of significant importance in several planning and technical applications where the environmental and climatological effects of extreme winds may be needed, particularly for essential social and economic development in the areas of structural and mechanical design of engineering structures, infrastructures, including telecommunications facilities, electrical power generation and transmission using wind-powered mechanical and electrical generating systems, and the effects of extreme winds on agricultural production.

Figure (6.2.1) shows the extreme wind distributions for 50-year mean return period or the basic wind speeds for all of West Africa. This map was prepared from available meteorological extreme wind data and processed as indicated above. The wind data are obtained from 3-second gust speeds calculated in knots at 33 feet (10 meters) at sites which are in open level country.

The principles of mathematical statistics of extremes developed here, or the use of the normal probability distribution function, can also be used to model the probabilistic distribution of the normal wind speed distribution in any region; from any available normal wind speed data.

7.0 CONCLUSIONS AND OUTLOOK FOR FUTURE

There is no doubt that wind-powered systems have provided a useful alternative, sustainable power in several countries which had suffered from the petroleum energy crisis which occurred about thirty years ago. However, in view of the practical experience gained during the last three decades, in both the economically developed countries of U. S. A and Western Europe, and the several economically less developed countries particularly in Africa, it has become necessary to set new guidelines for the engineering-economic planning, design and maintenance of wind-powered systems in order to achieve higher mechanical and electrical efficiencies, and system reliability; including lower maintainability for wind-powered systems. The following major points will therefore be discussed briefly here.

ENGINEERING-ECONOMIC PLANNING ANALYSIS AND DESIGN

Thorough analysis and evaluations of the potential wind resources, including the reliable designs of the various components of the entire systems, are extremely important factors which should be considered in the engineering-economic planning of viable and reliable wind-powered systems. In view of the variability of wind speeds and the intermittent nature of the wind, an area which has become important, is the need to make engineering design provisions for interconnections and supplementation to other generating systems. This is necessary in order to have dependable and interrupted sources of wind-powered mechanical and electrical generating systems, which can be utilized for industrial, commercial, domestic and other sectors of social and economic development, including the development of essential rural infrastructures. In order to achieve this desirable objective, the economic viability of several technical options should be considered in the general planning and design of new wind-powered generating systems. Some of the possible viable technical options include the following possible provisions: the generation of additional power from solar energy or other existing power generating sources. The economic viability of the generation of supplementary power from coal or gas may also be considered separately, on their own individual merits.

NEEDS OF DEVELOPING COUNTRIES

The need to produce wind-powered generating systems and indeed other associated useful machines for power generation and utilization, particularly in the developing countries of Africa, will always present tremendous opportunities for several years to come in view of the huge possible demands for viable systems for power generation and utilization in these countries. However, careful considerations should be given to the design of relatively simple and reliable engineering components of power generating systems which can be easily maintained in these countries. In other words, social factors need to be seriously considered in order to ensure that the design can be successfully executed for the good and benefit of all those who should derive benefits from the project.

STATISTICAL DATA OF WIND RESOURCES

The statistical data of wind resources of possible project areas should be continuously collected, updated and reviewed as new statistical data are made available by the Department of Meteorological Services or any other agencies. The successful engineering - economic planning and design of efficient wind-powered systems depend on the reliability of the statistical data of wind speeds, which are relevant to the project sites. This is the very important reason why serious efforts should be focused on the accurate collections and appropriate statistical analyses of such data.

SOME POSSIBLE NEW RESEARCH CONSIDERATIONS

The need to develop new accurate simplified methodologies, which are relevant to the environmental and climatological conditions of the project area has become an important consideration in recent times. In this respect, one of the important areas new research should focus on is the multiparameter modeling and validation of wind speed as functions of the several physical, environmental and climatological factors, which can affect wind speed model in any project area. This will provide better understanding, which will be necessary for the assessment and quantification of the magnitude and variability of wind speeds for any project area. Another area which needs further investigation is the need to establish models to quantify the relationship between the variabilities in wind speeds of any two project sites. This would enable accurate comparisons in wind speeds between two project sites.

REFERENCES

[1] T. R. Hiester and W. T. Pennell, 1981, "The Meteorological Aspects of Siting Large Wind Turbines" U.S. Dept of Energy, Report No. PNL-2522.

[2] A.B.O. Soboyejo, M.A. Foster, C. Mercer, J.C. Papritan and W.O. Soboyejo (1999) "A New Multiparameter Model for the Prediction of Fatigue Crack Growth in Structural Metallic Materials". Fatigue and Fracture Mechanics; 30th Volume, American Society for Testing and Materials (ASTM), Special Publication (STP) 1360, edited by: Paul. C. Paris and K. Jerina; ASTM.

[3] W.O. Soboyejo, Y. Ni, Y. Li, A.B.O. Soboyejo and J. F. Knott, 1998, "A New Multiparameter Approach to the Prediction of Fatigue Crack Growth" International Journal of Fatigue and Fracture of Engineering Materials and Structures. Vol. 21, pp541-555.

[4] D.M. Eggleston and F. S. Stoddard, 1987, Wind Turbine Engineering Design, Van Nostrand Reinhold.

[5] R.J. Gustafson, 1988, "Fundamentals of Electricity for Agriculture" 2nd Edition, American Society of Agricultural Engineers, St. Joseph, Michigan, USA.

[6] J.S. Rohatgi and V. Nelson, 1994, "Wind Characteristics-An Analysis for the Generation of Wind Power" Published by Alternative Energy Institute, West Texas A&M University, Canyon, Texas, USA.

[7] P. C. Paris, M. Gomez and W. E. Anderson, 1961, "A Rational Analytic Theory of Fatigue" Trends in Engineering, Vol. 13, pp9-14.

[8] P.C. Paris and F. Erdogan, 1963, " A Critical Analysis of Crack Propagation Laws" Journal of Basic Engineering, American Society of Mechanical Engineers, Vol. 85, pp528-534.

[9] A.G. Devenport, 1961, "Use of Application of Statistical Concepts to the Wind Loading on Structures." Proc. Inst. Of Civil Engineers, Vol. 19, August 1961, pp449-472.

[10] A.B.O. Soboyejo, 1998, "Simplified Model for Wind Speed/Height Relationship and Design of High Rise Structures." Long-Span and High-Rise Structures. Publication of International Association of Bridge and Structural Engineering, Zurich, Switzerland, Vol.79, pp575-580.

[11] A.B.O. Soboyejo, 1997, "Sustainable Wind Powered Generation and Utilization for West Africa" Proceedings: American Power Conference at Illinois Institute of Technology, Chicago, Illinois. American Society of Mechanical Engineers (ASME); American Society of Civil Engineers (ASCE). Pp59-64.

[12] Schueller. G.I., 1995, "On Procedures to Quantity the Reliability of Structures and Mechanical Components" Proceedings; American Soc. of Mechanical Engineers, International Mech. Engineering Conference, Safety Engineering and Risk Analysis. Vol. 4, pp 119-125.
[13] J.R. Benjamin and A.C. Cornell, 1970, "Probability, Statistics and Decisions for Civil Engineers" McGraw Hill, Inc. San Francisco, California.
[14] R. W. Thresher, W.E. Holley, C. E. Smith, N. Jafarey and S.R. Lin, 1981, "Modeling the Response of Wind Turbines to Atmospheric Turbulence" U.S. Dept of Energy, Report No. ET/23144-81/2
[15] D. E. Newland, 1975, "An Introduction to Random Vibrations and Spectral Analysis" Longman Group, London, United Kingdom.
[16] R. H. Kirchoff, F.C. Kaminsky, and J.S. Woo, 1988, " A Maker Chain Analysis of Wind Speed at Winsor, Massachusetts" 7th ASME, Wind Energy Symposium, pp9-16.
[17] "Minimum Design Loads for Buildings and Other Structures" December 1995, American Society of Civil Engineers, ASCE 7-9, Revision of ANSI/ASCE 7-93.
[18] "British Standard Code of Practice CP3, 1972, Code of Basic Data for the Design of Buildings." Chapter 5. Loadings; Part 2; Wind Loads: British Standard Institution, London.
[19] Soboyejo. A.B.O. "Structural Wind Loadings in West Africa" Journal of Construction in Nigeria, Vol.7, No.6, pp25 to 38, June 1973.
[20] Soboyejo. A.B.O. "Wind Loadings on Building Facades in West Africa" Building Envelope Systems and Technology, Editor: Edmund Choi, Nanyang Technological University, Singapore. Pp342-347. December 1994.
[21] Soboyejo. A.B.O., 1997, "Probabilistic Structural Mechanics and Design for the High Wind Environment" Proceedings American Institute for Aeronautics and Astronautics (AIAA), 38th Structures, Structural Dynamics and Materials Conference. Paper No. 97-1027, Kissimmee, Florida
[22] C.W. Newberry and K.J. Eaton, 1974, "Wind Loadings Handbook" Dept. of the Environment, Building Research Establishment, London, U.K.
[23] Golding, E.W., 1976, "The Generation of Electricity by Wind Power." E & F.N. Spon Ltd., London, 1955, Reprinted by Halsted Press.

Renewable Energy: Trends and Prospects. Edited by S.K. Majumdar, E.W. Miller and A.I. Panah

Chapter Twenty

IMPACTS OF RENEWABLE ENERGY DEVELOPMENTS ON REGIONAL ECOLOGY

FRED J. BRENNER, KIMBERLY D. KOSICK and CORRIE A. GARDNER

Biology Department
Grove City College
Grove City, PA 16127

INTRODUCTION

According Cooney (1999), renewable energy technologies produce an excess of 450 billion kilowatt hours (kWh) of electricity, or approximately 12 percent of the total power production in the United States. The term "green energy" generally refers to nonfossil fuel power plants using wind, solar, biomass, geothermal, and hydropower, including large dams, which produce approximately 10 percent of the renewable energy in the United States (Cooney 1999). Although this "green energy" is considered to produce less air and water pollution than fossil fuel plants, all power generation facilities will impact local environments and biodiversity. In the development of sustainable energy systems, their impact on local ecology and communities may not necessarily affect those that benefit from the generation and transmission of electric power. Although so called "green energy" includes small and large scale hydroelectric dams, their adverse impacts on Pacific salmon (*Oncorhynchus spp.*) populations in the Columbia and Snake river systems and elsewhere in the Northwest has resulted in many of these populations being placed on the endangered species list (Cederholm *et al.* 1999, Maser 1999). These effects are not only limited to the northwest since hydropower has also adversely impacted migratory species of Atlantic salmon

(*Salmo salar*), shad (*Alosa spp*) and sturgeon (*Acipenser spp*) in northeastern mid-Atlantic rivers. The basic question being addressed in this chapter is how can we maintain sustainable renewable energy developments while at the same time protecting local ecology and biodiversity.

BIOMASS

Biomass, especially wood, which has been used by man for over 5000 years as a source of heat, cooking and, at least since the bronze age, by industry to produce energy, is far from a new idea. The term "biomass" encompasses an array of materials either directly or indirectly produced by photosynthesis. These starting materials can be consumed directly, as in wood, or converted to a more useful secondary fuel, such as ethanol or methane (Hollander *et al.* 1980). Benefits of using biomass to produce energy include: reduced volumes of solid wastes requiring disposal in landfills, reduced quantities of nutrients being added to waterways, and reduced odor and pest problems (Hollander *et al.* 1980). Environmental impacts can incur, however, during growth and harvesting of biomass, during conversion into fuels, and in the consumption of these fuels (Hollander *et al.* 1978).

Organic urban and industrial wastes or refuse derived fuels (RDF)

Organic matter available for energy production comprises approximately 65% of urban refuse, over half of which is cardboard and paper. Other sources of RDF include solid wastes and sewage sludge. Refuse can be converted into energy either by burning to produce heat and electricity, pyrolyzed to produce low-medium calorific value gas, or digested to form methane or alcohol (OECD 1988).

Agricultural and Forestry Wastes

These materials include manure, straw and other crop residues from agricultural production, biogas and forestry wastes. These materials are naturally produced organic materials that currently have little or no commercial value. For example, unmarketable portions of trees are generally either burned or left on the ground to biodegrade. The harvesting of these sources allows a greater portion of the plant or tree to be harvested (OECD 1988, Hollander *et al.* 1981). Manure can be used to produce biogas or methane by natural microbial processes. The product, after slurry digestion, is high in nutrients and low in odor emissions, which enables it to be recycled as natural fertilizer (OECD 1988). Also, existing uncultivated stands of trees, shrubs, bracken, heather, and naturally or anthropogenically degraded forests, including those damaged by fire, insects, or diseased trees, may provide a source of biomass for energy production. Another potential source of biomass energy would include high caloric crops of sugar cane or other cash crops, such as corn, that can be utilized for energy production.

Environmental Impacts

Although the handling, transportation, and storing of biomass has only minor environmental impacts, local air quality may be impacted by the release of particulates during combustion. Due to the high moisture content, small amounts of incorporated nitrogen, and the low burning temperature of biomass, results in relatively low levels of NOx and SOx emissions (Swezey *et al.* 1995, Hollander *et al.* 1981, OECD 1988, Shinn *et al.* 1982). Thus, emissions from biomass combustion does not significantly contribute to acid rain (Hollander *et al.* 1981). Moreover, carbon monoxide emissions produced during combustion are generally more concentrated than those from fossil fuel plants (OECD 1988). In addition, the combustion of biomass does not release geologic carbon. Therefore, the release of CO_2 into the atmosphere is less than that from the combustion of fossil fuels (OECD 1988, Hollander *et al.* 1980, Hollander *et al.* 1981). Another factor is that during combustion of biomass products, such as ethanol, no or low net CO_2 and CH_4 emissions are released into the atmosphere (Swezey *et al.* 1995). On the other hand, Ebenhack (1995) suggests that the combustion of ethanol in internal combustion engines produces toxic aldehydes and may only be marginally or questionably more beneficial to local environments than gasoline.

Particulate emissions, however, are the greatest concern during biomass combustion, and, therefore, will require emission controls (Hollander *et al.* 1981, Swezey *et al.* 1995). For example, wood fibers are completely oxidized at high burning temperatures resulting in lower particulate emissions, while dry recovery in multicyclones, dry scrubbers, and electrostatic precipitators can collect up to 99% of particulate emissions. This combination allows compliance with all EPA regulations (Hollander *et al.* 1981). However, the current technology available for controlling these emissions are designed for large-scale industrial plants, and are not cost-effective for reducing emissions from residential wood burning units that burn at lower temperatures and cannot achieve complete combustion of wood fibers. The release of particulates from residential wood stoves may contain a variety of toxic irritating, carcinogenic and cocarinogenic agents including phenols, formaldehyde, acetaldehyde, benzo(a)pyrene and methylbenznthracene (Hollander *et al.* 1981, Ebenhack 1981). The incineration of urban refuse can produce dioxins during the combustion of plastics and chlorinated compounds. Although these emissions are 4-5 magnitudes below the current minimum chronic toxicity as established by the EPA (OECD 1988), these organic compounds have be shown to be both carinogenic and mutagentic.

Production processes using biomass such as pyrolysis, gasification, and alcoholic fermentation result in little air pollution, but the particulates and organic compounds produced by these processes are channeled into the waste-water or solid waste streams which may adversely impact local environments. Processed fuels such as methanol, synthetic oil, and charcoal burn relatively cleanly, producing less air emissions compared to the burning of fossil fuels (Hollander *et al.* 1981).

Hollander *et al.* (1981) indicated that approximately 2% of the original wood mass remains after combustion in the form of ash. This non-hazardous ash can be used as liming agents and partial fertilizers for acidic soils (Swezey *et al.* 1995), but the disposal of large quantities of ash poses a problem to both the industry and the environment. Wind blown ash from temporary storage piles has been shown to cause eye and respiratory irritation (Hollander *et al.* 1980, Hollander *et al.* 1981) that may result in health problems. As with fossil fuel generation facilities, waste water discharges may represent a more serious environmental concern. Leachates from wood chip storage piles are high in biological oxygen demand (BOD) and other organic materials, usually high alkaline (pH =13), but low in heavy metal concentrations (Hollander *et al.* 1980, 1981). These leachates can be collected and processed by secondary water treatment plants to neutralize pH and lower the BOD, thereby reducing the environmental impact (Hollander *et al.* 1981). Gasification of biomass also produce waste waters high in organic material, tars, phenols, acids, and aldehydes, and require secondary treatment to reduce their impacts on health and the environment (Hollander *et al.* 1981).

Biomass energy plantations have impacts similar to those of conventional agriculture including a net loss of groundwater resources due to irrigation from underground aquifers combined with slow recharge rates. Harvested biomass may be a source of leachate during storage, creating a need to control decomposition, and a reduction in tannic acid leachate, and odors associated with decomposition. The removal of crop residues for energy production will reduce soil organic and humus content, increase wind and water erosion, decrease water infiltration and moisture retention, reduce the cation exchange capacity and decrease a source of energy for micro-organisms essential for nutrient and mineral recycling and soil productivity (Hollander *et al.* 1981). During harvesting for biomass energy production, the entire crop is removed from the field, thereby reducing the nutrient content of the soil by as much as 10% and leading to reduced productivity and soil degradation (OCED 1988). The loss of soil nutrients, especially nitrogen, creates the need for increased commercial fertilizer applications and irrigation, which in turn leads to increased nutrient rich leachate, increased costs, and energy required for the production of commercial fertilization (Hollander *et al.* 1981).

Likewise, when wood is harvested for energy production, the environmental impacts are also quite extensive. Removal of well-established trees from forests results in soil compaction, soil acidification, and erosion. The degree of these impacts depends on the soil texture and moisture content, equipment used, and the extent of the harvest. Compaction of soils results in decreased water infiltration and increased overland flows, as well as increased soil erosion. Soil compaction may also inhibit the re-establishment of vegetation by reducing soil water recharge, limiting soil aeration, and resisting root penetration. All but the most severely impacted areas can recover. However, these effects may require years to several decades for the soils to return to their pre-impact conditions by natural processes (Hollander *et al.* 1981; Shinn *et al.* 1982). Sound management practices can alleviate some of these impacts, including soil erosion and re-vegetation (Hollander *et al.* 1980, 1981; OECD 1988).

Increased soil erosion, along with leachate production, will have an adverse impact on water quality within a watershed. Selected harvesting of forest resources can result in a 2-3 percent increase in runoff compared to a 40 percent increase in runoff from clear-cuts (Hollander *et al.* 1981), resulting in increased sedimentation, dissolved solids and nutrient loading in receiving streams. The deterioration of water quality adversely impacts aquatic communities, especially fish and macroinvertebrates. Hollander *et al.* (1981) suggests that these changes in water quality are temporary and generally most pronounced during the first 1-2 years following harvest, but recovery will vary among geographic regions and local environmental conditions.

In addition, the type and degree of harvest of forest resources affects community development during forest succession, as well as local wildlife populations (Hollander *et al.* 1981). Forest ecosystems are more vulnerable to disruption than established agrosystems since natural ecosystems are dynamic and always evolving through a series of sere stages. Each sere stage is comprised of specific species of flora and fauna, and harvesting disrupts this evolutionary process which, depending on the species, may be beneficial or detrimental to local wildlife communities. Although clear-cutting is generally considered the most detrimental to ecosystems, select harvesting of a particular species may result in a homogeneous mature stand which decreases the diversity and stability of the ecosystem (Shinn *et al.* 1982, Hollander *et al.* 1981). However, both methods of harvesting are likely to reduce small mammal and avian populations, along with game species, migratory birds species, and endangered species (Hollander *et al.* 1981).

But of all the biomass sources, plantations have the greatest potential to impact the local environments (Hollander *et al.* 1980). On a large scale, biomass energy systems can affect local or regional climate conditions by altering temperature and humidity patterns, aerodynamic roughness, moisture-transfer processes, and surface reflectivity (Hollander *et al.* 1980,1981). This, in turn, can affect the existing vegetation, atmosphere, soil (aridity, erosion, albedo, etc.), and hydrology of the watershed.

GEOTHERMAL POWER

The major sources of geothermal energy include hydrothermal, geopressurized, and hot dry rock, with the latter being the most effective. The ultimate source of geothermal energy is the hot molten core of the earth (Rosenberg 1993). Geologic activity within the earth's crust produces geysers, hot springs, and volcanoes that emit heated rock and water with the resulting heat and steam being used for a variety of purposes, including natural spas, systems to melt snow on roads, development of agriculture, horticulture, space heating, and hot water supplies (International Energy Agency 1998). But, more importantly, electricity can be generated from the heat released from faults where the molten mantle is near the surface, but this semi-renewable resource provides less than one percent of the world's electricity (Ramage 1997). This statistic, however, does not do justice for the potential development of geothermal energy. Due to the tendency of this resource to be found only in widely dispersed geographical areas, this statistic does not hold true for those regions near potential geothermal energy sources. For example, in the Philippines 25 percent of the electricity is provided by geothermal sources. In general, geothermal sources are more efficient than fossil fuel generation since less heat is lost during energy production (Ramage 1997).

Of the three sources of geothermal energy, hydrothermal discharges offer the greatest potential for energy development. Hydrothermal discharges originate at depths exceeding 3000 m and vary from warm water pools to geysers releasing superheated water vapor, which is condensed into pressurized steam once it reaches the atmosphere. The release of this superheated water is currently being harnessed for electric generation in New Zealand, Iceland and Japan (International Energy Agency 1998). In the United States, the only known geysers occur in Northern California and in Yellowstone National Park (Kleinbach and Salvagin 1986).

Although hydrothermal sources provide the greatest potential for energy developments, superheating water using hot dry rocks may also be used to generate geothermal energy. This technology allows for water recycling, thereby providing for water conservation. All three sources of geothermal energy have environmental concerns, but hot dry rock systems have fewer impacts since they are essentially a closed system (Kleinbach and Salvagin 1986).

There are, however, numerous environmental and health issues associated with geothermal energy development, including their potential impact on endangered species, water and air pollution, and alteration of land use patterns, all of which would affect local biodiversity. In addition, the potential increase in H_2S and particulate emissions, along with increased seismic activity along fault lines causing landslides and land subsidence, are a few of the human health and safety issues that must be addressed when considering the development of geothermal energy sources (Boyle 1996, Kleinbach and Salvagin 1986; Ramage 1997).

Both the quantity and quality of water required by geothermal developments remains a major issue in the development of these energy sources. Depending on the type of development, they may require an excess of 100,000 liters/day for cooling and evaporative water loss. Since many geothermal deposits are located in arid areas, they have an adverse impact on ground water aquifers while increasing the risk of land subsidence. For example, the extraction of groundwater resulted in a 7.6 m subsidence of a geothermal energy plant in Wairakei, New Zealand. But even in nonarid regions, the prolonged exploitation of water resources has resulted in a decline in water reserves and associated decreases in the reservoir pressure (Woods 1999). On the other hand in dry steam plants with condensing turbines, water is often produced in excess of what the region can recharge into the ground water or reservoirs. The disposal of these waters may result in the alteration of land use and pollution of aquatic ecosystems by increasing heavy metal concentrations.

The release of elevated concentrations of H_2S, SO_2 may increase the formation of acid rain, which will not only impact local forest communities but will affect the ecosystems of a wide geographical area as occurred with European forests in the 1980's (Ramage 1997) when particulate matter was being released into the air. These impacts may be mitigated to some extent by the removal of sulfur compounds and heavy metals from the emissions at the site of production, but although emissions from geothermal energy developments are less than those from fossil fuel plants, these issues still must be addressed because of their potential impact on local and regional ecology.

HYDROELECTRIC POWER

At the present time, approximately 20 percent of the world's electricity is produced by hydroelectric power. In fact, hydropower generation is the second largest renewable energy resource with a potential output of 2,400,000,000kWh or approximately twice the current demand for electric power. This, of course, depends on the availability to transmit this energy from the site of production to areas of high usage. In order to generate electricity using hydropower, water must be available in sufficient flow, volume and pressure to power the generators. This may be either free flowing water such as Niagara Falls or by constructing of hydroelectric dams. Generally dams are considered a better choice because of their storing capabilities, but they are more expensive both in terms of dollars (Rosenberg 1993) and environmental impacts.

In the era of construction and expansion of hydroelectric dams in the western United States and elsewhere in the world during the 1920's and 1930's, the environmental effects of the construction and operation of these facilities was not given adequate consideration by proponents of hydropower generation. Likewise, the total cost of construction and operation of these facilities is not being born by the user, but is spread out over the entire tax base of the nation. In other words, the users of electric power in Las Vegas, Nevada are not paying the full cost of the electricity being generated by Hoover Dam, but this applies to all dams constructed for the generation of electric power. But more important, however, is the numerous adverse environmental and sociological problems associated with the construction and maintenance of hydropower generation facilities, including the alternation in stream flow regimes, interference with the migration and spawning of anadromous fish, groundwater and surface water ecosystems and displacement of local citizens or, in some cases entire communities, such as was the case during the development of the Tennessee Valley Authority.

Many environmental groups question the concept that the development of relatively inexpensive electricity is offset by the loss of biodiversity and adverse impacts on local communities. If these habitats and communities are irreplaceable, as Kleinbach and Salvagin (1986) suggests, then this renewable energy source is not worth the money and environmental cost associated with hydropower generation. Rosenberg (1993) argues that although species and human communities are displaced, this does not necessarily mean extinction for the given species and may be a temporary inconvenience for the citizens being displaced. Parris (1997) supports the views of Rosenberg (1993) by referring to the success of the salmon migration program in the Columbia, but he does not indicate that large numbers of young salmon are killed in the turbines as they move downstream to the ocean. Moreover, many migratory salmon and steelhead populations are declining to the point of being endangered throughout the northwestern United States and Western Canada (Cederholm *et al.* 1999, Maser 1999). According to Kleinbach and Salvagin (1986), the fact that wildlife habitats are permanently lost by dam construction is a valid reason to argue against their construction and continual use for hydropower generation. The construction of hydroelectric dams also have been shown to impact the diversity of fish communities above and below the structure. In the Senegal River Project, there was an increase in the number of fish species and populations above the dam, whereas the diversity and size of the populations in the stream below the dam were drastically reduced (Roberts 1988).

The development of large scale hydroelectric dams will reduce downstream flooding, thereby protecting human life and property and providing water for irrigation and drought emergencies. However, at the same time, they adversely impact stream systems since periodic flooding is necessary to maintain channel morphology and diverse wildlife and fishery habitats. In tropical countries, these dams may increase the health risk to local peoples. For example, the damming of the Senegal River provides an ideal habitat for snails, which increased the incidence of Schistosomiasis, along with dysentery and malaria in the region (Roberts 1998). Although the project provided more water during drought conditions, it was not used effectively to improve agriculture, and malnutrition remains a major health problem in the region. Both the Senegal and Aswam dam projects adversely impacted agricultural production in their respective flood plains, and the lack of nutrients associated with their silt loads reduced fishery production in their deltas, a major source of protein for Senegal and Egypt. Although the era of large scale dam building for hydroelectric power generation may be over in the United States, it remains a threat to biodiversity in other parts of the world, especially in tropical countries. Let us hope that the lessons learned from the Senegal and Aswam Dam projects will be applied elsewhere in the world so biodiversity can be maintained in all the nations of the world.

RENEWABLE ENERGY DERIVED FROM THE OCEANS

The oceans of the world provide a tremendous potential for renewable energy developments. The mechanical energy of the ocean is derived from three main sources: solar radiation, the gravitational pull of the moon, and wind. Technology is currently available to harness and convert this mechanical energy into electric power. But, as with all energy development, there are environmental issues that must be addressed.

Wave Generated Energy

The mechanical energy of waves is derived from wind being produced by the differential heating of the atmosphere by solar radiation. Like solar and wind energy, wave energy is diffuse and requires large facilities to harness the available energy. Since wave energy is created by winds, it tends to be intermittent, however it is not subject to the rapid variations of wind energy. The ability of waves to persist long after the winds have ceased tends to make wave energy more constant than winds alone (Newman 1983, Ebenhack 1995). Although technology is currently available to harness wave energy, economics dictates that at present this type of development would not be cost effective. The environmental impacts of wave generation systems are unclear at this stage in their development (Ebenhack 1995).

Tidal Generated Energy

Tides are generated by the gravitational pull of the moon and although they are intermittent, they occur at regular intervals and thus provide a source of energy 10 times more concentrated than wind energy. As with all fluids, the concentration of energy in tides is directly proportional to its density (Fay 1983) and provides mechanical energy available for large scale electric power generation. Currently, two plans have been proposed to harness tidal power for electric generation. The first of these involves the construction of small-scale tidal dams at the mouth of estuaries, some of which are presently in use in Europe (Daborn 1987). The second proposal involves the construction of underwater "windmills" to harness the tidal currents flowing into and out of an estuary. These smaller underwater "windmills" would deliver more energy than larger land based structures since the energy of fluids is proportional to its density.

The conventional design for power plants operates on similar principles as hydroelectric dams but may have less of an environmental impact because the flora and fauna are already adapted to periods of inundation (Ebenhack 1995). The main component of such designs includes concrete dams built across the mouth of a bay or estuary. Water enters the head pond through sluice gates during high tides. When sufficient head has built up, turbine gates are opened and the water fuels the turbines for up to 6 1/2 hours after high tide (Daborn 1987). Another less conventional design involves the construction of devices similar to windmills beneath the surface of the water which harness tidal currents flowing in and out of an estuary. These underwater "windmills" deliver more energy than land based windmills because the energy of a fluid is proportional to its density. This design has the advantage of being smaller than a wind turbine that produces comparable amounts of energy (Fay 1983, Daborn 1987).

In most dams, less then 50 percent of the water entering the head pond is released back across the dam. The retention of large amounts of water above the dam inundates intertidal habitats, effectively converting them to subtidal habitats (Daborn 1987, Mitchell and Probert 1984). Hence, the existing water flow patterns in the estuaries and the interconnected coastal waters will be altered by the construction and operation of tidal dams. The construction of these facilities in macrotidal estuaries such as Fundy and Severn would result in large scale environmental impacts. These systems are tide dominated with the complete mixings of the water column along with high sediment loads that are balanced by intertidal and subtidal deposits. The construction of a barrier at the mouth of the estuary will change the patterns of water flow, resulting in decreased turbulent mixing in the head pond. Additionally, in systems with large freshwater inputs into the estuary, stratification of the water column will occur leading to sedimentation, increased light penetration, and conversion of firm and sandy substrates into mud (Daborn l987, Mitchell and Probert 1984).

Since estuaries are physically dominated systems, all the above physical impacts will have drastic biological effects. In the clear, nutrient rich waters of the head pond, phytoplankton communities flourish, compensating for the decline of productivity in the salt marsh and benthic diatom communities. The shift towards increased pelagic productivity favors the development of planktivoric species, such as herring, and benthic suspension feeders, such as mussels. The abundant benthic resources of bays and estuaries are utilized extensively by over 50 species of fish, some of which have commercial importance, including herring, cod and halibut. Changing currents result in the redistribution of mud and sand flats that may require 10-20 years to stabilize. The reduction of intertidal zones and the redistribution of mudflats will significantly affect migratory birds that utilize these areas as resting and feeding grounds during migration (Daborn 1987).

In addition to the adverse impact on estuarian species, tidal dams would interfere with the spawning migration of andronomous species such as salmon, striped bass (*Roccus saxatitis*), and shad, that require passage from the estuaries into the river systems. Although entrance into the head pond through the sluice gates during high tide allows fish access to the estuary, the reverse movement to sea requires passage through turbines, resulting in mortalities as high as 20% per passage. Some investigators (Daborn 1987, Mitchell and Probert 1994) suggest that increased phytoplankton production will compensate for turbine mortality by increasing the survival and growth of the remaining fry, but there is no substantial evidence to support that an increase in phytoplankton production offsets turbine mortality.

In addition to these impacts, thermal effluents from power plants will also alter the ecology of estuaries along with a variety of physiological effects, including diseases due to thermal stress occurring in species inhabiting these ecosystems. If temperature changes are gradual, then the organisms have an opportunity to acclimate to higher temperatures.

In the normal operation of power plants, toxins may be released into the water column. Chlorine is often used as an antifouling agent in power plant operations and may alter the ecology of estuaries, including reducing the productivity of phytoplankton communites causing a biomagnification effect throughout the food web. In addition, the release of heavy metals reduce the ability of fish to survive thermal shock, as well as being toxic to aquatic microorganisms (Hollander *et al.* 1978). Mitchell and Probert (1984) suggest that the bioaccumulation of heavy metals often reach lethal and sublethal concentrations in terminal consumers of the food chains.

Ocean Thermal Electric Conversion (OTEC).

In the view of some proponents of OTEC, this technology may provide a benign power production since ammonia is the only hazardous material involved in power production and no noxious by-products are produced (Bisio and Boots 1996, Hollander *et al.* 1980). OTEC uses large ducts to draw water from the mixed layer and the deep ocean and returns it to the mixed layer near the thermocline (Bisio and Boots 1996, Hollander *et al.* 1980). The same thermodynamic principals that conventional steam power plants operate on are applicable to OTEC technologies, and environmental impacts can be avoided by proper design (Basio and Boots 1996). However, several environmental concerns have been expressed concerning the operation of these power plants. The potential impacts are generally limited to the creation of rapidly flowing seawater streams with flow rates approaching those of flowing rivers that can affect thermal stratification and the entrapment of organisms on the intake screens.

WIND POWER

Once construction has been completed, wind power is often viewed as a renewable energy source with, except for noise and visual impacts (Freris 1990), few, if any, environmental concerns. The more populated the area, the greater the need for more windmills to fulfill the energy requirements of the area, and the greater the concern expressed for visual and noise pollution. Moreover, we do not know what affect noise pollution will have on local wildlife populations. Opponents of wind power generation argue that amount of electricity produced by the wind energy conversion systems (WECS) is not cost effective when considering the value of the land required for wind farms. Once the land is converted to wind power generation, it is effectively lost as wildlife habitat, food production or other alternative land uses. Since geographical variation exists in the effectiveness of WECS, their impact on local land values must be addressed and compared with alternative technologies for generating electric power. If wind power generation is considered to be a cost effective land use, the profit generated must exceed the theoretical cost of the land. Once WECS are in operation, concern has been raised in regard to the increasing mortality rate of migratory birds. Although the possible interference with electromagnetic waves may not be considered an environmental impact, the interference with communication systems may become a concern for local communities. Moreover, transmission of electromagnetic waves may represent environmental health problems for both wildlife and humans alike. Unfortunately, data bases are not available as to long term impacts of WECS on local wildlife and human populations. Before we enter into extensive development of these systems, all environmental issues must be addressed.

SOLAR ENERGY

Compared to fossil fuel power plants, solar energy is often considered as an environmentally safe source of electrical and heat production (Lawrence 1979, Hollander *et al.* 1978). Regardless of the ultimate use of solar energy, it involves harnessing solar radiation via solar collectors that convert solar energy into a more useful form. Currently, the uses of solar energy technology involve passive heating and cooling, active heating and cooling, thermal-electric power generation, and photovoltaic power generation (OECD 1988, Shinn *et al.* 1982, Hollander *et al.* 1980, Ebenhack 1995).

Passive Solar Systems

The use of passive solar energy has been shown to reduce conventional energy requirements for residential heating and cooling by at least 50 percent when integrated right into the building design. For maximum efficiency, the solar panels should face in a southerly direction for maximum solar radiation and should not be shaded by trees or other buildings.

Since solar panels are integrated into the building design, additional land will not be required for their construction (OECD 1988, Ebenhack 1995). For the effective use of solar energy, the buildings must to be well insulated, which may reduce air exchange and result in potential health concerns, high humidity and condensation, especially in the ventilation system. Depending on the type of insulation, toxic fumes may be released, especially in case of a fire in the building (Ebenhack 1995, OECD 1988).

Active Solar Systems

Active solar technology utilizes a fixed orientation of solar collector panels which transform solar radiation into a heated fluid. For residential heating and cooling, low to medium heat panels can be installed directly on the roof of the building or other surfaces exposed to solar radiation. For higher temperature functions, the solar collectors require larger amounts of land and materials than passive technology.

Operational environmental impacts are minimal with the majority of environmental impacts being incurred at the end of the useful life of the solar collectors. These collectors have a coolant that is replaced approximately every three years to avoid potential fire hazards. Although the coolant is water based, it may contain anti-freeze, biocides, rust inhibitors, and substances leached from the solid materials of the solar collectors, which could include nitrates, nitrites, chromates, sulphites, sulphates, alcohols, oils, fluorocarbons, and liquid sodium. As in all technologies, there is a danger of an accidental release of coolant fluid through leakage or fire (Hollander *et al.* 1980, OECD 1988).

Thermal-Electric Solar Systems

Thermal-electric technology is designed to convert thermal energy from the sun into mechanical energy using turbines to produce electricity. Currently, two technologies are being used for solar power plants. The first involves a central receiver system using an array of sun tracking mirrors (heliostats) that reflects radiation onto a receiver mounted on a central tower. The energy absorbed is then transferred as heat into fluids and converted into electrical energy via turbines (Ebenhack 1995, Hollander *et al.* 1980, OECD 1988). The second technology involves a distributed collecting system of parabolic collectors that focus the radiation into a receiver located on individual collectors which are interconnected to allow the heated fluid to flow into generators (OECD 1988, Hollander *et al.* 1980).

Generally, under normal operating conditions, air pollution is not a major concern, but, in the event of a fire or a rupture in a high pressure valve, products from heat transfer fluids or storage products may be released into the atmosphere. These potentially toxic compounds include nitrogen oxides, sodium monoxide or peroxide, and sodium hydroxide. Since large amounts of water are necessary for washing the heliostats and parabolic collectors, steam condensation, and boiler make-up purposes, the effects on water quality are debatable.

Some researchers say that the impact is very minimal, while Ebenhack (1995) suggests that 23.6 liters of potentially contaminated water on average are discharged for every giga-joule of electrical output. This is a non-negligible impact. The flushing of heat transfer fluids and storage products may release substances such as oils, corrosion inhibitors, biocides, and glycols into the waterways (Ebenhack 1995, Hollander *et al.* 1980, OECD 1988). Proper disposal plans should alleviate these concerns. Noise pollution has not been associated with solar technology (OECD 1988, Hollander *et al.* 1980).

The environmental impacts of most concern associated with solar-thermal power generation are the large land use requirements and the possible alteration of the microclimate. Solar energy is a diffuse resource, and, therefore, it requires large amounts of land for the generation facility and all the necessary collectors. The size of the plant and the number of collectors depends on the electrical demands of the community. Extensive land use includes removal of trees and other structures that can shadow the collectors and the destruction of wildlife habitat. Wind deflection and soil compaction can lead to increased erosion, which in turn affect water quality. Shading of the ground and deflection of wind by solar panels may be sufficient to lead to small changes in the microclimate, including decreased temperature. Glare from the solar panels is more problematic in these systems because the radiation is concentrated to an extent that the glare can be blinding to organisms or workers (Hollander *et al.* 1980, OECD 1988).

Photovoltaics

Photovoltaic cells directly transform the protons from incident radiation into electrical current, via the use of semi-conductors. These cells require material which can be explosive (silane), toxic (cadmium sulfide), or produce toxic substances during refining (mercuric effluents from gallium extractions). The manufacture of thin-film photovoltaic cells requires the use of toxic (AsH_3, PH_3, SiF_4, B_2H_6), pyrophoric (SiH_4) or flammable (H_2, CH_4) compounds. In addition, polycarbonated biphenols (PCBs) may be released from the outgassing of power conditioners (Hollander *et al.* 1980, Ebenhack 1995, OECD 1988). Polycarbonated biphenols have been shown to bioaccumulate in the food chain in aquatic ecosystems. Both centralized and decentralized technologies exist, therefore, as land use requirements may vary according to the design of individual facilities. Overall, since photo-voltaic cells have multiple uses, and with proper environmental controls, the environmental impacts on the local ecology will be minimal (OECD 1988)

LITERATURE CITED

Bisio, A. and S. Boots. 1996. The Wiley Encyclopedia of Energy and the Environment, NewYork: John Wiley and Sons, Inc. 2:1162-1515.

Boyle, G. 1996. Renewable energy: power for a sustainable future. Oxford University Press, Oxford. 477pp.

Center for the Analysis and Dissemination of Demonstrated Energy Technology (CADDET). 1995. The world's first off-shore wind farm. 13:4.

Cederholm, C.J., M.D. Kunze, T. Murota, and A. Sibatani. 1999. Anadromous Pacific salmon represent a major vector for transporting marine nutrients across boundaries. Fisheries 24:6-23.

Calvert, N.G. 1981. Windpower principles, their application on the small scale. Charles Griffin & Co., London. 122pp.

Cooney, C.M. Can renewable energy survive deregulation. 1999. Environmental Science and Technology. 33: 494-499.

Daborn, G.R. 1987. Potential impacts of hydro and tidal power developments on the ecology of bays and estuaries. pp 335-348. *in* Environmental Consequences of Energy Production: Problems and Prospects. S.K. Majumdar, F.J. Brenner, E.W. Miller (eds). Pennsylvania Academy of Science. Easton, PA.

Ebenhack, B.W. 1995. Nontechnical Guide to Energy Resources: Availability, Use, and Impact. Tulsa: PennWell Publishing Company. 290pp.

Fay, J.A. 1983. Harnessing the Tides. Technology Review. July: 55-61.

Freris, L.L. 1990. Wind energy conversion systems. Prentice-Hall, Englewood Cliffs, NJ. 388pp.

Hollander, J.M., M.K. Simmons, and D.O. Wood. 1978. Annual Review of Energy. Annual Reviews, Inc. Palo Alto, CA: 3:101-429.

Hollander, J.M., M.K. Simmons, and D.O. Wood. 1980. Annual Review of Energy. Annual Reviews Inc. Palo Alto, CA: Annual Reviews, Inc. 5:241-283.

Hollander, J.M., M.K. Simmons, and D.O. Wood. 1981. Annual Review of Energy. Annual Reviews Inc. Palo Alto, CA. 6:139-170.

International Energy Agency. 1998. The largest geothermal power station in Japan and effective use of hot water. CADDET 13:3.

Johnson, G.L. 1985. Wind energy systems. Prentice-Hall, Englewood Cliffs, NJ. 360pp.

Kleinbach, M.H. and C.E. Salvagin. 1986. Energy technologies and conversion systems. Prentice-Hall, Englewood Cliffs, NJ. 452pp.

Lawrence, K.A. 1979. A Review of the Environmental Effects and Benefits of Selected Solar Energy Technologies. Solar Energy Research Institute. SERI/ TP-53-11 pp. 1-10.

Maser, C. 1999. Ecological Diversity in Sustainable Development: The Vital and Forgotten Dimension. Lewis Publishing, Boca Raton, FL. 401pp.

Mitchell, R. and P.K. Probert. 1984. Environmental and Nature Conservation

Aspects of Tidal Power Proposals for the Severn Estuary (UK). Wetland Science and Technology. 16: 269-279.

Newman, J.N. 1983. Power from the Ocean's Waves. Technology Review. July: 52-69.

Organization for Economic Cooperation and Development (OECD). 1988. Environmental Impacts of Renewable Energy: The OECD Compass Project. Paris. 97pp.

Parris, T.M. 1997. Power and salmon along the Columbia River. Environment. 39:3-4.

Ramage, J. 1997. Energy: a guide book. Oxford University Press, Oxford. 894pp.

Roberts, L., 1998. World resources:A Guide to the Global Environment. The World Resources Institute, Oxford. 369pp.

Rosenberg, P. 1992. The alternative energy handbook. Fairmont Press, Lilburn, GA. 266pp.

Shinn, R.A., J.H. Freeman, and L.J. Hoch. 1982. Pennsylvania Renewable Energy Resource Assessment. Governor's Energy Council. Harrisburg, PA. 137pp.

Swezey, B.G., K.L. Porter, and J.S. Feher. 1995. Biomass and Bioenergy Vol. 8. Great Britain: Elsevier Science Ltd. 8:207-220.

Woods, A.W. 1999. Liquid and vapor flow in superheated rock. Annual Reviews Fluid Mechanics. 31:171-199.

Renewable Energy: Trends and Prospects. Edited by S.K. Majumdar, E.W. Miller and A.I. Panah

Chapter Twenty-One

REEVALUATING LATENT ETHANOL PRODUCTION

BROOKS C. PEARSON[1]
and
SARA C. PRYOR[2]

[1]Geography Program
Geosciences Department
State University of West Georgia
Carrollton, GA 30118-3100
and
[2]Atmospheric Science Program, Geography
Indiana University
Bloomington, IN 47405-7100

This chapter demonstrates that the technology represented by current installed fermentation facilities is capable of extracting fuel ethanol from within the present agricultural system without disrupting livestock or human food supplies and without expanding production. Most previous studies have suggested that large scale ethanol production is not feasible, but they have based their conclusions overwhelmingly on the assumption that this ethanol must be based on normally marketed corn, the cultivation of which cannot be expanded significantly without increased production costs and intrusion onto marginal lands. This is why the present study's latent ethanol provides such an important avenue for capturing the fuel from a portion of the biomass fed to livestock, thereby extracting ethanol from within the current system.

x.1. INTRODUCTION

Liquid biofuels, particularly alcohols, represent one of the renewable energy technologies most easily integrated into current energy consumption frameworks and offer potential additional benefits associated with clearer burning characteristics (leading to the potential for improved air quality) and reducing net carbon emissions. Ethanol has several advantages over other alternative fuel competitors in terms of safety and experience with production and use. However, widespread adoption of ethanol has been hampered by relatively low oil prices and competition for valuable cropland. Here we demonstrate that sufficient latent production potential lies within the present agricultural system to provide feedstocks for greatly increased fuel ethanol fermentation. The potential for expanded ethanol production exists because the byproducts of current ethanol milling technologies are high-protein animal feeds. The fermentation agents which excrete ethanol feed only on sugars (and mostly on simple sugars). Therefore, ethanol distillation realizes no protein loss from the original crop input. The waste product of ethanol production then can be fed as a major food source to most animals (Harsch, 1993; International Energy Agency, 1994; Powers, et al., 1995; Sosulski, et al., 1997; United Nations Environment Programme, 1991; Ward and Matsushima, 1980).

Since the process of ethanol refining depletes most if not all of the sugars in the mash, ethanol coproducts do not represent a balanced diet for livestock. Ethanol production thus cannot completely monopolize feed crops because stillage-fed animals need additional fresh foods (Harsch, 1993; Klass, 1998; Solar Energy Research Institute, 1979). Distiller's dried grains, DDG, and other feed byproducts of ethanol production coproducts are, however, suitable as high-protein supplements (replacing bulkier grain protein sources) for cattle, swine, sheep, and poultry (Powers, et al., 1995; Ward and Matsushima, 1980). Recent studies have further demonstrated that some species of fish can thrive on stillage-only diets (Wu, et al., 1995; Wu, et al., 1996). This seems to indicate that the market for ethanol coproducts could benefit from future expansion. The development and widespread application of coproduct-based livestock feeding could free large quantities of grain for ethanol production.

Under ideal conditions, as long as coproducts are priced in relation to their relative nutritional value (i.e. par with an equal volume of grain protein), they can be brought on-farm without an increase in real feeding costs. In practice it is unlikely that this can be fully accomplished. Additional costs will be incurred for transporting crops from and coproducts back to the farm. Furthermore, it is likely that actual distillery operations would result in at least some loss of biomass, since spillage, spoilage, and pests will eliminate some crops from the system. However, proper handling and storage should mitigate most loss and other waste. Likewise, the delivery of coproducts to the farm and the increased demand for feedstock crops will undoubtedly have some repercussions on farm economics. The effect here will be assumed negligible since it is beyond the scope of this study to quantify the intricacies of agricultural economics as impacted by the crop diversion strategy. For the purposes of calculation in the present study, it will be assumed that this feedcrop diversion system can be undertaken without any additional costs to the agricultural system.

x.2. CAPTURING LATENT ETHANOL FROM WITHIN THE CURRENT FARM SYSTEM

x.2.1. Estimating Potential Ethanol Production

A portion of livestock grain rations normally fed for protein could be converted to ethanol without a resultant rise in grain or meat prices so long as an equal quantity of protein is returned to the farm in the form of stillage co-products. Information from studies by Lipinsky, et al. (1979) and the Solar Energy Research Institute (1979) can be combined with other sources (Dowd, et al., 1993; "Family Chicken," 1966; Irish, et al., 1995; Sosulski, et al.,1997; Ward and Matsushima, 1980) to develop an estimate of the percentage of livestock feed for protein that can be converted to ethanol without disturbing animal nutritional requirements and thus without endangering animal health. These data are shown in the "Feed Ration Convertible" column in Table 1 below. The proportion of all feed consumed by each livestock category can be estimated from thirty year production and disappearance data contained in Economic Research Service (1997a; 1997b; 1997c) and in National Agricultural Statistics Service (1998), respectively, to obtain the fractions listed in the third column of Table 1. This information is combined with harvest statistics in Section x.2.3 to estimate annual feed consumption by each livestock species and then to estimate the portion of that feed available for ethanol conversion. Note: beef cattle are

TABLE 1.
Livestock Feed Available for Ethanol Feedstocks

Animal	Feed ration convertible	Proportion of all crop
beef (dormant)	24%	13%
beef (fattening)	35%	50%
chicken	10%	2%
dairy	10%	26%
hogs	25%	4%
sheep	20%	2%
turkey	10%	4%

Source: Dowd, et al. 1993; Economic Research Service 1997 a, b, c; "Family Chicken" 1966; Irish, et al. 1995; Lipinsky, Scantland, and McClure 1979; National Agricultural Statistics Service 1998; Solar Energy Research Institute 1979; Sosulski, et al. 1997; Ward and Matsushima 1980.

referred to as "dormant" when not being actively fattened for market. Error in the estimation of feedstock supply for latent ethanol production potential is primarily determined by reporting error in the agricultural statistics here employed. It is assumed here that the portion of each crop available for diversion does not vary among producing states, although in practice, the crop livestock systems in each state differ.

x.2.2. Estimating Croppage and Ethanol Conversion Factors

The 1997 USDA-NASS Crops Production Data (1998) provide agricultural production statistics for key crops. Suitable ethanol conversion crops included in these statistics are barley, edible dry beans, corn, hay, oats, peanuts, rice, grain sorghum, soybeans, sunflower, and wheat. Potatoes-another good ethanol feedstock-are absent from this data set. Biomass varies widely in its ethanol conversion efficiency. Nevertheless, it is possible to assign average ethanol yields to a typical unit of marketed crop based on conversion efficiency information listed in the following sources: Drinkwater (1984); Henk and Linden (1994); International Energy Agency (1994); Johnson and Anth (1951); Lipinsky, et al. (1979); Putnam, et al. (1991); Rendleman and Hohmann (1993); Solar Energy Research Institute (1979); Sosulski, et al. (1997); Sosulski and Sosulski (1994); Thomas, et al. (1995); Thomas and Ingledew (1995); United Nations Environment Programme (1991); World Bank (1980). These data are summarized in Table 2 below. The table also shows the fraction of each crop's production that is

TABLE 2.
Sample Crops and Ethanol Production Characteristics

Crop	Ethanol	Yield	Production for feed
barley	2.05	gal/bu	56.4%
beans, edible dry	1.5	gal/bu	23%
corn	2.7	gal/bu	22.9%
hay	0.02	gal/lb	99%
oats	1.05	gal/bu	29.1%
peanuts	1.24	gal/bu	25%
rice	2.3	gal/bu	0%
sorghum, grain	2.6	gal/bu	97.5%
soybeans	2.0	gal/bu	50%
sunflower	0.92	gal/bu	25%
wheat	2.6	gal/bu	23.9%
gal = gallon (U.S.)	*lb = pound*	*bu = bushel (U.S.)*	*1 bu = 35.21*

Source: Drinkwater 1984; Henk and Linden 1994; International Energy Agency 1994; Johnson and Anth 1951; Lipinsky, Scantland, and McClure 1979; National Agricultural Statistics Service 1998; Putnam, et al. 1991; Rendleman and Hohmann 1993; Solar Energy Research Institute 1979; Sosulski, et al. 1997; Sosulski and Sosulski 1994; Thomas, et al. 1995; Thomas and Ingledew 1995; United Nations Environment Programme 1991; World Bank 1984.

consumed as animal feed based on average data accounting the disappearance of each crop for the period 1985 to 1994 derived from the National Agricultural Statistics Service (1998).

Conversion efficiencies do not vary much with source. In those few cases where sources differed, the average ethanol yield was selected. To be accepted for the study, conversion information had to be representative of currently installed technology. When a source listed rates based on several technologies, the yeast-based batch process figure was selected since this is the traditional alcohol fermentation system. This underestimates latent ethanol production potential relative to that obtainable from high-tech processing facilities, but more accurately reflects installed capacity.

x.2.3 Estimating Latent Ethanol

Using the information discussed in this section, latent ethanol production potential was calculated for each relevant state following the formula detailed below:

$$Eth_{st} \{(C_x f_x D r_y e_x) + (C_x f_x BF r_y e_x) + (C_x f_x BD r_y e_x) + (C_x f_x S r_y e_x) + (C_x f_x CH r_y e_x) + (C_x f_x T r_y e_x) + (C_x f_x H r_y e_x)\} \quad \text{Eq. 1}$$

where,

Eth_{st} = ethanol production for each state st
C_x = production of crop x
f_x= feed diversion fraction
r_y = feed ration diversion fraction for livestock category y
e_x = ethanol yield per unit of crop x
D = dairy fraction
BF = beef (fattening) fraction
BD = beef (dormant) fraction
S = sheep fraction
CH = chicken fraction
T = turkey fraction
H = hog fraction

Livestock fractions are the feed allocation percentages by animal from Table 1, f_x is Table 2's feed diversion fraction, and e_x is the appropriate yield information, also from Table 2. Results of these calculations are seen in Table 3. These figures detail the latent ethanol production potential for the forty-eight conterminous states. To further illuminate these findings, relative production potential for each state is shown graphically in Map 1. Annual latent ethanol production potential by crop is seen in Table 4. These results are discussed in greater detail in the following section.

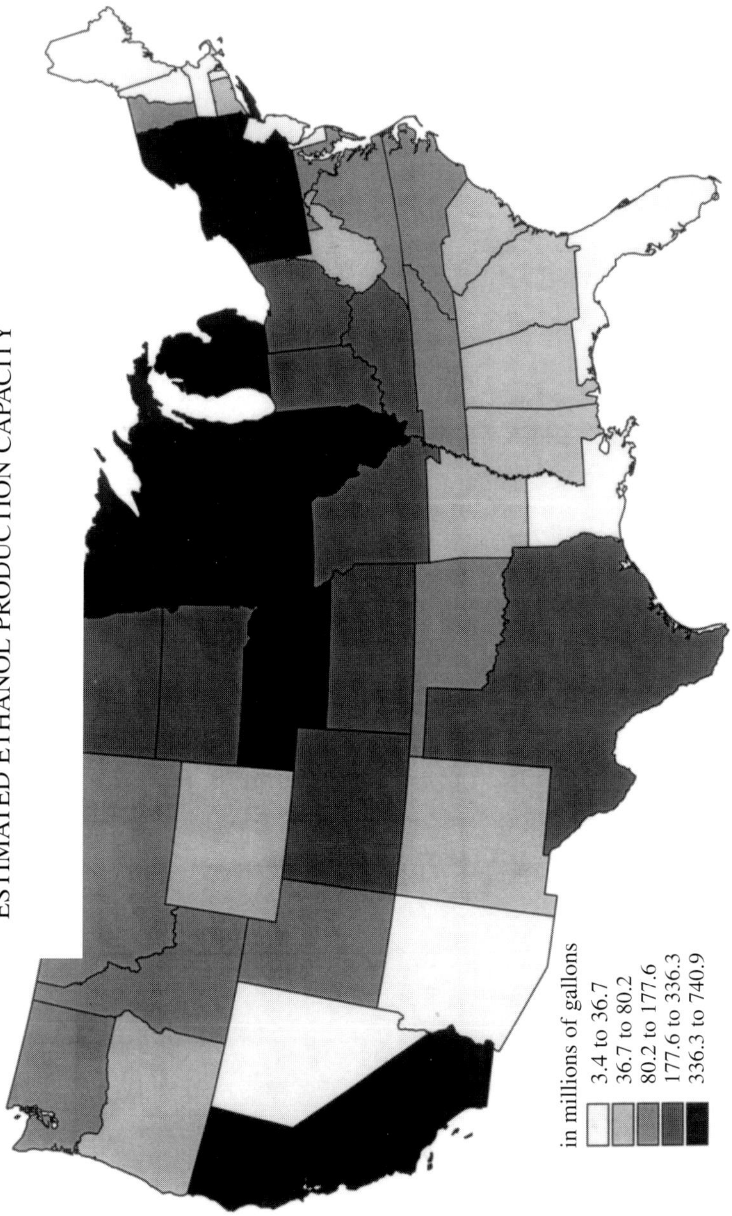

MAP 1.
ESTIMATED ETHANOL PRODUCTION CAPACITY

TABLE 3.
Estimated Latent Ethanol Production By State
in million gallons

State		State		State	
Alabama	44.8	**Maryland**	88.3	**Oregon**	80.2
Arizona	30.9	**Massachusetts**	36.7	**Pennsylvania**	421.9
Arkansas	54.4	**Michigan**	346.8	**Rhode Island**	3.4
California	376.7	**Minnesota**	694.5	**South Carolina**	39.9
Colorado	244.0	**Mississippi**	61.8	**South Dakota**	336.3
Connecticut	46.9	**Missouri**	222.2	**Tennessee**	146.7
Delaware	12.1	**Montana**	148.3	**Texas**	186.8
Florida	26.5	**Nebraska**	508.1	**Utah**	88.1
Georgia	79.8	**Nevada**	15.8	**Vermont**	83.1
Idaho	168.6	**New Hampshire**	21.4	**Virginia**	177.6
Illinois	485.6	**New Jersey**	28.8	**Washington**	128.0
Indiana	288.8	**New Mexico**	46.6	**West Virginia**	38.4
Iowa	740.9	**New York**	520.1	**Wisconsin**	740.4
Kansas	295.1	**North Carolina**	125.4	**Wyoming**	65.6
Kentucky	196.4	**North Dakota**	231.3		
Louisiana	32.9	**Ohio**	316.7		
Maine	31.1	**Oklahoma**	90.9		

Total 9,194.6
Figures may not add to total due to rounding

TABLE 4.
Estimated Latent Ethanol Production by Crop
in million gallons

Crop	Ethanol Production
barley	129.38
beans, edible dry	5.14
corn	6,601.02
hay	1,551.50
oats	31.09
peanuts	14.13
rice	0.00
sorghum, grain	35.05
soybeans	468.24
sunflower	1.11
wheat	357.97

Total 9,194.6
Figures may not add to total due to rounding

X.3. DISCUSSION OF FINDINGS

x.3.1 National Latent Ethanol Production: Discussion

In showing that 9.2 billion gallons of ethanol can be produced annually through latent potential in domestic agriculture, this study has successfully demonstrated that current U. S. agricultural output can provide large quantities of fuel ethanol. To properly establish a context for this statement, however, it is necessary to compare these estimates of annual ethanol production to current gasoline consumption. In recent years, the United States has consumed approximately 743 million gallons of gasoline per day (Energy Information Administration, 1997). The ratio of Btu's per gallon for ethanol and gasoline is 0.6763. The number here is used to convert hypothesized ethanol production for a more effective comparison to gasoline. Latent ethanol production could thus displace 6,218.3 million gallons of gasoline each year on an energy-equivalent basis, providing the nation's automotive fuel needs for approximately 8.4 days: clearly a small contribution. Latent ethanol production could, therefore, potentially keep America rolling for a mere 202 hours; obviously minimal in terms of sustainable alternative fueling. It is clear that agricultural expansion would be necessary to support a fuel ethanol industry of more than localized importance if we assume use in an ICE.

x.3.2 Regional Variation in Ethanol Production Potential

So far the discussion of latent ethanol production has focused on the national aggregate. However, since agricultural activities are not uniformly distributed across the landscape, it is reasonable to assume that fuel alcohol production potential as calculated will show marked regional variation. This conclusion is supported by the information in Table 3 as well as in Map 1.

This map reveals that the raw materials for ethanol production are most abundant in the Midwest (in agreement with information presented by Keeney and DeLuca (1992)), the Plains, Texas, and California. Of the primary petroleum-producing states, only Texas currently demonstrates notable potential for fuel ethanol production. One can conclude from this that the centers of production in a large-scale fuel ethanol industry would be located largely outside the region of traditional fuel extraction and refining. It should be noted, however, that this study's calculations represent conservative estimates for two potentially active ethanol production areas. Since data were not examined for sugar cane production, ethanol potential in the South, particularly in Louisiana and Florida, may be seriously underestimated. The Pacific Northwest is another potential producing region for which results are likely underestimated since Idaho's potatoes as well as the farmlands of Oregon and eastern Washington could prove well situated to serve the ethanol needs of the region's remote markets.

From the perspective of a flexible-fueled transportation future, these findings may be somewhat heartening since the locational advantages for ethanol production are spatially auto-correlated with the majority of the nation's transportation fuel demand. This fact is something of a rarity among alternative energy sources which are, like solar and aeolian sources, often greatly removed from demand centers. The Southeast's increasing transportation market could be at least partially offset through the local production of energy crops on the region's largely idled farmlands.

Thus, while latent ethanol production is not sufficient to meet the entire nation's demand for transportation fuel, it does indicate that the fuel's production could be locally important and can be produced in large volumes from within the current agricultural system.

X.4. POSSIBLE AVENUES FOR FEEDSTOCK CROPPAGE EXPANSION

Traditional crops are not the only possible feedstocks for ethanol; other crops perform better in high-tech continuous batch processing fermentation systems, or they can be more sustainably cultivated and thus are more suitable for fuel dedication. Sugar beets produce nearly as much ethanol per bushel as corn, are suitable for many types of marginal lands, and could easily be integrated into crop rotation schemes on farms throughout the country. Grains, such as oats and barley, are tolerant to marginal soil and climatic conditions but yield only slightly less ethanol than corn, especially when their reduced cultivation intensities are considered. Sorghum has nearly as high a yield per bushel as does corn and has fewer cultivation requirements. Readily fermentable semi-arid crops such as buffalo gourd could help extend the range of energy croppage (Icerman and Schultz, 1985). Jerusalem artichokes are a perennial tuber that provide excellent stillage biomass through harvested leaves (Baker, et al., 1993). Since the long-lived root is undisturbed, the field can remain productive for years without cultivation thereby allowing the production of stillage crop on lands prone to erosion. Sustainable cultivation of alternative feedstocks for ethanol production, therefore, exist and are a subject worth further investigation.

X.5 CONCLUSIONS

This chapter has demonstrated that the technology represented by current installed ethanol fermentation facilities is capable of extracting large volumes of fuel ethanol from within the present agricultural system. This is accomplished by diverting some livestock feedcrops to ethanol production and then returning the protein-rich distillery coproducts to the farm. The 9.2 billion gallons of ethanol which this strategy could produce would displace 6.2 billion gallons of gasoline annually. Although small relative to the aggregate national transportation fuel demand, these volumes could be locally important in a flexible-fueled environment for those regions with a competitive advantage for feedstock production.

REFERENCES

Baker, R., et al. 1997. Classification Criteria for Solid Oxide Fuel Cell Electrode Materials. *J of the Electrochemical Society* 144 (7): 2427-33.

Dowd, Michael K., et al. 1993. Low Molecular Weight Organic Composition of Ethanol Stillage from Corn. *Cereal Chemistry* 70 (2): 204-09.

Drinkwater, Al. 1984. *Arkansas Energy Future: Alcohol.* Arkansas Department of Energy, Little Rock.

Economic Research Service. 1997. *Food Grains: Background for 1997 Farm Legislation.* U. S. Department of Agriculture. GPO, Washington.

Economic Research Service. 1997. *Soybeans and Peanuts: Background for 1997 Farm Legislation.* U. S. Department of Agriculture. GPO, Washington.

Economic Research Service. 1997. *U.S. Feed Grains: Background for 1997 Farm Legislation.* U. S. Department of Agriculture. GPO, Washington.

Energy Information Administration. 1997. *Petroleum 1996:Issues and Trends.* U. S. Department of Energy, Washington.

n.a. 1966. *A Family Chicken Flock for Appalachi*a. Leaflet #541. U. S. Department of Agriculture. GPO, Washington.

Harsch, J. 1993. *New Industrial Uses, New Markets for U. S. Crops.* Cooperative State Research Service, U. S. Department of Agriculture. GPO, Washington.

Henk, L.L. and J.C. Linden. 1994. Silage Processing of Forage Biomass to Alcohol Fuel. *ACS Symposium Series.* American Chemical Society, Washington. pp 391-410.

Icerman, L. and E.B. Shultz, Jr. 1985. A Preliminary Economic Analysis of Buffalo Gourd as a Diesel Fuel and Ethanol Feedstock in the High Plains of New Mexico, Texas, and Oklahoma. *Energy* 10 (8): 901-09.

International Energy Agency, 1994. *Biofuels.* Policy Analysis Series. IEA/OECD, Paris.

Irish, G. G., et al. 1995. Removal of the -Galactosides of Sucrose from Soybean Meal Using Either Ethanol Extraction of Exogenous -Galactosidase and Broiler Performance. *Poultry Science* 74: 1484-94.

Johnson, A.J. and G.H. Anth. 1951. *Fuels and Combustion Handbook.* McGraw-Hill, New York.

Keeney, D. R. and T. H. DeLuca. 1992. Biomass as an Energy Source for the Midwestern U. S. *American Journal of Alternative Agriculture* 7 (3): 137-44.

Klass, D.L. 1998. *Biomass for Renewable Energy, Fuels, and Chemicals.* Academic P, San Diego.

Lipinsky, E. S., et al. 1979. *Systems Study of the Potential Integration of U.S. Corn Production and Cattle Feeding with the Manufacture of Fuels via Fermentation.* Vol. 1. W-7405-ENG-92-077. Department of Energy, Washington.

National Agricultural Statistics Service. 1998. *Agricultural Statistics*, 1997-98. U. S. Department of Agriculture. GPO, Washington.

Powers, W. J., et al. 1995. Effects of Variable Sources of Distillers Dried Grains

Plus Solubles on Milk Yield and Composition. *J of Dairy Sci* 78: 388-96.
Putnam, Daniel, et al. 1991. A Comparison of Sweet Sorghum Cultivars and Maize for Ethanol Production. *J of Production Agriculture* 4: 377-81.
Rendleman, C.M. and N. Hohmann. 1993. The Impact of Production Innovations in the Fuel Ethanol Industry. *Agribusiness* 9: 217-31.
Solar Energy Research Institute. 1979. *Fuels From Farms: A Guide to Small-Scale Ethanol Production*. EG-77-C-01-4042. Department of Energy, Washington.
Sosulski, K., et al. 1997. Preprocessed Barley, Rye, and Triticale as a Feedstock for an Integrated Fuel Ethanol-Feedlot Plant. *Applied Biochemistry and Biotechnology* 63-65: 59-70.
Sosulski, K. and F. Sosulski. 1994. Wheat as a Feedstock for Fuel Ethanol. *Applied Biochemistry and Biotechnology* 45/46: 169-80.
Thomas, K. C., et al. 1995. Production of Fuel Alcohol from Hull-less Barley by Very High Gravity Technology. *Cereal Chemistry* 72 (4): 360-64.
Thomas, K. C. and W. M. Ingledew. 1995. Production of Fuel Alcohol from Oats by Fermentation. *J of Industrial Microbiology* 15: 125-30.
United Nations Environment Programme. 1991. *Green Energy: Biomass Fuels and the Environment*. United Nations, New York.
USDA-NASS. 1998. *USDA-NASS Crop Production Data, By State*. HTTP://usda.mannlib.cornell.edu: 70/00/data-sets/crops.
Ward, G. M. and J. K. Matsushima. 1980. *Feeding Value of Dried Distillers' Grains (DDG) in Beef Feedlot Rations: Final Report*. Golden: Solar Energy Research Institute.
World Bank. 1980. *Alcohol Production from Biomass in the Developing Countries*. World Bank, Washington.
Wu, Y. V., et al. 1995. Evaluation of Corn Gluten Meal as a Protein Source in Tilapia Diets. *J of Agricultural and Food Chemistry* 43: 1585- 1588.
Wu, Y. V., et al. 1996. Effect of Diets Containing Various Levels of Protein and Ethanol Coproducts from Corn on Growth of Tilapia Fry. *J of Agricultural and Food Chemistry* 44: 1491-1493.

Renewable Energy: Trends and Prospects. Edited by S.K. Majumdar, E.W. Miller and A.I. Panah

Chapter Twenty-Two

RENEWABLE ENERGY AND CLIMATE CHANGE

DWAYNE S. BREGER

Department of Civil and Environmental Engineering
Lafayette College
Easton, PA 18042

1. OVERVIEW OF THE RELATIONSHIP BETWEEN ENERGY AND CLIMATE CHANGE

The earth's climate is an immensely complex system of solar flux, the atmosphere, ocean and land formations, and the biosphere. Climate science and modeling has advanced substantially over the past few decades, and will continue to improve our ability to understand the system and predict cause and effect of system changes. Historical and geological records show that climate changes, such as the ice ages, are normal over very long periods of time, and that short term perturbations (such as volcanic eruptions) are normally assimilated by the system.

Atmospheric chemistry plays a critical role in the system by allowing solar radiation to penetrate and by absorbing its heat to provide an insulating blanket and a warm temperature for earth relative to space. This function of the atmosphere is analogous to the glass in the greenhouse, and hence the gases within the atmosphere responsible for this heat trapping effect are often referred to as greenhouse gases. The primary greenhouse gases are carbon dioxide, methane, nitrous oxide, and hydroflourocarbon, each of which exhibits different concentrations and heat trapping capabilities. The amount and role of CO_2 is especially important and its concentration in the atmosphere is an important element in the climate system, and hence to the life which has evolved and adapted to these conditions.

Over the past few decades, the atmospheric concentration of CO_2 has increased remarkably, and longer trends shown in Figure 1 indicate this increase coincides with the human industrial revolution and increased use of fossil fuel energy. Although this trend by itself does not prove a cause and effect relationship, the large majority of scientists and the Intergovernmental Panel on Climate Change (IPCC) have concluded, "the balance of evidence suggests that there is a discernible human influence on global climate" (1). Moreover, carbon-based fossil fuels represent carbon sequestered from the present balance of the earth system, and their reappearance upon human extraction and combustion over a relative second of geological time is a highly probable cause of the increase in CO_2 in the atmosphere.

The most direct effect of the increase in greenhouse gases is an increase in average global temperatures, often referred to as global warming. However, the more general term "climate change" captures the anticipated range of temperature changes across the globe, as well as changes in the frequency and intensity of precipitation and drought, and other climate patterns.

While the biosphere has adapted to slow changes in climate during ice ages, the impact of human induced climate change is of a pace too fast for ecosystem adaptation and migration, and hence serious and unpredictable impacts on ecosystems are anticipated. Human economic infrastructure, notably coastal development and agriculture, are also particularly prone to disruption by climate change.

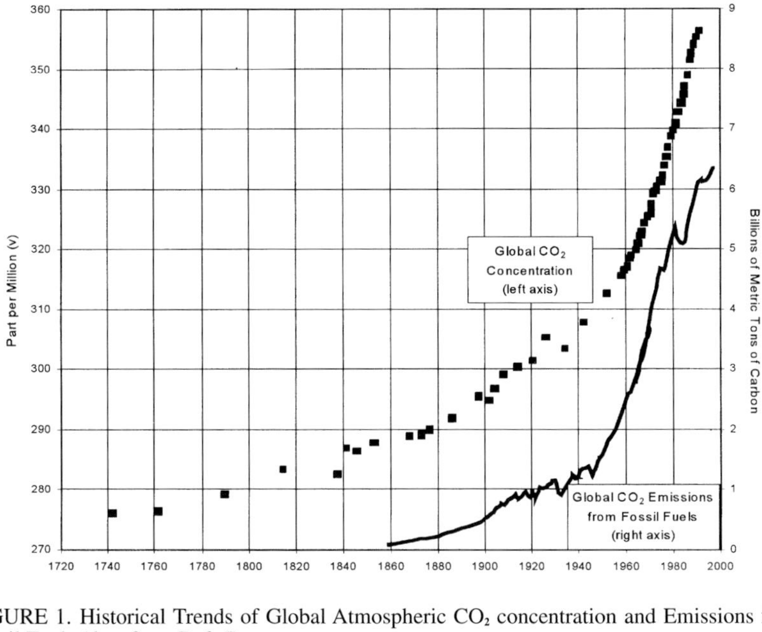

FIGURE 1. Historical Trends of Global Atmospheric CO_2 concentration and Emissions from Fossil Fuels (data from Ref. 5).

Over 98 percent of the CO_2 emissions in the United States result from the use of fossil fuel energy that drives our lifestyles and economy (2). Emission reductions can come from increased energy efficiency, economic transitions to less energy intensive activities, or the substitution of energy technologies that emit less (such as natural gas) or no greenhouse gases (such as renewable energy and nuclear). However, with the foreseen growth in the global economy over the next few decades, clearly all strategies need to be embraced to limit the growth rate of greenhouse gas emission and eventually level and reduce the emissions.

2. PROJECTIONS OF ENERGY USE, ECONOMIC DEVELOPMENT, AND GREENHOUSE GAS EMISSIONS

As discussed later in this chapter, the international policy of climate change is confounded by the rapid industrialization of the global economy. While industrialized nations (notably the U. S. and Europe) are responsible for the large majority of greenhouse gas emission to date and will continue to dominate in terms of per capita emissions, developing countries are projected to be the dominant source of emissions by the 2030-2050 time frame. How energy needs of these emerging economies are met will be critical for future greenhouse gas concentrations. Dominant future economies such as China and Russia have vast coal reserves, which provide low cost energy for economic development, but huge potential CO_2 emissions.

An obvious link exists between economic growth and energy consumption. However, the strength of this correlation and the ability of economic transitions and policy to reduce the energy intensity of economic activity are not as clear. When economic growth in the U. S. returned after the energy crises of the 1970s and 1980s, new energy conservation measures and the strength of the service industries reduced energy use per unit of GDP.

Official projections for energy use and greenhouse gas emission for the U.S. (3) are shown in Figure 2, which indicates a steady rise in energy consumption, a continued small market share for renewable energy, and little change in the rising greenhouse gas emissions. Without any current climate change policy, these projections can be viewed as "business as usual" scenarios.

Other energy and emission futures under alternative economic and policy scenarios are discussed later in this chapter.

Global projections for greenhouse gas emissions (4) by country groups are shown in Figure 3, which indicates the immensity of the challenge and threat we currently face in developing international agreements and mechanisms to effectively and equitably reduce these impending greenhouse gas emissions. A "safe" concentration of CO_2 is not adequately know, though long term scenarios by climate scientists suggest a target of stabilizing concentrations at about double the pre-industrialized levels (or about 550 parts per million). According to the Intergovernmental Panel on Climate Change, this target would require eventual reduction of global carbon emissions by at least 60 percent of current levels (5). Such a reduction will require significant departure from the rising global projections given above.

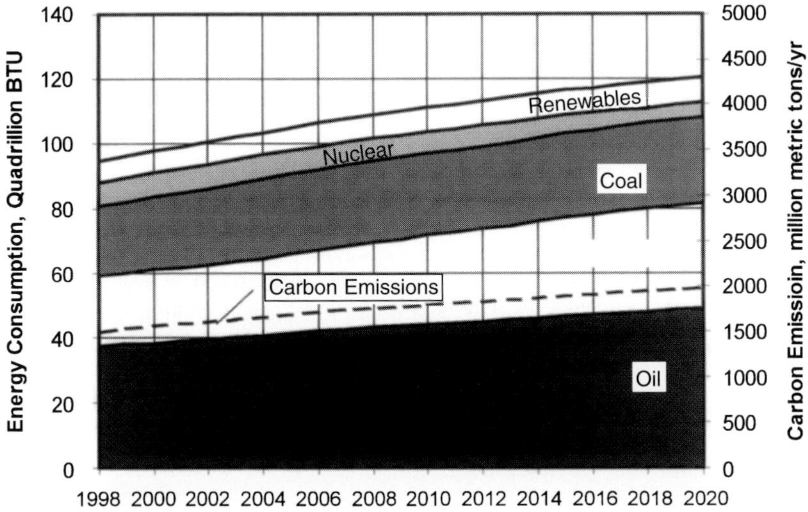

FIGURE 2. Projections for U.S. Energy Consumption and Greenhouse Gas Emissions (data from Ref. 3).

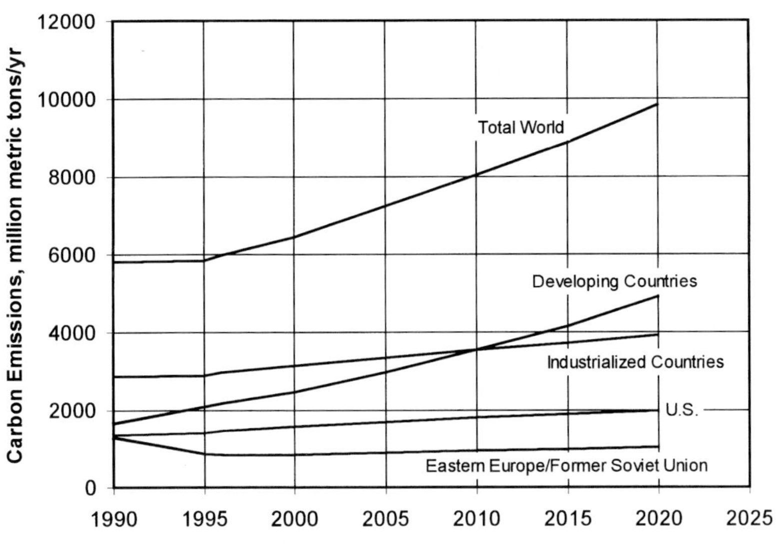

FIGURE 3. Projections for Greenhouse Gas Emissions by Country Groups (data from Ref. 4).

3. CLIMATE CHANGE POLICY ISSUES RELATED TO RENEWABLE ENERGY

International negotiations to address climate change formally began under the United Nations Environmental Program with the adoption of the Framework Convention on Climate Change at the Earth Summit in Rio de Janeiro in 1992. A discussion of the climate change policy process is beyond the scope of this chapter, but is often regarded as the most complex and challenging high level international negotiations ever faced. The Kyoto Protocol of 1997 signed by 84 countries, including the United States, marks the first binding commitment by the developed countries to specific timetables and targets for greenhouse gas emission reductions. The treaty calls for the U. S. to reduce its emissions by 7 percent below their 1990 emission level on average over the 2008 to 2012 time period. Overall, the industrialized nations would be required to reduce emissions by 5.2 percent by this time period. The treaty ratification, though, is unlikely by the U. S . Senate based on concerns of the impacts on the national economy and on the treaty' s non-binding commitments of the developing countries. The treaty enters into force only after at least 55 countries accounting for at least 55% of 1990 CO_2 emissions fully ratify the agreement.

The reduction of greenhouse gas emissions and the impact on the economy depend strongly on the emerging role of renewable energy technologies. The

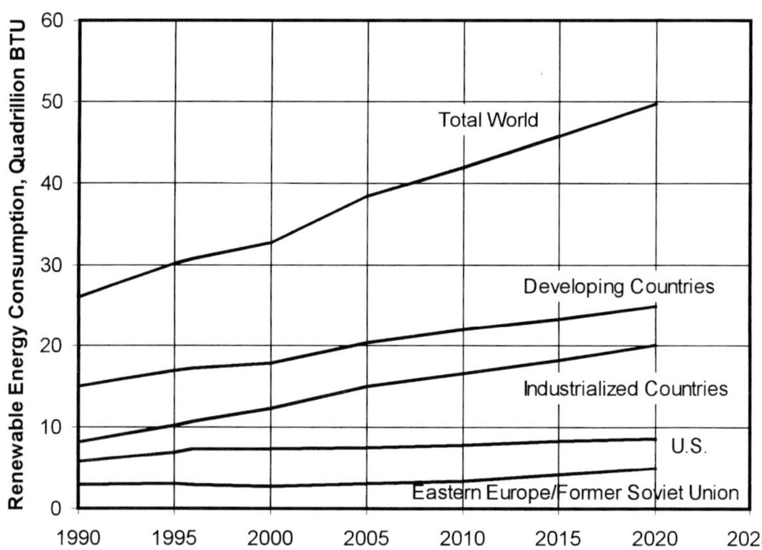

FIGURE 4. Projections for Renewable Energy Consumption by Country Groups (data from Ref. 4).

development and market penetration of renewable energy technologies in developed countries will also strongly influence the opportunities for developing countries to adopt renewable energy technology to meet their burgeoning energy needs.

The Kyoto Protocol does not prescribe policies for countries to meet greenhouse gas targets and timetables; each country is to pursue their own path. However, the treaty does begin establishing international mechanisms by which nations can negotiate agreements to trade emission reductions. The treaty and negotiators recognize the economic importance of minimizing the cost of greenhouse gas mitigation, and that the cost of reducing greenhouse gases can be lowered by targeting emissions which can be reduced or eliminated at low cost. Often, greenhouse gas emissions in another country may be easier (or cheaper) to reduce than reducing emissions domestically. This opportunity exists particularly in increasing energy efficiency in developing countries and often it may be less costly for developed nations to pay for energy efficiency improvement in developing countries than take steps to reduce their own greenhouse gas emissions. The treaty has established the Clean Development Mechanism that provides the structure and rules by which developing countries can meet their own targets by reducing emissions in developing countries. The rules and limitations of this highly complex and contentious aspect of the international negotiations are still being debated.

Within a country, a host of policy measures can be adopted to reduce greenhouse gas emissions, and these policies influence the opportunities for renewable energy technologies. The government can impose a tax on fossil fuel energy, or the carbon content of the specific fuel, to increase the cost of energy use and reduce demand. The induced reduction in fossil fuel energy use (and greenhouse gas emissions) will depend on consumers' price elasticity of demand for energy, the turnover rate of capital stock in the energy use sector, and the substitutability of energy sources with lower or no tax applied. The tax on conventional fuels will enhance the financial attractiveness of renewable energy technologies and trigger more aggressive market activities to develop and supply the technology. Proponents of energy or carbon taxes, especially in the United States which has relatively low taxes on energy currently, argue that the taxes reflect the true cost of energy use and the absence of such taxes distort the market away from an efficient equilibrium. Further, the taxes could be gradually imposed to avoid market shocks and that other taxes could be simultaneously reduced to avoid any net increase in the economy's tax burden. The political climate in the U. S. has not allowed serious consideration of any new taxes over the past few administrations, and there was a very quick rejection of the "BTU Tax" proposed by President Clinton early in his administration. Studies on structuring an environmental tax system and reducing taxes on economic goods (e.g., employment, income, profits) have been made and for the U. S. and are being more seriously considered on a state tax level by several states (6).

Closely related to energy taxes is the policy of establishing a cap and trade program for CO_2 or other greenhouse gases. Under this policy, the government would set a cap for total emissions and then issue tradable permits to emitters. The permits take on an economic value and a trading program is established. This system of tradable permits has been in operation for SO_2 emissions by large electric utilities in the eastern U. S. and is designed to induce the economically efficient (cost minimizing) allocation of emissions and mitigation efforts.

The government policy of supporting research and development is also essential in mitigating greenhouse gases and developing and demonstrating renewable energy technologies. The fluctuating renewable energy R&D budget of the U. S. Department of Energy over the past decades is discussed in the following section. Despite efforts of the Clinton Administration, the budget has not seen significant change as a result of new climate change goals.

In the U.S. and other countries, the deregulation of the electric utility sector has resulted in conflicting new market conditions for renewable energy. On the one hand, the promise of deregulation to reduce electric rates to consumers will make it more difficult for renewable energy technologies to compete. On the other hand, consumers are promised the opportunity to choose the source of their electric supply and green marketers can provide electricity derived from renewable energy to consumers who are willing to pay more. Already there is empirical evidence that a green market and a willingness-to-pay do exist (7). Additionally, the federal and state rules defining the electric deregulation often provides measures to assure an initial and growing market for renewable energy. Most often this is defined in terms of a Renewable Energy Portfolio which stipulates a percentage of the electric supply that must come from renewable sources (normally set at a few percent increasing over time to about ten percent).

4. TECHNOLOGICAL CHANGE AND MARKET PENETRATION OF RENEWABLE ENERGY

As the climate change debate swings from the scientific recognition of the problem to the economic cost of mitigation, the issue of how quickly and cost effectively renewable energy can deliver future energy demands becomes essential.

Over the next few decades, it is generally assumed that renewable energy technologies will supply an increasingly large amount of the national and global energy demand. Some of these projections will be reviewed later in this chapter. The process by which these technologies develop and penetrate the markets depends on research and development effort, relative price signals, capital stock turn over, consumer acceptance, and other things including education and political leadership. Predicting technological change and market penetration is an area of economic research based largely on analysis of historical cases of technological change and extension to current technologies and market conditions (8, 9). The process of renewable energy market penetration is not well understood theoretically and many factors make this case different from historical cases. For example, energy itself does not hold economic value; the value resides only in the services it provides - building comfort, motors turning, lights illuminating. Hence, unlike new technologies such as color television or airplane travel, renewable energy technologies do not provide new (or clearly superior) services to consumers. Therefore, market penetration need not follow such past examples, and likely will depend more on price signals, government incentives, and public interest.

Although the use of renewable energy technologies date back to the very earliest days of human history, the modern era of technological development might be considered since the 1960s and 1970s, when the world's burgeoning energy use exhibited vulnerable dependence on fossil fuel supplies. Issues of resource depletion, environmental damage, and economic dependence on the OPEC oil cartel prompted social and government action to support research and development of renewable energy technologies. At the time, projections for depleting oil reserves and escalating energy prices were the driving forces behind the government investment, public concern, and new business marketing the new energy technologies.

The renewable energy research and development budget spent by the U. S. government grew from less than $1 million per year in the early 1970s to over $1.2 billion per year by the end of that decade characterized by the energy crises. From that time, and resulting from the Reagan Administration's change of energy policy, the renewable energy budget declined steadily to $119 million per year by 1990 (10). The budget has since climbed back to about $336 million per year in 1999 under the Clinton Administration, though larger increases requested have been thwarted by the Senate claiming the Administration was seeking to implement the Kyoto agreements before ratification.

As R&D continues to improve renewable energy performance and cost, the private sector will continue to play an increasing role in its commercialization and market penetration. In the private market sector, price signals become fundamental in defining opportunities for renewable energy firms to sell products and sustain their operations. Private firms also become critical players in R&D (sometimes in public/private partnerships), especially in methods to reduce costs of production through manufacturing technology and other efficiency measures.

Market imperfections exist when price signals do not reflect true costs of production and use and such imperfections lead to market inefficiencies. External environmental costs of using fossil fuels, and other externalities, have always led to market prices that are unfavorably weighted against cleaner renewable technologies. The more recent and broader environmental issue of climate change adds to the externalities of fossil fuels and to a greater recognition of market inefficiencies.

The speed at which market penetration of renewable technologies can occur is limited by the turnover of the energy sector capital stock. Investments in energy consuming capital equipment often are long term commitments.

Automobiles produced today will consume fuel for the next ten years or more, industrial plants are operated for several decades, and utility power plants for up to fifty years. While attractive incentives can be used to stimulate early retirement of capital equipment, the transition period for significant market penetration of new energy technologies is on the order of decades to a half-century. This inertia in energy capital stock constrains the rate at which greenhouse gas emissions can be reduced, and must enter into policy decisions concerning optimal timing of actions.

5. ECONOMIC EVALUATION OF CLIMATE CHANGE POLICY AND INCORPORATION OF RENEWABLE ENERGY TECHNOLOGICAL CHANGE

At the crux of the climate change debate is the need to compare cost of actions taken in the present with avoided costs in the future to adapt to change or compensate victims unable to adapt. The fact that these economic impacts concern agents across geo-political regions and through future generations suspends perfect market assumptions of perfect information and access to the market. In the U. S. the cost of taking action (for example, accelerating market penetration of renewable energy technologies) is typically based on economic models of the national economy and projections for change in the gross domestic product. These results have typically shown a negative impact of a few percent GDP after a few decades due to meeting Kyoto timetables and targets. However, these models have been criticized, for their assumptions about the economic efficiency of the status quo and the limitations of modeling transitional impacts of growing energy efficiency and renewable energy industries and economic development.

The process by which renewable energy technologies penetrate the energy markets depends on research and development, relative price signals, capital stock turn over, consumer acceptance, and other things. Given the effectiveness of these zero-emission energy technologies to mitigate greenhouse gas emissions and the sensitivity of technological change and market penetration to government policy, the explicit consideration of the technological change process is critical to predict future emissions and to set optimal policy strategies. Projected cost and effectiveness of greenhouse gas mitigation strategies generally result from macroeconomic and/or general equilibrium modeling of the national (or global) economy, with detailed attention to the energy sector. For good historical reasons, current models are structured to address change within the conventional energy markets and industries, but are not specifically designed to track the diffusion process of new energy technologies and the resulting environmental and macroeconomic impacts. Two reasons may underlie this deficiency. First, energy economic projections over a period of a few decades to over a century are somewhat unique to the climate change arena and only over such a long run do the issues of technological change become critical. Second, the process of technological change is not well understood, inherently uncertain and, hence, not well suited to modeling.

The influence of technological change typically was estimated in early modeling efforts by the assumption of some autonomous process by which energy use becomes more efficient and renewable technologies experience price change and enter the market (11). While convenient as a modeling approach, the autonomy of technological change is a vast simplification of a process that is well recognized to be largely influenced by market signals and government policy – both of which are critical components of greenhouse gas mitigation strategies.

Under the autonomous treatment of technological change, the effect of market or regulatory based policies on the diffusion process of renewable energy sources is not captured. To the extent that policies directed towards mitigating conventional energy emissions will concurrently enhance the market competitiveness and penetration of energy efficiency and renewable energy technologies, modeling results will tend to overestimate mitigation costs and underestimate emission reductions. Furthermore, evaluation of policy options, such as government R&D and market incentives to accelerate the rate of technological change, may be overlooked simply due to the limitations of available model structure and capabilities.

The need to incorporate induced (or endogenous) technological change within greenhouse gas mitigation policy modeling is well recognized and is now being pursued (12, 13). How this enhancement will alter projections for emission levels and mitigation costs is difficult to predict.

It seems reasonable that economic signals or policies that stimulate technological change and market penetration of renewable energy sources (either directly or through the increased cost of conventional sources) will produce a reduction in emissions. The net cost of the emission reduction is not as clear. Market penetration of renewable energy technologies that would naturally result due to optimal market reaction to changes in relative price signals would tend to reduce the projected cost of mitigation. The cost of public expenditures directed at accelerating the use of renewable energy technologies would have to be compared to costs expended in other ways to meet specific emission targets. Presumably, there may be some optimal mix of policies to minimize costs. Such optimization is impractical without an endogenous treatment of technological change.

The economic and policy analysis of renewable energy technological change and market penetration must also consider other positive benefits to the economy beyond primary net energy costs and greenhouse gas reductions. Secondary benefits are especially important when considered on a regional or national scale when conventional energy is largely imported. Renewable energy offers new industry and market opportunities for regions without indigenous fossil fuel resources, producing employment, indirect economic activities through the multiplier effect, and greater gains in local tax revenues. Such an analysis for the

U.S. by the Union of Concerned Scientists and others (14) indicates the importance of these macroeconomic impacts in guiding policy. Their scenario in which U. S. renewable energy consumption is nearly doubled by 2010 relative to their status quo scenario produces a $2.8 billion gain in GDP, a $14 billion increase in income, and a net gain of 773,000 jobs.

Finally, another criticism of U. S. policies towards renewable energy in the new context of climate change has been recently raised by the Renewable Energy Policy Project (15). Given the limited political action allowed in the U. S. by the Senate, and the relatively aggressive policy taken by European nations to embrace renewable energy, the U. S . industry is at risk of losing substantial share of their own domestic and the immense global market demand. Despite leading contributions to the technological development of most of the renewable energy technologies, the U. S. may lose its opportunity to benefit from their research investments if early industry leaders emerge from Europe and elsewhere and capture huge new demands for renewable energy especially in developing countries.

6. EXPECTATIONS FOR RENEWABLE ENERGY IN REDUCING GREENHOUSE GAS EMISSIONS

Projections for renewable energy contribution to the U.S. and global energy supply all indicate growing opportunities, and generally point to biomass and wind as dominant technologies over the first few decades of the transition. Figure 4 shows reference projections by the U. S. Department of Energy for renewable energy consumption by different country groups. Under current policy, these data indicate a slower rate of growth in renewable energy for the U. S. than other industrialized nations and developing countries. In Figure 5, the U.S. Department of Energy reference projection is compared with projections made by the Union of Concerned Scientists and others under their innovative path scenario, which leads to the macroeconomic gains described above. Along with other studies, the range of these projections varies widely, but nonetheless indicated under appropriate policy, the potential huge market is about to unfold. Although relatively new and small firms have generally dominated the growth of the renewable energy industries to date, the huge market potential is beginning to be recognized by large energy corporations that may seek to dominate market activities in the future.

7. THE TRAGEDY OF THE COMMONS AND THE ROLE OF GOVERNMENT IN CLIMATE CHANGE

Without ownership rights to the earth's atmosphere, users of the atmosphere to dispose of greenhouse gases are expected to behave in the manner described by the *tragedy of the commons*. Under this market structure, no individual has the incentive to reduce the use of the commonly owned resource because the cost of doing so is borne by the individual taking action, while the benefits are spread equally to all users. Instances of the tragedy of the commons (water rights, urban air quality) have typically been resolved through government intervention to regulate behavior or restructure market signals.

Extending the analogy to climate change is appropriate in the need for policy to avoid the tragedy. However, in the case of climate change, the 'users' of the common atmosphere are considered the nations themselves and the governments that represent them. The United States, for example, balks on the notion of reducing their own greenhouse gas emission unless other nations (including developing nations) commit to reductions as well - for fear of hurting their own economy for benefits distributed globally. Hence, an international level of government is needed to resolve the tragedy, and while the United Nations is in (and has taken) this position, the complexity of the issue and the stakes involved are stretching the role of this body way beyond previous experiences. Additionally, the U.N. does not have the legislative command and enforcement over member nations that national governments have over their citizens and industries. Therefore, the tragedy of the common global atmosphere has the structure of the classical tragedy of the commons, but not the same political scheme to approach the resolution of the issue.

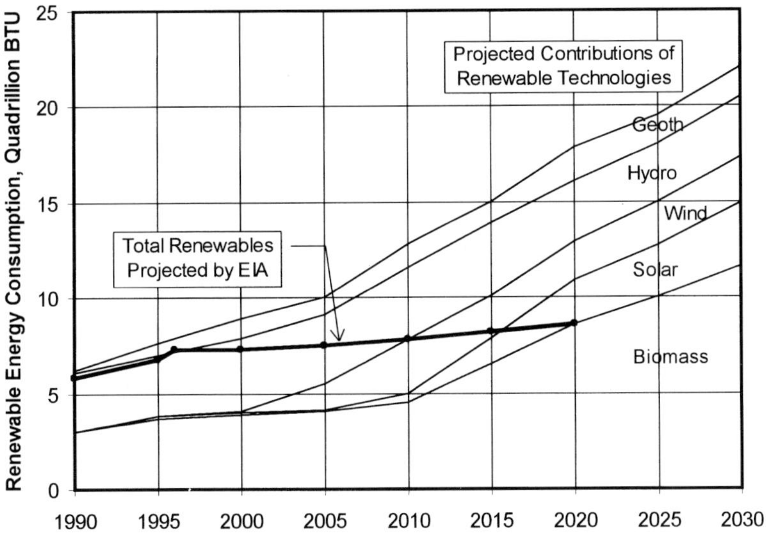

FIGURE 5. Projections for U.S. Energy Consumption (data for technology contributions from Ref. 14, EIA projections from Ref. 4).

In the United States, the inability of the government to consider the Kyoto Protocol and embrace climate change action is seemingly in stark contrast to citizens' expressed concern and their willingness to pay for prudent action (16). The public believes the government should be taking prudent actions, especially where benefits result in reducing other environmental impacts and promoting economic development, such as support of renewable energy R&D and commercialization.

8. ETHICAL ANALYSIS OF CLIMATE CHANGE

Given the issues at hand, it is somewhat surprising that the ethical dimension of climate change has not been more at the center of policy or academic debate. Some significant attention has been given to the ethical analysis (17). Also, the role and association of the religious leaders and organizations in discussing climate change as a moral and social issue, is an important means of public engagement.

The debate or predicament of climate change is essentially an ethical one. Given uncertainty in all aspects of the debate, given the responsibilities of human beings to their children and future generations, given our shared presence of the earth with other co-evolving life and aesthetic beauty, the question that begs answering is what we "ought" to do. A decision not burdened by rigorous calculations of limited scope and accuracy, nor by the whims or persuasion of individuals or corporations, but by the ethos of the human society and condition. The role of ethics in defining the U. S. Constitution, and the ongoing struggle between individual rights and utilitarian needs, would lead one to think that many politicians would engage an open and active discussion of what we *ought* to do. Given that our ability to limit the extent of climate change is reduced by delaying action, the inattention of the U. S. government could be itself ethically questioned. The public dialogue and the role climate change takes in national politics are much further advanced in many European nations.

The policy debate on climate change typically concerns whether (or how much) humans should expend resources and economic costs today, for benefits to future generations in avoiding, or more likely, adapting to climate change. These semantics are often used, and leads one to suppose that current society is sacrificing their own well-being for the betterment of the future, concealing the issue of whether the well-being was ethically gained and the resulting condition of the future illegitimately eroded.

The ethical question better stated is whether (or how much) the current generation should decline benefits in order to reduce costs borne upon future generations of adapting to climate change including irreversible changes. Human beings show a condition for such sacrifice when it comes to raising children, but not so much when it comes to economic principles and market activities. Sacrifice in economic terms is reflected in investment, which takes place only if the payback is worthwhile and quick relative to the investor.

An argument is often put forth that current economic growth is essential to provide the wealth to future generations to dedicate to climate change adaptation. That is, instead of bearing costs today by mitigating greenhouse gases, we could alternatively invest those savings to encourage economic growth and have greater wealth in following decades for adaptation. While one can not argue economically against this strategy given adequate parameters, little attention is given to the ethical issue of intergenerational utilitarian calculations and the rights of unborn generations to the planetary conditions from which we have evolved.

9. DISCUSSION AND CONCLUSIONS

The future of renewable energy and climate change is inextricably connected. Although renewable energy can find market share without demand driven by climate change, the ability of our global economy to develop while reducing fossil fuel use over the next century demands major contribution by renewable energy technologies. Energy efficiency, fuel switching to natural gas, and new schemes for carbon sequestration will also play essential roles, but little doubt remains of the significant increase in renewable energy over the next few decades. Already, rapid growth in wind and biomass energy in Europe and the U. S. is largely a result of climate change policies and pressures.

The rate at which renewable energy penetrates markets will depend on government policies and leadership, public interest and willingness to pay, and the role of businesses in marketing, financing, and adopting the technologies. As international climate change agreements translate to national policies and strategic plans, new and large markets for renewable energy must be opened to meet greenhouse gas targets and timetables. Clean Development Mechanisms promise opportunities for industrialized nations to market renewable energy in developing countries. This market growth may, in fact, help to mature renewable energy technologies and industries, making them more attractive for domestic markets as well.

With limited government leadership or action in the U. S., the role of the public and business sector may be especially important. Public scrutiny and ethical principles may stimulate public action to democratically pressure government action, or new methods of public support may come forth, either through bottom-up grassroots activities or through top-down demonstration of public will. The business sector clearly shows its propensity to connect its public image with environmentalism in advertising, which implies potential schemes for businesses to provide support for greenhouse gas mitigation and renewable energy activities.

Unlike the call for renewable energy during the 1970s based on concern of oil dependence and resource depletion, the climate change issue is likely to be a major international energy issue for the next century. While other energy sources, such as fusion and new natural gas resources, and schemes to sequester huge amounts of carbon might come forth, renewable energy technology has matured over the past decades and the public is prepared to embrace its large role in the new century of growth and global change.

REFERENCES

1. Intergovernmental Panel on Climate Change (IPCC). 1996. Summary for Policy Makers: The Science of Climate Change. Report of the IPCC Working Group I, p. 22.
2. U. S. Environmental Protection Agency. 1999. Inventory of U.S. Greenhouse Gas Emissions and Sinks: 1990-1997.
3. U. S. Energy Information Administration. Annual Energy Outlook 2000. U. S. Department of Energy, Washington, D.C.
4. U.S. Energy Information Administration. International Energy Outlook 1999. U.S. Department of Energy, Washington, D.C.
5. MacKenzie, James, J. 1997. Climate Change and the National Interest. World Resources Institute, Washington, DC.
6. Muller, Frank and J. Andrew Hoerner. 1994. Greening State Energy Taxes: Carbon Taxes for Revenue and the Environment. Pace Environmental Law Review.
7. Farhar, Barbara C. and A.H. Houston. 1996. Willingness to Pay for Electricity from Renewable Energy. National Renewable Energy Laboratory, NREL-TP460-21216.
8. Greening, L. A. and M. Khrushch. 1996. Modeling the Process of Technological Innovation and Diffusion: An Overview of Issues. Lawrence Berkeley National Laboratory, Berkeley, CA.
9. Grubler, Arnulf, Nebojsa Nakicenovic, and David G. Victor. 1999. Dynamics of Energy Technologies and Global Change. *Energy Policy*. 27: 247-280.
10. Sissine, Fred J. 1995. Renewable Energy: A New National Commitment? Science Policy Research Division, Congresssional Research Service, Library of Congress, Washington, D.C.
11. Nordhaus, W. D. 1994. *Managing the Global Commons: The Economics of Climate Change*. MIT Press, Cambridge, MA.
12. Laitner, S. 1997. WYMIWYG (What You Measure is What You Get): The Benefits of Technology Investment as a Climate Change Policy. Proceedings of the 18th Annual North American Conference of the USAEE/IAEE, San Francisco, 7- 10 September.
13. Breger, D. S. 1998. A System Dynamics Model of Technological Change and Market Penetration of Renewable Energy Technologies as a Tool for Greenhouse Gas Mitigation Policy. Final Report to the U. S. Environmental Protection Agency, Washington, D.C.
14. Energy Innovations. 1997. Energy Innovations: A Prosperous Path to a Clean

Environment Alliance to Save Energy. American Council for an Energy Efficient Economy, Natural Resources Defense Council, Tellus Institute, and Union of Concerned Scientists, Washington, D.C.

15. Moore, Curtis and Jack Ihle. 1999. Renewable Energy Policy Outside the United States. Renewable Energy Policy Project, Washington, D.C.
16. Krosnick, Jon A., Penny S. Visser, and Allyson L. Holbrook. 1998. American Opinion on Global Warming: The Impact of the Fall 1997 Debate. *Resources*, Issue 133, Resources for the Future, Washington, D.C.
17. Coward, Harold and Thomas Hurka (editors). 1993. Ethics and Climate Change: The Greenhouse Effect. Wilfrid Laurier University Press, Waterloo, Ontario, Canada.

Renewable Energy: Trends and Prospects. Edited by S.K. Majumdar, E.W. Miller and A.I. Panah

Chapter Twenty-Three

SUSTAINABLE AGROECOSYSTEMS: PRODUCTIVITY, ENVIRONMENTAL QUALITY AND RENEWABLE ENERGY

FRED J. BRENNER,
SHAWN T. ROBINSON
and
JOY H. YATES

Biology Department
Grove City College
Grove City, PA 16127-2104

INTRODUCTION

For agroecosystems to be truly sustainable, soil quality and crop production must be maintained without the use of chemicals or the need for supplemental energy and economic resources. In order to provide a conceptual setting for the definition of sustainable agroecosystems, it is necessary to review the evolutionary trends leading toward development of sustainable agriculture. In the early 1980s, the concepts of regenerative agriculture (Rodale 1983) and the articulation of sustainable agriculture (Jackson 1980) began to emerge. These early concepts evolved into a construct of agriculture based on principles of ecological interaction which form the philosophical basis for sustainable agriculture. Beginning around 1987, the term sustainable referred to a stable agriculture in the global sense involving all facets of agriculture, and its interaction with society began to emerge (Harwood 1990). In this context, sustainable agroecosystems are defined as an agricultural system that can evolve indefinitely toward greater human utility, greater efficiency of resource use, and a balance with the environment that is favorable both to humans and to most other species (Harwood 1990). Sustainable agroecosystems should therefore resemble natural ecosystems rather than traditional agricultural systems that relay on auxiliary economic and energy sources to enhance productivity. Also, because of the intense human involvement in management and artificial selection to maximize yields, species and genetic diversity are reduced in traditional agrosystems. In addition, traditional agroecosystems tend to be goal orientated in terms of maximizing profits rather than maintaining natural ecosystems.

In order to develop and maintain sustainable agroecosystems, they must be based on proven ecological and economic principles, as well as being socially acceptable. In the view of the many supporters of sustainable systems, a new agricultural revolution will be required that offers social, economic and environmental stability. But to comprehend the need for sustainable agroecosystems, we must understand the history of traditional agriculture as we know it today. Approximately 8,000 years ago, portions of the human population abandoned their nomadic life style, settled in communities and began domesticating animals and cultivating plants (Meadow 1992), thereby forming the basis of agroecosystems. In the United States, many communities originated to provide goods and services to farmers and ranchers. For the most part, these early farms were self sustaining, utilizing natural sources of energy and fertilizers and providing food for their families while trading their surpluses for goods and services within the evolving communities.

At the beginning of the 20th century, the United States was basically an agrarian economy with approximately 40 percent of the population being involved in agriculture, compared to less than three percent today. As the Industrial Revolution proceeded, agriculture became more mechanized and specialized resulting in increased productivity and lower production cost, while becoming more dependent on fossil fuels. In order to increase productivity, additional lands were placed under cultivation requiring larger equipment and an increased dependency on commercial fertilizers, herbicides and insecticides, all of which increase the dependency of agriculture on fossil fuels. At the time of the civil war, agriculture productivity in the United States was less than 2000 kg/ha when compared to approximately 8000 kg/ha today (Nebel 1996). But, unfortunately, this increased productivity has adversely impacted terrestrial and aquatic environments. Numerous studies have demonstrated that discharges from agricultural lands are the principal cause of the nutrient and bacteriological loading of aquatic ecosystems (Brenner and Mondok 1995). In the view of many, grain yields have reached a plateau which appears to be related to lack of response of crops to commercial fertilizers, the selectivity of pest species to pesticide applications and increased soil degradation.

MANURE MANAGEMENT

Livestock and human manure can not only be used as natural fertilizer, but, by the use of methane generators, as source of energy as well. Brenner and Mondok (1995) estimated that a total of 36,204 animals units (one animal unit = 400 kg live weight) on 950 farms in Mercer County, Pennsylvania produced 435,715 metric tons of manure per year which represented a nutrient value of 2,447, 1,198, and 1,869 metric tons of nitrogen, phosphorus and potassium, respectively. Based on interviews of 190 farmers, 51 percent indicated that manure was accounted for in their fertilizer applications, but they did not have the manure analyzed for nutrient availability. Biosolids (composed sewage sludge) are another potential source of fertilizer available at relatively low cost from municipal sewage treatment plants, but there are environmental concerns including heavy metals and potential pathogenic organisms, especially spore forming bacteria that must be addressed prior to biosolids application. Although natural fertilizers represent a potential source of organic matter and nutrients, run off from agricultural lands have been demonstrated to be a major contributor to the degradation of surface and groundwater systems in the United States, as well as elsewhere in the world (Brenner and Mondok 1995, Clark *et al.* 1985, Diebel *et al.* 1992, EPA 1983, 1987, Hill 1985, Macharis 1985, OECD 1985, Reay *et al.* 1992, Schaller and Bailey 1985, Tim *et al.* 1992). But, the use of constructed and natural wetlands in conjunction with the restoration and maintenance of riparian buffers along stream channels will reduce and/or eliminate the adverse impact of agricultural drainages into stream systems (Brenner *et al.* 1991, Brenner and Johns 1996, and Brenner and Brenner (1998). To large commercial farming operations with large animal concentrations such as dairy, beef cattle and hog feed lots and poultry, the short and long term manure storage represents a continual potential source of surface and groundwater contamination. In addition to wetlands and riparian buffers, the risks of contamination of these ecosystems may be reduced through the use of rotational grazing systems which allows for the natural distribution and biological decomposition of manure.

SOIL QUALITY

For agriculture to be truly sustainable, soil quality and productivity must be maintained without the use of commercial fertilizers which require energy for their production, distribution and application. Soil quality has been defined as the capacity of a specific soil to function within natural or managed ecosystem boundaries, to sustain plant and animal productivity, maintain or enhance water and air quality, and support human health and habitation (Karlen 1997). Smith (1994) relates soil health to its ability to provide a media for organic decomposition and nutrient recycling, as well as the degradation of toxic compounds, including herbicides and insecticides. Sustainability of the soil ecosystem is directly related to soil management and soil quality should be a measure of how well the soil performs its natural functions and not by crop production alone.

Nutrient management is a key ingredient of maintaining a quality soil ecosystem which requires that nutrient applications be correlated with the nutrient requirements of each crop and the amount of naturally available nutrients. It is imperative, therefore, that soils be analyzed for total and available nitrogen and phosphorus, but this is often difficult because these nutrients, especially nitrogen, occur in different forms that exhibit seasonally shifts in nutrient availability. In the majority of cases, spring fertilizer applications are based on nutrient analyses undertaken the previous fall, but this is not an indication of nutrient availability during the growing season since additional nitrogen becomes available via the action of N-fixing bacteria. Any fertilizer application in excess of that required to achieve optimum yield either will be leached into groundwater or discharged to surface water ecosystems.

In addition, using nutrient soil analyses in conjunction with nutrient management, cover crops, and crop rotation will incorporate nitrogen into soil, thereby reducing or eliminating the use of commercial fertilizers. The dependency on commercial fertilizers may also be reduced by alternating crops with a high nitrogen demand, such as corn, with legumes, especially alfalfa, and clover associated with nitrogen fixing rhizobum bacteria. For examples, corn planted after a multiyear alfalfa stand may not require additional fertilizer to obtain maximum yield. Although soybeans are legumes, they do not provide sufficient residue to reduce erosion, thereby, requiring the planting of a fall cover crop after harvest. For example, winter rye and barley will utilize 108 kg and 1 19 kg of nitrogen/ha, respectively. But, the National Research Council (1993) reports that less than 50 percent of the nitrogen absorbed by cover crops is used by subsequent crops. At present it is not known whether cover crops introduce nitrogen into a continual recycling system, or if they act as a temporary nitrogen sink that eventually leaches into the ground water. On the other hand, cover crops have been shown to be effective in reducing soil phosphorous loss. For example, in Georgia, the use of winter rye as a cover crop in corn fields reduced phosphorous loss by 70 percent. Likewise in Quebec, the use of an alfalfa-timothy combination as a cover crop with no-till corn resulted in a 94 percent reduction of phosphorous loss in comparison to conventionally tilled fields. Numerous studies have documented that runoff and accompanying erosion of agricultural lands is a major factor in the nutrient enrichment of aquatic ecosystems. In this regard, nutrient and soil management practices may have a positive or negative effect on sediment and nutrient transport to receiving streams. For example, conservation tillage will reduce the nutrient transport of sediment bonded nutrients, but may increase the leaching of nutrients into the groundwater. Although grass buffer strips may reduce nutrient loading in overland flows by as much as 70%, their effectiveness drastically declines with increased flow conditions, as well as seasonally (Brenner *et al.* 1990). Moreover, during high flow conditions in the spring or intense summer storms, the sediment and nutrient previously retained by filter strips may flush into streams (Brenner *et al.* 1995, 1998). On the other hand, riparian and constructed wetlands have been shown to be the most effective systems for nutrient removal from agricultural lands and, therefore, are the most beneficial and cost effective systems for water quality improvement in aquatic ecosystems (Brenner *et al.* 1991, 1995, 1996, 1998).

PESTICIDE MANAGEMENT

Over last the five decades, insecticide and herbicide applications have increased at an accelerated rate on agricultural lands at not only a tremendous cost in terms of energy, but at an adverse impact on both environmental and human health. Brenner *et al.* (1995) summarized the adverse impacts of pesticides on ecosystem functions and human health. The action of these agricultural chemicals on aquatic and terrestrial ecosystems depends on a variety of factors, including their solubility and ability to be absorbed by soil particles, their dissipation into the atmosphere and their persistence in the environment. In a survey of the South Platte River in Colorado, Kimbrough and Litke (1996) detected 30 different pesticides in streams draining agricultural areas and, in urban areas, 22 different pesticides were detected in one or more samples. Hiligren (1993) reported that the average farm uses over 260 kg of pesticides annually which contribute to the accumulation of toxic materials in terrestrial, as well as surface and groundwater ecosystems. Graham (1991) reported that on farms that use a high volume of agrochemicals, that these chemicals are delivered to groundwater systems via leaching after their application. Graham (1991) further reported that pesticide concentrations in groundwater increase over time in a manner analogous to nitrate concentrations. The adverse environmental impacts is not restricted to North America, but is also a global concern. Muir *et al.* (1992) reported varying concentrations of polychlorinated biphenyls (PCB's), and organochlorine pesticides along with mercury and lead in arctic marine fish and mammals. Organophosphorous insecticides are currently being used at a rate that adversely impact sensitive marine ecosystems and the continued use of defoliants represents a threat to the Amazon basin and other tropical rainforests around the world (Brenner *et al.* 1995).

In addition to the adverse impacts for pesticides and herbicides on natural ecosystems, these chemicals also pose a serious threat to human health. According to a report issued by the Council of Scientific Affairs of the American Medical Association (JAMA 1988), agricultural workers exposed to pesticides have increased mortality rates of leukemia, all lymphopoietic cancers and cancers of the stomach, rectum and eyes. The United Farm Workers of America estimates that approximately 300,000 U. S. Farm Workers suffer illnesses resulting from pesticide exposure. On a global scale, the World Health Organization estimates that over a million people suffer from acute pesticide poisoning resulting in over 20,000 deaths annually. For example, exposure to the herbicide 2,4 D (2-4 dichlorophenoxyacetic acid), commonly referred to as agent orange, significantly increases the risk of non-Hodgkin's disease (NRL) (Siberner 1986). Several studies have demonstrated a correlation between the exposure to paraquat and other herbicides and the incidence of Parkinson's disease (Lewin 1985, Cunningham 1999). It is of interest to note that Parkinson's disease was unknown prior to the industrial revolution and its incidence has increased during the 20th century leading Lewin (1985) to conclude that Parkinson's and some auto-immune diseases are environmentally induced. A large number of pesticides have been shown to be carcinogenic in animals, but only arsenic and vinyl chloride compounds have been linked to carcinogenics in man (JAMA 1988). In 1987, the Environmental Protection Agency (EPA) reported that a correlation exists between the exposure to organophosphate pesticides and a variety of neuropsychological conditions, including depression, irritability and memory loss. Because of the adverse impact of these pesticides on natural ecosystems and human health and welfare, it is imperative that society address these problems on a global scale.

The U. S. Department of Agriculture estimates that over a billion dollars are expended annually on agrochemicals (Cramer 1991), but recent studies suggest that the target species are developing a resistance to pesticides. The United Nations Environmental Program reported that over 250 plant pest species and plant pathogens and 500 insects have acquired resistance to herbicides and insecticides. In 1946, an effective pesticide application rate was 0.17 kg/l0,000 bushel of corn and in less than 30 years the effective application rate increased to 10.7 kg/l0,000 bushel of corn. It is apparent that pesticide applications affect not only the evolution of target pest species, but the beneficial non target species as well. The ultimate goal of sustainable agroecosystems should be the total elimination of pesticides and, if this is not possible, their use should be restricted to problems that impact the health of the ecosystems being managed. In recent years, integrated pest management (IPM) has been used to reduce the dependence of agroecosystems to maintain or increase crop yields. Integrated Pest Management evaluates agriculture from an ecological rather than an industrial viewpoint and crop lands are viewed as components of a larger dynamic ecosystem. The goal of IPM is not to eliminate the agricultural pests, but to reduce and maintain crop damage below the economic threshold where earnings would be affected. For IPMs to be effective, crop lands must be monitored and when the economic threshold is reached, the recommended amount of pesticides is applied to the portion of the field being impacted. If IPM is applied correctly, the reduction in pesticide applications will not only result in higher crop yields and profits, but will reduce the dependence of agriculture on external energy. In 1986, Indonesia eliminated the use of pesticides that had been in effect on rice plantations for over 15 years, resulting in a saving of over $120 million in production costs while producing higher yields (Nebel 1996).

Another ingredient of IPM is the cultural control of crops as demonstrated by crop rotation and planting schemes. Cultural control is hereby defined as any nonchemical factor used to manage agricultural pests. A key ingredient of controlling agricultural pests is the selection, placement and rotation of diversified crops (Bender 1994). For example, the planting of soybeans for two consecutive years will disrupt the life cycle of root cyst nematodes. The planting of soybeans adjacent to wheat in place of sorghum will reduce chinch bug damage to wheat since sorghum is also a natural host for this species which increases chinch bug survival (Bender 1994). The larva of many insect pest species feed on the roots of corn and small grains and since many of these species overwinter as larva, delaying planting until after emergence will reduce damage from these pest species.

Numerous predatory insects, insectivorous birds and mammals provide a natural biological control of agricultural pest species but, unfortunately, pesticides adversely impact the food supply and habitats of these species. For the predator-prey system to be effective, food and nesting cover for the predatory species must be maintained adjacent to the crop fields. In addition, it is possible to use other species to control weeds in crop fields. For example, domestic geese have been used to control grass and broadleaf plants in pick your own commercial strawberry fields eliminating the use of herbicides. The geese avoid the leaves of the strawberry and only feed on the berries after the other food species have been eliminated.

It is possible to develop pest resistant crops through genetic breeding and/or genetic engineering. For example, the fungus that produced the Irish potato blight has been eliminated by breeding strains of potato resistant to the fungus. Similar success has been reported for corn, wheat and other cereal grains (Nebel 1996). Although plant breeding has succeeded in increasing crop productivity around the world, these homogenous genetic strains of corn and other grains may not be resistance to existing plant pathogens, especially strains of plant viruses and bacteria that may evolve in the future. In recent years, the consumer has been become concerned about genetically engineered species being used for food production.

Insect pheromones have also been used to control agricultural pest species. Pheromones placed on traps have been effective in attracting and controlling Japanese beetles. In other species, males are captured and sterilized by irradiation and released. The sterile males will mate with females and once mating has occurred, the females will not mate with other males. But to be effective, a sufficient number of males must be irradiated annually to reduce the population below the economic threshold for the particular crop being impacted by the pest species.

ENERGY AND SUSTAINABLE AGRICULTURE

In sustainable agricultural ecosystems, it may not be possible to eliminate the reliance on external energy sources, but it can be reduced by using a combination of renewable energy sources. For example, the anaerobic digestion of both human and animal manure produces a biogas that is approximately 67 percent methane with the remaining 33 percent consisting of carbon dioxide and organic gasses. By passing biogas through a water column, the water soluble carbon dioxide is removed with the end produce being almost pure methane, which can be used to supplement natural gas supplies for heating and cooking. In the Central American county of Belize, many facilities solely rely on biogas for cooking and electrical generation. In addition, the nutrient rich organic material remaining after anaerobic digestion can be used as fertilizer. In the case of human production of biogas, the heat (104° F) of digestion largely, if not entirely, eliminates pathogens and, therefore, may be used as fertilizer, especially on forage crops. In many parts of the world, dry animal dung is used as fuel for heating and cooking.

In many parts of the world, including the United States, dairy barns and other farm buildings may be designed using solar panels for electric generation and, if necessary, heating the facility. Solar energy can also be used to power water distribution systems and wind power has been used for centuries to generate power to draw and distribute water from wells. Water powered pumps (RAMS) have also been designed for water distribution systems for rotational grazing systems.

SUMMARY AND CONCLUSIONS

Although agroecosystems can be sustainable and compatible with the environment, it will require a new revolution in agricultural practices on a global scale. For agriculture to be sustainable, it must reduce or eliminate its dependence on commercial fertilizers and pesticides which require energy for their production and distribution while maintaining soil quality and production, along with environmental integrity. At the present time, the majority of agricultural enterprises are completely dependent on auxiliary energy sources. But this dependence may be reduced by employing methane generation technology and designing more energy efficient buildings using solar panels for heating and electrical generation. Auxiliary energy demand may also be reduced by using solar energy, along with wind and water power to obtain and distribute water to live stock. To be successful, this revolution must provide social and economical stability for an ecological sustainable system to which sustainable agriculture is an integral part of natural ecosystems.

LITERATURE CITIED

Bender, J. 1994. Future Harvest. University of Nebraska Press, Lincoln.

Brenner, F.J. and E.K. Brenner. 1998. A watershed approach to agriculural nonpoint source pollution. pp 203-220 *In.* Watershed Management: Practice, Policies, and Coordination. R.J. Reimold (ed) McGraw-Hill, New York.

Brenner, F.J. and R. Johns. 1996. Riparian vegetation, stream systems and non-point source pollution abatement. pp 275-285. *In.* Forests - A Global Perspective. S.K. Majumdar, E. W. Miller and F. J. Brenner (eds). Pennsylvania Academy of Science. Easton.

Brenner, F.J. and J.J. Mondok. 1995. Nonpoint source pollution potential in an agricultural watershed in northwestern Pennsylvania. Water Resources Bulletin 31:1101-1112.

Brenner, F.J., J.J. Mondok, and R.J. McDonald, Jr. 1991. Impact of riparian areas and land use on four nonpoint source pollution parameters in Pennsylvania. J. Penn. Acad. Science 65:64-70.

Brenner, F.J., R.P. Steiner, J.J. Mondok, and R.J. McDonald, Jr. 1990. Nonpoint source pollution: A model watershed approach to improve water quality. pp. 483-498. *In.* Water Resources in Pennsylvania: Availability, Quality and Management. Pennsylvania Academy of Science. Easton.

Brenner, F.J., R.P. Steiner. and J. J. Mondok. 1995. Groundwater-surface water interaction in an agricultural watershed. J. Penn. Acad. Science. 70:3-8.

Brenner, F.J., H.J. Whitbeck, and M.M. Langdon. 1998. Agricultural impacts on headwater streams and associated wetlands. pp 432-446. *In.* Ecology of Wetlands and Associated Systems. S. K. Majumdar, E. W. Miller, F. J. Brenner (eds). Pennsylvania Academy of Science. Easton.

Brenner, F.J., S.D. Yoder, and R.J. Blair. 1995. Impact of nonpoint source contaminants of ecosystems and human health. pp 87-99. *In.* Environmental Contaminants, Ecosystems and Human Health. S.K. Majumdar, E.W. Miller, F.J. Brenner (eds). Pennsylvania Academy of Science. Easton

Clark, E.H., H.H. Haverkamp, and W. Chapman. 1985. Eroding Soils: The Off Farm Impacts. The Conservation Foundation. Washington, D.C.

Council on Scientific Matters. American Medical Association 1988. Cancer risk of pesticides in agricultural workers. J. Amer. Med. Assoc. 260:959-963.

Cramer, C. 1991. Controlling Weeds with Fewer Chemicals. Emmaus: Rodale.

Cunningham, W. 1999. Barbara Woodworth Saigo. Environmental Science A Global Concern. McGraw-Hill, Boston.

Diebel, P.L., D.B. Taylor, S.S. Batie, and C.D. Heatwole. 1992. Low-input agriculture as ground water protection strategy. Water Resources Bulletin 28:755-762.

Environmental Protection Agency. 1983. Chesapeake Bay: A Frame Work for Action. Chesapeake Bay Program. Environmental Protection Agency Chesapeake Bay Liaison Office. Annalpolis, MD.

Environmental Protection Agency. 1987. Nonpoint Source Guidance. Nonpoints Source Branch, Office Water Regulations and Standards. U. S. Environmental Protection Agency, Washington, D.C. 35 pp.

Graham, J.A. 1991. Monitoring groundwater and well water for crop protection chemicals. Anal. Chem. 63:613-621

Harwood, R.R. 1990. A history of sustainable agriculture. pp 3-19. *In.* Sustainable Agricultural Systems (C.A. Edwards, R. Lal, P. Madden, R.H. Miller, and G. House (eds)). Soil and Water Conservation Society. Ankeny, IA.

Hill, P.L. 1987. Monitoring the effects of land management practices on water quality in Gulford County, North Carolina. pp 137-148. *In.* Monitoring, Modeling and Mediating Water Quality. S.J. Nix and P.E. Black (eds). American Water Resources Association. Herndon, VA.

Hiligren, S. and B. Mills. 1993. Manure handling spread to wetlands (man-made marshes filter livestock waste). Farm Journal. Dec. pp. 1-4.

Karlen, D.L., M.J. Mausbach, J. W. Doran. R.G. Cline, R.F. Harris, and G.E. Schuman. 1997. Soil Quality: A concept, definition, and framework for evaluation. Soil Science Society of America.

Kimbrough, R. A. and D. W. Utke. 1996. Pesticides in streams draining agricultural and urban areas in Colorado. Environmental Sciences and Technology. 30:908-916.

Jackson, W. 1980. New roots for agriculture. Friends of the Earth. San Francisco, CA.

Lewin. R. 1985. Parkinson's Disease an environmental cause. Science 229:257-258.

Macharis, J. 1985. Chesapeake Bay Non-Point Source Pollution. *In* Perspectives on Non-Point Source Pollution. United States Environmental Protection Agency. EPA 44015-85001.

Meadow, D. D. Meadows, and J. Randers. 1999. Beyond the limits. Chelsea Green Publishing Company. Burlington, VA.

Muir, D.C. G. R. Wageman, B. T. Hargrove, D. J. Thomas, B. D. Peakall and R. J. Norstrom . 1992. Arctic marine ecosystem contamination. Science for the Total Environment. 122:75-134.

National Research Council Committee on Long Range Soil and Water Conservation. 1993. Soil and Water Quality: An Agenda for Agriculture. National Academy Press. Washington, D. C.

Nebel, B. and R. Wright. 1996. Environmental Science, The Way the World Works, Upper Saddle River: Prentice Hall. Princeton, N.J.

OCED. 1985. The State of the Environment. Organization for Economic Cooperation and Development. Paris, France.

Reay, W. G., P. L. Gallagher, and G. M. Simmons, Jr. 1992. Groundwater discharge and its impact on surface water quality in a Chesapeake Bay Inlet. Water Resources Bulletin 28: 1121-1134.

Rodale, R. 1983. Breaking new ground: The search for sustainable agriculture.

The Futurist 1:15-20.

Schaller, F.W. and G.W. Bailey. 1985. Agricultural Management and Water Quality. Iowa State University. Ames.

Smiths M. 1994. The Real Dirt. Northeast Region Sustainable Agricultural Research and Eduction Program. Burlington, VA.

Tim, U. S. S. Mostaghim, and V. O. Shanholtz. 1992. Identification of critical nonpoint pollution source areas using geographic information systems and water quality monitoring. Water Resources Bulletin 28: 877-887.

Renewable Energy: Trends and Prospects. Edited by S.K. Majumdar, E.W. Miller and A.I. Panah

Chapter Twenty-Four

POWER FOR THE PEOPLE: A COMPARISON OF THE U.S. AND GERMAN COMMITMENTS TO RENEWABLE ENERGY[1]

MICHAEL K. HEIMAN

Dickinson College
Carlisle, PA 17013-2896

INTRODUCTION

Given the unreliability and rising cost of foreign energy supplies, and growing concern over the social and environmental impact of conventional fossil and nuclear power, the case for renewable energy has never been more compelling. Following the 1973 Arab oil embargo, the United States was an early leader.[2] By the early 1980s, the U.S. had installed ninety percent of the world's wind power capacity and led in the manufacture of photovoltaic (PV) cells–a technology that it pioneered. This early commitment came to an abrupt halt by the mid '80s, however, during an era of cheaper fossil fuel and outright hostility to renewable energy from the energy and utility industries and under political leadership during the Reagan era. As a result, since the 1980s the renewable share for electric generation, excluding large hydroelectric provision, has stagnated in the U.S. at approximately two percent, mostly biomass and geothermal, with solar and wind power accounting for just one-tenth of one percent of the total 1998 electric generating capacity (Brown 1999, USDOE 1999a, 2000). Despite a modest increase in wind and solar generating capacity, the renewable share of total generating capacity declined slightly in recent years as new fossil fuel plants (i. e., natural gas) were built (USDOE 1998, 2000). The immediate future for renewables in the U.S. does not look any brighter, at least according to the Department of Energy. It recently concluded that contribution from renewable sources would remain nearly stable over the next 20 years, while the proportion of generation from coal and natural gas was projected to rise 25 percent and 300 percent respectively (USDOE 1999b).

As the U.S. commitment to renewable energy was collapsing, a remarkable transformation was occurring across the Atlantic in Europe, particularly in Germany. Despite being a latecomer, within the last decade Germany became a world leader with installed wind capacity, PV manufacturing, cogeneration, and other efforts to develop a renewable electricity supply. In wind alone, the nation went from just a few turbines generating 10 megawatts (mw) in 1990 to 7,900 turbines responsible for 4,400 mw by 2000–some 50 percent of the European total wind power and one third of world capacity (Wagner 1999, 2000).

In this paper the factors favoring Germany's assent with, and hindering the U.S. commitment to, renewable energy will be discussed. The comparison is informative, for just as the German renewable energy industry once learned from pioneering American technology and experience, energy advocates in the U.S. are now turning to Germany for guidance on policy to promote renewable energy. Particular attention will be paid to utility deregulation, a recent reform that may restrain the struggling U.S. renewable energy industry.

RENEWABLE ENERGY IN THE UNITED STATES: THE UNFULFILLED PROMISE

Modern photovoltaic systems were first developed in the U. S. for satellites and global telecommunication and then for other remote uses where transmission costs were prohibitive and economies of scale from conventional power plants unattainable. From 1974 to 1984 the U.S. PV market soared from 1.3 million to 17.2 million square feet installed, and even President Carter placed a thermal water heater on the White House roof (Moore and Ihle 1999, Weltman 1999). Modern wind generation had a similar trajectory, particularly in California where ninety percent of the global capacity was installed during the 1980s prior to European initiatives. No doubt federal and state tax credits, direct grants for demonstration projects, and public funding for research and development helped spur this growth in the aftermath of the oil price hikes. The federal Public Utility Regulatory Policies Act (PURPA) of 1978 also encouraged development, as it required utilities to purchase electricity from independent renewable providers at the avoided cost of self generation–at first a high price given the cost for new power plants during the era of lofty interest rates and rapid energy cost escalation during the late '70s.

In time, as the tax credits expired in the 1980s and the price of oil and natural gas fell, most manufacturers were forced to pull out of solar and wind production or go bankrupt, as did Kenetech, supplier for the 420-mw Altamont Pass (CA) wind farm, and Luz International, developer of a 345-mw solar thermal installation in the Mojave Desert. Many were prone to blame President Reagan, who pulled the solar heater off the White House roof and under whose administration the tax credits ended while federal spending for research and development was cut to just one-sixth of its 1980 level. But the U.S. manufacturers themselves share fault. They could not agree on uniform quality control or promotional activities, had faith in rising electricity demand and oil price projections, were unprepared for the dramatic drop in natural gas prices, and became over-extended through high-interest loans (Kozloff 1994, Weltman 1999; Peter Varadi, personal communication, 1999). Moreover, environmental advocates, at the time focused largely on shutting down nuclear plants, failed to coalesce behind renewable energy.

In the 1990s, natural gas replaced coal and oil as the desired fuel for new power production. As the more expensive alternative, solar installment was particularly hard hit. Between 1984 and 1986 alone, the number of U.S. PV manufacturers fell by over half, while in California the domestic installed solar industry declined from 500 million dollars in sales to just 20 million as homeowner tax credits for solar installation ended (Berman and O'Connor 1996, p. 33). The two major surviving manufacturers, Solarex and Arco Solar, were eventually acquired by the European energy conglomerates, British Petroleum/Amoco and Siemens respectively, both eager to cover all bases as the rest of the world turned toward renewable sources. Some saw this turn toward market domination as a dangerous trend. As solar and wind power are inherently decentralized common goods, they felt that public access must not be channeled through a few multinational energy companies and large utilities treating renewable energy as a sideline to conventional power provision for private profit (Berman and O'Connor,1996).

In 1992 the federal Energy Policy Act brought dramatic change to the U.S. utility industry. States were now encouraged to open up electricity provision to market competition. Thus, generation, transmission, and distribution need no longer be supplied by the same regulated monopoly service provider. The states complied, under pressure from large industrial users seeking to wheel electricity from low-cost natural gas-based suppliers. By fall of 1999, 23 states had enacted deregulation legislation, with the rest expected to follow soon (Bensinger 1999).

Even though a growing market share is now open to "green energy" providers, many energy advocates feel that deregulation threatens instead to delay, rather than promote, renewable energy. Chief among the flaws cited with deregulation in its present form is the expectation that voluntary consumer choice can be a substitute for legislation requiring a designated renewable mix from certified renewable sources, providing a guaranteed price for renewable energy providers, and installing a carbon tax to account for the negative externalities of conventional power production–all reforms now common in Europe. These advocates feel that such legislation is warranted given the public good in renewable energy, be this through environmental, political, or long-term economic security.

Instead of a strong mandate for renewable energy, the compromise reached in most states allows the utilities to pass on so-called "stranded costs" for their now non-competitive nuclear power plants–estimated at 200-300 billion dollars over the next few decades. These surcharges are assessed to all consumers as transition costs, regardless of the individual energy source chosen (Bensinger

1999). According to Wasserman (1998), rate restructuring allowing for stranded cost bailout results in public refinancing of nuclear power plants on behalf of their now-deregulated private owners. Taken together with low-price electricity from coal-fired power plants built before the 1977 Clean Air Act required installation of expensive flue gas desulfurization (making these plants a valued asset in a deregulated market), conventional sources of power remain cheaper as the full costs are written off to the environment or passed on to all users. Finally, in a deregulated market, there is less incentive for utilities to provide renewable energy and promote energy efficiency since direct requirements are usually dropped and linked rate hikes are no longer guaranteed.

Home to the nation's largest electricity market, and leader in renewable solar and wind power installation, California provides an example of what can go wrong with deregulation. Here the 1996 enabling legislation only allows municipalities to sign up individual consumers through an "opt-in" process, making it much more difficult to collect the numbers necessary for wholesale rate reduction. In effect, only the large industrial users enjoy a significant rate reduction (Fenn 1999). Moreover, with residential use just one-third of total consumption (the rest taken by industry and commercial users who usually go with cheap, rather than "green," power), one-half of California's residential customers would have to sign up for one-hundred percent renewable energy just to subscribe to the installed capacity already in place–an unlikely scenario, with just one percent enrolling since the market opened two years ago (Booth 1999, Fenn 1999).The result is that most renewable energy providers are merely reselling power from existing facilities at a premium to the few customers volunteering to pay more to be green. Many fear that when the modest state subsidies provided to renewable energy providers end in 2001–mostly used for advertising rather than new construction–the market will collapse in the state that pioneered renewable energy during an earlier era of market intervention (Rader 1998). Elsewhere, in Minnesota, Arizona, Texas, and lowa for example, the situation is not as bleak due to energy tax credits and a mandated renewable portfolio standard in power provision. But even with these reforms in place, the commitment is very modest (e.g., just one percent required renewable in Arizona and 2000 mw or 3 percent of the state's electricity capacity in Texas over the next 10 years) (Weltman 1999, Brown 1999, American Wind Energy Association 2000). As a result, the entire U.S.-added wind capacity in 1998, for example, was less than that added in Denmark (235 mw) and under a third of Germany's addition (790 mw) (Flavin 1999). Despite a surge to 732 mw new wind power installed in 1999, thanks in large part to extension of the federal (wind energy) Production Tax credit as developers raced to complete projects before the new termination deadline, the prospects for 2000 do not look as promising, with only 50-200 new mw anticipated (American Wind Energy Association 2000).

According to researchers at Resources for the Future, the failure of renewable energy to become a major component of the U.S. energy mix was not due to any intrinsic problem with the technology–where every major expectation with cost reduction has been met and even exceeded–but rather with broader market forces beyond the manufacturers' control. Chief among these were the declining cost of conventional fossil fuels (in particular natural gas), inaccurate electric consumption forecasts in the 1980s, and continuing market volatility in the current era of rate deregulation (Burtraw, et. al. 1999). Since the early '80s, the installed price/watt for PVs fell on average more than ten-fold, from 70 dollars to just 7, while the price to install wind fell from 2.2 dollars/watt in the early 1980s to under 1 dollar today (USDOE 1999a, McVeigh, et. al. 2000). This is a level that now makes wind installation–at least in areas with good potential such as the mountain passes of California and the upper Great Plains–competitive with coal, oil, and perhaps soon natural gas.

These reductions in generation costs notwithstanding, the average non-subsidized cost per kilowatt hour (kw hr) for PV electricity is approximately four times that from the new generation of combined-cycle natural gas plants, while unsubsidized wind (running at 25 percent maximum rated capacity) comes in at approximately twice the cost (Taylor and VanDoren 1999, Wasserman 1997). However, when the price for grid connection and transmission is included, decentralized renewable supply is already the preferred choice in isolated rural settlements, for highway lighting, telecommunications, and for other remote requirements such as for the two billion, mostly rural, inhabitants of the world still without access to electricity. Indeed it is booming foreign demand that has kept the U.S. industry alive, even as conventional residential and industrial use has stalled. Today, despite the highest prices for oil since the Gulf War, there remains no broad public discussion about a renewable energy future, at least when compared to the commitment unfolding in Europe.

RENEWABLE ENERGY IN GERMANY: GUARANTEED RATES, GUARANTEED FUTURE?

Reinforced through directives of the European Union (EU), the strong German involvement with renewable energy stems from a variety of factors. Compared to the U.S., there is a greater collective concern over urban smog and acid deposition, as well as a pledge to significantly reduce carbon emissions under the Kyoto Protocol (which the U.S. Senate has yet to ratify) (USDOE 1999). Furthermore, the ruling government coalition is committed to phasing out nuclear power, the main issue now in negotiation being the timetable for such and the level of dependence on inexpensive Eastern European and Russian natural gas as a transition fuel. With the residential rate for electricity in Germany almost twice that for the U.S., renewable sources are already competitive with nuclear and oil-based power.

Germany's rapid assent to renewable dominance began with the 1990 Electricity Feed Law (EFL) requiring utilities to purchase electricity from independent renewable sources at 80 to 90 percent of the retail price they themselves charged (Wagner 1999, USDOE 1999). In 1990 the federal government also initiated a major 250-mw demonstration program for wind energy, and the national banks were directed to offer loans at below market rates for renewable energy projects (Wagner 1999). Furthermore, when rate deregulation finally reached Germany in 1998, stranded cost surcharges were not allowed (Andrews 1999). Taken together, the guaranteed retail price, existing European Union carbon tax, high level of government support for research and development, low-interest loans, and now lack of market distortion from stranded costs, led to dramatic results—all the more so for a nation that has much weaker solar resources, and far less wind potential than found throughout much of the U.S.

About to surpass Japan, Germany will soon be the world's largest PV market. Demand is supported through a "100,000 roof program" providing low-interest loans for residential installation and guaranteed prices for the power produced (Moore and Ihle 1999). Direct government purchase for public buildings, such as the new Reichstag complex and central train station in Berlin, help guarantee a stable market for a technology that is not yet competitive, on its own, with fossil fuel-derived power.

At 800 dollars/kw installed and dropping, wind is already competitive with conventional power in many regions. Generating capacity has grown by over 40 percent per annum since 1993. The windy northern state of Schleswig Holstein derives more than 15 percent of its electricity through wind (Flavin 1999). This market is in the hands of thousands of individual investors and landowners, rather than a few companies operating large wind farms as in the U.S. Leases are popular among farmers who can still plow around the new 60-meter, 1.5-mw turbines. With the EU spending ten times as much for wind research and technology as the U.S., these turbines employ technology pioneered in Denmark and Germany now that the U.S. has lost its lead (Andreas Wagner, personal communication,1999).

In February, 2000 the Electricity Feed Law was amended by the Renewable Energy Law (REL). The national target now is to double the renewable supply electricity from 5 to 10 percent by 2010. Distortions in the conventional electricity market will be addressed by basing the price paid renewable suppliers of the actual cost of generation rather than a fixed percent of the retail rate charged consumers. Thus the act accounts for instability in the retail market as it affected price support under the old EFL while also encouraging renewable energy development beyond the choice coastal areas and solar regimes with the lowest generation costs (Andreas Wagner, personal communication, 2000).

To be sure, problems remain. These include multi-billion-dollar subsidies to maintain thousands of threatened coal miner jobs, resistance from large utilities fearing less need for their transmission service with decentralized power supply, EU-enforced imports of inexpensive natural gas and (French) nuclear power, and aesthetic concerns over large wind turbines (particular along the norther coast where steady winds favor installation). Nonetheless, with all major projections and goals for solar and wind-derived power met and surpassed, continued steady double-digit annual growth for renewable energy in Germany and throughout the

EU appears attainable, particularly if the more successful German model of minimum price support is adopted instead of the capacity-support (minimum percent obligation–or renewable portfolio standard) employed in some U.S. states, Britain, and France. With capacity support, the quota set tends to be the limit reached, even in areas with more favorable wind and solar regimes (e.g. in Britain and France). As a result, market penetration is a small fraction of that achieved under price-support (Moore and Ihle 1999, Wagner 1999).

SUMMARY: FROM GREEN MARKETING TO GREEN POLITICS

Many promoters predict that renewable energy is about to take off in the decentralized and deregulated market, eventually replacing large-scale conventional power plants, much as the personal computer has replaced the mainframe for many uses (Flavin 1996). However, this potential will not be reached under the green-marketing, consumer-choice approach to power provision now popular in much of the U.S. Without acknowledgment that renewable sources deserve compensation for conventional market distortions—preferably through price guarantees as found in Germany, instead of more-limited production quotas (or renewable energy portfolios)—it is unlikely that a significant number of consumers, on their own, will choose higher-priced green power.

For the immediate future, market protection for renewable sources is warranted, not only on environmental grounds, but also to offset decades of regulated price support, tax relief, funding for research and development, and other forms of assistance to conventional power providers. This assistance was even expanded through recapture of "stranded-costs" that continue to distort what is euphemistically referred to as a deregulated and free market. With U.S. solar and wind production now growing again, even as domestic installation is relatively stagnant, evidence suggests that the market is there, in this case for export, once the playing field is leveled. As the German experience demonstrates, assistance should move toward price support for all providers of renewable energy based on the cost of electricity produced, and away from production quotas or, worse, grants for facility construction–both more likely to be captured by companies with vested interest in optimizing return on existing, conventional power sources.

Finally, for renewable energy to succeed, it has to be taken out of the hands of the large-scale energy and utility companies that have, up to now, thwarted democratic local control of service provision while protecting private monopoly profits–a history well-covered by Berman and O'Connor (1996). Local economic control is a necessary, not merely a sufficient, condition for renewable goals to be met. The required transformation, at least in terms of technology and production efficiency, is already well underway. What remains is the political will to prevent private forces, now deregulated, from capturing and then stifling a sustainable, renewable energy future.

NOTES

1. Research in Germany was supported through the (U.S./ German) Fulbright Commission. The author gratefully acknowledges the thoughtful comments of Barry Solomon and Andreas Wagner on earlier versions of this chapter.

2. Renewable energy sources are taken here to include small-scale hydro (under 50 mw), biomass, wind, geothermal, and solar (photovoltaic and thermal) electricity generation. Moreover, as the concern is with comparison of renewable energy procurement in the U.S. and Germany, the focus is on wind- and solar- derived power. Other renewable sources are either geographically circumscribed (e.g., geothermal), or in areas where there has not been a comparable record of commitment (e.g., specific forms of biomass).

REFERENCES

American Wind Energy Association. 2000. *Wind Energy Outlook 2000.* Washington, DC. Available: www.awea.org

Andrews, Edmund. 1999. "Cheaper Power to the People: In Germany, a Radical Deregulation Benefits Households." *The New York Times* October 28, 1999, p. C1.

Bensinger, Charles. 1999. "What is Electric Utility Restructuring? What's at Stake?" *The Workbook* (published by the Southwest Research and Information Center, Albuquerque, NM) 24(3):66-70.

Berman, Daniel and John O'Connor. 1996. *Who Owns the Sun?: People, Politics, and the Struggle for a Solar Economy.* White River Junction, VT: Chelsea Green.

Booth, William. 1999. "'Green Power,' an Electric Idea?" *The Washington Post* October 17, 1999, p. A1.

Brown, Kathryn. 1999. "Bright Future—Or Brief Flare—For Renewable Energy?" *Science* 285(5427):678-680.

Burtraw, Dallas, et. al. 1999. "Renewable Energy: Winner, Loser, or Innocent Victim?" *Resources* (Resources for the Future, Washington, DC) Issue # 135: 9-13.

Fenn, Paul. 1999. "Deregulation in California, One Year Later: Without Community Choice, No Consumer Choice." A position paper of the American Local Power Project, Oakland, CA. Available: www.local.org/califor3.html

Flavin, Christopher. 1996. "The Next Energy Revolution." *World Watch* 9(1):10-19.

Flavin, Christopher. 1999. "Bull Market in Wind Energy." *World Watch* 12(2): 24-27.

Kozloff, Keith. 1994. "Renewable Energy: An Urgent Need, A Hard Sell." *Environment* 36(9): 4-9, 25-30.

McVeigh, James, et. al. 2000. "Renewables from Another Angle." *Electric Perspectives* 25(2): 10-20.

Moore, Curtis and Jack Ihle. "Renewable Energy Policy Outside the United States." Issue Briefing # 4, September 27, 1999 for the Renewable Energy Project, Washington, DC. Available: http://www.repp.org/articles/issuebr14/ index.htm

Rader, Nancy. 1998. "Green Buyers Beware." Position paper prepared for the Public Citizen Critical Mass Energy Project, Washington, DC. Available: www.citizen.org/cmep/renewables/greenbuy.htm

Taylor, Jerry and Peter VanDoren. 1999. "The Soft Case for Soft Energy." *Journal of International Affairs* 53(1):209-235.

United States Department of Energy. 1998. *Renewable Energy Annual 1998 With Data for 1997*. Washington, DC: U. S. Department of Energy, Energy Information Administration, Office of Coal, Nuclear, Electric and Alternate Fuels. Available: http://www.eia .doe .gov/cneaf/solar. renewables/ rea_data/ html/front-1. html#waste

United States Department of Energy. 1999a. *International Energy Outlook 1999*. Washington, DC: U. S. Department of Energy, Energy Information Administration, International Energy Office. Available: http://www.eia.doe.gov/oiaf/ieo99/

United States Department of Energy. 1999b. *Choices for a Brighter Future: Perspectives on Renewable Energy*. Oak Ridge, TN: U. S. Department of Energy, National Renewable Energy Laboratory, Office of Science and Technical Information.

United States Department of Energy. 2000. *Renewable Energy Annual 1999 With Data for 1998*. Washington, DC: Energy Information Administration, Office of Coal, Nuclear, Electric and Alternative Fuels.

Wagner, Andreas. 1999. "Prospects for Wind Energy in Europe." *Renewable Energy World*, January 1999, pp. 38-44.

Wasserman, Harvey. 1997. "Inherit the Wind: Will Clean Energy Pass Us By?" *The Nation* June 16, 1997, pp. 11 -15.

Wasserman, Harvey. 1998. "The Last Energy War: The Mega-Battle Over Utility Deregulation." *The Nation* March 16, 1998, pp. 11-15.

Weltman, Eric. 1999. "Here Comes the Sun: Whatever Happened to Solar Energy?" *In These Times* February 7, 2000 p. 14 ff.

PERSONAL COMMUNICATION

Varadi, Peter: President of Photovoltaic Enterprises and founder of Solarex (Frederick, MD). In Bonn, Germany on June 7, 1999.

Wagner, Andreas: Executive Director of the German Association for Wind Energy Promotion. In Hamburg, Germany on June 11, 1999, and in Berlin, Germany on June 19, 2000.

Renewable Energy: Trends and Prospects. Edited by S.K. Majumdar, E.W. Miller and A.I. Panah

Chapter Twenty-Five

RENEWABLE ELECTRIC GENERATION IN COMPETITIVE MARKETS

MARWAN MASRI, SULAYMAN AL-QUDSI
and
MICHAL C. MOORE

California Energy Commission
Energy Technology Development, MS-45
1516 Ninth Street
Sacramento, CA 95814

INTRODUCTION

California's electric generation system is one of the most diverse in the world. Its electricity is supplied by a blend of fuels and technologies that include fossil generation, nuclear generation, renewable[1] generation, large hydro, and imported electricity from both the Northwest and Southwest, made up of nuclear, large hydro, and fossil generation. California's electricity accounts for 10 percent of the energy consumed in the state and for nearly 50 cents out of every dollar spent on energy. The state's network for providing electricity is elaborate: 1,013 power plants, 2,500 substations, and 40,000 miles of transmission lines. Power in the state is provided by five investor-owned utilities, 26 municipal utilities, four irrigation districts and five rural electric cooperatives.[2]

This chapter examines the regulatory and market factors that have helped California to become a world leader in the development and use of renewable electric generation. The period under review here begins in the 1970s and covers twenty-five years. In the late 1970s, fossil generation dominated the electricity system in California. Heavy dependence on fossil generation, up to 70 percent of total generation, coincided with a period in which oil and gas were perceived to be in short and unreliable supply, raising concern about the longterm security of fossil supplies and associated cost and price escalations. This concern about cost and security of supply of fossil fuels prompted policy responses to reduce these long-term risks. The search for alternative, less risky supplies intensified, and policy incentives and regulations prompted market participants to explore alternatives. Renewable electric technologies offered an attractive option because they were produced domestically and thus were not subject to the vagaries and risks of political and economic instability overseas. These technologies were relatively clean, with little or no emissions associated with their use. Their reliable operation had been proven through technological demonstration. On the down side, however, renewable technologies were perceived to have high costs relative to conventional alternatives.

Methodologically, this chapter applies quantitative and institutional analysis to characterize the respective role of market forces and policy regulation in the development of renewables in California. The quantitative analysis includes an empirical estimation of a cost function utilizing panel data techniques and the application of the Hodrick-Prescott method to differentiate between *actual* and *potential* renewable electricity generation in California. The quantification portion also includes the application of a simple logit model to gauge the relative contribution of economic forces to the survival of renewable facilities in California. In the institutional portion, this chapter offers an analytical and historical account of the policy environment, including both policy challenges and policy responses, which has been conducive to the evolution of renewable energy in California. The quantitative and institutional framework enables us to highlight lessons learned and to offer an assessment of the future of renewables in the generation of electricity in California.

The remainder of this chapter is organized as follows: section two surveys the historical development of renewables in California; section three empirically addresses the role of market forces in the development of renewable-based electricity generation in California, and also applies a simple logit model to estimate the chance of renewables surviving under a rapidly changing market structure; section four is a discussion of the policy measures that have nurtured the development of renewables in California; and finally, the chapter concludes with a summary of findings and policy lessons.

2. CHARACTERIZATION OF RENEWABLE ELECTRICITY IN CALIFORNIA

Several observations can be made regarding the time-series data on the historical profile of electricity generation that are shown in Table 1. First, growth rates of electricity generation varied during the 1980s and 1990s. High rates of growth were recorded in the early eighties relative to the rates in the nineties. The deceleration in growth rates resulted from the introduction of energy conservation, enhanced efficiency, and saturation of electricity use. In 1990, for instance, regulators in California allowed utilities to share the savings they helped customers to achieve.[3]

Second, the share of renewables in electricity generation has greatly increased between 1983 and 1998. This remarkable growth is due to a variety of factors that include incentive regulation and education programs that expanded the demand for environmentally clean generation sources. On the supply side, technological progress caused the cost of generation by renewable sources to decline geometrically over time. For instance, the cost of generating electricity from wind has fallen dramatically since the mid-1970s, when the U.S. Department of Energy estimated that it cost $1 per kilowatt-hour to generate electricity from wind in the United States. By 1996, that cost had dropped dramatically to only 5 cents per kilowatt-hour.[4]

Third, despite the remarkable growth that renewable energy has achieved, areas of malaise remain. Specifically, there are marked time-dependent variations in the growth of renewables. In other words, California's progress toward a renewable electricity market has been volatile. The degree of volatility varies according to utility and independent suppliers. During the period under consideration, the logarithm of the variance of renewables' supply by utilities was 0.053, while the corresponding statistic for independent suppliers was 0.59; that is, more than eleven times as large. The higher volatility of electricity generated by independent suppliers of renewables is indicative of the higher risks that circumscribe the supply of renewables through independent suppliers. Utilities have integrated facilities and access to financing at favorable credit terms.

3. MARKET FORCES: THE ROLE OF COSTS

In general, costs can be divided into three categories: fixed capital costs, fuel costs, and operating and maintenance costs. Renewables are characterized by high capital costs relative to nonrenewables. We attempt to explain the observed variations in generation according to source, renewable versus nonrenewable, by applying a simple regression model to California's electricity generation and cost data.[5] The data are panel covering the period 1990 to 1995 for nine technologies: five renewable (small hydro, solar, wind, organic waste, and geothermal), three fossil-fueled (coal, gas, oil), and nuclear. The model examines annual variations in electricity generated by each technology type during the six-year period and attempts to establish the relative contribution of its capital costs, fuel costs, and operating and maintenance costs. The analysis yields crude estimates of the elasticity of output - electricity generation - with respect to capital, fuel and operating and maintenance costs. Algebraically, the model is

$$\mathbf{F}_{it} = \alpha_o + \mathbf{X}_{it}\beta + \mathbf{a}_i + \varepsilon_{it} \qquad i = 1,\ldots 9,\ t=1990,\ldots,\ 1995.$$

Where F_{it} connotes the annual generation by the ith technology, X_{it} contains

TABLE 1.
California Electricity Generation 1983-1998

	1983	1985	1989	1990	1991	1992	1993	1994	1995	1998
TOTAL GENERATION	199,609	210,172	238,567	252,355	242,434	245,535	242,026	257,799	256,367	268,136
HYDRO-ELECTRIC	59,351	33,898	32,742	26,092	23,244	22,373	41,595	26,706	51,665	48,462
NUCLEAR	6,738	18,911	33,803	36,586	37,167	38,622	36,579	38,828	36,186	41,565
COAL	17,564	14,977	19,702	21,402	23,442	32,435	22,907	25,095	17,925	29,043
OIL	6,535	2,790	9,275	4,449	523	107	2,085	1,954	489	123
GAS	45,486	69,771	78,916	76,082	75,828	87,032	70,715	95,025	78,378	79,616
GEOTHEMAL	7,020	10,957	15,247	16,038	15,566	16,491	15,770	15,573	14,267	12,554
ORGANIC WASTE	731	1,171	5,204	6,644	7,312	7,362	5,760	7,173	5,969	5,368
WIND	52	655	2,139	2,418	2,669	2,707	2,867	3,293	3,182	2,776
SOLAR	2	33	471	681	719	700	857	798	793	839
OTHER	0	0	4	4	0	2	0	0	0	230
ENERGY IMPORTS	56,130	57,009	41,064	61,959	55,873	37,704	42,892	43,354	47,514	47,559

Source: California Energy Cornrnission, 1999.

the observed explanatory variables of technology i in period$_t$. The variables are capital, fuel and operating costs. All variables are transformed into logarithmic scale. The a_i's are the fixed effects that may vary by technology and reflect unobserved technology-specific characteristics that may be correlated with X_{it}. The ε_{it} are typical disturbance terms, assumed to be identically and independently distributed, with a zero mean and a constant variance.[6] Table 2 contains a summary of the empirical findings.

The results indicate that the single most important variable in explaining technology-specific generation variations is the cost of capital. On average, a ten percent change in capital costs is associated with an eight percent change in technology-specific generation. Since renewable technologies are highly capital intensive, this finding implies that market participants in general and policy makers in particular must find appropriate ways and means that can effectively contribute to lowering the capital costs, thereby enhancing the competitiveness of renewables in the electricity market.

Judged by the size of its coefficient, fuel cost is a strong determinant of the variations in technology-specific generation in California. The negative parameter estimate of the fuel costs variable, 0.59, suggests that lowering the fuel cost by 10 percent can increase the generation by about 6 percent. This result corroborates the powerful role that fuel prices have played and continue to play in the development of renewables. The low fossil fuel prices that prevailed during most of the 1990s constrained development of renewable electricity technology in California. Although the costs of installing and generating electricity with renewable resources continued to decline and technological advances continued to improve generating efficiencies, these factors historically have been unable to keep pace with the declining costs of energy from fossil fuels, making it difficult for the use of renewables to increase as a share of total electricity generation and use.

TABLE 2.
Fixed-Effects Model: Generation and Costs by Technology

Variable	Coefficient	T-value	Mean[a]
Logarithm of fuel cost ¢/kWh	-0.59*	-3.53	0.39
Logarithm capital cost ¢/kWh	-.810*	-4.03	1.74
Logarithm O&M ¢/kWh	-.354*	-2.67	-.103
Constant	10.17*	26.63	
R-squared within	0.453*		
Between	0.97		
Overall	0.33		
Sample size	54.0		

[*] Means the variable is significant at the 95 percent level.
[a] Logarithm of costs.

Operating and maintenance (O&M) costs are least important as determinants of the historical variations in electricity generation in California; the approximate measure of elasticity of negative 0.35 implies that reducing O&M costs by 10 percent can cause a generation increase of nearly 3.5 percent. It should be noted, however, that for most technologies the cost share of O&M is low and therefore the significance of this variable is in fact lower than what is implied by the parameter estimate. Finally, we note that the value of the Hausman test statistic leads us to accept the current specification of the model. The value of the χ^2 test statistic is 19.7. Therefore the hypothesis of systematic difference in coefficients is supported.

3.1 The Potential for Renewables

To arrive at the potential of renewable electricity generation, we utilize the Hodrick-Prescott de-trending method. The historical fluctuations evident in the time series of electricity generation can result from seasonal, cyclical, and random or trend factors. For our purposes, it is desirable to separate the trend that underlies the renewable time series from its other components to arrive at the potential production of renewables. That is, we filter out undesirable elements (noise) while extracting the trend. The Hodrick-Prescott (H-P) filter uses a method of constrained optimization to fit the trend path to the time series.

In the context of renewable electricity generation, the historical long-term pattern depicted by its time series can be viewed as the sum of cyclical (short term) and growth components. The cyclical components arise because of short-term irregularities due, for example, to spells of drought in the case of small hydro, periods of random low winds in the case of wind power, or otherwise when renewable facilities are not operational because of contractual or maintenance reasons. The conceptual framework is that the time series of renewable electricity generation, F_t, is the sum of a growth component, g_t, and a cyclical component, c_t: $F_t = g_t + c_t$ for t=1983 to 1998. One measure of the smoothness of g_t path is the sum of the square of its second difference. The c_t are deviations from g_t and the H-P method considers that over long time periods, their average is zero. Therefore, the H-P method boils down to minimizing the following relationship for determining the growth components:[7]

$$\mathbf{Min}\ \{g_t\}^{T}{}_{t=-1}\ \{\Sigma^{T} c_t^{2} + \lambda\, \Sigma^{T}\ [\,(g_t - g_{t-1}) - (g_{t-1} - g_{t-2})\,]^{2}$$

The parameter λ is a positive number that penalizes variability in the growth component series. The larger the value of λ, the smoother is the solution series. Following H-P, the value of (symbol) that was selected for our annual series is 800.

Applying the H-P method, we arrive at estimates of the potential generation of electricity using renewables. The time series data cover the period 1983 to

1998. The series for California is shown in Figure 1. Clearly, the curve of electricity generation using renewable energy is concave and rising and therefore when random forces that influence the series are de-trended, the renewable electricity profile progresses upwards.

The thick segment connoting "actual" generation by non-utility or independent suppliers in Figure 1 illustrates the S-shaped pattern of renewables development: high rates of penetration and growth until the end of the 1980s. It peaks at the beginning of the 1990s, but falls-off after 1995. By contrast, the dashed curve depicting the potential of renewables reveals that their potential is robust and suggests that when cyclical trends are removed, a steady and sustainable level of renewables-based electricity generation is occurring.

For comparative purposes, Figure 2 shows the potential and actual utility profile of renewables. Despite the fact that the level of actual generation by

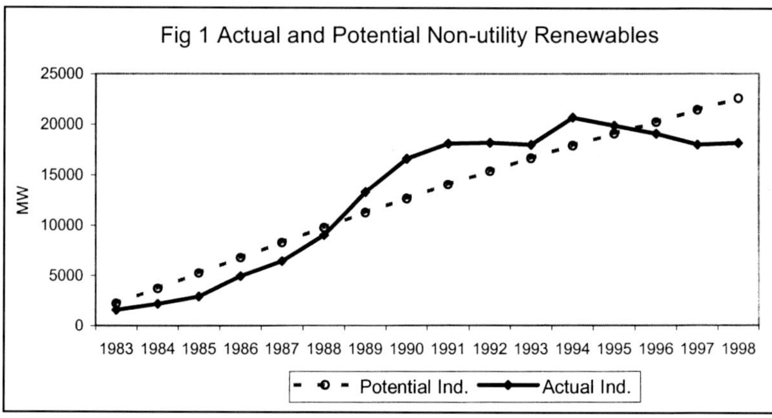

FIGURE 1. Actual and Potential Non-utility Renewables

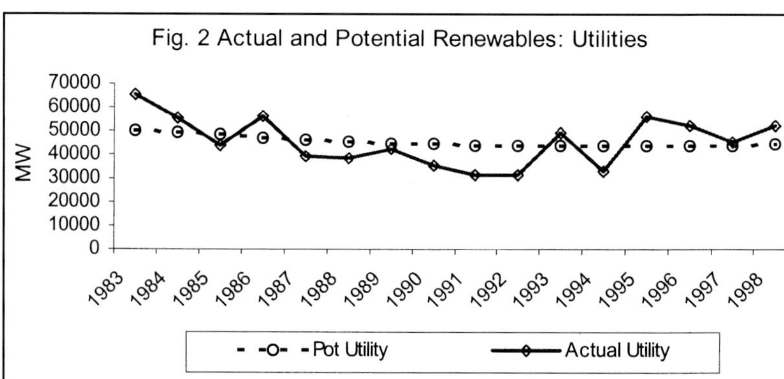

FIGURE 2. Actual and Potential Renewables: Utilities

utilities is much higher than that of independent suppliers, the time-path of its growth varies from that of independents. More noteworthy, there is a clear tendency for the potential of utility-generated renewables to taper off dramatically more than renewables generated by independents, as Figures 1 and 2 suggest.

3.2 Survival of Renewable Projects

This section applies a two-step model: first, a dichotomous model, and second, a Weibull survival model. Formally, the two-step model consists of two equations. The first is a binary equation that shows the supplier is either operating ($Y= 1$) or is not ($Y= O$). A set of factors, such as nameplate capacity, quarter of the year, contract type and type of and so on, are gathered in a vector x and explain the decision to operate the specific independent supplier facility, so that[8]

$$\text{Prob}\ (Y= 1) = F\ (\beta x)$$

$$\text{Prob}\ (Y= O) = 1 - F(\beta x)$$

The set of parameters β, reflect the impact of changes in **x** on the probability that the project will be operating. For example, among the factors that might interest us is the marginal effect of nameplate capacity on the probability of operating status.

The second equation is a Weibull equation that describes the duration or survival of independent renewable projects. Duration is represented by the random variable *T*. The Weibull distribution has a nonnegative random variable (in our case the positive survival of renewable projects); that is, it is monotonically increasing. In the Weibull equation, each individual renewable project is assumed to be at risk at every instant, and that risk is summarized in the instantaneous failure or hazard rate function,

$$h(t) = h_o(t)e^{Bo+B_1 I_1 +\ldots+B_k I_k}$$

For exponential regression, $h_o(t) = 1$. For Weibull regression, $h_o(t) = pt^{p-1}$ where p is the shape parameter to be estimated from the data.

Two data sets are deployed in order to assess the chances of survival of renewable projects. The first set is a cross-section survey that contained information on the hurdle rate and prospective future operating plans of projects.[9] The second is cross-section time series panel data that were collected by the utilities on independent renewable projects in California on a quarterly basis from 1988-1996. The two surveys collected information on location (city, code), contractual agreement (SO1, S04,[10] etc.), technology type, fuel type, project capacity, the year the project was commissioned, and the current operating status. Table 3 shows our empirical findings.

The results of the first equation indicate that projects with large nameplate capacity have higher prospects of remaining operational than projects with small nameplate capacity. The first derivative of the function with respect to capacity is positive, but the coefficient of the variable gauging capacity squared is negative, suggesting that capacity positively influences the operating status of projects. Second, the findings support the hypothesis that the contractual arrangement aided the survival of projects. That is, some contracts reduced uncertainty about revenue streams since the revenue per unit of energy per kWh was to be known in advance for extended periods, up to 10 years. Third, capacity utilization is positively associated with the operating status of projects. That is, projects that have a high capacity utilization are more likely to be operating during the sample period than projects with small average capacity utilization. This result is in agreement with the conventional wisdom that low capacity factors, and low dependable on-peak capacity factors, are sources of renewable power cost disadvantage. California's wind capacity, which represents about 90 percent of the U.S. wind capacity, operated at only 23 percent realized average capacity factor in 1994,[11] and thus is a prime illustration of this relationship.

Table 3.
Regression Results of the Two-Step Model:
Operation and Survival

Equation/Variable	Marginal effect	Z-value
Operating Status Equation		
Nameplate capacity	.0101572*	-2.12
Capacity squared	.000107**	-1.78
SO No. 1	.194906	3.08*
SO No. 2	.024350	0.30
SO No. 4	.088349	1.45
Capacity utilization	.002739	2.18*
Change in utilization	.042630	-2.57*
Log likelihood	-83.6066	
Survival Equation		
Nameplate capacity	.004302*	1.901
SO No. 1	.268330*	2.339
SO No. 2	-.21242*	-1.803
SO No. 3	-.86762*	-2.902
SO No.4	-.34323*	-2.864
O&M costs	-.0040	-1.238
Predicted operation status	-1.3825	-1.238
Constant	3.3577*	10.983
Log likelihood	-103.254	
No. of observations	163.	

* Indicates the variable is significant at the 5 percent level.
** Indicates the variable is significant at the 10 percent level.

In the second equation dealing with survival of renewable projects over time, it is noteworthy that the operational status of renewable projects is time-dependent. This time-dependency is due to variations in the seasons and weather conditions (sun intensity, wind speed, availability of water or wood). We also found that the expected probability of survival declines slightly over time. The supplier's fuel type also affects the probability of survival. For instance, geothermal projects have an expected probability of survival of 0.93, and hydroelectric achieves an expected probability of 0.91. The expected probability of renewable facilities' survival also varies among the three investor-owner utilities. Renewable projects selling to Pacific Gas & Electric encounter an expected probability of 0.81. The corresponding probability for renewable projects selling to San Diego Gas & Electric is 0.87 and for Southern California Edison, it is 0.92. These variations are partly a reflection of composition (differences among utilities with respect to technology and contractual mix) and partly a reflection of differential utility load requirements.

4. THE STATE'S POLICY ROLE

While market forces influenced the relative cost-effectiveness of renewable electricity in California, the favorable regulatory environment greatly accelerated their penetration by reducing risk as well as by offering monetary incentives to develop and use electricity generated by renewable sources. Specifically, over the past 25 years California has employed a large number of policy measures aimed at pursuing the public policy goal of advancing the development and use of renewable electricity generation. California's stakeholders realize that to be effective, renewable public policy must be consistent with the market environment in which it operates. Until 1996, the California electricity system was characterized by vertically integrated utility monopolies, controlling the production, transmission, and distribution of electricity. The state regulated these monopolies. In California, these are the California Public Utilities Commission (CPUC), and the California Energy Commission (CEC), each with a different set of responsibilities relating to various aspects of the electricity market and its players.

The implementation of renewable public policy formulated in the environment of a regulated monopoly consisted of using a public process to issue a directive to utilities to carry out the elements of public policy. For example, utilities were directed to spend certain levels of funds on research and development to improve the efficiency and cost of renewables and other technologies, with regulators giving the criteria by which these funds are to be used and the projects on which they are to be expended.

The regulated market for electric generation saw the beginning of the end in 1978, with the passage of the federal law known as the Public Utilities Regulatory Policies Act (PURPA). That landmark law opened the generation market to non-utility entities, by requiring utilities to purchase power from generating facilities that meet certain qualifying criteria. These facilities later came to be known as qualifying facilities, or QFs. The other provision of PURPA that led to the development of non-utility generation in California was the requirement that utilities pay for the power generated by these non-utility projects and delivered to the utility system the so-called "avoided cost," which in theory was to leave the ratepayer indifferent.

The avoided cost was to be equal to the cost of the power that the utility would have had to generate itself, or purchase from other sources, but for the production of the QF power. To the extent that this condition is met, the ratepayer should see no difference in cost of power whether the utility generated it or the QF. Thus, the criteria of ratepayer indifference would be preserved.

In practice, however, the development of methodologies for estimating avoided cost were not so straightforward and often conflicted with each other. In fact, the principle of avoided cost was used to guide policy development that led to policy goals that encouraged renewable power generation.

The federal law left it up to the states to implement the details of the power purchase arrangements between utilities and independent power projects. In California, the development of contracts that governed the purchase of power by utilities from independent power projects was completed in 1983 and 1984. Some of these contracts have proven to be very lucrative, and within a year or two after the contracts were made available, led to the announcement of plans to develop almost 15,000 MW of independent power capacity.

This huge market response to the development and availability of power purchase contracts from the utilities was the result of a provision in the contracts that made the revenue stream that would result from the sale of power known to the project once the amount of energy production was known. That is, the electricity production from the project was the variable that projects needed to control in order to have certain revenue. The revenue per unit of energy or kWh was fixed in advance for up to 10 years.

In addition to known and certain energy prices for these projects for up to 10 years, these contracts also offered fixed and known capacity prices for up to 30 years. As a result of these contracts, about 10,000 MW of renewable generation and gas-fired cogeneration were added to the system between the years 1985 and approximately 1990. About half of that capacity, approximately 5,000 MW, was from renewable power generation projects. The 1980s and early 1990s saw California as the leader in the development of renewable generation in the U.S. and beyond.

In 1996, the California legislature passed the electricity restructuring legislation known as Assembly Bill 1890 (AB 1890) that began the end of the integrated utility environment in electricity markets in California. It opened the market for retail competition, during a four year transition period beginning January 1998 and ending in 2002. Prior to the passage of this bill, the CPUC had already begun investigating the approaches to restructuring the electricity market with less regulation and more competition. Private markets are very good at accounting for private costs and benefits, but fail to adequately account for external costs and benefits of electricity production and consumption.

The restructuring legislation, AB 1890, directed the three largest investor-owned electric utilities in California, Pacific Gas & Electric Company (PG&E), Southern California Edison (SCE), and San Diego Gas and Electric (SDG&E), to collect $540 million from their ratepayers over the period 1998-2002. This $135 million per year would be deposited in a fund to be administered by an agency selected by the state Legislature.

The same legislation, passed in September 1996, directed the California Energy Commission to prepare a report to the Legislature by the end of March 1997 recommending how this money should be allocated among different categories of renewable projects. The report was to observe broad constraints imposed by the Legislature, such as not less than 40 percent of the fund was to be used to support existing renewable projects, and not less than 40 percent was to be used to support new and emerging renewable projects.

An intensive effort by the Commission, including a series of public workshops and hearings to collect input on the preparation of the this report to the Legislature, culminated in March 1997 with the publication and transmittal to the Legislature and the Governor of a report entitled *Policy Report on AB 1890 Renewables Funding*, which carried out the legislative directive by making recommendations on how to allocate the money among different categories of renewables and went further, by describing a distribution mechanism designed to disperse the allocated funds to the designated category.

The California Legislature then codified the report submitted by the Commission into law by passage of Senate Bill 90 (SB 90) in October of 1997. This bill also established the Renewable Resources Trust Fund into which the money collected from ratepayers is deposited, and gave the California Energy Commission the responsibility to administer this fund largely along the design principles that were established in the policy report.

4.1 California's Energy Program Under Restructured Electricity Markets

In designing a public program for support of renewables under a restructured electricity market, the Commission relied on the following major design principles for such a program:

- The program must be simple to administer. This program was not to be a complex, bureaucratic structure that adds to the cost of supporting renewables and creates a new bureaucracy to administer the program.
- The program would be performance-based; that is, to the extent possible, funds should be used to increase production and consumption of renewable energy and therefore payments should be made only when production and/or consumption of renewable energy is made.
- The program must rely on market forces to the extent possible.

- The program design should strive for economic efficiency. This meant that the amount of funding spent to accomplish the goal of supporting renewable energy in competitive markets should be minimized.
- The program should create incentives for and pressure on renewable energy projects to increase their competitiveness in the open electricity market.

The renewable strategy underlying the design and implementation of California's Renewable Energy Program can be considered a total market strategy, in the sense that it operates on both the demand and supply sides of the market. The renewable funding flows to the market through five accounts. Each account was designed to reflect the needs of the particular category or segment of the renewables market that it was designed to serve.

On the supply side, the Existing Renewable Resources Account was allocated $243 million to support existing renewable projects. The New Renewable Resources Account, allocated $162 million, provides financial incentives for the development of new renewable projects that are generally more efficient and lower costing than existing generation that operated up to this point.

On the demand side, the Customer Credit Account was allocated $75.6 million and issues cents per kilowatt-hour credits to customers who opt to purchase renewable energy in the open market. The Emerging Renewable Resources Account, allocated $54 million, was created to encourage customers to install their own renewable generating systems on-site, primarily to offset their own load rather than selling to the grid. Finally, the last account on the demand side was the Consumer Education Account, which strives to educate users of renewable energy about the availability of such an opportunity and the financial support available to them if they opt to purchase renewable energy. The Consumer Education Account was allocated $5.4 million.

5. CONCLUSIONS AND POLICY IMPLICATIONS

The analysis in this paper has shown that the share of renewables in California's power sector has increased steadily. The growth has been volatile, however, due to the vagaries of fossil fuel markets and the phasing out of financial incentives at the federal and state levels. The economics of renewable electricity generation would improve if and when capital costs are significantly lowered. However, the continued low price of fossil fuels is adversely impacting the penetration of renewables, which may continue to overwhelm the competitive position of renewable generation to fossil generation, even in the face of declining capital costs.

California's support for renewables culminated in the enactment of California's AB 1890, which required the California Energy Commission to develop a program to encourage renewable electricity generation technologies. It directed the collection of $540 million from investor-owned utility ratepayers from 1998 to 2002 to support existing, new, and emerging renewable electricity generation technologies. Unless legislation to extend the program is enacted in time, renewable project revenues will be based on significantly lower energy prices than previously available. This will have a serious financial impact on current renewable suppliers, as market prices may not cover the independent suppliers' current long-run marginal generation costs, and could deter the entry of other small renewable electricity generators.

The state of California is likely to continue its policies to promote renewable-based electricity production. Justifications for this continuation include the clear links between renewables and air quality and environmental objectives. In addition, renewable electricity generating technology benefits California's export capabilities and helps the state's private sector to secure dynamic niche markets as well as links to the international energy markets.

REFERENCES

1. California defines renewables as those that are not fossil, not nuclear, and not large hydro, large hydro being defined as over 30 MW.
2. Little Hoover Commission, State of California. 1996. "When Consumers Have Choices: The State's Role in Competitive Utility Markets" p. 12.
3. Hirsh, R. and A. Serchuck. 1999. "Power Switch: Will the Restructured Electric Utility Help the Environment?", *Renewable Energy Policy Project.* Helen Dwight Reid Education Foundation.
4. Department of Energy. 2000. "Wind Energy: From Coast to Coast Wind Turbines Are Generating Electricity" www.doe, April.
5. The data are derived from the files that accompanied the study titled Feldman M. 1998. *Diversity and Risk Analysis* Final Report Submitted to CEC Contract #500-96-019.
6. Baltagi, B. 1995, *Econometric Analysis of Panel Data. Wiley, London.*
7. Hodrick, R. and E. C. Prescott 1997 "Postwar U.S. Business Cycles: An Empirical Investigation" *Journal of Money, Credit and Banking* Vol.29(1) February: 1-25.
8. Greene, W. 1997. Econometric Analysis: Third Edition. Prentice-Hall, New Jersey.
9. Owners or their representatives were questioned about future plans under alternative market structures. The responses were used to determine whether the project would continue to operate or not.
10. These independent projects sell power to utilities under four contract types: Standard Offer No. 1 (SO1), which provides for variable energy and capacity prices; Standard Offer No. 2 (S02), which provided for variable energy and firm capacity prices; Standard Offer No. 3 (S03), which is a simplified version of SO1; and Standard Offer No. 4 (S04), which provided firm energy and capacity prices.
11. CEC Wind Project Performance: 1994 Summary, Sacramento 1995.

Renewable Energy: Trends and Prospects. Edited by S.K. Majumdar, E.W. Miller and A.I. Panah

Chapter Twenty-Six

THE IMPACT OF CLEAN ENERGY POLICIES ON RENEWABLE ENERGY TECHNOLOGIES[1]

MARILYN A. BROWN[2]
and
WALTER SHORT[3]

[2]Oak Ridge National Laboratory
P. O. Box 2008
Oak Ridge, TN 37831-6186
and
[3]National Renewable Energy Laboratory, MS-2721
1617 Cole Boulevard
Golden, CO 80401-3393

ABSTRACT

This paper summarizes several scenarios that assess how energy-efficient and clean energy technologies can address key energy and environmental challenges facing the United States. A particular focus of this chapter is on the impact of public policies and programs on the market penetration and performance of renewable energy technologies. The study concludes that policies exist that can significantly reduce oil dependence, air pollution, carbon emissions, and

[1]Support for this analysis was provided by the U.S. Department of Energy's Office of Energy Efficiency and Renewable Energy. This chapter summarizes the findings of many contributors to the study entitled Scenarios for a Clean Energy Future. These contributors include: David Greene and Stan Hadley (Oak Ridge National Laboratory), Steve Plotkin (Argonne National Laboratory), Mark Levine, Jon Koomey, Ernst Worrell and Lynn Wright (Lawrence Berkeley National Laboratory), and Paul Lemar (Resource Dynamics Corporation).

inefficiencies in energy production and end-use systems at essentially no net cost to the U.S. economy. The study also shows the important role that renewable energy resources can play in every sector of the economy in addressing these multiple energy objectives.

INTRODUCTION

The goal of this study by five U.S. Department of Energy National Laboratories is to produce well-documented scenarios of public policies and programs that could foster efficient and clean energy technologies to meet a number of energy-related challenges facing the United States. These challenges include global climate change, air pollution, oil supply vulnerability, and inefficiencies in energy production and end-use systems. Some of these concerns are visible today and are clear public priorities; others are emerging as issues or are possible outcomes of an uncertain future. How the nation responds to them will affect the prosperity and well-being of future generations.

This study builds upon the results of a previous report, *Scenarios of U.S. Carbon Reductions*-also known as the "Five-Lab study" (Interlaboratory Working Group, 1997; Brown, et al., 2001). This earlier report quantified the potential for energy-efficient and clean energy technologies to reduce carbon emissions in the United States by the year 2010. Key sectors of the economy were examined independently: buildings, transportation, industry, and electric generators. The report identified a set of technologies with the potential to restrain the growth in U.S. energy consumption and carbon emissions so that levels in 2010 could be close to those in 1997 (for energy) and 1990 (for carbon).

The Clean Energy Future study goes one step further by identifying specific policies and programs needed to motivate consumers and businesses to purchase such technologies. Specifically, it examines the potential impacts of different packages of public policies and programs in an effort to identify feasible, low-cost policy pathways to a clean energy future. As such, the CEF study responds to a recommendation by the President's Committee of Advisers on Science and Technology (PCAST), Panel on Energy Research and Development (1997), that the nation identify and adopt a commercialization strategy to complement its national energy R&D portfolio.

THE CEF SCENARIOS

The study develops three scenarios - Business-as-Usual (BAU), Moderate, and Advanced. The BAU scenario assumes a continuation of current energy policies and a steady, but modest, pace of technological progress. In contrast, the Moderate and Advanced scenarios are defined by policies that are consistent with increasing levels of public commitment and political resolve to solving the nation's energy challenges. Some of the public policies and programs that define the scenarios are cross-cutting; others are designed individually for each sector (buildings, industry, transportation, and electric generators) and assessed the impacts out to 2020.

Numerous policies are examined, including fiscal incentives, voluntary programs, regulations, and research and development. Most of the policies were selected on the basis of their potential to reduce carbon emissions.[2] Others were designed specifically for air quality (e.g., reducing SO_2 emissions in the electric sector), oil security (e.g., alternative fuels research), and economic efficiency (e.g., restructuring of the electric sector). Regardless of the driving force behind them, almost all reduce carbon emissions and improve air quality. Policies are generally stronger in the Advanced than in the Moderate scenario, with larger expenditures on public-private RD&D partnerships, stricter standards, higher tax incentives, and greater government investment in programs that promote efficient and clean energy technologies. Some policies are assumed to begin in 2000; others are assumed to be enacted in subsequent years. Their impacts tend to be gradual, as stock turnover and other factors dampen initital responses. Later implementation would "miss" earlier stock turnover opportunities reflected here.

Two key differences for all of the sectors is the addition of a domestic carbon trading system to the Advanced scenario and increased R&D resources in both the Moderate and Advanced scenarios.

- **Domestic carbon trading system.** Emissions trading programs work by allocating allowances that permit the emission of limited quantities of pollutants during a specified period (e.g., annually). They allow sources to comply with the cap by reducing emissions or purchasing permits for emissions above the cap. A firm's response will depend on its costs of control compared with the market price of allowances.

  We assume that a domestic carbon trading program is announced in 2002 and is implemented in 2005. Each year, beginning in 2005, permits are sold in a competitive auction run by the federal government. The carbon pollutants limit is set so that the permit price equilibrates at $50/tC (in $1997) throughout the study period.[3] The federal government collects all of the carbon permit revenues and transfers them back to the public. The idea of the carbon allowance rebate is to leave people's "incomes" intact while changing the relative price of carbon.

- **Increased R&D resources.** The Moderate scenario assumes a 50 percent increase in federal government appropriations for cost-shared research

[2]Throughout this report, emissions are reported in terms of metric tons of carbon. To convert metric tons of carbon to metric tons of carbon dioxide, multiply the amount of carbon by 44/12 or 3.67.

development, and demonstration (RD&D) of efficient and clean-energy technologies. This includes increased research on energy-efficient end-use technologies as well as power generation technologies using renewable resources, natural gas, coal, and nuclear energy.[4] Since these resources are spent in public/private RD&D partnerships, they are matched by private-sector funds.

Altogether, the Moderate scenario assumes an increase of $1 .4 billion per year by 2005 (half as federal appropriations and half as private-sector cost share). This increment continues through 2020. The Advanced scenario assumes that the federal government doubles its appropriations for cost-shared RD&D, resulting in an increase of $2.8 billion per year. Both scenarios assume a careful targeting of funds to critical research areas and a gradual, 5-year ramp-up of funds to allow for careful planning, assembly of research teams, and expansion of existing teams and facilities. Much of the increased RD&D funding on electric technologies (in both the Moderate and Advanced scenario) is devoted to renewable energy technologies.

The policies identified as most important in the Advanced scenario are summarized in Table 1.

METHODOLOGY

A scenario-based approach is used to allow examination of alternative portfolios of public policies that address energy-related challenges. Scenarios are stories - not predictions - of how the future might unfold. They are useful for organizing scientific insight, gauging emerging trends, and considering alternative possibilities.

Various assessment methods, analytic tools, and expert judgements are used to analyze the impacts of individual policies. The CEF-NEMS model – based on the Energy Information Administration's (EIA) National Energy Modeling System (NEMS) - is then employed to quantitatively integrate the impacts of each scenario's policies. The integration step of CEF-NEMS allows the estimated effects of changes in energy use in each sector to be taken into account in the resultant energy use patterns of the other sectors. The CEF-NEMS also assesses additional changes in energy demand where new policies or technologies affect

[3]$50 per tonne of carbon corresponds to 12.5 cents per gallon of gasoline or 0.5 cents per kWh of electricity produced from natural gas at 53 percent efficiency (or 1.3 cents per kWh from coal at 34 percent efficiency).

[4] These budget estimates are based on a 1997 report by the President's Committee of Advisors on Science and Technology (PCAST, 1997), entitled "Federal Energy Research and Development for the Challenges of the Twenty-First Century." This report recommended that the United States double its federal energy R&D expenditures by the year 2003.

TABLE 1.
*Key Policies in the Advanced Scenario**

Buildings	Industry
–Efficiency standards for equipment	–Voluntary programs
–Voluntary labeling and deployment programs	–Voluntary agreements with individual industries and trade associations
Transportation	**Electric Generators**
–Voluntary fuel economy agreements with auto manufacturers[a]	–Renewable energy portfolio standards and production tax credits
–"Pay-at-the-pump" auto insurance	–Electric industry restructuring
Cross-Sector Policies	
–Doubled federal research and development	–Domestic carbon trading system

*The scenarios are defined by approximately 50 policies. The 10 in this table are the most important ones in the Advanced scenario. Each policy is specified in terms of magnitude and timing. For instance, "Efficiency standards for equipment" comprise 16 new equipment standards introduced in various years with specific levels of minimum efficiencies.

[a]These voluntary agreements, because they are met in the Advanced scenario, would have the same effect as a corporate average fuel economy (CAFE) standard of the same level.

energy prices. Macro-economic impacts and feedback are separately assessed through an analysis of previously published modeling results.

We employed the EIA's reference case from the *Annual Energy Outlook 1999* (EIA, 1998) as the starting point for the CEF Business-As-Usual (BAU) forecast. Thus the EIA's reference case assumptions on fossil fuel supplies, world oil prices, energy transport, end-use service demands, and macro-economic growth underlie the three CEF scenarios. To capture the policies of the Moderate and Advanced scenarios, CEF-NEMS inputs associated with process economics, stock turnover rates, consumer discount rates, and fuel prices are changed from the BAU scenario (and therefore from the EIA's Reference case). The translation of these policies to the inputs required by CEF-NEMS was conducted through off-line analysis, reference to past studies, expert judgment, and outside review. This process has the essential advantage that it quantitatively estimates the impacts of key non-fiscal policies.

OVERVIEW OF RESULTS

Key results of the study derived through the above approach are presented in Table 2 for the Moderate scenario and the Advanced scenario with a domestic carbon trading system that equilibrates at a permit price of $50/tC. Results are also shown for one of the numerous alternative policy sets that are examined in the full study (Interlaboratory Working Group, 2000)- in this case, the Advanced scenario with a permit price of $25/tC.

Energy Use. The reductions in energy use shown for these scenarios result from the many CEF policies that are directed at the adoption of energy-efficient technologies in the end-use sectors. Efficiency standards are shown to be highly effective in the buildings sector. Voluntary agreements with industries and voluntary labeling and deployment programs are also key to the substantial demand reductions of these scenarios. Such efficiency improvements are generally most economic when capital stock turnover occurs; they therefore take time to materialize. In the Advanced scenario the nation consumes 20 percent less energy in 2020 than it is predicted to require, based on the BAU forecast. This savings of 23 quadrillion Btu in the Advanced scenario is equal to almost one quarter of the nation's current energy use. It is enough to meet the current energy needs of all the citizens, businesses, and industries located in the top three energy consuming states (Texas, California, and Ohio) or the combined current energy needs of the 30 lowest consuming states.

The 2020 reductions are also a function of the continuing technology improvements that result from accelerated RD&D. For example, in the transportation sector much of the stock turns over relatively quickly, but the reductions in energy use are even larger in the second decade as research and voluntary agreements with automobile manufacturers make more efficient vehicles available. Reductions in primary energy use in all sectors can be expected to grow even larger after 2020 as RD&D improves technology further and additional stock turns over.

Energy use reductions in the Advanced scenario (with carbon permit values of both $25/tC and $50/tC) are more than twice those of the Moderate scenario because of two types of policy changes. First, the policies of the Moderate scenario have been strengthened in the Advanced scenario. For example, RD&D has been further expanded and performance standards in the buildings sector have been applied to more end uses. Second, additional policies are applied in the Advanced scenario. Key amongst these are the domestic carbon trading policy, voluntary agreements to improve the fuel economy of light-duty vehicles, and pay-at-the-pump automobile insurance. An off-line analysis of combined heat and power in industry suggests that policies tackling barriers to this technology could increase energy savings by an additional 5 to 10 percent.

There are several notable observations about the fuel mix in the BAU forecast compared with the Moderate and Advanced scenarios (Fig. 1): First, fossil fuel consumption is reduced in both the Moderate and Advanced scenarios, compared with the BAU scenario, while a higher proportion of nuclear power is retained and renewable energy grows more rapidly. However, the magnitude and composition of these trends differ across the two policy scenarios. For example, coal consumption is impacted much less in the Moderate than in the Advanced scenario. In the Moderate scenario, coal consumption increases from 1997 levels in both 2010 and 2020. Relative to BAU, coal consumption declines by about the

TABLE 2.
*Selected Results for 2010 and 2020**

			2010			
	1990	**1997**	**BAU**	**Moderate**	**Advanced ($25/tC)[a]**	**Advanced ($50/tC)[b]**
U.S. Primary Energy Use in Quadrillion Btu (Percent Change from BAU)	84.2 –	94.0 –	110.4 –	106.2 - 106.5 (-4%)	101.0 (-9%)	98.2 - 99.3 (-11% to 10%)
U.S. Energy Bill in Billion 1997$ (Percent Change from BAU)	516 –	552 –	651 –	595 (-9%)	598[c] (-8%)	634[c] (-3%)
U.S. Carbon Emissions in Million Metric Tons (Percent Change from BAU)	1,346 –	1,480 –	1,769 –	1,679 - 1,684 (-5%)	1,539 (-13%)	1,437 - 1,463 (-19 to -17%)
			2020			
	1990	**1997**	**BAU**	**Moderate**	**Advanced ($25/tC)[a]**	**Advanced ($50/tC)[8]**
U.S. Primary Energy Use in Quadrillion Btu (Percent Change from BAU)	84.2	94.0 –	119.8 –	109.6 - 110.1 (-9% to -8%)	98.8 (-18%)	94.4 - 96.8 (-21% to -19%)
U.S. Energy Bill in Billion 1997$ (Percent Change from BAU)	516 –	552 –	694 –	594 (-14%)	541[c] (-22%)	572[c] (-18%)
U.S. Carbon Emissions in Million Metric Tons (Percent Change from BAU)	1,346 –	1,480 –	1,922 –	1,730 - 1,740 (-10% to -9%)	1,472 (-23%)	1,307 - 1,347 (-32% to -30%)

*A number of key technologies were not modeled within the CEF-NEMS framework, including: combined heat and power (CHP), solar domestic hot water heaters, and fossil fueled on-site generation in buildings. An off-line analysis of policies to tackle barriers to CHP in industry was completed. It produced estimates of energy and carbon impacts for the Moderate and Advanced Scenarios. These estimates are included in the lower numbers in the ranges shown in this table. Estimates of impacts of CHP policies on the U.S. energy bill are not available.

[a] This variation of the Advanced scenario has a domestic carbon trading system that equilibrates at a carbon permit value of $25/tC.

[b] The Advanced scenario includes a domestic carbon trading system that equilibrates at a carbon permit value of $50/tC.

[c] The energy prices used to calculate this energy bill include the cost of the carbon permit.

same magnitude as natural gas and petroleum in both 2010 and 2020 - on the order of 5 to 8 percent from 1997 levels. However, in the Advanced scenario with a $50/tonne carbon permit price, coal use declines to 77 percent of 1997 consumption in 2010 and 60 percent of 1997 consumption in 2020.

Even with the significant decline in coal consumption in the Advanced scenario, the growth in demand for natural gas is lower than in the BAU forecast.

This is because the increased energy savings from efficiency investments, the increased use of renewable energy, and the maintained use of nuclear power in the Advanced scenario are greater in magnitude than the decline in coal use.

The use of renewable energy sources increases above BAU by 10 percent in the Moderate scenario and by 31 percent and 27 percent in the Advanced scenario for 2010 and 2020, respectively. In 2020, non-hydro renewables double from 2.3 quads in the BAU scenario to 4.6 quads in the Advanced scenario. Such contributions, consistent with cost projections for renewables in this time period, are especially notable for their longer-term role. This analysis suggests that the 20-year CEF scenario horizon could see the beginning of a significant growth in renewables.

Another implication of the fuel use results is that growth in petroleum consumption slows in both the Moderate and Advanced scenarios (by 9 percent to 21 percent in 2020 compared with BAU). Nuclear power retirements continue in all cases, but at much lower rates in the Advanced scenario than in BAU (6.4 quads of nuclear power consumed in 2020, compared with 5.6 quads in BAU).

The off-line analysis of CHP policies suggests that increased CHP in industry would result in the following adjustments to the scenario results, both in 2010 and 2020. It would increase natural gas consumption, decrease petroleum-based industrial boiler fuels, decrease coal in both the electricity and industrial sectors, and slow the growth of wind and biopower, especially in the Advanced Scenario in 2020.

Carbon Emissions. The difference between the Moderate and Advanced scenarios is even more pronounced with respect to carbon emissions (see Figures 2 and 3). In both 2010 and 2020, the carbon emission reductions of the Advanced

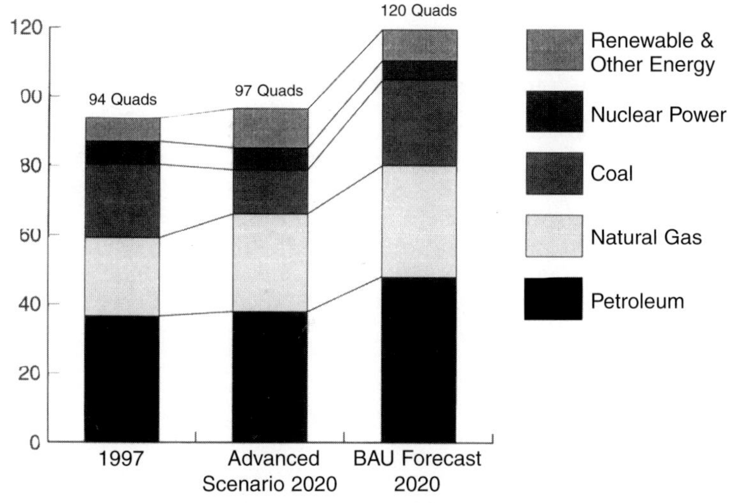

FIGURE 1. Energy Consumption by Source of Energy in the Advanced Scenario.

scenario are more than three times those of the Moderate scenario. By 2020, carbon emissions in the Moderate scenario are 9 to 10 percent lower than in the BAU forecast, while in the Advanced scenario carbon emissions are 30 to 32 percent lower than in the BAU forecast.

This much stronger performance of the Advanced scenario results from the focus of many of its policies on carbon emission reductions, particularly in the electric sector. These policies include domestic carbon trading, expansion of the production tax credit for renewables, restrictions on emissions of particulate matter, and restructuring of the electricity industry that allows cost-effective options to be introduced more quickly. Given the assumptions of the Advanced scenario, such policies produce a 2020 reduction from BAU carbon emissions in the electric sector of 47 percent, versus only 12 percent in the Moderate scenario. The largest portion of these electric sector reductions come from the repowering or replacement of coal-fired power plants by natural gas-fired power generation as well as wind, biomass, and geothermal power. An off-line analysis of combined heat and power in industry suggests that policies tackling barriers to this technology could reduce carbon dioxide emissions by an additional 5 to 8 percent throughout the period.

Overall, the Moderate scenario brings CO_2 emissions 20 percent of the way back to 1990 levels by 2010; the Advanced scenario with a carbon permit value of \$25/tC brings them 54 percent of the way down; and the Advanced scenario at \$50/tC closes 72 percent of the gap. In the context of the Kyoto goal of reducing greenhouse gas emissions to roughly 3 percent below 1990 levels by

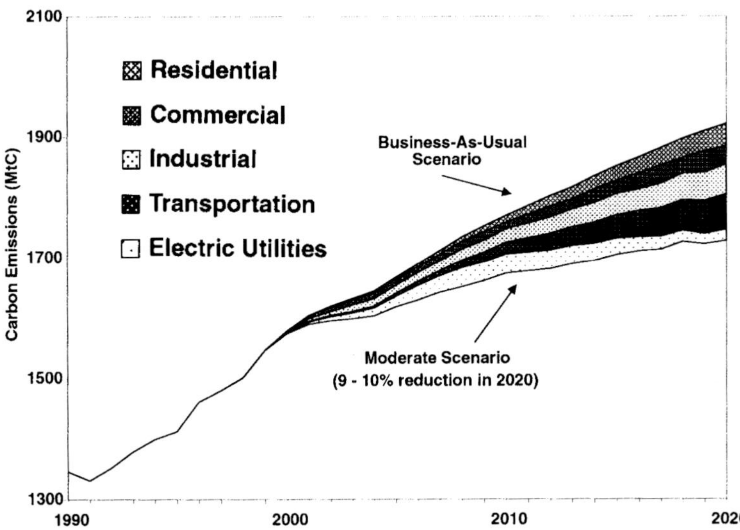

FIGURE 2. Carbon Emission Reductions, by Sector, in the Moderate Scenario Carbon Emissions.

2010, the CEF policies would need to be supplemented by other means such as international carbon trading, reductions in other greenhouse gases, and stronger domestic policies. In the Advanced scenario, carbon emissions drop to 1990 levels by the year 2020.

Costs and Benefits. In both the Moderate and the Advanced scenarios and in both time frames (2010 and 2020), the national energy bill is lower than in the BAU forecast (Table 2). This is largely due to the accelerated development and deployment of energy-efficient technologies, which reduces primary energy use. In 2010 the Advanced scenario bill is higher than the Moderate scenario bill because energy producers increase energy prices to recover their cost of purchasing carbon permits. Energy prices in the Advanced scenario drop between 2010 and 2020 due to the increased use and improved performance of efficient and low-carbon energy technologies. The net effect is that by 2020 the Advanced scenario's energy bill is $23 billion lower than that in the Moderate scenario and $124 billion lower than in the BAU forecast.

While consumers see the energy bill and must pay for any incremental investment costs of the energy-efficient and clean energy technologies they purchase, there are also public costs to implement policies and programs and to recycle revenues to the public from the auction of carbon permits. In both the policy scenarios, the energy bill savings from investments in energy efficiency, in combination with recycled revenues from the cap and trade system, exceed the annualized direct costs of the technologies and policies (including an estimate of incremental investment costs and policy implementation costs). The annualized net direct benefits are estimated to be at least $40 billion dollars for both

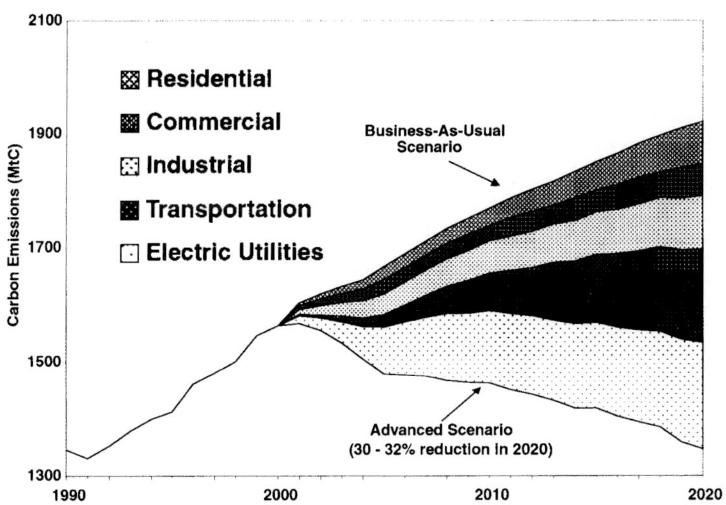

FIGURE 3. Carbon Emission Reductions, by Sector, in the Advanced Scenario.

scenarios in 2010, and growing to more than $100 billion per year in the Advanced scenario by 2020.

We assessed the macroeconomic costs of a $50/tonne carbon allowance price by examining the Energy Modeling Forum's recent compilation of results from simulations using seven of the leading energy/economic models (Weyant and Hill, 1999). These seven models provide alternative scenarios for achieving the carbon emissions targets of the Kyoto Protocol.

The estimated 2010 GDP losses (in $1997) associated with $50/tonne carbon permit price range from $4 billion to $66 billion. These can be compared with the $48 billion in net direct benefits estimated for the Advanced scenario in 2010. The combined impact of these direct benefits and the lowest and highest estimates of GDP losses is a range of total impacts of -$ 18 billion to +$44 billion in 2010.

In addition to reducing carbon emissions and cutting energy bills, a clean energy future could produce important collateral benefits: reduced threat of disruptions in oil supply, cleaner air, and increased opportunities to export energy technologies. For example, in the Advanced scenario in 2020:

- SO_2 emissions from the electric sector decline from 8.2 million metric tons in the BAU forecast to 4.1 million metric tons in the Advanced scenario, resulting in substantial environmental and health benefits.

- Petroleum consumption decreases 23 percent below the BAU forecast, thereby strengthening oil security. In addition, the U.S. benefits from significant reductions in annual wealth transfers from U.S. oil consumers to world oil exporters.

In spite of the overall net economic and environmental benefits of the scenarios, implementation of the policies in the CEF scenarios could produce important transition impacts and dislocations. The impact of the Advanced scenario on the coal industry is of particular note. Overall coal production in the United States decreases by 2020 to 50 percent of the BAU estimate for 2020, causing significant adverse impacts on that industry. On the other hand, the growth of strong domestic wind, bioenergy, and other "green" industries envisioned in this scenario would bring new employment opportunities to many regions and could lead to a revitalization of the economies of rural America. Efficiency technologies could boost output over a range of industries located throughout the United States, such as agriculture and bioprocessing, lightweight materials fabrication, sensor and control systems, and energy service companies, contractors, and consultants.

POLICIES THAT PROMOTE RENEWABLE ENERGY RESOURCES

The policies shown in Table 1 are found to be the most influential at promoting energy efficiency and clean energy technologies overall in the two policy scenarios. In many cases there are synergies between the policies. For example, RD&D breakthroughs from additional Federal funding reduce the cost impact of the Renewable Portfolio Standard. Interestingly, the policies and technologies sometimes compete with each other as well. For example, increased efficiency in the end-use sectors reduces demand for new electric capacity and therefore the opportunities for new clean renewable energy capacity.

Additional policies play a particularly influential role in accelerating the development and market deployment of renewable energy resources. These policies are listed in Table 3. They stimulate utilization of renewable energy resources in all four of the sectors of the economy, as illustrated in the following four sections. For electric power, we describe the role of wind generation technologies, in industry we look at biomass gasification, in transportation we examine cellulosic ethanol, and in buildings we discuss the prospects for photovoltaic systems.

WIND POWER

In the electric power sector, a diverse array of policies stimulate the rapid expansion of wind power in the two policy scenarios. In addition to the two key cross-sector policies (i.e., expanded cost-shared, federal R&D expenditures and the domestic carbon trading system), the following policies have significant impact on the development of wind power.

- **Policies to Facilitate Wind Deployment.** There are a number of issues associated with wind deployment and operation within a competitive electric market that could be mitigated through focussed policies. One policy is to facilitate siting on Federal land by, for example, reducing the National Environmental Protection Act (NEPA) filing requirements which currently require avian, archeological, and floral/fauna studies. Other policies involve expediting challenge procedures, limiting liabilities for all sites, and designing independent system operator protocols to accommodate wind intermittence.

- **Production Tax Credit.** In the Advanced scenario, a production tax credit of 1.5¢/kWh (1992$) is assumed for the first 10 years of operation from all non-hydro renewable electric generators installed through 2004. In the Moderate scenario, only wind and biomass power qualify, consistent with the President's Climate Change Technology Initiative proposals. In addition, for both scenarios a 1¢/kWh credit is given for cofired biomass during the years 2000-2004.

- **Renewable Portfolio Standard (RPS).** President Clinton's proposed (April 1999) legislation on competition in the electric sector included a

TABLE 3.
Additional Policies to Promote Renewable Energy Resources.

Moderate Scenario	Advanced Scenario
Wind deployment facilitation (e.g., facilitate siting on Federal land, design operator protocols to accommodate wind intermittency)	Same
1.5¢/kWh production tax credit (PTC) for the first 10 years of operation for wind and biomass power installed through 2004.	Same, for all non-hydro renewable electricity options
1¢/kWh credit for biomass cofiring during the years 2000-2004	1¢/kWh credit for biomass cofiring during the years 2000-2014
	Renewable portfolio standard-represented by 1.5¢/kWh PTC in 2005-2008 to signify cap in Clinton Administration proposal
Up to 1% net metering.	Up to 5% net metering
Full national restructuring of the electricity industry in 2008 resulting in marginal cost pricing, lower reserve margins, etc.	Same
	SO_2 ceiling reduced in steps by 50% between 2010 and 2020 to represent tighter particulate matter standards
CHP Policies: tax credits as specified in the Administration's 1998 Climate Change Technology Initiative, expedited siting and permitting, interconnection standard in 2002	CHP Policies: Extend tax credits beyond 2003, increase state grants through Clean Air Partnership Fund, further reduce expense associated with interconnection
Program to promote investment in cellulosic ethanol production.	Same

mandate to generate 7.5 percent of all electricity sales from either wind, biomass, solar, or geothermal for the years 2010 through 2015. However, a 1.5¢/kWh cap on the price premium for the renewable power was included. If the price difference between renewable energy and other alternatives is more than the cap, then it could come into play and lower the portfolio percentage, which could end up less than 7.5 percent. Although CEF-NEMS has the capability to include an RPS, it cannot directly model the 1.5 cents/kWh cap and it has problems combining RPS with marginal-cost-based rates. As a surrogate to the CEF-NEMS method of modeling the RPS, we extended the PTC of 1.5¢/kWh to capacity added between 2005 and 2008. Because the biomass cofiring tax credit only applies in the years specified, as opposed to the following ten years, it was extended to 2014. We calculated the added cost and carbon saved due to the tax credit extension and determined it to be between $60 and $70/tC. For this reason, the credit was only applied in the Advanced scenario.

- **Restructuring**. This policy assumes that all states implement competitive wholesale markets for electric power by 2008 in the Advanced scenario and the Moderate scenario. This translates to pricing based on real-time marginal costs instead of regulated, average-cost-based rates. This is as opposed to the BAU forecast, in which marginal cost-based pricing is applied in the five regions of California, New York, New England, the Mid-Atlantic Area Council (consisting of Pennsylvania, Delaware, New Jersey, and Maryland), and the Mid-America Interconnected Network (consisting of Illinois and parts of Wisconsin and Missouri). Restructuring can cause other changes to the market, such as higher costs of capital, lower reserve margins, and flatter load shapes. It also allows the non-quantified benefits of choice of supplier and competition. This may create dynamic efficiencies that spur development of lower cost and higher value energy services to customers. A recent study by the Northeast Midwest Institute gives more details on the potential for efficiency improvements in a restructured market (Kaarsberg et al., 1999). Market forces are already at work in today's environment, changing the generation mix to more efficient and cleaner plants. For example, the top two types of plants built in recent years have been combined cycle gas turbines and wind plants.
- **Stricter Particulate Matter (PM) Emission Standards.** This policy assumes that PM standards are tightened in response to increasing concerns of their impact on health and the environment. The CEF-NEMS does not include PM emissions; however, one of the major precursors to the formation of small (< 10 microns) particulates is SO_2, which can be constrained in CEF-NEMS. Following the example of the EPA's analysis of mercury and particulate emissions (EPA, 1999), we restricted $S0_2$ emissions to 50 percent below the current requirements. However, we delayed the ramping down to between 2010 and 2020, in part to shift policy impacts to the latter part of the study period.

Of the renewable technologies, wind received the most benefit from improvements in technology and other policies. Its capacity in 2020 grows from 4 GW in the BAU scenario to 15 GW in the Moderate scenario to 47 GW in the Advanced scenario (Figure 4). There is a large growth of wind through 2008 because of the PTC and the RPS (to 11 GW in the Moderate and 39 GW in the Advanced scenarios). In the Advanced scenario, economics (and the carbon permit) cause wind to continue to grow beyond these levels in later years.

The renewable portfolio standard creates strong incentives for growth in wind energy during the first decade of analysis. By encouraging the development of a strong capacity base, wind generation continues to grow even after production tax credits and renewable portfolio standards expire.

In the Advanced scenario, wind generation grows from 7.1 TWh in 2000 to 129 TWh in 2008, as a result of the production tax credits and renewable portfolio standard, with help from the carbon allowance penalty on other technologies and advances in technology (Table 3). This 18-fold increase would require an unprecedented growth in production capacity of suppliers of wind generation equipment. In the Moderate scenario, with its shorter schedule for the production tax credits and no renewable portfolio standards or carbon permit price, wind quadruples by the time the tax credits expire and the carbon trading system is fully in place (2005).

Geothermal also contributed substantially in the Advanced scenario. Figure 5 shows that generation from all renewable resources increases over time in the Advanced scenario. Once the RPS levels are reached in 2010, growth continues at a more moderate pace led by new geothermal installations.

COMBINED HEAT AND POWER SYSTEMS

Recent studies have identified combined heat and power (CHP) to be one of the most important technologies to improve energy efficiency and reduce GHG emissions in the United States (DOE, 1997; Onsite, 1998; Kaarsberg and Elliott, 1998). Combined heat and power, also known as cogeneration, is a system that generates electricity and captures the waste heat to meet other energy needs. Each system has at least one prime mover (e.g., a reciprocating engine, gas turbine, or fuel cell), fueled by fossil energy or renewable resources such as

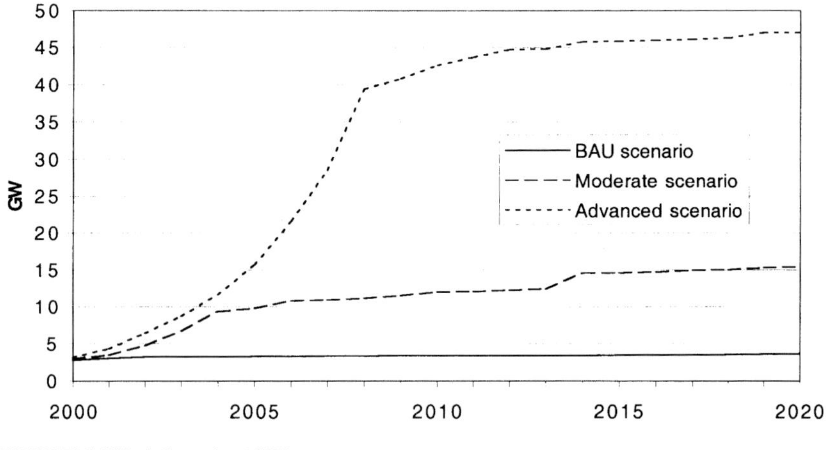

FIGURE 4. Wind Capacity (GW).

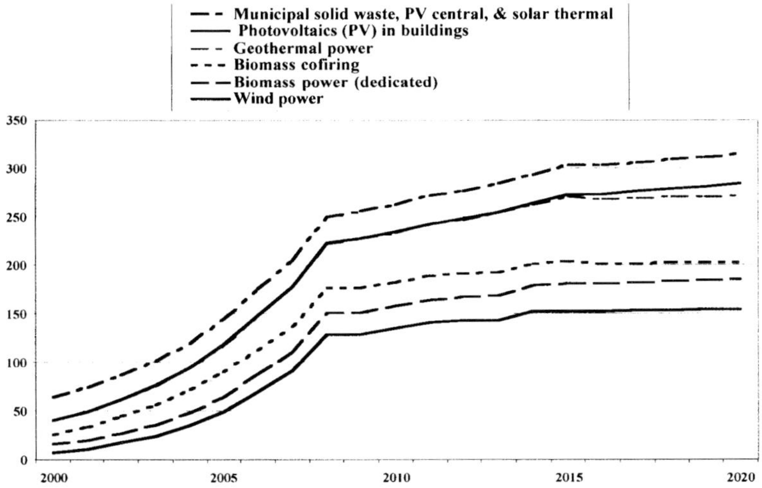

FIGURE 5. Renewable Electric Generation in the Advanced Scenario (TWh).

biomass feedstocks. The energy needs met by recovered waste heat include hot water, process heat, dehumidification, refrigeration, air conditioning and motor drive. The primary energy savings from small-scale CHP gas turbine units can be about 30 percent when replacing a conventional coal-fired power plant and about 15 percent when replacing combined cycle natural gas power plants. The potentials for CHP vary by sector, and even site, as they depend on site specific technical characteristics (e.g., steam load, demand pattern, and heat to power ratio) and on non-technical issues (e.g., regulation, buy back tariffs, and standby contracting). Most of the potential sites for new CHP units can be found in the 30 to 75 MW range (Khrushch et al., 1999).

Combined heat and power is currently used extensively in the pulp and paper industry with over 30 TWh of electricity produced. The bulk of this power is used on site with the remainder supplied to the grid. The power is generally produced in systems using Tomlinson boilers. Significant opportunity exists to replace these boilers with more efficient combined cycle and cogeneration systems that use gasified black liquor or hog fuels.

In this study we examined the potential of several policies to encourage all CHP systems, including the following.

- **Investment Tax Credit for CHP Systems.** This policy would establish an 8 percent investment credit for qualified CHP systems with an electrical capacity in excess of 50 kilowatts or with a capacity to produce mechanical power in excess of 67 horsepower (or equivalent combination of electrical and mechanical energy capacities). A qualified CHP system would be required to produce at least 20 percent of its total useful energy in the form of thermal energy and at least 20 percent of its total useful energy in the form of electrical or mechanical power (or a combination thereof). In the moderate scenario, it is expected that the tax credit scheme is maintained at the 2000-2002 level throughout the scenario period (2020). In the advanced scenario it is expected that the tax credit scheme is started at a level of $100 million for the period 2000-2005, and maintained at a higher level of $200 million from 2005 till 2020. The investment credit remains at 8 percent for qualifying CHP units. After 2005, higher credit levels are available for advanced cogeneration systems, including advanced turbines, gas turbines for industrial furnaces, high efficiency systems using waste gases, and for industrial applications of fuel cells. The program is maintained throughout the modeled period, and is expected to contribute to the expansion of CHP in industry, as well as third party (merchant) producers at industrial sites.

- **Expedited Siting and Permitting for CHP Projects.** In the Moderate scenario, this includes providing guidance to state agencies on establishing faster CHP permitting processes and encouraging arrival of new capacity online earlier (EPA Handbook and workshops). In the Advanced scenario, this policy involves appropriations for a Clean Air Partnership Fund to (1) increase state grants to encourage streamlined CHP siting and permitting, and (2) provide grants to states with accelerated CHP siting and permitting.

- **Remove Utility Barriers.** This policy is the same for both the Moderate and Advanced scenarios. It specifies enactment of national interconnection standards for CHP and other distributed generation projects by 2002.

The CHP analysis was performed using Resource Dynamics Corporation's DISPERSE model.[5] DISPERSE estimates the achievable economic potential for CHP applications by comparing on-site generation economics with competing grid prices. The model not only determines whether on-site generation is more cost effective, but also which technology and size appears to be the most economic. As a result, double counting of market potential for a variety of competing technologies is avoided. This model has been developed over the past five years, and has been applied on a variety of projects for utilities, equipment manufacturers, and research organizations. Fuel and electricity prices are based on those of the CEF scenarios. The overall steam demand for the industrial sub-sectors is taken from the results of the CEF baseline and policy scenarios. For modeling purposes it is assumed that steam use in each site follows the national developments. Various financial parameter assumptions are taken into account, including depreciation periods, tax rates, and insurance.

By permitting retirement of existing boilers where economically feasible, the model estimates CHP potential for the year 2020 ranging from 46 to 107 GW, depending on the timing and impact of CHP policies designed to remove

[5]Distributed Power Economic Rationale Selection (DISPERSE) model.

technical and market barriers. These estimates include both traditional (where all unit output is used on-site) and non-traditional (where sales of electricity to the grid is permitted) applications of CHP.

Our off-line analysis of combined heat and power in industry suggests that policies tackling barriers to this technology could increase energy savings by an additional 5 to 10 percent. In the BAU scenario, 8.8 GW of new CHP is projected, based on a continuation of current market penetration trends. Several technical and market barriers stand in the way of further use of CHP, as evidenced by the fact that over 80 percent of the potential capacity is projected as untapped. Most potential for CHP can be found in the paper, chemical, food and the non-energy-intensive manufacturing sub-sectors.

In the Moderate scenario, the projected additional CHP capacity grows to approximately 14 GW by 2010 and 40 GW by 2020. This includes 3 GW of integrated black liquor gasification cogeneration by 2020. It is expected that

TABLE 4.
Estimated Market Penetration and Impacts of Industrial CHP

Market Impact	**Year 2010 Impacts**			**Year 2020 Impacts**		
	BAU	Moderate	Advanced	BAU	Moderate	Advanced
New Installed Capacity (GW)	**4.4**	**14.1**	**28.9**	**8.8**	**40.1**	**76.2**
– natural gas	4.4	12.3	24.5	8.8	34.9	63.6
– black liquor gasifier combined cycle	0	1.1	2.6	0	3.1	7.5
– biomass gasifier combined cycle	0	0.7	1.8	0	2.1	5.1
Generated Electricity (TWh)	31	98	201	62	278	539
Fuel Consumed by CHP Systems (TBtu)	**274**	**901**	**1,853**	**551**	**2,542**	**4,985**
– natural gas	274	793	1,595	540	2,232	4,237
– biomass	0	108	258	11	310	747
Net Energy Reduction, relative to BAU Forecast (TBtu)	**NA**	**298**	**1,073**	**NA**	**450**	**2,367**
Net Carbon Reductions, relative to BAU Forecast (MtC)	**NA**	**4.9**	**26.1**	**NA**	**9.7**	**39.7**

expanded research and development will result in black liquor gasifier combined cycle technology by 2010, which will result in several demonstration projects by 2010 and an installed base of 3.1 GW by 2020. In addition, this expanded R&D will result in the emergence of high efficiency gas turbines, which are expected to increase CHP capacity in under 5 MW unit size ranges. Furthermore, policies designed to remove financial barriers, expedite siting and permitting, improve grid sell back price, and reduce interconnection costs are expected to contribute significantly to the expanded market potential and penetration.

In the Moderate scenario, newly installed CHP consumes almost 2 quads of energy (principally natural gas) more than in the BAU forecast, by 2020. This is offset by 0.9 quads of boiler fuel that is displaced by the CHP systems and 1.6 quads of energy that is displaced in the power sector. The net impact in 2020 is an energy savings of 0.5 quads and a reduction in carbon dioxide emissions of 9.7 MtC. (In 2010, the net reductions are 0.3 quads of energy and 4.9 MtC of carbon.)

In the Advanced scenario, the projected level of new CHP reaches approximately 29 GW by 2010 and 76 GW by 2020. Accelerated development of black liquor gasifier combined cycle units, as well as cost and efficiency improvements, in 5 MW-and-under gas turbines contribute significantly to the 107 GW of market potential. This includes 64 GW of natural gas based cogeneration, 7.5 GW of black liquor gasifier combined cycle capacity, and 5.1 GW of biomass gasifier combined cycle capacity. More aggressive policies designed to remove financial barriers, expedite siting and permitting, improve grid sell back pricing, and reduce interconnection and backup power costs all contribute to improved market penetration levels, as well as reduce the costs of the Advanced Turbine System. This leads to an accelerated implementation of CHP, despite the lower steam demand due to energy efficiency improvement.

In the Advanced scenario, newly installed CHP consumes 4.4 quads of energy (3.7 quads of natural gas and 0.7 quads of biomass) more than in the BAU forecast. This is offset by 2.1 quads of boiler fuel that is displaced by the CHP systems and 4.7 quads of energy that is displaced in the power sector. The net impact in 2020 is an energy savings of 2.4 quads and a reduction in carbon dioxide emissions of 39.7 MtC. (In 2010, the net reductions are 1.1 quads of energy and 26.1 MtC of carbon.)

CELLULOSIC ETHANOL AS AN ALTERNATIVE TRANSPORTATION FUEL

About one billion gallons of ethanol produced from corn is currently used annually in U.S. transportation markets as a blend stock for gasoline (Davis, 1998). Although the efficiencies and fuel choices used over the fuel cycle in producing this ethanol vary widely (e.g., fuel choices for powering the distillery can be corn stover, natural gas, or coal), recent studies show that the use of this ethanol provides a moderate GHG advantage over gasoline of approximately 20 percent. Processes to produce ethanol from cellulose - from woody biomass or municipal wastes - for use as a gasoline blending agent or neat fuel offer to reduce greenhouse gases about 80 percent compared to gasoline (Wang et al., 1999). A DOE program at the National Renewable Energy Laboratory has been working intensively to improve the efficiency and cut the costs of producing ethanol to about half of the current level (NREL, 1999). Program success could allow significant quantities of cellulosic ethanol to enter the market over the next decade, though a critical factor in its commercialization will be the world market price of oil. Land requirements could ultimately limit cellulosic ethanol production. About 15 billion gallons of ethanol (1.2 Quads) could be produced annually by converting municipal and agricultural wastes with minimal land requirements (Lynd, 1997). If about 35 million of the roughly 60 million acres idled by Federal programs were used for energy crops, about 25-32 billion gallons, or about 3-4 Quads, of ethanol could be produced annually (assumptions: 8.4 dry tons/acre/year crop productivity, 107.7 gallons of ethanol/ton yield) (Lynd, 1997). If only 10 billion gallons of ethanol were produced annually, this would leave 200 million dry tons of biomass for other uses, such as biomass power. The CEF-NEMS model explicitly treats these resource limitations and the trade-offs between different uses of biomass energy.

In the two policy scenarios, it is assumed that increased R&D funding yields a 50% decrease in overall ethanol production costs. This assumed level of cost reductions is consistent with the findings of Bowman and Leiby (1998); NREL (1999); and NRC (1999). It presumes a focused program of research on ethanol production processes associated with increases in transportation R&D funding in both scenarios. A recent assessment of DOE's bio-ethanol research program by the National Research Council (1999) made several recommendations for redirecting and expanding the effort, including:

1. improvement in pretreatment efficiency to minimize enzyme requirements,
2. development of less expensive, more effective digestive enzymes,
3. improved design to minimize costs, improve efficiency, and maximize coproduct value,
4. research in bioengineering and genomics to improve yield, pest resistance and stress resistance, and
5. research into feedstock changes to improve processing and conversion efficiency.

In addition to enhanced cost-shared R&D, the market penetration of cellulosic ethanol is fostered in the CEF scenarios by the assumption of a cellulosic ethanol commercialization program.

- **Cellulosic Ethanol Commercialization Program.** The use of cellulosic ethanol in vehicles would be extremely useful in reducing GHG emissions

> because, over the fuel cycle, such use generates a small fraction of the emissions generated by the equivalent use of gasoline - about 10 or 20 percent depending on assumptions. The DOE has an active research program aiming to develop commercializable processes for producing ethanol from energy crops, forest and agricultural residues, and municipal wastes (NREL, l 999). If successful, cellulosic ethanol would first be used primarily as a blending agent with gasoline, with added value as an octane enhancer and oxygenate.

Current market incentives for ethanol use include exemption from most federal excise taxes on gasoline for the ethanol portion of gasoline/ethanol blends, mandated oxygenate levels in Federal reformulated gasoline (RFG) required in ozone non-attainment areas (and other areas that have joined the RFG program), alternative fuel fleet requirements mandated by EPACT, and CAFE credits associated with the sale of alternative fuel vehicles. The latter two incentives affect vehicle production and sales only, and do not currently have any effect on ethanol production.

In the scenarios, we use the 1990 AEO assumptions about extension of the current "gasohol" tax break: that it will not expire in 2007 but continue at the nominal level of $0.51 per gallon. The value of the credit will thus be eroded by inflation over time to $0.27 per gallon in 2020. We also assume a program of loan guarantees, tax breaks, or subsidies to reduce or eliminate the added risk of investment in new ethanol plants.

In the Advanced scenario, use of ethanol for motor fuel increases from 1.1 billion gallons in 1999 to 3.9 billion in 2010, to 6.4 billion gallons in 2015, and 7.3 in 2020 (Figure 6). Ethanol's share of the gasoline market also increases from 3.8 percent in 2015 to 4.7 percent in 2020 (5.3 percent to 6.6 percent, by volume). The much lower demand for gasoline in the Advanced scenario depresses demand for ethanol. (Gasoline use in the Advance scenario is down 12 percent in 2010 and 24 percent in 2020 versus the Moderate scenario.) Ninety-five percent of the demand for ethanol as a motor fuel is for gasoline blending. Once again, these projections do not account for the potential impact of an MTBE ban, which would tend to increase demand for ethanol as a blending stock.

BUILDING INTEGRATED PHOTOVOLTAIC SYSTEMS

Photovoltaic cells convert light energy into electricity, at efficiencies that range from 7 percent to 17 percent. Building integrated photovoltaic systems can be combined with roof tiles or other parts of building structures to supplement grid-supplied power, reduce energy costs, and provide emergency backup power during utility power outages (NREL, 1998). In addition to the policies that support wind generation and other renewable power options, DOE's Million

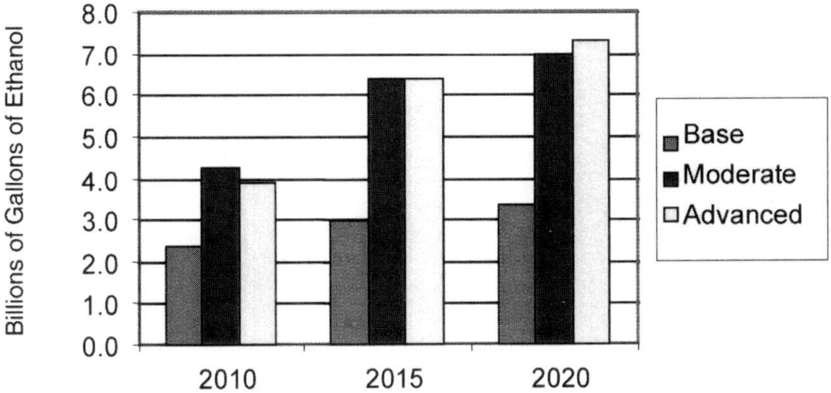

FIGURE 6. Ethanol Use for Motor Fuel

Solar Roofs Initiative is instrumental in accelerating the use of photovoltaic systems in the policy scenarios.

- **Million Solar Roofs Initiative.** This initiative is designed to address barriers to two types of solar energy technology - photovoltaics that produce electricity from sunlight and solar thermal systems that produce heat for domestic hot water, space heating or heating swimming pools. The initiative involves the development of new and existing financing options, leveraging resources, coordinating Federal agency support and sharing information. This initiative has already collected commitments for over 900,000 roof-top photovoltaics and active solar hot water units by 2010. These commitments are a reflection of the public's interest in green power, a range of benefits associated with distributed generation, and the continuing improvement in the economics of solar technologies.

Other policies having a significant impact on PV include:

- **Net Metering.** Consistent with the President's recently (4/19/99) proposed legislation on competition in the electric sector (DOE, 1999), this policy assumes a minimal level of net metering is allowed by the states. It is applicable only to systems of 20 kW or less in residential and commercial applications. Net metering means that on-site generation exceeding site loads can be fed back to the grid at values equal to the purchase price, i.e., the meter can be run "backwards" when on-site generation exceeds on-site loads. Net metering creates incentives for distributed generation which can have environmental and reliability benefits through higher efficiencies and reduced transmission and distribution requirements. Allowing customers to resell power at the retail price means that distribution costs are not recovered by the distribution company, requiring those costs to be recovered from sales to other customers. For this reason, net metering may face resistance and limits are often placed on the maximum amount of net metering allowed. The current analysis allows net metering of only residential buildings using PV. The high value afforded PV in buildings through this policy significantly impacts its market penetration.

- **Enhanced RD&D.** Increased cost-shared, Federal RD&D expenditures are assumed to lead to cost reductions that yield PV rooftop power for about 10 cents/kWh in 2020.

In the CEF Advanced scenario, the economics of PV are improved by 2020 to the point that over 2.6 million PV rooftop systems are estimated to generate approximately 17 TWh/year. This trend could become a significant factor in U.S. carbon reductions after 2020 as the technology continues to improve.

SUMMARY

This analysis has documented the valuable role that efficient, clean energy technologies could play in meeting the nation's many energy-related challenges, and describes portfolios of policies that could motivate consumers and businesses to invest in them. These policies could significantly reduce oil dependence, air pollution, carbon emissions, and inefficiencies in energy production and end-use systems at essentially no net cost to U.S. energy consumers. The analysis also shows the important role that renewable energy resources can play in every sector of the economy in addressing these multiple energy objectives.

While not directed specifically at renewable energy resources, the CEF study illuminated many opportunities for renewables:

- Renewable energy can make significant contributions towards the reduction of U.S. carbon emissions.
- Of all the renewable energy electric technologies, wind has the potenetial to contribute the most to carbon reductions.
- Effective policies for expanding the role of renewables include the renewable portfolio standard, carbon trading, and enhanced RD&D.
- If efficiency policies and technologies are not effective, renewables could make up much of the shortfall in anticipated carbon reductions.
- Cellulosic-based ethanol can make a significant contribution to carbon reductions as a blended fuel.

REFERENCES

Bowman, D. and P. Leiby, 1998. *Methodology for Constructing Aggregate Ethanol Supply Curves*, Draft, Revision 3, Oak Ridge National Laboratory, Oak Ridge, Tennessee, August 24.

Brown, M.A., M.D. Levine, W. Short, and J.G. Koomey, 2001. *Scenarios for a Clean Energy Future*, Energy Policy, 29 (14): 1179-1196.

Davis, S.C., 1998. *Transportation Energy Data Book, Edition 18*, Center for Transportation Analysis, Oak Ridge National Laboratory, ORNL-6941, September.

Energy Information Administration. 1998. *Annual Energy Outlook 1999: With Projections to 2020*, DOE/EIA0383 (99) (Washington, DC: U.S. Department of Energy), December.

Environmental Protection Agency. 1999. *Analysis of Emissions Reduction Options for the Electric Power Industry*, Office of Air and Radiation, U.S. Environmental Protection Agency, Washington, DC, http://www.epa.gov/capi/multipol/mercury.htm, March.

Interlaboratory Working Group. 2000. *Scenarios for a Clean Energy Future*, Lawrence Berkeley National Laboratory, Berkeley, CA (LBNL-44029), and Oak Ridge National Laboratory, Oak Ridge, TN (ORNL/CON-476), April, draft.

Interlaboratory Working Group. 1997. *Scenarios of U.S. Carbon Reductions: Potential Impacts of Energy-Efficient and Low-Carbon Technologies by 2010 and Beyond*, Lawrence Berkeley National Laboratory, Berkeley, CA (LBNL-40533), and Oak Ridge National Laboratory, Oak Ridge, TN (ORNL/CON-444), September.
URL address: http://www.ornl.gov/ORNL/Energy_Eff/labweb.html.

Kaarsberg, T., and N. Elliott, 1998. "Combined Heat and Power: How Much Carbon and Energy Can it Save for Manufacturers?" *Intersociety Engineering Conference on Energy Conversion* (IECEC-98), Colorado Springs, CO, August 2-6, 1998.

Khrushch, M., E. Worrell, N. Martin and D. Einstein, 1999. "Carbon Emissions Reduction Potential in the US Chemicals and Pulp and Paper Industries by Applying CHP Technologies," *Proc. 1999 ACEEE Summer Study on Energy Efficiency in Industry*, Washington, DC: ACEEE.

Lynd, L.R., 1997. "Cellulose Ethanol: Technology in Relation to Environmental Goals and Policy Information," in DeCicco, J. and Delucchi, M. (eds.) *Transportation, Energy and Environment: How Far Can Technology Take Us?* ACEEE, Washington, DC.

National Renewable Energy Laboratory, 1999. *Bioethanol Multi-Year Technical Plan: Fiscal Year 2000 and Beyond*, Office of Fuels Development, U.S. Department of Energy, Washington, DC, July.

National Renewable Energy Laboratory, 1998. *Photovoltaics and Commercial Buildings -A Natural Match*. Golden, CO: Produced for the U.S. Department of Energy by the National Renewable Energy Laboratory. September.

National Research Council, Board on Energy and Environmental Systems, 1999a. *Review of the Research Program of the Partnership for a New Generation of Vehicles: Fifth Report*, National Academy Press, Washington, DC.

Onsite Energy Corporation, 1998. "Pulp and Paper Industry. On-Site Power Market Assessment" Onsite Energy Corporation, January 1998.

PCAST (President's Committee of Advisors on Science and Technology). 1997. *Federal Energy Research and Development for the Challenges of the Twenty-First Century*, Executive Office of the President, Washington, D.C., November.

U.S. Department of Energy (DOE). 1999. *DOE Research and Development Portfolio, Energy Resources*, Washington, DC: U.S. Department of Energy, Volume 2 of 5, April.

U. S. Department of Energy, 1997. *OHVT Technology Roadmap*, Office of Heavy Vehicle Technologies, Office of Transportation Technology, DOE/OSTI-11690, October.

Wang, M.Q., Saricks, C.L., and D.J. Santini, 1999. *Effect of Fuel Ethanol Use on Fuel Cycle Energy and Greenhouse Gas Emissions*, Center for Transportation Research, Argonne National Laboratory, ANLESD-38, January.

Weyant, J. P., and J. N. Hill, 1999. "Introduction and Overview," in *The Costs of the Kyoto Protocol: A Multi-Model Evaluation,* Special Issue of The Energy Journal.

Renewable Energy: Trends and Prospects. Edited by S.K. Majumdar, E.W. Miller and A.I. Panah

Chapter Twenty-Seven

RENEWABLE ENERGY: A Cartographic Overview of State Programs and Regulatory Policies

MARK MONMONIER[1]
and
GEORGE A. SCHNELL[2]

[1]Department of Geography
Maxwell School of Citizenship and Public Affairs
Syracuse University
Syracuse, NY 13244-1020
[2]Department of Geography
State University of New York at New Paltz
New Paltz, NY 12561-2499

INTRODUCTION

"Energy, Environment, and Economy," the first of the Pennsylvania Academy of Science's 25 books, contained the following simple but prophetic sentences: "Are there solutions to the energy problem that will still allow a viable environment? The answer is 'yes,' but the solution can be painful to a society brought up in luxury consumption. The short-term solution is to make best use of various sources of energy." The long-term solution is described as a reduction in entropy, and continues that "that amounts to energy conservation. Although conservation can significantly reduce energy demands, it is not popular with the power industry or government… because a reduction in energy consumption is equated with lower economic growth" (Smith 1981, 58). The passages cited were written at a time when renewable energy development enjoyed progress, but eventual price stabilization for petroleum products and government disregard resulted in serious reductions in funding for research. In the late 1990s, though, renewable energy technology regained popularity — witness revived interest in power from the sun, wind, and water, along with less utilized sources such as those derived from biomass and geothermal operations.

This chapter tracks the status of renewable energy efforts by way of state programs and regulations. Recent rapid development of renewable technologies has been fostered by government funding incentives along with programs to regulate production of energy. The presentation is essentially cartographic and the primary sources of data are a survey and summary by Larsen, Rogers, and Shirley (1998), and of related data at the North Carolina Solar Center's homepage at http://www-solar.mck.ncsu.edu (April 21, 1999). In addition to describing variations among states in energy-related policies, programs, and financial incentives, this chapter uses principal-components analysis to search for broad regional trends in political efforts to promote renewable energy.

DISTRIBUTION OF STATE PROGRAMS AND POLICIES

Figures 1 and 2 illustrate the states' strategies for promoting renewable energy by way of an array of programs and regulations (Figure 1) and financial incentives (Figure 2). The following section serves to summarize the distribution of these activities across the nation.

Programs and Regulations

Renewable Energy Portfolio Standards and Set Asides are similar insofar as Portfolio Standards require that, before an established deadline, a percentage of capacity be derived from renewables, whereas Set Asides deal with like requirements for new capacity (Larsen, et al. 1998, 19). The map (Figure 1, top row, left) shows seven states in the programs, four with Portfolio Standards (Arizona, Maine, Massachusetts, and Nevada) and the remainder with set-aside rules. System Benefits Charges are universal fees on electricity use or, more commonly, fees as part of legislation calling for restructuring (Larsen, et al., 19). The states with these levies (top, center) include a cluster in the Northeast (Connecticut, Massachusetts, New York, and Rhode Island) and three others located from Illinois westward. Twelve states have Green Pricing Programs (top, right), the option offered, at a premium, to consumers who want all or part of their power from renewables (Larsen, 31). States with this option form two groups, the larger extending from Texas to the north and west as far as Oregon, and the other including Indiana, Michigan, Wisconsin, and Minnesota. Otherwise, only Florida and Hawaii are shown. Legislatures in ten states passed laws regulating Disclosure or Certification of the fuel mix used in power generation (second row, left). Certification, operating only in California, requires that "green" power sources provide a given percentage from solar production (Larsen, 51 and 52). Allowing electricity to flow through one meter to and from consumers with their own generators and using renewable resources is termed Net Metering. Savings

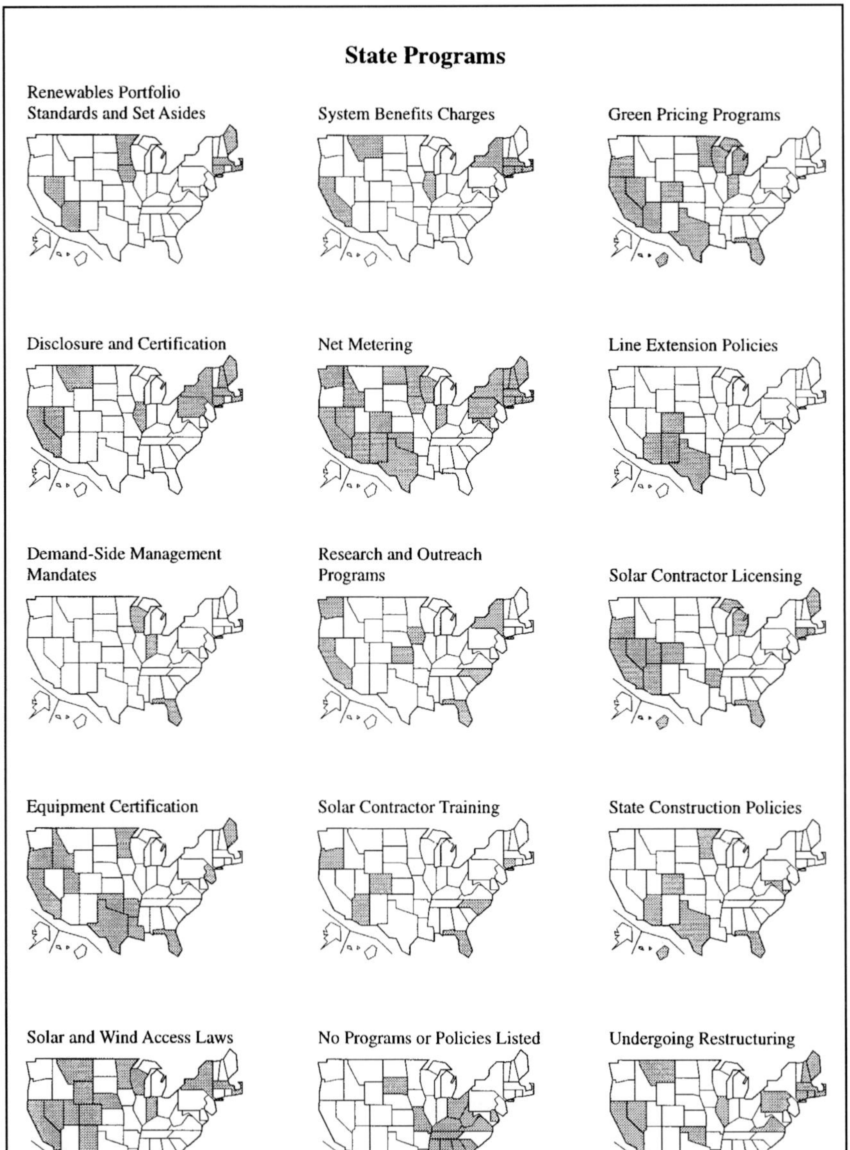

FIGURE 1. Array of maps describing states' policies and programs for promoting renewable energy. Compiled by authors from data at the North Carolina Solar Center's homepage, *http://www-solar.mck.ncsu.edu* (April 21, 1999).

accrue for industrial, commercial, or residential customers because any excess power by other users can offset the cost of the electricity that would have been purchased outside the system. Twenty-three states (second row, center) have Net Metering, making it the most widespread of the policies available (Larsen, 61). States involved include much of the Northeast and Southwest, and part of the Midwest.

Four states, Arizona, Colorado, New Mexico, and Texas (second row, right), comprise the extent of Line Extension Policies in which renewable energy resources are viewed as alternatives to the very expensive power line extension to remote areas (Larsen, 75). Just Florida, Indiana and Wisconsin are shown with Demand Side Management Mandates (third row, left), a strategy that reduces the burning of fossil fuel by requiring the use of renewable energy sources (Larsen, 79). Seven states, North Carolina and six others scattered across the country (third row, center), have established Research and Outreach centers to educate consumers and encourage adoption by residents and manufacturers. Solar

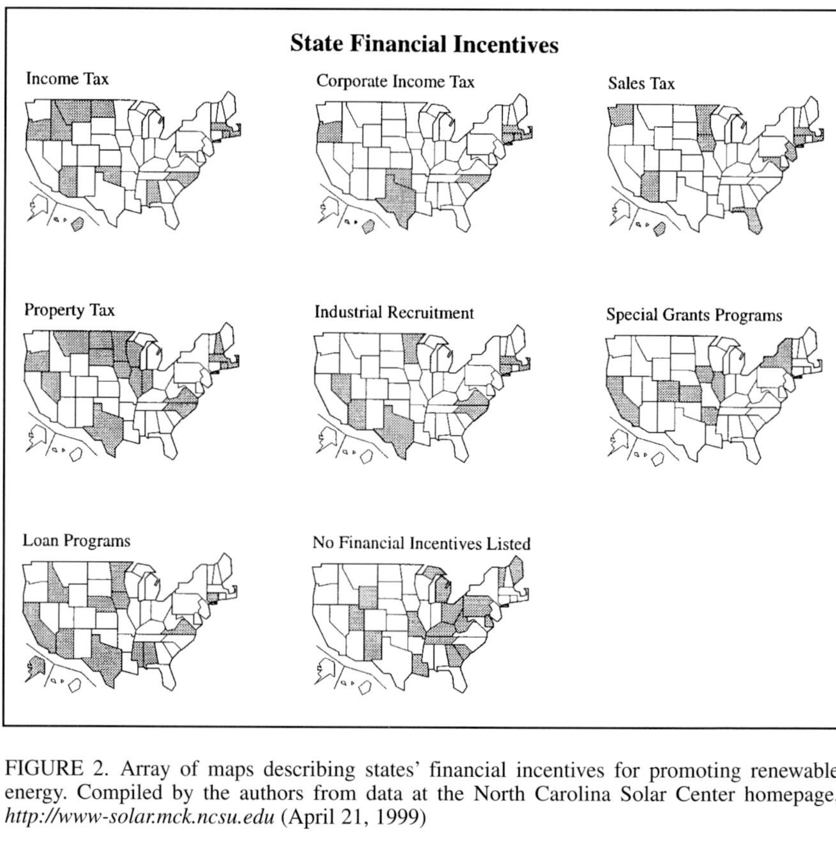

FIGURE 2. Array of maps describing states' financial incentives for promoting renewable energy. Compiled by the authors from data at the North Carolina Solar Center homepage, *http://www-solar.mck.ncsu.edu* (April 21, 1999)

Contractor Licensing (third row, right) is required in thirteen states. These range across the country from New England to Hawaii, with the largest group in the West. Solar Equipment Certification (fourth row, left) appears in twelve states, eight of which are in the Southwest and West, with the other four scattered as shown. The purpose is to provide safeguards for consumers via state standards within which financial incentives are offered. Solar Contractor Training is offered in six states, located from coast to coast. Curricula to meet the need are available in a community college in Oregon, through training and apprentice programs in Connecticut, and those sponsored by solar energy centers in Florida and North Carolina (fourth row, center). State Construction Policies, intended to promote the use of renewable energy in public buildings and other projects (fourth row, right), are found in Arizona, Colorado, Florida, Hawaii, Maryland, Minnesota, and Texas. The fourteen states with Solar and Wind Access Laws (fifth row, left) are principally in the western half of the nation, with groups of three states in the Midwest, and eight in the West. Massachusetts and New York complete the array. The thirteen states without programs or policies (fifth row, center) are largely clustered in the Southeast and Appalachia, with notable outliers in Alaska and South Dakota. The map of states actively restructuring their policies and programs (fifth row, right) in early 1999 indicates a broad interest, largely outside the South, in evaluating and enhancing legislative solutions.

Financial Incentives

The state financial incentives arrayed on the maps in Figure 2 are more easily understood by their titles than those in Figure 1 and require little or no explanation beyond the description of the distribution by state of each incentive. To begin, personal income tax incentives (first row, left in Figure 2) are offered in Hawaii and ten of the conterminous states scattered from New England across the South to Arizona in the Southwest, and an additional cluster extending from North Dakota to Oregon. Corporate Income Tax (top row, center) is offered as a stimulus in eight states mostly dispersed from New England to Hawaii. Sales Tax as an impetus (top row, right) appears mainly scattered, once again, but sited around the periphery and in Hawaii as well. Exemption from Property Tax as a motivation to use renewable energy (second row, left), is the most widespread incentive, found in 16 states with a huge concentration extending from Indiana to Montana and smaller clusters in New England, the South, and the West. Industrial Recruitment (second row, center) exists in eight states with pairs in New England, the Carolinas, and the West, along with Texas and Minnesota standing alone. Special Grants Programs (second row, right) feature the largest populations, California and New York, and a five-state cluster in the nation's center. Twelve states are listed with Loan Programs (third row, left). Mostly found west of the Mississippi, this incentive extends otherwise from Mississippi to Virginia, and includes Connecticut.

Massachusetts leads the states with five incentive programs, followed by six with four (Arizona, Connecticut, Hawaii, Oregon, Rhode Island, and Virginia), and four with three (Iowa, Minnesota, North Carolina, and Texas). New England is well represented in this group with three states listed. The final map showing states lacking financial incentive programs (third row, center) contains 17 states, five of which are west of the Mississippi, whereas most of the remainder occupy a large contiguous area extending northward from Georgia-South Carolina to include Michigan, Ohio, and Pennsylvania and westward to include Kentucky and Tennessee. Otherwise, only Maine and Vermont are contained.

PRINCIPAL COMPONENTS ANALYSIS

Principal-components analysis (Davis 1986, 527-545) affords a concise assessment of accord among the states in their strategies for promoting renewable energy. Widely used as a data-reduction technique, principal-components analysis collapses a set of m variables into a smaller set of orthogonal (uncorrelated) measures that account for much of the variation in the original data. The importance of each component is revealed by its *eigenvalue*, which when divided by m yields the proportion of the total variation accounted for by the component. For example, in an analysis involving ten variables, a component with an eigenvalue of 3.0 accounts for 30 percent of the variation, whereas a component with an eigenvalue of only 0.5 accounts for a mere 5 percent. Note that each individual variable adds one unit of variation to the data set. Thus, if two of the ten variables are perfectly correlated (r=1.0), one principal component should be able to account for at least 20 percent of the total variation. But, because other variables are probably correlated at least minimally with these two variables, the technique should extract at least one component that accounts for more than 20 percent of the total variation. The analysis also yields a matrix of *loadings*, which show the correlation of each original variable with each component. Because the square of a bivariate correlation coefficient measures the extent to which the variation of one of the variables "explains" or "accounts for" variation of the other, a component's eigenvalue equals the sum of the squares of its loadings.

Principal-components analysis extracts components in a parsimonious sequence, so that the first component always accounts for the greatest amount of common variation that a single component can explain, the second component represents the greatest amount of the remaining variation that can be accounted for by an index uncorrelated with the first component, the third component reflects the greatest amount of the remaining variation that can be accounted for by a composite variable uncorrelated with the first two components, and so forth. Thus, if the variables in a data set are highly similar — and thus able to yield maps similar in pattern—a small number of components, perhaps only one or two, will account for most of the total variation. By contrast, if the areas are highly dissimilar, the first eigenvalue might not be much greater than 1.0 and the next few principal components might account for little more variation than an individual variable.

Intended for interval and ratio data, principal-component analysis can also be used with ordinal (rank) and binary data, for which Spearman rank-order and phi coefficients, respectively, provide comparable measures of similarity (Burt and Barber 1995, 399; Griffith and Amrhein 1991, 97-101). For the thirteen policy and program variables described cartographically in Figure 1, phi coefficients range from a high of 0.66 to a low of -0.20. The 0.66 correlation, for Systems Benefits Charges (top row, center) and Disclosure and Certification (second row, left), reflects six states with both kinds of policy and five states with one kind but not the other. In contrast, a -0.20 correlation between Disclosure and Certification (second row, left) and State Construction Policies (fourth row, right) identifies a pair of policy categories that no states share despite 17 states with one kind of policy but not the other. Most of the other 76 pairings are only minimally similar, and only three have correlations greater than 0.5. Not surprisingly, the first principal component, with an *eigenvalue* of 3.08, accounts for only 23.7 percent of the total variation, while the second component, with an *eigenvalue* of 2.25, accounts for another 17.4 percent, for a cumulative explanation of 41.1 percent.

Figure 3 is a concise summary of the analysis. Orthogonal axes represent the first two principal components, and the points represent the thirteen programs and policies, plotted according to their loadings, which describe their correlations with each component. Variables that plot relatively far from the origin, like Disclosure and Certification programs and Green Pricing policies, are better accounted for by the pair of principal components than are variables that plot

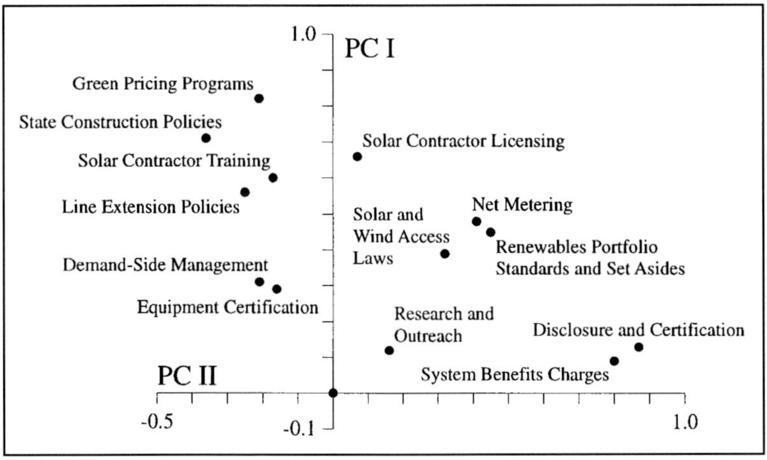

FIGURE 3. Relative similarity of the thirteen policies and programs, as depicted by the first two principal components, which account for 41 percent of the variation among the states.

closer to the origin, like Research and Outreach Programs. Thus, the relative closeness of Disclosure and Certification programs and System Benefits Charges policies is more meaningful than the even greater proximity of New Metering policies and Solar and Wind Access Laws. Overall, the dearth of significant clusters of variables reinforces the observation that states are highly diverse in their approach to energy conservation and renewable resources.

The states are no more similar in their use of financial incentives. For the seven strategies described cartographically in Figure 2, phi coefficients range from 0.56 to -0.21. The 0.56 correlation reflects six states offering both Income Tax (first row, left) and Corporation Tax (first row, center) incentives and seven states with one of these incentives but not the other. At the other extreme, the -0.21 correlation between Income Tax incentives (first row, left) and Special Grants Programs (second row, right) epitomizes 18 states with one incentive or the other but none with both. All of the other 19 pairwise correlations are less than 0.5, and as with the programs and policies data set, principal-components analysis confirms minimal similarity. The first component, with an eigenvalue of 2.28, accounts for only 32.5 percent of the total variation, while the second component, with an eigenvalue of 1.31, represents another 18.7 percent, for a cumulative explanation of 51.2 percent. Although this latter result might suggest that the states are more similar in their adoption of financial incentives than in their choice of programs and policies, a pair of components with eigenvalues totaling 3.59 and based on seven variables need not reflect greater similarity than two components with eigenvalues totaling 5.34 for thirteen variables.

As Figure 3 revealed diversity in programs and policies, Figure 4 reflects substantial variation among the states in their use of financial incentives. In addition to confirming the aforementioned positive correlation between income and corporation tax incentives and the negative correlation between income tax incentives and loan programs, the graph suggests that states that favor tax incentives of any kind (property or sales as well as income or corporate) tend not to offer special grants. Even so, the overall picture is a pattern of strong diversity.

CONCLUDING REMARKS

The foregoing analysis reveals little regional structure or geographic coherence to states' efforts to promote the use of renewable energy resources through regulation or financial incentives. Although resistence is concentrated in a cluster of Southeastern-Appalachian states, legislatures elsewhere are experimenting with a vast variety of structures and inducements. Their experiences — in particular their relative successes and failures — during the first decade of the twenty-first century might usefully inform the development of regional and national energy policy.

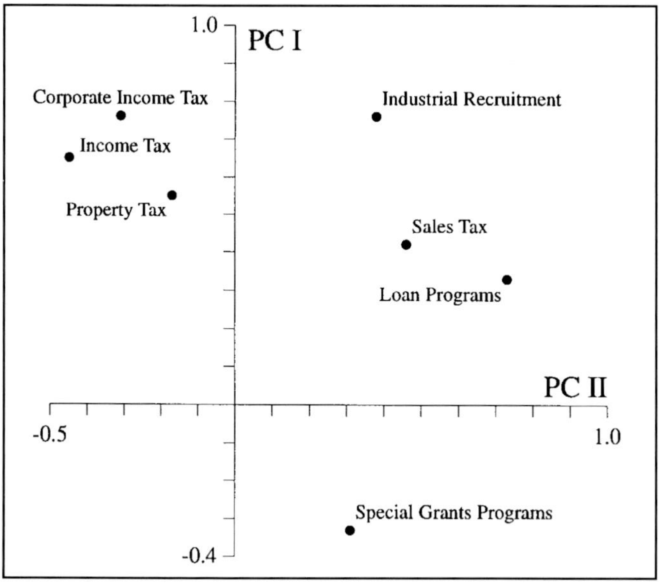

FIGURE 4. Relative similarity of the seven financial incentives, as depicted by the first two principal components, which account for 51 percent of the variation among the states.

REFERENCES CITED

Burt, James E. and Gerald M. Barber. 1995. *Elementary Statistics for Geographers*, 2nd ed. New York: Guilford Press.

Davis, John C. 1986. *Statistics and Data Analysis in Geology*, 2nd ed. New York: John Wiley and Sons.

Griffith, Daniel A. and Carl Amrhein. 1991. *Statistical Analysis for Geographers.* Englewood Cliffs, NJ: Prentice-Hall.

Larsen, Chris, et al. 1998. *National Summary Report on State Programs and Regulatory Policies for Renewable Energy.* Raleigh: North Carolina Solar Center, Industrial Extension Service, North Carolina State University. See also North Carolina Solar Center's homepage at http://www-solar.mck.ncsu.edu (April 21, 1999)

Smith, Robert Lee. 1981. Energy and the Environment, in *Energy, Environment, and the Economy*, S.K. Majumdar (ed.). Easton, PA: Pennsylvania Academy of Science.

Renewable Energy: Trends and Prospects. Edited by S.K. Majumdar, E.W. Miller and A.I. Panah

Chapter Twenty-Eight

RENEWABLE ENERGY: POLICY, PUBLIC/PRIVATE PARTNERSHIPS, AND PROMISE

NANCY McCREARY WATERS

Nancy McCreary Waters
Department of Biology
Lafayette College, Easton,
Pennsylvania 18042 USA

INTRODUCTION

Too often scientists concentrate on communicating with one another at the expense of making their message understandable to the general public. The patterns over the last 30 years by which the public has alternately viewed renewable energy sources-with skepticism, then enthusiasm with unrealistic expectations, followed by disappointment and ultimately realistic cautious optimism-exemplify our shortcomings in effectively translating our scientific knowledge base to policy makers in various levels of government and individuals who drive those policy decisions. This chapter examines the application of renewable energy technologies to the public sector: it begins with a brief overview of the timeline for policy decisions. It then moves to case studies, placing examples of successful renewable energy policies in the context of how general policies function, as well as identifying specific renewable energy technology policies which have demonstrated uniquely favorable outcomes. Finally, it suggests the guidelines by which we might enact effective policy changes in pursuit of comprehensive renewable energy development.

DEVELOPMENT OF RENEWABLE ENERGY POLICIES: STRUCTURING FORCES

Assessing the ebb and flow of renewable energy policy requires an understanding of the global forces over the last half century which have generated interests in renewables. Within the United States, policies governing alternative renewable energy sources (ARES) have been tightly linked to changes in cost and availability of fossil fuels. Oil production and consumption trends fall into four major time frames, upon which we can superimpose dramatic events on the global stage (Nebel and Wright 2000). Prior to 1970, oil consumption in the United States demonstrated essentially linear increases, rising at a rate of @ 28% over the thirty-year period from 1950 to 1970 (Nebel and Wright, 2000). This was somewhat higher than continental US production (@12%), which climbed predominantly due to discovery of new reserves. That discovery, however, peaked in 1970 to mark the end of the first time frame, after which domestic oil production declined, but consumption accelerated. In this second time period, rapid consumption outpaced production, and imported oil became increasingly necessary. That heavy reliance contributed to the massive impact of two important concurrent events in 1973-the Arab oil embargo and the US Mineworkers Strike. These mark the onset of a third phase of the oil timeline, with oil prices skyrocketing by nearly ten-fold (Nebel and Wright, 2000), priming the economic recessions of the mid-70's.

While some analysts suggested the crisis was the beginning of the end for global oil, scientific estimates placed global consumption of oil at only about one-eighth of accessible crude (Campbell and Laherrere, 1998). The embargo was successful for the five OPEC nations, not due to scarcity of oil internationally, but rather due to their ability to corner over a third of the global oil market (Campbell and Laherrere, 1998). The national response to this fossil fuel energy shortage included (1) attempts to reduce consumption by increasing efficiency and raising environmental standards, (2) attempts to increase exploration and production of Alaskan oil, and (3) government support for research, industrial tax breaks and depreciation schedules, and personal tax incentives for development and use of ARES. Indeed, the establishment of the Department of Energy by President Carter was referred to by some as the energy equivalent of the recent "war on drugs" (Wasserman, 1997).

This third phase ended in 1986 when oil prices declined by almost 50% to nearly $10 per barrel, a decline generated by the combined efforts enumerated above. While a successful resolution would have been to continue to reduce consumption, market forces rebounded. Lowered prices reduced conservation efforts, while tax incentives and government subsidies were systematically reduced and eliminated, bringing the ARES industry nearly to a standstill. To provide a dollar figure comparison, the federal renewable energy budget in 1981 was $500 million, but the Reagan-proposed fiscal year 1984 allocation was 20% of that figure (Jeffries, 1983). In this fourth and last phase of the timeline, consumption rates have gradually increased again and the outlook remains uncertain—we have become victims of our own success, and have generated an energy nonchalance which makes us ripe for a repeat of the fossil fuel energy shortages. Conventional wisdom couches the energy problem in terms of how to increase supply to meet projected demand (Lovins, 2000).

Some political theorists suggest that the Gulf War of 1991 was little more than a preventative measure to maintain United States supply lines for foreign oil, and economic estimates of the military costs associated with the Persian Gulf average around $50 billion annually. When viewed in the context of the 635 million barrels of oil imported by the US from the Persian Gulf (in 1997), this military cost adds a subsidy of between $24 to $80 a barrel to the existing price of $14 per barrel (Hubbard, 1991; Nebel and Wright, 2000). While some geologists contend that the world is not yet running our of petroleum, they do caution that we enter a new century with perhaps only a decade before we face the end of cheap and abundant oil on which our economies depend (Campbell and Laherrere, 1991). Kerr (1998) provides data which predict the next oil crisis before 2010. Prudent economists then can argue that a paradigm shift away from fossil fuel-based economies toward less environmentally-hazardous economies makes good money sense, in addition to sound ecological sense (Brown, 2000).

From this timeline, it is evident that the policies of increased support and tax relief for alternative energy development, which were adopted in the late 1970's, subsequently eliminated in the early 1980's, and partially restored through the late 1980's, played a critical role on the global energy stage. Some companies managed to weather the bracing economic storm of shifting policies, others flourished, while still others were not sufficiently well-developed to survive at all and were eliminated from the energy arena. Over a span of perhaps a dozen years, changes in the renewable energy industry proceeded at a rate equaled only by the development in the last 2 decades of the information industry. In fact, experts contend that it will be highly developed information technology capabilities which ultimately will enable alternative energy forms to supplant traditional energy sources (Dunn and Flavin, 2000). These diverse responses by the renewables industry suggest that an examination of the impacts of different policies on research and development of ARES would be useful.

RENEWABLE ENERGY POLICIES: ECONOMIC IMPACTS OF DIFFERENT POLICY DECISIONS

Elsewhere in this volume, the intricacies of how different forms of renewable energy work are painstakingly explained. There can be no question that technological capability drives the approach to economic feasibility for these ARES. A more fundamental question, though, is what enables our technological capability? Irrespective of the source of renewable energy, all of these industries must develop technological sophistication, which arises only from application of ingenuity along with extensive capital expenditures in research and development (Gibbons, *et al.,* 1990). The ability to make those expenditures is largely a function of political willingness and economic opportunities. Once technical sophistication is achieved, economic feasibility must be attained, and this step presents an additional suite of challenges to renewable energy companies. There exists a clear "catch-22" in which renewable energy companies find themselves-small companies cannot justify investment in larger entities without market demand, itself undermined by high prices, which slows further investment ad infinitum (Owen and Chiras, 1995). This section reviews the impact of some types of policy action, as well as specific legislation, on renewable energy development.

Policies governing renewable energy can be divided into categories which reflect emphasis on different stages of the energy industry. Perhaps most familiar to the consumer are initiatives to increase energy efficiency, thereby reducing conventional energy consumption. An early example was the National Appliance Energy Conservation Act (NAECA) of 1987. US homeowners spend an average of $1330 per household annually to operate home appliances—kitchen appliances alone consume about one-fifth of household energy (EREN, US DOE, 1997). NAECA was embraced by appliance manufacturers who faced the burden of trying to comply with variable state standards. One of the few promising moves of the late 1980's, NAECA established federally mandated efficiency standards for most household appliances. It called on manufacturers to produce appliances that use 20% less energy, eliminating nearly 22,000 megawatts of electricity for a proposed savings of nearly $48 billion in electric bills over 20 years (Owen and Chiras, 1995). Current estimates based on newly amended standards place direct savings to consumers at nearly $2 billion, energy savings to date at $20 billion, and projected energy savings of an additional $58 billion by 2015 (EREN, US DOE, 1998).

On a statewide basis, California enacted legislation that cut maximum allowable energy use in new refrigerators in half, beginning in 1992 (Owen and Chiras, 1995). Residents of the state of Oregon who buy designated energy-efficient appliances for their new homes are eligible for a maximum $1000 state income tax credit, initiated in 1998 under an innovative new law targeting residential tax credits (Ohrenschall, 1997a). At the local level, Portland, Oregon has adopted some of the nation's toughest building codes requiring higher insulation in both new homes and existing office buildings—resulting in nearly one-third less energy consumption in the city in 1995 (Owen and Chiras, 1995). Local electric and water utilities in the Pacific Northwest have responded to increasing consumer demand and regional initiatives by offering supplemental rebates ranging from $50-$200 to customers purchasing clothes washers compliant with the "Wash Wise" Program—Wash Wise promotes energy- and water-saving horizontal axis washing machines and associated appliances (Ohrenschall, 1997b).

These examples demonstrate how governmental initiatives at every level can set and favorably impact energy policies targeting decreased consumption through increased efficiency. The resultant energy conservation makes good economic sense as well. Consider that improving efficiency of an appliance may cost between 1 and 2 cents for each kwh saved, but a coal-fired power plant generates energy at 5-7 cents per kwh, while nuclear power may cost 10-12 cents per kwh; with this typical scenario, the cost of conserving traditional fuels may generate 60-90% savings in energy (Owen and Chiras, 1995).

Unfortunately, savings associated with improved efficiency in a free market may undermine overall energy conservation measures by promoting lowered energy costs and increased energy availability, thereby leading to higher consumption rates, as occurred in the early 1970's. Yet, it is still reasonable to suggest that in a better educated 21st century, a reduced demand scenario to govern energy policy is possible (Lovins, 2000). This is evidenced by the rebound from 1989 to 1993 in funding proposed by former President George Bush for the National Renewable Energy Lab (Amato, 1993), as well as proposals by the Clinton administration to increase renewable energy research by 30%—reduced consumption is cautiously though clearly desirable, from ecological, economical and political vantages.

At its most simplistic, a market economy functions with supply and demand being related inversely, though not linearly, to one another. The price of a good or service, coupled with availability, drives the supply/demand relationship such that lowered prices can increase demand, thereby reducing availability and subsequently driving price higher. One assumption for this relationship is that price includes all of the costs for making a good or service available; for example, the cost of a pack of cigarettes includes the cost to the farmer for growing and harvesting the tobacco, as well as costs associated with transport, manufacturing and distribution of the product, with a profit margin included. These costs are considered to be *internalities*, costs which are accounted for in the pricing of an item, but do not include any *externalities*, costs not borne directly by those involved in the industry and not part of the price structure. In this example, externalities include lost work due to smoking-related illness and increased health care and insurance costs, estimates for which range from $1.25 to $3.15 per pack, over and above manufacturing costs (Miller, 1988)! The challenge then becomes converting externalities into quantifiable costs which can be accounted for in the pricing/availability structure which drives supply and demand (Hubbard 1991). Nowhere is this challenge more evident, and more difficult, than in setting energy industry policies beyond increased efficiencies.

Governmental policies operate in the realm of what are currently considered externalities, and include subsidies, or direct grants to private enterprise considered by the government to be of public benefit, as well as indirect costs borne by the government on behalf of consumers. Direct subsidies may be considered a form of incentives, mechanisms which stimulate or motivate a desirable action. Other incentives take the form of tax credits, or reduced tax liability for companies investing in renewable energy, while active disincentives generally take the form of added taxes or penalties. Both subsidies and incentives function at the corporate level, but are also present for the individual consumer, as evidenced by increased tax credits to reduce federal income tax burdens for individuals who install direct solar heating capabilities in existing homes. More importantly, subsidies and incentives operate strongly to select for renewable energy research and development when they are in place, yet their absence in the development of ARES, coupled with their continued presence in the traditional (*e.g.*, fossil fuel and nuclear) energy industry generates market forces which actively discourage investment, research and development of ARES.

The concern of how conservation affects consumption in the supply/demand model has led to some questions regarding what steps are needed to promote renewable energy, or if all energy subsidies should be eliminated to level the playing field and permit true free market forces to function (Hubbard, 1991, Wasserman, 1997; Nebel and Wright, 2000). For example, Lovins (2000) cites estimates that subsidies to conventional fuel and power industries in the United States alone exceed $10 billion annually! Although deregulation of the electricity industry theoretically ought to generate a truly competitive marketplace for energy, practical estimates are that at least half of the nation's 107 nuclear reactors could not compete with wind energy sources (a viable and promising renewable technology) in a completely subsidy-free open market (Wasserman, 1997). The traditional energy industry receives a direct annual subsidy averaging $424 per US household (ASE, 1993). However, that does not include the costs of externalities in the traditional energy industries, costs such as environmental degradation due to acid deposition, atmospheric deterioration and warming, crop losses due to landscape disturbance, public health costs associated with air pollution and occupational risks to employment, or waste disposal and decommissioning costs for closing nuclear power plants (Hubbard, 1991; Pimental, *et al.,* 1994; Casey, 1995).

Promoting development of renewable energy, the Public Utility Regulatory Policy Act (PURPA) of 1978 created a market by requiring utilities to buy power from independent companies capable of producing it with better cost-effectiveness (UCS, 1999). PURPA is credited for contributing to the now 7% of the entire US power market derived from non-utility producers, which are predominantly renewables like solar and wind, and natural gas co-generation plants (UCS, 1999). However, there are no provisions in PURPA to incorporate the costs of negative externalities for traditional energy and corresponding positive benefits of ARES into the "cost-effectiveness" equation. Since PURPA, the Energy Policy Act of 1992 is federal legislation which has encouraged wholesale power competition, while recent Federal Energy Regulatory Commission rulings have opened transmission lines to all generators equally, thereby permitting, but not requiring, free market competition (UCS, 1999).

Market issues are at the crux of attempts to value externalities in the various subsidy programs which favor traditional energy and inhibit development of ARES. The gap between sale prices and actual costs are so large for traditional fuels that it distorts the choices of consumers. While carbon taxes in the United States are deemed political suicide and comprise less than 20% of the cost per gallon of gasoline, they are more accepted as an essential part of marketing elsewhere in the globe; for example, in Germany, 60% of the price for gasoline reflects governmental-imposed taxes, while in France the proportion is

74% (Gibbons, *et al.,* 1990). Unmasking these costs and internalizing them will provide huge incentives for policymakers to support a transition toward ARES (Hubbard, 1991).

While consumer taxes may be a simple and direct method to internalize unaccounted-for indirect costs, they may not always be the best method. Integrated resource planning is a process that attempts to meet energy demands by combining factors that provide the lowest overall cost (Hubbard, 1991). Thus far, 18 states have mandated that utilities provide integrated resource plans (IRP's) which account in some way for externalities, and as many as 26 are in the process of doing so (Hubbard, 1991). In these IRP's, preferential technologies are identified and percentage adders are incorporated to factor in costs such as workforce generation and environmental impacts—the percentage adders tend to be additional costs for traditional energy, while they are often negative (or are credits) for ARES (Hubbard, 1991). Some additional innovative state policies for assessing environmental costs include that in New York, where each kilowatt hour of energy is assessed a penalty based on the source of the energy—a coal-burning plant lacking scrubbers might be penalized by an additional 25%, while cleaner energy sources incur proportionately less penalty (Hubbard, 1991).

Another category of subsidy enjoyed by traditional energy sources is research funding. Fossil fuels supply 86% of US power consumption—with petroleum leading at 39%, natural gas at 24%, and coal at 23%—leaving 14% of consumption to be met by nuclear energy (@6%), hydropower (@5%) and ARES, (@ 3%) (EREN, US DOE 1997). Of the estimated $50 billion of tax credit research funding, $26 billion goes to fossil fuels, another $19 billion subsidizes nuclear power research, and only @ $5 billion is earmarked for research in all forms of renewables (Hubbard, 1991). The research investment is a critical component of generating economic feasibility-in the absence of research and development, it is difficult for some ARES to make the technological advances that will insure they are truly cost-competitive with traditional energy sources. Photovoltaic solar power demonstrates this case in point.

The first large scale solar modules used by the Vanguard I space satellite in 1958 cost more than $600 per watt of generating capacity (Jeffries, 1983). At the onset of the 1970's prices averaged around $100 per watt, but the favorable research investment climate generated by the Carter administration resulted in an order-of-magnitude decline in costs (Jeffries, 1983). By reducing costs to @ $10 per watt prior to the Reagan administration cuts in 1981 (Quinn, 1997), the solar industry was able to offset reduced governmental investment with private investment opportunities, and continued to improve cost effectiveness.

Presently, photovoltaics hover around $4.50 per watt of new generating capacity (Kane, 1993). This is equivalent to a price of nearly $0.30 per kwh of electricity (Kane, 1993; Pimental, *et al.,* 1994); in order to compete effectively in the United States with traditional energy sources, photovoltaics are still about 4 to 6 times more expensive than conventional energy sources, and costs need to decline further to about $0.08 per kwh (Kane, 1993; Pimental, *et al.,* 1994).

Alternatively, photovoltaic efficiency needs to increase from the existing 7-8% energy conversion in commercial units to the 21% achieved in prototype test cells currently being developed (Moore, 1992). Interestingly, global energy costs more accurately reflect hidden costs of fossil fuels, and have resulted in a more favorably competitive environment for photovoltaics worldwide (Quinn, 1997). From 1995 to 1997, global demand for photovoltaic energy surged from 90 to 127 megawatts (Nebel and Wright, 2000). As a result, photovoltaic energy is expected to exhibit growth of 15% or more annually (Nebel and Wright, 2000), converting itself from primarily a microapplications specialty market (*e.g.*, solar digital watches, harbor buoys, highway emergency telephones, *etc.*) and niche market (*e.g.,* centralized desert field generation, inaccessible transmission power grids, *etc.*) to a fully competitive energy player (Gibbons, *et al.,* 1990; Quinn, 1997; Wasserman, 1997). In addition to promising increases in efficiency, cadmium-telluride thin-films, new solarized roofing shingles and integrated solar windowglass loom on the horizon to propel projected growth (Quinn, 1997).

The continued success story of photovoltaic energy might seem more an exception than the rule in ARES policy decisions. Evidence lies in the dramatic upsurge and downward spiral of American wind generated power. The first of nearly 20,000 wind turbines began operating in 1981 in California, where extensive state tax credits eventually contributed to that state bearing 90% of the global wind source market (Wasserman, 1997). In spite of an increase in wind-generated electricity sales of $40 million to $70 million from 1985 to 1986, that favorable tax structure was eliminated in 1986 (Shea, 1988), contributing to the demise of multinational wind energy companies like Luz (a US-Israel venture) and the largest turbine producer in the world, Kenetech Windpower, which filed for bankruptcy protection in 1996 (Wasserman, 1997).

Those companies that have survived have done so primarily through innovative private partnerships with traditional energy companies: the turbine marketing powerhouse Zond is a subsidiary of the Houston-based Enron, a natural-gas supplier (Wasserman, 1997). Enron also owns major interests in Solarex, one of the largest US solar-cell manufacturers (Flavin, 1997). Other traditional energy companies are likewise diversifying into ARES: in 1995, the nuclear-based Bechtel Enterprises teamed with PacifiCorp (coal-fired energy) to form EnergyWorks for investment in solar power, among other renewables (Flavin, 1997). DaimlerChrysler is incorporating a Canadian-developed hydrogen fuel cell design into fleets expected to be marketed in Iceland by 2004 (Dunn and Flavin, 2000), and in 1999 Shell Oil opened its first hydrogen fuel station in Hamburg, Germany (Brown, 2000).

On a less global scale, West Texas Utilities has teamed with Solamotor of Texas to replace commercial diesel- and propane-powered water pumps with solar-powered pumps (Lamarre, 1995). The new cost-competitive systems for ranchers and farmers are relatively trouble free and highly reliable (Lamarre, 1995). The New York Power Authority is engaged in a 9 megawatt wind-power project on Plum Island, while installing a 300 kilowatt photovoltaic system atop a NYC Transit Authority bus garage (Lamarre, 1995). The Solar Two project, with support from the Department of Energy, employs advanced molten salt solar-thermal technology in an experimental online supply to Southern California Edison's power grid (EPRI, 1995). Furthermore, utilities like the Sacramento Municipal Utility District are targeting customer responsiveness, where nearly 70% of surveyed customers were willing to pay \$4-\$6 more monthly for renewable-based electricity (Lamarre, 1995). These conventional utilities are being economically prudent, since experience with ARES will likely be a deciding factor in successful implementation and maintenance of these new technologies when they are viable energy sources (Lamarre, 1995).

RENEWABLE ENERGY POLICIES: GUIDELINES, RECOMMENDATIONS AND CONCLUSIONS

Given the diversity of successful public-private initiatives in renewable energy, the conventional argument that policies promoting environmentally benign practices are bad business no longer have credibility. Renner (2000) cites three major reasons for this: first, projected high job losses associated with increased environmental policy regulations have failed to materialize in any energy sector; second, forcing companies to become technologically advanced actually can confer a competitive advantage in the marketplace; and, third, policies targeting environmentally friendly business practices and regulations have themselves generated a rapidly growing industry employing millions globally. So, while the AFL-CIO argues against alternative energy policies based on projections of devastation in US coalworker jobs, the limitations of fossil fuel reliance have already been dramatically evident in the decline from 242,000 workers in 1980 to 90,000 in 1998 (Renner, 2000). Delaying or avoiding altogether a shift to a renewable energy economy is likely only to further economic declines. A comprehensive renewable energy plan is likely to create more, not fewer, employment opportunities, generating a strengthening, not weakening, of national and global economies.

Among the guidelines for enhancing and furthering policies, using the existing marketplace tools which proved beneficial to conventional energy is essential to ease the perceived abruptness of economic impacts associated with renewable energy. Rather than advocating elimination of all government-based subsidies, this author recommends extension of existing forms of energy subsidies to ARES at the same or higher level as their application in the traditional energy industries. Enacting policies which mandate state or local integrated resource plans provides the opportunity for decentralized, micropower-based renewables efforts which are tailored to the specific needs and concerns of a geographic area (Dunn and Flavin, 2000). IRP's can then incorporate the advances in internalizing the economically and environmentally damaging impacts of conventional energy along with the comparably favorable aspects of renewables and shift the education base for consumers in the direction of the latter.

And make no mistake, Americans do indeed like renewable energy (CSG, 1998). In a survey by the Safe Energy Communication Council, 6 out of 10 voters desire renewable energy and increased efficiency as top priorities for federal research and development funding, irrespective of their affiliation as Independents, Republicans or Democrats (CSG, 1998). Increasingly well-educated consumers see nuclear and fossil fuel power as having had more than their fair share of subsidy, and encourage a shift to renewables research (CSG, 1998). By leveling the energy playing field and identifying promising investment opportunities, policies can enhance the continued growth rates of renewables (Brown, 2000). To promote dissemination of information on governmental incentives and financing opportunities, the US Department of Energy established in 1995 the Database of State Incentives for Renewable Energy (DSIRE), which contains current and extensive information on financial incentives, state programs, regulatory policies, utility programs, even schools going solar programs, all available via traditional and internet access (Rogers, 1998). Examples of private investors interested in advancing renewable energy business capability abound on the internet: the San Francisco-based Investors' Circle has not only provided $44 million dollars for over 100 companies, simply gaining attention (not funds) from the nonprofit generates additional interest from private investors (GEFT, 1999).

Continuing emphasis on improvements in efficiency, like the EPA sponsored Energy Star Program and Green Lights Program (Nebel and Wright, 2000) need to be viewed as efforts which are shorter term—they buy the time needed to make the transition to renewable energy which is essential to a sustainable economy (Pimental, *et al.,* 1994). Brown (2000) cites leadership and time as the desperately scarce resources needed to enable the paradigm shift toward a responsible, renewable and sustainable energy base.

Even as the final costs may remain unknown, the benefits are too large to ignore (Hoffert and Potter, 1997).

LITERATURE CITED

Alliance to Save Energy (ASE). 1993. Federal Energy Subsidies: Energy, Environmental, and Fiscal Impacts. ASE, Washington, DC.

Amato, I. 1993. A New Life For a National Clean Technology Workshop. *Science* 260: 1889-1890.

Brown,L.R. 2000. Challenges of the New Century. pp. 3-21. In: L. Starke (Ed.) *State of the World 2000: A Worldwatch Institute Report on Progress Toward a Sustainable Society*. W.W. Norton and Company, New York, NY, pp 276.

Campbell C.J. and J. H. Leherrere. 1998. The End of Cheap Oil. *Scientific American* 278:78-83.

Council of State Governments (CSG). 1998. Americans Like Renewable Energy. *Spectrum: The Journal of State Government.* 71:28.

Casey, J. 1995. Here Comes the Sun. *Civil Engineering* (October):54-57.

Dunn, S. and C. Flavin. 2000. Sizing up Micropower. pp. 142-161. In: L. Starke (Ed.) *State of the World 2000: A Worldwatch Institute Report on Progress Toward a Sustainable Society.* W.W. Norton and Company, New York, NY, pp 276.

Electric Power Research Institute (EPRI). 1995. Working Together For Renewables: Joint Projects of the Electric Power Research Institute and the Department of Energy. *Electric Power Research Institute Journal* 20:20.

Energy Efficiency and Renewable Energy Network (EREN). 1997. About the Office of Codes and Standards. US Department of Energy (US DOE), retrieved 13 January 2000 from <http://www.eren.doe.gov/buildings/codes_standards/aboffbck.html>.

Energy Efficiency and Renewable Energy Network (EREN). 1998. About DOE's Appliance Standards Program. US Department of Energy (US DOE), retrieved 13 January 2000 from <http://www.eren.doe.gov/buildings/ consumer_information/about.html>.

Flavin, C. 1997. Clean As A Breeze. *Time* (November 1997):46-49.

Gibbons, J.H, P.D. Blair and H.L. Gwin. 1990. Strategies for Energy Use. pp. 85-86. In: J. Piel (Ed.) *Managing Planet Earth.* W.H. Freeman and Company, New York, NY, pp. 146.

Global Environment and Technology Foundation (GEFT). 1999. *Sustainable Business Insider: Business Activism.* pp. 1 -4, retrieved 13 January 2000 from <http>//www.sustainablebusiness.com/insider/april99/busactivism.cfm>

Hubbard, H.M. 1991. The Real Cost of Energy. *Scientific American* 264:36-42.

Hoffert, M.I. and S.D. Potter. 1997. Beam It Down: How the New Satellites Can Power the World. *Technology Review* (October 1997):30-36, adapted from P. Glaser (Ed.) *Solar Power Satellites: A Space Energy System for Earth,* Oxford University Press, New York, NY.

Jeffries, J.P. 1983. The New Alchemy of Photovoltaics. *Sierra* 68:45-49.

Kane, H. 1993. Photovoltaic Sales Growth Slows. pp. 52-53. In: L. Starke (Ed.) *Vital Signs 1993: The Trends That Are Shaping Our Future.* W.W. Norton and Company, New York, pp. 150.

Kerr, R.A. 1998. The Next Oil Crisis Looms Large-and Perhaps Close. *Science* 281:1128-1131.

Lamarre, L. 1995. Renewables in a Competitive World. *Electric Power Research Institute Journal* 20:16-24.

Lovins, A.B. 2000. Soft Energy Paths: Toward a Durable Peace. pp. 119-125. In: T.D. Goldfarb (Ed.) *Notable Selections in Environmental Studies, Second Edition.* Dushkin/McGraw Hill, Guilford, CT, pp. 390.

Miller, J.T. 1988. *Living In the Environment, Fifth Edition.* Wadsworth Publishing Company, Belmont, CA, pp. 603, plus 65 pp. appendices.

Moore, T. 1992. High Hopes for High-Power Solar. *Electric Power Research Institute Journal* 17:16-25.

Nebel, B.J. and R.T. Wright. 2000. *Environmental Science, Seventh Edition.* Prentice Hall, Upper Saddle River, New Jersey, pp. 664.

Ohrenschall, M. 1997a. Making a Mark in the Market. pp.4-6. In: M. Ohrenschall (Ed.) ConWeb: *Pacific Northwest Energy Conservation and Renewable Energy Newsletter* Volume 20, pp. 18. Retreived 13 January 2000 from <http:www.newsdata.com/enernet/conweb20.html>

Ohrenschall, M. 1997b. Oregon Credits Efficient Appliances. pp. 13-15. In: M. Ohrenschall (Ed.) ConWeb: *Pacific Northwest Energy Conservation and Renewable Energy Newsletter* Volume 20, pp. 18. Retreived 13 January 2000 from <http://www.newsdata.com/enernet/conweb20.html>

Owen, O.S. and D.D. Chiras. 1995. *Natural Resource Conservation: Management for a Sustainable Future*. Prentice Hall, Englewood Cliffs, NJ, pp. 586.

Pimental, D., G. Rodrigues, T. Wang, R. Abrams, K. Goldberg, H. Staecker, E. Ma, L. Brueckner, L. Trovato, C. Chow, U. Govindarajulu and S. Boerke. 1994. Renewable Energy: Economic and Environmental Issues. *Bioscience* 44:536-547.

Quinn, R. 1997. Sunlight Brightens Our Energy Future. *The World and I* (March 1997):156-163.

Renner, M. 2000. Creating Jobs, Preserving the Environment. pp.162-183. In: L. Starke (Ed.) *State of the World 2000: A Worldwatch Institute Report on Progress Toward a Sustainable Society*. W.W. Norton and Company, New York, NY, pp 276.

Rogers, H. 1998. *Database For State Incentives for Renewable Energy (DSIRE)*. Retrieved 13 January 2000 from <http://www.ncsc.ncsu.edu/ dsire.html>.

Shea, C. P. 1998. Harvesting the Wind. *Worldwatch* 1 :12-17.

Union of Concerned Scientists (UCS). 1999. Seven Ways to Switch America to Renewable Energy. pp. 1-3, excerpted from Executive *Summary, Powerful Solutions*, Union of Concerned Scientists, Cambridge, MA. Retrieved 13 January 2000 from <http://www.uscusa.org/energy/brief.7ways.html>.

Wasserman, H. 1997. Inherit the Wind: Will Clean Energy Pass Us By? *The Nation* (June 1997):11-15.

Renewable Energy: Trends and Prospects. Edited by S.K. Majumdar, E.W. Miller and A.I. Panah

Chapter Twenty-Nine

THE INTEGRATED SCIENCE AND TECHNOLOGY PROGRAM: Offering a New Paradigm for Teaching Energy Concepts and Technologies

JONATHAN J. MILES[1]
and
CHARLES E. ISKANDER YUOSIF[2]

[1]College of Integrated Science and Technology
James Madison University
Harrisonburg, VA 22807
[2]Institute for Energy Technology
University of Malta
Roman Port Street
Marasaxlokk, ZTN 09, Malta

INTRODUCTION

Today's university graduates are entering a working environment that is becoming more and more interdependent. As the world becomes smaller, and communications and commerce stretch farther across international boundaries, a growing segment of industry strives to capitalize. The energy industry is no exception. All nations share a common problem - to provide adequate energy resources to a demanding population and, in the developed world, to the commercial sector. Although the makeup of energy resources, infrastructures, and available technologies differs between each nation, especially between those developed and those developing, every nation must support education in the energy field in order to maintain or improve its standard of living.

The state university system in Virginia is committed to preparing graduates to interact professionally and personally in the global community. In an effort to provide an international element to higher education in Virginia, James Madison University[1] in 1990 established an International Internship Program with ten positions available in Great Britain and France. Today the program offers more than fifty internships, each semester, worldwide. The basis for these offerings are primarily in business, the arts, government, communications, and social services. The programs are carefully planned, closely supervised, and provide participating students a clearer view of their place in the world, a more thorough understanding of other cultures, and a stronger sense of their potential to contribute in the global marketplace. However, since none of these opportunities deal directly with energy technology and the environment, a program was established in 1998 to fill this void as will be described later in this chapter.

The Program of Integrated Science and Technology[2] (ISAT) at James Madison University was established in 1993. This program is best described as a "unique enterprise among science and technology programs within higher education" that provides students a broad, liberal, science and technology-based education to prepare them to enter technology-based industry in the 21st century. The Blue-Ribbon Panel, appointed in 1989 to develop the overall philosophy of the College of Integrated Science and Technology, recommended a curriculum that would provide:

- a broad, versatile understanding of the world;
- the quantitative and analytical skills necessary for experimentation;
- experience in team investigations;
- understanding of the economic, societal, and ethical contexts of technology decisions;
- solid oral and written communications skills;
- sensitivity to a multi-cultural environment;
- exposure to multi-national influences.

The ISAT program embraces all these elements and aims to produce a science generalist broadly acquainted with the basic sciences, engineering, and business.

All ISAT majors pursue a program that, during the first two years, encompasses the study of the role of science and technology in society, discrete and continuous mathematics, information and knowledge-based systems, statistics, management, organizational behavior, chemistry, physics, and biology. All students then continue with a deeper study of applied topics such as energy, environment, engineering and manufacturing, biotechnology, knowledge management, health systems, and telecommunications. Each student focuses on three of these areas, plus laboratory work in instrumentation and measurement as juniors. Students select a concentration in any one or two of these areas as seniors. The senior year also requires independent study in the concentration culminating in a senior project and thesis.

ISAT Energy Curriculum

The Mission of the *ISAT Energy Sector*[3] is to provide students a broad-based set of skills and a comprehensive body of knowledge pertaining to energy and energy-related business. The student who pursues studies within the energy concentration acquires an understanding of the vast technical, policy, and business aspects of the energy industry, as well as the tools needed to contribute to these aspects of the industry. An emphasis is placed on critical skills of analysis and judgment. Students acquire a solid technical and scientific foundation, enabling them to work with consideration to energy production, supply, consumption, and conservation. This includes conventional, advanced, and renewable energy systems.

Those students who concentrate in energy also develop a foundation in energy policy and economics, providing them with the means to comprehend the structure and dynamics of energy policy on regional, national, and global bases. This knowledge, coupled with technical competence, provides students the ability to develop strategies and make prudent decisions related to energy technologies and problems. Students with an energy concentration possess the knowledge, understanding, and confidence necessary to evaluate situations, make intelligent business decisions related to energy supply, production, and end-use technologies, and appreciate the significance and impact of energy as related to the national economy and the environment.

The Energy Sector can best be summed up by sixteen goals described below by category:

Science and Technology

Goal #1: Students will demonstrate a technical and scientific foundation incorporating the knowledge and understanding of physics and engineering principles needed to work with energy systems.

Goal #2: Students will demonstrate a broad array of skills and tools of experimental sciences as applied to energy systems and technologies.

Goal #3: Students will demonstrate computational and mathematical methods of data analysis as applied to energy systems and technologies.

Goal #4: Students will understand the fundamentals of energy - its forms, transformations, and interactions.

Goal #5: Students will be able to apply the principles of thermodynamics and heat transfer to numerous energy systems and demonstrate appropriate problem-solving skills.

Goal #6: Students will be able to plan and conduct research pertaining to energy systems and technologies.

Goal #7: Students will understand methods for developing dynamic models of energy processes.

Energy, Economics, and Policy

Goal #8: Students will possess a basis for the understanding and critical evaluation and interpretation of energy policies.

Goal #9: Students will understand the relationships, conflicts between energy policy, economics and other societal needs, wants, or desires.

Goal #10: Students will have an understanding and ability to evaluate analytically the various energy types, supplies, and distribution methods.

Goal #11: Students will acquire the basis to understand and critically evaluate the regulation of the energy industry and energy supplies.

Goal #12: Students will acquire a basis to understand the politics of energy.

Goal #13: Students will understand and critically evaluate the economics of energy supply, production and distribution.

Business Decision Making

Goal #14: Students will use economic tools to evaluate energy technologies and energy projects.

Goal #15: Students will demonstrate the ability to analyze, evaluate, and make a sound business decision.

Personal and Professional Skills

Goal #16: Students will acquire a set of personal and professional skills that makes them valued members of a profession or organization.

The ISAT energy curriculum composes a series of courses described below. All ISAT majors take the first course listed (ISAT 212) as part of their general second-year program. Majors who elect energy as one of their three junior *sectors* (areas of more in-depth study) will take the three 3XX courses listed. Majors who elect energy as their senior *concentration* (area of most in-depth study) will take at least four of the 41X courses listed and will develop and carry out a three-semester senior project and thesis that focuses on energy.

Energy Issues in Science and Technology (ISAT 212): The objective of this course is to introduce important issues and scientific concepts relevant to the use of energy in modern society. The purpose of this course is to (a) introduce students to the broad field of energy and its importance in modern technological society; (b) develop a basic understanding of important scientific concepts in preparation for upper-level energy courses; and (c) develop an appreciation of the environmental and economic issues associated with the different energy technologies and the challenges of sustainable energy development.

Instrumentation and Measurement in Energy (ISAT 301): Instrumentation is used to acquire data from representative systems that include mechanical, thermal, solar, chemical and nuclear energy. Students analyze the data to enhance understanding of these forms of energy. Energy transport processes are characterized. Computer-based data acquisition is emphasized.

Energy Fundamentals (ISAT 310): The objective of this course is to introduce the fundamentals of energy, its forms and transformations, and the interactions of energy with matter. The course is intended to provide: a solid framework for the study of energy related subjects such as heat transfer and power generation; an understanding of how different forms of energy differ in quality and how that quality of energy affects the efficiency of energy-consuming devices; guidance in developing sound problem-solving skills, especially for systems involving steady flow; an introduction to a powerful software tool for performing thermodynamic analyses.

Role of Energy in Modern Science (ISAT 311): This course covers the role of energy in the U.S. and world economies; geology of energy-valued natural resources; size, quality, and economics of domestic and world resource base. Models are developed for energy use by different sectors. The role of energy in global climate change, and other energy-related environmental concerns, and the implementations for national and international security are studied.

Sustainable Energy Development (ISAT 410): This course covers the science and the applications of solar and other renewable technologies, e.g. solar thermal electric, photovoltaics, wind power, biomass-derived alcohols, solar hydrogen, and ocean thermal energy conversion. Energy storage systems and materials, combined renewable-conventional systems for peaking and load management, and alternative energy sources for transportation are studied.

Energy Economics and Policy (ISAT 411): Methods for analyzing the economic, environmental, and societal benefits of energy technologies are considered. Topics include: optimization techniques, utility planning and finance, cost-benefit techniques, discounting for time and risk, econometric models, and input-output analysis. The role of government in determining energy costs, supply and markets is also considered.

Dynamic Control of Energy Systems (ISAT 412): This course considers methods for developing dynamic models of energy processes and technologies to achieve improved process control and increased efficiency, with applications of differential equations and discrete mathematical equations. Dynamic models are used to evaluate load management strategies and to develop computer-control algorithms for building energy systems.

Options for Energy Efficiency (ISAT 413): Detailed examinations are made of new technologies to increase the efficiency of energy conversion, transportation systems, and end-use technologies. Examples include MHD, combined-cycle systems, advanced nuclear reactors, intelligent transportation systems, high-efficiency lighting, energy management, and utilization of low-temperature heat. Socioeconomic and governmental barriers to energy efficiency are considered.

International Energy Studies (ISAT 416): A study-abroad course that examines international energy problems and provides a team-oriented, technical project experience. Energy issues associated with economic and social development are also addressed. Project participation, tours, and arranged meetings with local experts are used to illustrate energy-related problems that apply globally and which are compared and contrasted with those in the United States. Projects undertaken may be of national relevance to the host nation.

Thermal Processes in Energy Systems (in development)

Software Development and Applications for Energy Analysis (in development)

Each of the courses listed provides 3 credit-hours with the exception of ISAT 301, solely a laboratory course providing 1 credit-hour. ISAT 212 involves lecture sessions as well as laboratory and issues-based sessions. All other courses are lecture-based but benefit from hands-on activities available in a dedicated, comprehensive energy laboratory. This laboratory also provides additional resources for energy-based senior projects and theses.

ISAT Study-Abroad Program in Malta

The course listed as ISAT 416 in the previous section, *International Energy Studies*, and a corresponding study-abroad program[4] (described below) offered at JMU, was developed to address the demand among ISAT majors for an international educational experience that is science- and technology-based. James Madison University, like many universities, maintains a strong tradition of supporting international programs throughout the world. However, with few exceptions, offerings for semester- and summer-term programs are limited to the arts and humanities. The *ISAT in Malta Summer Program* was developed specifically to address the lack of opportunity for ISAT students to gain a study-abroad experience that focuses on applied science and technology. The program focuses on energy and environment and was offered for the fourth consecutive year during a five-week period in May and June 2001.

The ISAT in Malta Summer Program provides the following educational benefits to students:

- hands-on experience in the development and implementation of renewable-energy projects that are of importance to Malta and the Mediterranean region;
- access to a national energy picture small enough in scale to see all components working together;
- experience and understanding associated with living and working in an international setting - students live in university housing and are in close contact with study abroad students from other countries and the Maltese;
- applied, interdisciplinary, team project experience that integrates the technical, social, and economic dimensions of real-world problem-solving.

The program is conducted as a partnership between James Madison University and the University of Malta, but is essentially a joint effort of ISAT and Malta's *Institute for Energy Technology*[5] (IET), a center for teaching, research, and demonstration projects related to sustainable energy technologies and energy efficiency and conservation. The IET was established in 1988 with the following objectives:

- to analyze the use of energy;
- to determine feasible measures to conserve energy;
- to apply renewable sources of energy;
- to originate and participate in teaching and research projects;
- to collaborate with other universities, industries, and international bodies.

The program expanded in 1999 to include projects that focus on environment-related issues and concerns, and established ties with Malta's *Gaia Foundation for the Protection of the Environment* and again in 2000 to address biotechnology related issues. However, only the energy-related aspects of the program are discussed here.

The *ISAT in Malta Summer Program* comprises four components intended to provide students a complete and comprehensive international experience that stresses not only an increased technical competence, but also an enriched cultural perspective. These components include the development, completion, and presentation of a technical project, participation in a technical lecture/tour series, participation in a cultural lecture /tour series, and weekly planned extracurricular activities.

PROJECTS

The core of the *ISAT* in *Malta Summer Program*, and the basis for the majority of the 3 credit hours earned, are projects that are of particular national or academic interest to the Institute for Energy Technology. Students are selected during the fall semester prior to each program and assigned to a project group on the basis of their expressed interests. A project group works informally with a mentor during the spring semester prior to develop goals, objectives, and a work plan. A project group consists of 3 to 5 students working jointly.

Upon arrival at IET in May (the program takes place during JMU's first summer session following final exams in May) students implement their work plan over four 24-hour work weeks. Teamwork is considered of utmost importance, and supervision is provided without undermining the creativity and initiatives of the individual participants. The nature of projects changes from year to year according to the resources available at the institute; however, students are free to choose the area in which they prefer to work and provide considerable input regarding the details of the project.

In general, the first week of the program is dedicated to the teaching of theory and review of project objectives. In some cases, this involves some lecture, demonstrations, laboratory training, and field visits. The middle two weeks are spent on the formulation and execution of the project as well as the writing of a report. The final week is dedicated to the analysis of results, completion of a written report, and an oral presentation of the project to a general audience at the University of Malta. Each group presents their work orally a second time in the following September to an audience at JMU. The overall program time line is depicted on the following page.

A list of the projects carried out through 2000 is provided.

- Modelling and feasibility study of a hybrid wind/PV electricity generating system (1998).
- Modelling of active and passive solar water-heating systems using TRNSYS(tm) software (1998, 1999).
- Engineering design of an experimental hot box apparatus for thermal measurements of building materials (1998).
- Site analysis and visual impact assessment for wind power (three sites studied, 1998, 2000).
- Economics study of implementation of photovoltaics on rooftops (1999).
- Development and testing of data monitoring system using LabView(tm) software (1999).
- Installation and testing of a photovoltaic tracking system (2000).

Fall Semester prior to program	Students are selected to participate in program based on competitive application (October) and review (November). Project groups are established (December).
Spring Semester prior to program	Each group appoints a student coordinator and develops project goals, objectives, and work plan. This phase includes literature survey, reading, and other activities to ensure adequate knowledge and understanding for each project.
Week 1 of program	Review of project goals and objectives, training, initial field work.
Week 2 of program	Continued field work as required, execution of project work plan.
Week 3 of program	Continued field work and execution of project work plan, begin development of final written report.
Week 4 of program	Complete final project tasks and written report, prepare and deliver oral presentation.
September following program	Project groups present their experience and deliver oral presentations at JMU.

TECHNICAL LECTURE/TOUR SERIES

The *ISAT in Malta Summer Program* incorporates weekly lectures presented by noted experts in the field. The program also provides tours conducted at various plants that bear special significance with respect to energy production, consumption, and conservation. The plant visits are led by trained experts and provide students the opportunity to see large-scale, energy-intensive facilities in operation. The tours help reinforce the energy concepts taught and courses completed to date within the ISAT program, provide a more complete picture of the overall energy industry and burden in Malta, and motivate the students to consider real-world applications of the concepts learned. The 2000 program schedule of technical lectures and tours is provided below.

Lectures:

- *Renewables in Europe*, Mr. Mario Fsadni, Executive Director of the Institute for Energy Technology, Malta.

- *Renewable Energy in Developing Countries*, Professor Edward Scerri, Chairman of the Institute for Energy Technology, Malta.
- *The Energy Industry in Malta*, Professor Robert Ghirlando, Chairman of Enemalta.

Tours:

- **Delimara Power Plant** - The Delimara plant maintains two oil-fired boilers for base load operation, gas-turbine units for peak power production, and a newly-commissioned combined-cycle unit.
- **Pembroke Reverse Osmosis Plant** - Malta depends upon a network of four desalination plants located around the perimeter of Malta to provide over 50 percent of its potable water supply. Reverse osmosis accounts for 10-15 percent of the electricity demand in Malta. The Pembroke plant was commissioned during the 1990s and features two modern, independently-operating systems.
- **Farsons Brewery** - The national brewery added a facility during the 1990s that features building envelope technology designed specifically to minimize cooling costs. The design includes double-wall construction and computer-controlled dampers and windows to enhance passive cooling of the building at night and minimal infiltration of heat during the day.

CULTURAL LECTURE/TOUR SERIES

In order to enhance the appreciation of living and working in an international setting, a cultural program of lectures and tours was instituted for the first time in 2000. This cultural series provides students an expert view of features of Malta and the Mediterranean region, raising their awareness and appreciation of a very unique and rich civilization. The 2000 program schedule of cultural lectures and tours is provided below.

Lecture	*followed by*	**Tour**
Maltese History, Language, and Culture		Capital City of Valletta
Maltese Archeology		Neolithic Temples
Antiquity to the Late Middle Ages		Cities of Mdina and Rabat, Catacombs
Order of St. John in Malta		Inquisitor's Palace, Maritime Museum
Maltese Economy and Future Prospects		

EXTRACURRICULAR ACTIVITIES

Weekends during the *ISAT in Malta Summer Program* are occupied, in part, by organized group activities. These include day trips to sites of interest in Malta, an overnight trip with guided tour to the sister island of Gozo, snorkeling and scuba diving excursions, coastal cruises, and a theatrical production. Students are also allotted sufficient free time on weekends and evenings to explore independently.

DISCUSSION

The Program of Integrated Science and Technology at JMU was created to address the need within the State of Virginia, and across the United States and abroad, for qualified and motivated graduates to contribute in the industrial sectors pertaining to science and technology. Energy was identified at the inception of the program as a critical technology in the 21st Century, and thus a new and comprehensive energy curriculum was established and continues to develop. The *ISAT in Malta Summer Program* extends the energy curriculum across international boundaries, providing ISAT majors the opportunity to appreciate many additional aspects of energy technology and industry and to gain an international perspective.

Since 1997, eighty ISAT students and five ISAT faculty members have participated in this program. A number of those participating were influenced to the extent that they chose to concentrate in energy and seek work, following graduation, in the energy field. Two students who participated in the 1998 offering of the program are now employed with the United States Department of Energy. Their comments regarding the influence of the *ISAT in Malta Summer Program* on their careers are provided.

Mr. Brent Beerly, a program manager with the U.S. Department of Energy, states, "My experience in Malta stimulated my thoughts to take up the subject of renewable energy seriously and to strive to find a job in that field upon my graduation. Without doubt, the Malta experience (along with my senior thesis that was an extension of my work in Malta) was the key factor in gaining the position of Program Manager for the U.S. Department of Energy's Wind Powering America initiative." Mr. Beerley today is employed by Community Energy, Inc.

Mr. Joseph Mariano, also a program manager for the U.S. Department of Energy, states, "The Malta program was indeed a real-world experience that instilled in me the beauty and applicability of solar photovoltaic energy technologies. My employers were impressed with the fact that I had some experience in an international environment. It has also helped me in my current job with the United States Department of Energy, as a Program Manager for the Green Energy Parks Initiative, and in my contribution to Wind Powering

America, Million Solar Roofs and remote-powering programs."

The 2001 program added a new initiative through which the Institute for Energy Technology cooperates with Enemalta, the government entity responsible for electricity production and distribution in Malta - thus forging a new, permanent working relationship among energy professionals in Malta. This project involved the development of measures to enhance energy conservation in the commercial sector and to promote energy-saving techniques in Malta.

The development and application of renewable energy is very popular with the new generation of scientists and engineers. International collaboration enhances technology transfer and promotes raised awareness on the parts of both the students from the United States participating abroad and the sponsors and citizens of Malta who support the *ISAT in Malta Summer Program*. In the process of earning credit-hours, students are provided the opportunity to design and implement renewable energy applications, gain new knowledge of both scientific and cultural significance, and enjoy a number of unique recreational activities.

Most importantly, students return from the *ISAT in Malta Summer Program* with a fresh new perspective, having participated in an experience they regard unanimously as one of the most memorable of their college careers. Also, many will convert their experience to advanced study of energy technologies as seniors and a pursuit of graduate degrees and career goals in the field of energy.

REFERENCES

Iskander, C., E. Scerri, J. J. Miles. July 2000. Integrating Renewable Energy Studies In Under-Graduate Curriculum: The IET/ISAT Experience. Proceedings of the World Renewable Energy Congress - VI. Brighton, United Kingdom. Part IV, 2454-2457.

FOOTNOTES

1. http://www.jmu.edu
2. http://www.isat.jmu.edu/
3. http://www.isat.jmu.edu/energy/energy.htm
4. http://www.isat. jmu.edu/users/milesjj/malta/malta.htm
5. http://home.um.edu.mt/ietmalta/

Renewable Energy: Trends and Prospects. Edited by S.K. Majumdar, E.W. Miller and A.I. Panah

Chapter Thirty

UNITED STATES' RENEWABLE ENERGY LAWS AND REGULATIONS

E. WILLARD MILLER

E. Willard Miller
Department of Geography
The Pennsylvania State University
University Park, PA 16802

INTRODUCTION

The Industrial Revolution was founded on the availability of cheap, abundant fossil fuels. These vast resources have made their consumption nearly invisible and waste painless. As a consequence, the only renewable energy resource developed was hydropower. In 1952 the prestigious U.S. President's Material Policy Commission observed, "Efforts to date to harness solar energy are infinitesimal. It is time for aggressive research in the whole field of solar energy, an effort in which the United States could make an immense contribution to the welfare of the Free World." This recommendation was ignored not only by government agencies but also by public energy corporations.

In 1955 when the International Solar Energy Society was established, only twelve people were present at the meeting and only two solar equipment companies existed in the world. During the next decade, fewer than thirty solar energy homes were built in the United States. In the 1950s the photovoltaic cell was thought to be applicable for space use only. A number of windmill experiments in the Great Plains and New England failed in the 1950s as government sponsored, rural electrification programs developed. It was only in the energy crisis of the 1970s that it became evident to the general public that the fossil fuels were finite. The quadrupling of the price of oil and the temporary shortage of gasoline due to the embargo on oil imports had a dramatic impact on the general public.

In response, Congress recognized for the first time that renewable energy could play a role in the total energy system. To encourage development, Congress passed a number of acts considering the development of specific renewable energy resources. In addition, when the Energy Research and Development Administration was formed, renewable energy became one of its six programs. However, the renewable energy program was funded at a minimal level, receiving only $14.8 million out of a total budget of $1,034 billion. Nevertheless, this marked the initial federal effort. Most of the early federal activity involved research in three interrelated areas: technology development, breaking down commercialization barriers, and the establishment of small-scale demonstration programs.

President Carter placed a high priority on both conservation and the development of renewable energy programs. To illustrate, the Solar Energy Research Institute (SERI) set a goal of 2.5 million homes to be solar energy homes by 1985.

In response, two initiatives evolved:

1. The energy tax act provided federal credit to help reduce the initial cost barrier in the use of solar energy. Credits of up to $2000 were provided to homeowners who purchased an active solar or wind energy system for use in their principal residence. Ten to thirteen percent tax credits were also provided for business installations. The federal excise tax was also canceled on gasoline that contained 10 percent ethanol alcohol content.
2. The Public Utility Regulatory Policies Act (PURPA) required utilities to interconnect with small independent electrical producers and purchase their electricity at prices equal to that of the utilities' "full cost." This encouraged the development of the small solar and wind energy facilities that provided only a small amount of electricity. For the first time, these independent, small producers were guaranteed a market for their product.

By the late 1970s, governmental legislation provided the foundation for the growth of renewable energy. From 1976 through the early 1980s, renewable energy achieved real growth and wide expectations. To illustrate, the Federal Energy Regulatory Power Commission received many requests for permits to develop small hydropower sites across the nation. Wood used for energy increased by at least two-thirds. By 1981, nearly 200,000 residential solar heating panels had been installed in homes, generating over $600 million in sales. Almost all manufacturers of prefabricated housing offered passive solar options. By 1981, it was estimated that total renewable energy production totaled 5.0 quads (one quad equals one quadrillion Btus).

In the 1980s, however, with the decline in the price of a barrel of oil to around $20.00 or lower, the emphasis changed in the utilization of energy. Energy conservation became the key to energy utilization, and, because renewable energy was more costly than traditional fossil fuels, the use of renewable energy declined. Most significant were government policy changes. The Reagan administration feared that a change in energy types could cause an economic recession. Although Congress authorized $525 million for renewable energy development for 1981-83, the Reagan administration reduced the size of the appropriation by 95 percent and then impounded the remaining funds. In addition, the Reagan administration opposed the extension of the important tax incentive applied to the use of renewable energy enacted in the early 1980s. This abolishment of these incentives severely affected the renewable energy companies. As a result, many companies were forced into bankruptcy. Technological advances also essentially closed. In 1989 and 1990, Congress passed a number of acts to renew efforts to use renewable energy, but none received adequate funding for effective implementation.

After nearly ten years of decline, the renewable energy industry began its recovery in the early 1990s. The federal government has played no role in this recent development. The increased demand for energy in the Third World countries that were poor in the fossil fuels sparked the revival of the renewable energy industry. The increased demand provided the incentive for rapid technological innovation. While the growth of the use of renewable energy in the Third World in the 1990s has been most significant, the lowered costs have made it possible for renewable energy production to be cost effective in the Industrial World. Market forces have thus driven the recent developments in renewable energy.

RENEWABLE ENERGY LEGISLATION

Energy Security Acts

Biomass Energy and Alcohol Fuels Act of 1980 (Public Law 96-294, 94 Statute 683)

Renewable Energy Resources Act of 1980 (Public Law 96-294, 94N Statute 715)

Energy Security Act of 1980 (Public Law 96-294, 94 Statute 611)

This act was passed to guarantee the energy needs of the United States. It was based on the findings of Congress that:

1. The achievement of energy security for the United States is essential to the health of the national economy, the well-being of our citizens, and the maintenance of natural security.
2. Dependence on foreign energy sources can be significantly reduced by the

production from domestic resources of the equivalent of at least 500,000 barrels of crude oil per day of synthetic fuel by 1987 and of at least two million barrels of crude oil per day of synthetic fuel by 1992.

3. Attainment of synthetic fuel production in the United States in a timely manner and in a manner consistent with the protection of the environment will require financial commitments beyond those expected to be forthcoming from non-governmental capital sources and existing government incentives.
4. Establishment of an independent federal entity of limited duration which will provide additional financial assistance in conjunction with private sources of capital to assist the development of domestic non-nuclear energy resources for the production of synthetic fuel which will facilitate the expeditious achievement of synthetic fuel production from domestic resources.

The purpose of this act can be served by:

1. Demonstrating at the earliest feasible time the practicality of commercial production of synthetic fuel from domestic resources employing the widest diversity of feasible technologies.
2. Fostering the creation of commercial synthetic fuel production facilities of diverse types with the aggregate capability to produce from domestic resources in an environmentally acceptable manner.
3. Creating the United States Synthetic Fuels Corporation, a federal entity of limited duration formed to provide financial assistance to undertake synthetic fuel projects.
4. Providing for financial assistance to encourage and ensure the flow of capital funds to those sectors of the national economy which are important to the domestic production of synthetic fuel.
5. Encouraging private capital investment and activities in the development of domestic sources of synthetic fuel and to foster competition in the development of the nation's synthetic fuel resources.
6. Encouraging and supplementing, but not competing with or supplementing private capital investments in the development of domestic sources of synthetic fuel.
7. Fostering greater energy security and reducing the nation's economic vulnerability to disruptions in imported energy supplies.
8. Giving special consideration to the production of synthetic fuel which has national defense applications and expediting its initial development through the Defense Production Act of 1950.

Title I was devoted to the establishment of the United States Synthetic Fuels Corporation under the United States Fuel Corporation Act of 1980. Its primary purpose was to create the financial arrangements to develop the synthetic fuel industry.

Title II is cited as the Biomass Energy and Alcohol Fuels Act of 1980.

The purpose of this act was to develop:

> A national program for increased production and use of biomass energy that does not impair the nation's ability to produce food and fiber on a sustainable basis for domestic and export use must be formulated and implemented within a multiple-use framework.

Title III considered Energy Targets.

Title IV is cited as the Renewable Energy Resources Act of 1980. The purpose of this act was:

> To establish incentives for the use of renewable energy resources, to improve and coordinate the dissemination of information to the public with regard to renewable energy resources, to encourage the use of certain cost-effective solar energy systems and conservation measures by the federal government, to establish a program for the promotion of local energy self-sufficiency, to broaden the existing programs for accelerating the procurement and use of photovoltaic systems, and to provide further encouragement for the development of small hydroelectric power projects.

Title V is known as the Solar Energy and Conservation Act of 1980 and the Solar Energy and Conservation Bank Act of 1980.

Title VI was intended as the Geothermal Energy Act of 1980. This act was based on Congress's finding that:

1. Domestic geothermal reserves can be developed into regionally significant energy sources providing for the economic health and national security of the nation.
2. There are institutional and economic barriers to the commercialization of geothermal technology.
3. Federal agencies should consider the use of geothermal energy in government buildings.

Title VII considers Energy Conservation for Commercial Buildings and Multifamily Dwellings.

Title VIII amends the Strategic Petroleum Reserve portion of the Energy Policy and Conservation Act.

Regrettably, most of these acts were not funded, and implementation has been minimal.

Renewable Resources Extension Act

Renewable Energy and Efficient Technology Competitiveness Act of 1989 (Public Law 101-218, 103 Statute 1859)

This act provides federal assistance and leadership to a program of research, development, and demonstration of renewable energy and energy efficient technologies. The secretary of energy is empowered to develop this program to ensure a stable and secure future energy supply by:

1. Achieving, as soon as practicable, cost competitive use of those technologies without need of federal financial incentives.
2. Establishing long-term federal research goals and multi-year funding levels.
3. Directing the Secretary of Energy to undertake initiatives to improve the ability of the private sector to commercialize in the near term renewable energy and energy efficient technologies.
4. Fostering collaborative research and development efforts involving the private sector through government support of a program of joint ventures.

A number of specific programs included:

Wind — The goals for the Wind Energy Research Program included improving design methodologies and developing more reliable and efficient wind turbines to increase the cost competitiveness of wind energy. Research efforts shall emphasize solving of near-term technical problems, developing technologies such as advanced airfoils and variable speed generators to increase wind turbine output, increasing the basic knowledge of aerodynamics, structural dynamics, and electrical systems, and improving the compatibility of electricity produced from wind fans with conventional utility processes.

Photovoltaics — The goals for the Photovoltaic Energy Systems Program shall include improving the reliability and conversion efficiencies of, and lowering the costs of photovoltaic conversion.

Solar Thermal Energy — The goal of the Solar Thermal Energy Systems Program is to advance research and development to a point where solar thermal technology is cost competitive with conventional energy sources, and to provide the integration of this technology into the production of industrial process heat and the conventional utility network. Research and development shall emphasize development of a thermal storage technology to provide capacity for shifting power to periods of demand when full insolation is not available, improvement in receivers, energy conversion devices, and innovative concentrations using stretch membranes, levels, and other material; and exploration of advanced manufacturing techniques.

Other Renewable Technologies — These include Biofuel Energy, Hydrogen Energy, Solar Building Energy, Ocean Energy, Geothermal Energy, Low-Head Hydro, and Energy Storage Systems.

The act encourages joint ventures to improve coordination in technology development among industries attempting to commercialize renewable energy. There is also a provision to establish a program to inform other countries of the benefits of these policies and to encourage the export of renewable industries technology.

Solar, Wind, Waste, and Geothermal Power Production Incentives Act of 1990 (November 15, 1990, Public Law 101-575, 104 Statue 2834)

This act encourages, solar, wind, waste, and geothermal power production by removing the size limitations contained in the Public Utility Regulatory Policies Act of 1978.

For a complete listing to the renewable energy acts, see: *United States Code Annotated, General Index, Popular Name Table*. St. Paul, MN: West Publishing Co., Published yearly.

HYDROELECTRIC LEGISLATION

Federal Power Acts

The first Act was passed in 1920 and since then has been amended sixteen times.

Federal Waste Power Act (Chapter 285, 41 Statute 1063)

This act established the Federal Power Commission. It was authorized to:

1. Make investigations and to collect and record data concerning the utilization of the water resources of the region to be developed, the waterpower industry and its relation to other industries and to interstate and foreign commerce, and covering the location, capacity, development costs, and relation to markets of power sites; and whether the power from government dams can be advantageously used by the United States for its public purposes, and what is the face value of such power.
2 Make public, from time to time, the information secured.
3. Issue licenses to citizens of the United States for the purpose of construction, operating, and maintaining dams, water conduits, reservoirs, power houses, transmission lines, or other project works necessary or convenient for the development and improvement of navigation.

Each applicant for a license had to conform to a number of requirements.

These included:

1. The preparation of maps, plans, specifications, and estimates of cost as required for a full understanding of the proposed project.
2. A comprehensive scheme of improvement and utilization for the purposes of navigation of waterpower development, and of other beneficial public uses.
3. No substantial alteration or addition, not in conformity with the approved plans, shall be made to any dam or other project without approval of the commission.

4. Annual payment had to be made by the licensee.

This act initiated the federal government's involvement in the development and utilization of the water resources of the United States for production of electricity and navigability of rivers.

Solar Energy Research, Development, and Demonstration Acts

Solar Energy Research, Development, and Demonstration Act of 1974 (Public Law 83-473, 88 Statute 1431)

This act authorizes a vigorous federal program of research, development, and demonstration to ensure the utilization of solar energy as a viable source for our material energy needs.

The act was based on congressional findings that:

1. The needs of a viable society depend on an ample supply of energy.
2. The current imbalance between domestic supply and demand for energy is likely to be present for some time.
3. Dependence on non-renewable energy resources cannot be continued indefinitely, particularly at current rates of consumption.
4. It is in the nation's interest to expedite the long-term development of renewable and non-polluting energy resources, such as solar energy.
5. The various solar energy technologies are today at widely differing stages of development, with some already near the stage of commercial application and others still requiring basic research.
6. The early development and export of viable equipment utilizing solar energy, consistent with the established preeminence of the United States in the field of high technology products, can make a valuable contribution to our balance of trade.
7. The mass production and use of equipment utilizing solar energy will help to eliminate the dependence of the United States upon foreign energy sources and promote national defense.

This act establishes the Solar Energy Coordination and Management Project. This project has the overall responsibility for the provision of effective management and coordination with respect to a national solar energy, research, development, and demonstration project, including:

1. The demonstration and evaluation of this resource base, including temporal and geographic characteristics.
2. Research and development of solar technologies.
3. The demonstration of appropriate solar energy technologies.

Solar Heating and Cooling Demonstration Act of 1974 (Public Law 93-409,

88 Statute 1069)

This act provided for the early development and commercial demonstration of the technology of solar heating and/or combined solar heating and cooling systems.

In passing this act, Congress found that:

1. The current imbalance between supply and demand for fuels and energy is likely to be present for some time.
2. The early demonstration of the feasibility of using solar energy for the heating and cooling of buildings could help to relieve the demand upon present fuels and energy supplies.
3. The technologies for solar heating are close to the point of commercial application in the United States.

The act stated that the policy of the United States was to demonstrate within a three-year period the practical uses of solar heating technology.

Demonstrations were to be conducted on:

1. Heating systems to be used in residential dwellings.
2. Heating and cooling systems for commercial and residential buildings.

Solar Photovoltaic Energy Research, Development, and Demonstration Act of 1978 (Public Law 95-590, 92 Statute 2513)

This act provides legislation for an accelerated program of research, development, and demonstration of solar photovoltaic energy technologies leading to early competitive commercial applicability of such technologies to be carried out by the Department of Energy, with the support of the National Aeronautics and Space Administration, the National Bureau of Standards, the General Services Administration, and other federal agencies.

The act was prompted by the growing energy problem. Congress thereby found that:

1. The early development and widespread utilization of photovoltaic energy systems could significantly expand the domestic energy resource base of the United States, thus lessening its dependence on foreign supplies.
2. The establishment of sizable markets for photovoltaic energy systems will justify private investment in plant and equipment necessary to realize the economies of scale, and will result in significant reductions in the unit costs of these systems.
3. The use of solar photovoltaic energy systems for certain limited applications has already proved feasible.
4. There appear to be no technical obstacles to the widespread commercial use of solar photovoltaic energy technologies.

5. An aggressive research and development program should solve existing technical problems of solar photovoltaic systems and, supported by an assured and growing market for photovoltaic systems during the next decade, should maximize the future contribution of solar photovoltaic energy to the nation's future energy production.
6. It is the proper and appropriate role of the federal government to undertake research and development and demonstration programs in solar photovoltaic energy technologies, and to supplement and assist private industry and other entities and thereby the general public, so as to hasten the general commercial use of such technologies.
7. The high cost of imported energy sources impairs the economic growth of many nations which lack sizable domestic energy supplies and are unable to develop their resources.
8. Photovoltaic energy systems are economically competitive with conventional energy resources for a wide variety of applications in many foreign nations at the present time, and will find additional applications with continued cost reduction.
9. The early development and export of solar photovoltaic energy systems, consistent with the established preeminence of the United States in the field of high technology products, can make a valuable contribution to the well-being of the people of other nations and the nation's balance of trade.
10. The widespread use of solar photovoltaic energy systems to supplement and replace conventional methods for the generation of electricity would have a beneficial effect upon the government.
11. To increase the potential use of solar photovoltaic energy systems in remote locations, programs leading to the development of inexpensive and reliable systems for the storage of electricity should be pursued.

The objectives of this act were to:

1. Double the production of solar photovoltaic energy systems each year during the decade starting with the fiscal year 1979, so as to reach a photovoltaic energy production of approximately two million peak kilowatts and a total cumulative production of several systems of approximately five million peak kilowatts by fiscal year 1988.
2. Reduce the average cost of installed solar photovoltaic energy systems to $1 per peak watt by fiscal year 1988.
3. Stimulate the purchase by private buyers of at least 90 percent of all photovoltaic systems produced in the United States during fiscal year 1988.

The Secretary of Energy was to implement the program by acceleration of research, development, and demonstration. A Solar Photovoltaic Advisory Committee was nominated to advise the Secretary of Energy on all aspects of the project. The Secretary was authorized to encourage to the maximum extent practicable, international participation and cooperation in the development and utilization of photovoltaic systems.

Wind Energy Systems Act of 1980 (Public Law 96-345, 94 Statute 1139)

This act provided for an accelerated program of wide energy research development and demonstration, to be carried out by the Department of Energy and with the support of the National Aeronautics and Space Administration and other federal agencies.

Congress based passage of this act on:

1. The nation's best interests being served by providing opportunities for the increased production of electricity from renewable energy sources.
2. The early widespread utilization of wind energy for the generation of electricity and for mechanical power could lead to relief on the demand for existing nonrenewable fuel and energy supplies.
3. The use of large wind energy systems for certain limited applications is already economically feasible.
4. An aggressive research, development, and demonstration program to accelerate widespread utilization of wind energy should solve existing technical problems of converting wind energy into electricity and mechanical power.
5. It is the role of the federal government to undertake research and development, to participate in demonstration programs for wind energy systems, and to assist private industry and the general public in fostering the widespread utilization of such systems.
6. The widespread use of wind energy systems will have a beneficial effect on the environment.
7. The evaluation of the performance and reliability of wind energy technologies can be expedited by the testing of prototypes under carefully controlled conditions.

This act declared the policy of the United States to:

1. Reduce the average cost of electricity produced by installed wind energy systems by the end of fiscal year 1988 to a level competitive with conventional energy sources.
2. Reach a total megawatt capacity in the United States from wind energy systems, by the end of fiscal year 1988, of at least 800 megawatts, of which at least 100 megawatts are provided by small wind energy systems.
3. Accelerate the growth of a commercially viable and competitive industry to make wind energy systems available to the general public as an option in order to reduce national consumption of fossil fuels.

Under the act, the Secretary of Energy was to implement the program. As part of the project, the Secretary was to initiate a three-year national wind resources assessment program.

The program was to include:

1. Development of activities to validate existing assessment of known wind resources.
2. Performance of wind resource assessments in regions of the United States where the use of wind energy may prove feasible.
3. Initiation of a general site prospecting program.
4. Establishment of standard wind data collection and siting techniques.
5. Establishment, with other agencies, of a national wind data center which shall make public information available on the known wind energy resources of various segments throughout the United States.

Little has been accomplished to implement this legislation.

GEOTHERMAL ENERGY LEGISLATION

Geothermal Steam Act of 1970 (Public Law 91-581, 84 Statute 1566)

This act authorizes the Secretary of the Interior to develop geothermal steam and associated geothermal resources. A number of provisions are:

1. If lands are to be leased under this act and are within any known geothermal resource area, they shall be leased to the highest responsible qualified bidder by competitive bidding. Other areas may be leased without competitive bidding.
2. Mining claims are limited to 10,240 acres.
3. A royalty of not less than 10 percent or more than 15 percent of the amount of the value of steam, or any other form of energy, will be paid.
4. A royalty of not more than 5 percent of the value of any by-product derived from steam production will be paid.
5. Geothermal leases shall have a primary term of 10 years, and no longer than 40 years, before renewal.

Geothermal Steam Act Amendment of 1988

Geothermal Energy Research, Development, and Demonstration Acts

Geothermal Energy Acts

CONCLUSION

There is general agreement that there will be an increase in the use of renewable energy in the 21st century. However, as long as fossil fuels are available, renewable energy will be only a very small portion of the total energy requirements. Only after a national energy policy is established will renewable energy supply a significant portion of the nation's energy requirements. Only after an energy crisis, such as occurred in the 1970s, will a national energy policy evolve. Until then, all energy will be consumed as if none is finite.

Renewable Energy: Trends and Prospects. Edited by S.K. Majumdar, E.W. Miller and A.I. Panah

Chapter Thirty-One

RENEWABLE ENERGY: TRENDS AND PROSPECTS IN NEPAL

P.K. JHA, S.K. GHIMIRE
and
P. LACOUL[1]

Central Department of Botany,
Tribhuvan University, Kirtipur, Kathmandu, Nepal
[1]Biology Department, Kathmandu University, Dhulikhel, Nepal

ABSTRACT

The annual per capita energy consumption in Nepal is about 13 to 15 GJ, which is almost the lowest in Asia. About ninety percent of the population in Nepal who live in rural areas depend predominantly on conventional sources of energy, i.e. fuel wood, agriculture residue and animal waste. Even the urban population is not independent of these traditional energy sources. Of the total energy used, 88 percent is met from traditional sources and remaining 12 percent from commercial sources. Ninety percent of the total energy is used in the domestic sector for cooking, lighting, spaceheating, water heating, operation of household appliances and carrying out agro-processing activities. Biogas, micro-hydro, solar, and wind energy are alternative energy sources being utilised in Nepal. So far, the share of alternative energy supply is negligible in the national context, although they have good potential for utilisation in future. Hydropower development is a major element in government policies and strategies. Consequently, the hydropower sector has received high priority in the government's plan. The major objective of energy plans is to gradually replace imported fuels by indigenous energy sources. Recent policies and plans have considered energy development as an industry and have given more stress on the involvement of the private sector for development and promotion of the energy sector. His Majesty's Government of Nepal has provided incentives and subsidy policy in the alternative energy sector. However, the subsidy for alternative energy has been referred to as being inappropriate, inadequate, and inconsistent. Government line agencies, private entrepreneurs, INGOs and NGOs have been involved in the promotion and dissemination of alternative energy. However, there are socio-cultural, economic, technological and organisational constrains in the utilisation of alternative sources of energy.

INTRODUCTION

Energy is a basic input for development, and its demand is growing as the rapidly growing population and their changing life style has to deal with limited sources of energy. Nepal is not an exception from other developing countries, and is in a more critical situation, as she is more dependent on traditional sources of energy. About 90 percent of the Nepalese continue to live in rural areas, and have agrarian life styles relying predominantly on conventional sources of energy, i.e. fuel-wood, agriculture residue and animal waste. Even the urban population is not independent from these traditional energy sources. But, the growing number of cities extract heavy costs on the economy because the major portion of energy they use has to be imported with the limited purchasing capacity of the nation.

Nepal, a Himalayan country, is situated on the southern slopes of the central Himalaya. It has a land area of 147,181 sq km and stretches for over 800 km length from east to west forming one third of the entire 2500 km length of the Himalaya. About 83 percent of its total land area is occupied by high hills and mountains and the remaining 17 percent by flat lands of the Terai. The altitude varies from some 60 m above the sea level in the Terai to 8,848 m on Mt. Everest, which is the highest point of the world. The country is very rich in terms of natural wealth but is poor and underdeveloped in economic terms. Nepal had a total population of 22.4 million in 1999. Population growth rate was 2.08 percent between 1981-91, but recent assumption indicates the population growth rate has increased to 2.66 percent. Nepal is primarily an agricultural country which contributes about 41 percent to the national GDP and provides employment to more than 80 percent of the economically active population (CBS, 1998).

In Nepal, with the growth in population, the spread of urbanisation, the increase in the volume of construction works and rapid expansion of other economic activities, the consumption of energy is increasing in every day life. However, the annual per capita energy consumption in Nepal is about 13 to 15 GJ, which is almost the lowest in the world (HMG, 1998a). Even among the South Asian countries, Nepal's status of commercial per capita energy consumption level is very low (HMG, 1998a). The energy resource base of the country consists of a combination of traditional and commercial sources, including fuel-wood, agricultural residue, animal wastes and hydroelectricity. Biogas, micro-hydro, solar and wind energy are alternative energy sources being utilised in Nepal at realistically small scale. Historically, Nepal's rural populations have been meeting their energy needs from traditional sources such as fuel-wood and other biomass resources. The country lacks any proven deposits of conventional fossil fuels such as coal, lignite, gas and oil. These conventional sources of energy are imported from other countries.

ENERGY CONSUMPTION PATTERN

Two major characteristics of energy systems in Nepal are excessive dependence on traditional sources of energy and very low efficiency in their use. According to the statistics compiled by the Water and Energy Commission Secretariat, His Majesty's Government of Nepal (WECS, 1999) total energy consumption stood at 7.4 million tons of Oil Equivalent (TOE) with an annual increment of 3 to 4 percent. Of the total energy used, 88 percent is met from traditional sources (fuel-wood 78.6 percent, agricultural waste 3.53 percent and animal waste 5.55 percent) and the remaining 12 percent from commercial sources (petroleum fuel 8.53, coal 2.37 and electricity 1.11 percent, Fig 1). It has been recorded that the proportion of consumption of commercial energy has been increasing 3 to 4 percent every year, and there is a decline in the consumption of traditional energy (Fig. 2).

The sectoral distribution of energy consumption pattern in Nepal is that 90 percent of the total energy is used in the residential sector or for domestic use, 4.2 percent in the industrial sector, 2.9 percent in transport sector, 1.2 percent in

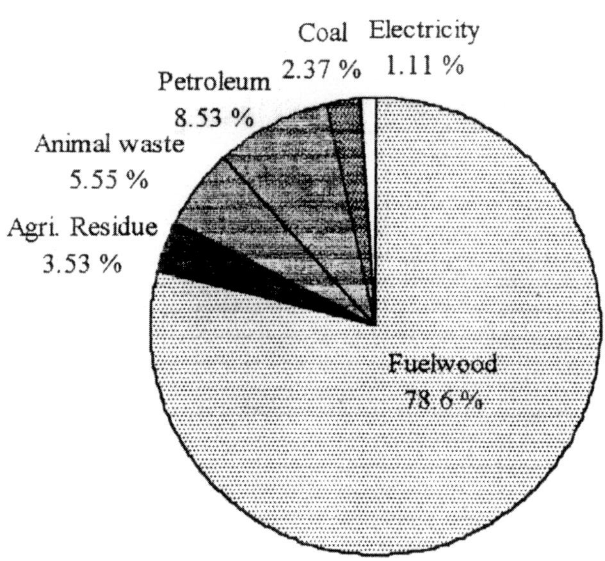

FIGURE 1. Energy consumption according to fuel type.

the commercial sector and 0.2 percent in the agricultural sector. The transport, commercial and industrial sectors are mostly dependent on imported commercial fossil fuels. The energy needs of the domestic sector are cooking, lighting, space-heating, water heating, operation of household appliances and carrying out agro-processing activities (Fig. 3). The bulk of consumption is for cooking and is based on non-monetised use of traditional fuels. In mountain areas, 32 percent of energy is used for cooking and 56 percent for heating, while in the hills the ratio of energy use is 40 percent for cooking and 36 percent for heating, the rest being consumed for lighting and operating electrical appliances.

The trend and pattern of energy use in Nepal illustrates that the nation relies predominantly on fuel-wood with its demand on the rise. Also, cooking takes a huge amount of energy. This reflects the poor economic situation of the nation, as developed economics consume less than 16 percent of the energy for the same purpose.

Although the imported fossil fuels represent only a small fraction of total energy sources, the country spent a total of Nepali Rupees (NRs) 7.5 and 10.4 billion (1 US $ equivalent to 68.40 NRs) for the import of fossil fuels in the year 1996/97 and 1997/98, respectively (HMG, 1998b). Since electricity contributes 1.11 percent of the total energy demand, it plays a minor role in sectoral energy

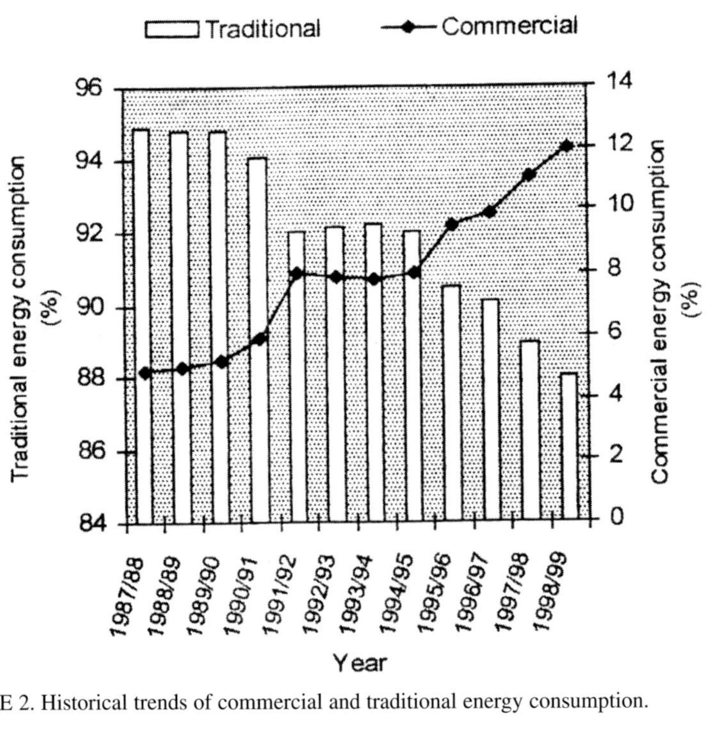

FIGURE 2. Historical trends of commercial and traditional energy consumption.

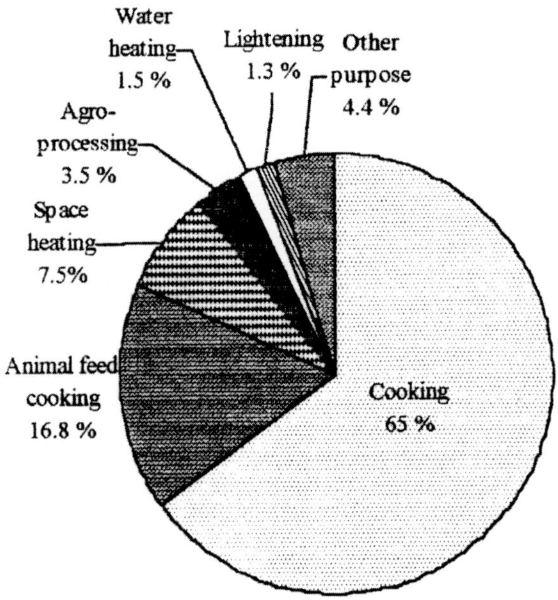

FIGURE 3. The energy needs of domestic sector.

needs. Industry is the main consumer of the total electricity supply followed by domestic and commercial sectors. Recently, the alternative (renewable) energy sources have been encouraged by government and non-government organisations because they are more environmentally friendly, but their contribution has been very limited due to the constraints of information on technology, high up-front cost, human resource and institutional setup (HMG, 1998a).

RENEWABLE ENERGY RESOURCES AND TECHNOLOGY

Given its size and population, Nepal is a country which possesses a considerable potential for indigenous energy resources development (Table 1). The most potential renewable or alternative sources of energy available for development in the context of Nepal are related to hydropower, biomass, solar and wind energy. Although there are indications of some potentials of geothermal energy, natural gas, lignite and draught animal power, their magnitude, as well as feasibility, are not well established.

Hydropower

Nepal has well-developed hydrological networks with abundant water resources representing an annual flow of about 224 billion cubic meters. In this respect Nepal has been regarded as the second richest country in the world, having nearly 2.27 percent of the world's water resources. There are about 6000 rivers and streams in Nepal with an approximate total length of 45,000 km (HMG, 1998a). The estimated surface water area is about 4063 sq km of which 97.3 percent is represented by large rivers (HMG, 1998a). The Koshi, Gandaki and Karnali form the three main river systems in Nepal which receive substantial portions of their water supply from the snow melt, glaciers and small tributaries. These large and many small rivers, rivulets and streams in Nepal, falling steeply from the high mountains (more than 8000 m) in the north to low laying Terai (less than 300 m) in the south, offer a great potential for hydropower development at different scales. Based on the average flow of the rivers, theoretical potentiality of hydropower generation is estimated to be 83,290 MW. Of this, 45,610 MW (54.8 percent) is technically feasible for development and 43,000 MW is estimated economically feasible (WECS, 1996, 1999). However, only about 262 MW of hydropower (0.3 percent of total potenıial) has been harnessed so far from 10 major (249.95 MW) and 37 small (11.808 MW) hydropower projects managed by Nepal Electricity Authority (Tables 2 & 3). Besides, thermal power stations are producing 57 MW of electricity. A total of about 1475 GWh of electricity is annually available in the country. Of the total availability, 76.2 percent is met from hydro-power, 8.05 from thermal power and 15.75 percent is met by importing from India.

Although the total installed generating capacity in the country is about 319 MW (both hydropower and thermal power), in winter the capacity decreases to 276 MW while the peak load demand is estimated at 326 MW. Thus, there is a power shortfall of about 50 MW in the present context (Shrestha, 1999). Therefore, the demand is being managed through increased import and power shedding. The power demand is increased by 8.6 percent per annum.

The present electricity supply meets only 1.11 percent of the total energy demand of the country. Moreover, only 15 percent of the total population have so far gained access to supply of electricity. The rural population, which comprise about 90 percent of the total population, has very limited access to electricity (4 percent).

To fulfill the increasing demand for electricity, large and medium sized hydroelectricity projects such as Kaligandaki A (144 MW), Khimti (60 MW), Bhote Koshi (36 MW), Chilime (20 MW), Modi (14 MW), Puwa Khola (6.2 MW) and Indrawati III (5 MW) are under construction totalling 285.2 MW. Negotiations between the officials of Nepal and India are underway for the implementation of large multipurpose hydropower projects such as Karnali (10,800 MW) and Pancheswor (6,480 MW). Besides these, 21 hydropower projects of different capacity (14 to 750 MW), totalling 4370 MW, are in various stages of energy development (NEA, 1999).

Nepal's rural economy and environmental considerations do not provide enough economic basis for large-scale investment for the exploitation of vast

TABLE 1.
Facts concerned to energy resources and consumption pattern in Nepal.

Variables	Value	Unit	Remarks
Total Area	147,181	sq km	
Total Population	22.4	million	
Forest Area Gross	4.27	milhon ha	
Total Forest Land	2.71	million ha	>10 % crown cover
Percentage of total land of the country	29.0	percent	>10 % crown cover
Total Shrub Land	1.56	million ha	
Percentage of total land of the country	10.7	percent	
Gross Fuel-wood Availability	13.9	million mt	
Sustainable Fuel-wood production	7.0	million mt	
Annual Deforestation Rate (forest)	1.7	percent	Last two decades
Annual Deforestation Rate (forest and shrubland)	0.5	percent	
Cultivation Area	3.0	million ha	
Crop Production	7.5	million mt	
Agricultural Residue Production	14.1	million mt	
Population of Livestock	10.0	million	
Annual Manure Production	12.0	million mt	
Collectable Manure	4.6	million mt	
Estimated Potential for no. of Biogas Plants	1.5	million	
Estimated Solar Energy Potential	26.6	million MW	
Estimated Theoretical Hydropower Potential	83,290	MW	
Estimated Economic Hydropower Potential	43,000	MW	
Total Installed Capacity (Hydro and Thermal Power)	318.8	MW	
Installed Capacity Major Hydropower	249.9	MW	
Installed Capacity Small Hydropower	11.808	MW	
Installed Capacity Thermal power	57.0	MW	
Per Capita Energy Generation	55	kWh	
Annual Energy Availability	1475	GWh	
Hydropower	1124	GWh	
Thermal Power	118.82	GWh	Increment of 11% over 1997/98
Electricity imported from India	232.39	GWh	Increment of 22% over 1997/98
Electricity Exported to India	60	GWh	Decrease of 7.4 % over 1997/98
Transmission lines:			
132 kV single circuit	1119.60	km	
132 kV double circuit	177.02	km	
66 kV single circuit	179.00	km	
66 kV double circuit	153.00	km	
33 kV single circuit	1348.70	km	
Energy Loss	23.4	Percent	
Annual Energy Consumption:			
Electricity Consumer: 15% of the total population	620000		Increment of 7.8% over 1997/98
Electricity Consumption (grid based)	1113	GWh	Increment of 5.8% over 1997/98
Average Consumer's Consumption	1698	kWh	Decrease of 0.8% over 1997/98
Peak load Demand	326.4	MW	Increment of 9.4% over 1997/98
Fuel-wood	11.0	Million mt	
Agricultural Residue	1.0 million	mt	
Animal Manure	1.7	million mt	
Coal	300,000	mt	
Liquified Petroleum Gas	28,950	mt	
Gasoline (Petrol)	34,550	kL	
Aviation Turbine Fuel	34,260	kL	
Kerosene	225,000	kL	
High Speed Diesel	246,730	kL	
Light Diesel Oil	930	kL	
Fuel Oil	930	kL	
Cost of Petroleum Products as a percentage of Total Merchandise export	40	percent	

Source: CBS (1998), HMG (1998a), NEA (1999), WECS (1992, 1994, 1995, 1996, 1999).

TABLE 2.
Existing Major Hydropower Projects in Nepal.

Name of the projects	Plant type[1]	Date of commission	Installed capacity (MW)	Owner[2]
Panauti	PROR	1965	2.40	NEA
Trishuli	PROR	1969	24	NEA
Sunkoshi	ROR	1973	10.05	NEA
Gandaki	ROR	1979	15	NEA
Kulekhani I	STO	1982	60	NEA
Devighat	ROR	1983	14.1	NEA
Kulekhani II	STO	1986	32	NEA
Marsyangdi	ROR	1990	75	NEA
Andhi Khola	ROR	1993	5.1	BPC
Jhimruk	ROR	1994	12.3	BPC
Total			**249.95**	

[1]ROR,Run-of-River; PROR, Peaking Run-of-River, STO, Storage
[2]NEA-Nepal Electricity Authority

TABLE 3.
Existing Small Hydropower Projects in Nepal.

SN	Name of the projects	Date of commission	Installed capacity (MW)	SN	Name of the projects	Date of commssion	Installed capacity (MW)
1.	Pharping	1911	0.500	20.	Manang	1988	0.080
2.	Sundarijal	1935	0.640	21.	Terhathum	1988	0.100
3.	Phewa	1967	1.088	22.	Chaurjhari	1989	0.150
4.	Dhankuta	1971	0.240	23.	Syarpudaha	1989	0 200
5.	Surkhet	1977	0.345	24.	Khandbari	1989	0. 250
6.	Tinao	1978	1.024	25.	Bhojpur	1989	0.250
7.	Baglung	1981	0.200	26.	Ramechhap	1989	0.150
8.	Doti	1981	0.200	27.	Bajhang	1989	0.200
9.	Phidim	1981	0.240	28	Bajura	1990	0.200
10.	Gorkha	1982	0 064	29.	Okhaldhunga	1990	0.125
11.	Jomsom	1982	0.240	30.	Arughat	1991	0.150
12.	Jumla	1982	0.200	31.	Tatopani	1991	1.000
13.	Dhading	1982	0.032	32.	Rupalgad	1991	0.100
14.	Syangja	1984	0.080	33.	Surnaiyagad	1991	0.200
15.	Seti	1985	1.500	34.	Darchula	1992	0.300
16.	Helambu	1985	0.050	35.	Namche	1993	0.600
17.	Salleri-Chialsa	1986	0.400	36.	Achham		0.400
18.	Chame	1987	0.045	37.	Dolpa	1999	0.140
19.	Taplejung	1988	0.125				
	Total capacity				**11.808 MW**		

hydro-potentials. Similarly the grid electricity can not be made available to remote places. In this context, micro-hydro (up to 100 kW) as alternative energy has a great potential for fulfilling the energy requirements of rural Nepal. Small hydro-projects such as micro-hydro have the advantages of being environmentally superior, requiring no resettlement or leading to deforestation or submergence, and have reduced transmission and distribution losses. Over the past few decades, Nepal has made significant progress in developing the micro-hydro sector in harnessing water resources for producing power. In fact, its importance has been long realised and reflected in the large number of traditional water wheels (ghattas) of less than 1kW found throughout the country. According to an estimate there are over 25,000 such traditional ghattas in hilly and mountainous regions of Nepal which provide agro-processing energy for local people (GATE, 1983). But recent technological development has resulted in the more efficient use of water resources for power generation. With the addition of cross flow turbines, these traditional ghattas might be made to provide power not only for the traditional functions of agro-processing activities but also for generating up to 10 kW at each site, making the total potential to be as much as 250 MW (WECS, 1994). However, only about 350 units of traditional ghattas have been improved to a power output in the range of 1-3 kW (Shrestha and Amatya, 1998). Currently, there are over 1349 units of micro-hydropower turbines and 156 peltric sets installations in remote hilly and mountainous areas of Nepal (HMG, 1998a). Most of the micro-hydro installations are primarily used for agro-processing activities such as grinding, rice dehusking and dehulling, oil expelling, rice beating, etc. Some of the micro-hydro installations are being used for electrification (Shrestha and Amatya, 1998). The total generating capacity from micro-hydro installation in Nepal has been estimated as over 10 MW. In most of the micro-hydro plants, generating electricity lighting still remains the predominant end-use. However, some efforts have been made for the promotion of electric cooking and water heating systems, beating and digesting pulp and drying paper, pumped irrigation, operation of chilling plant, producing ice blocks, etc.

Biogas

Since time ımmemorial fuel-wood has become the major source of biomass energy for the rural people in Nepal. At present, fuel-wood accounts for about 80 percent of the total energy sources and is derived from public, community and private forests and agricultural land. Besides fuel-wood, agriculture residue and animal waste are increasingly being used as a source of energy. The implication of the use of biomass energy has been quite serious environmentally, with soil loss, decline in agricultural productivity, and occasionally the onset of the process of desertification (Bhadra, 1993). Thus, His Majesty's Government (HMG) Nepal has started a village electrification programme in Terai. Efforts have been made to substitute the traditional use pattern of biomass energy by appropriate technologies. Biogas, briquettes and improved cooking-stoves are some of the alternative energy technology used to harness biomass energy more efficiently and sustainably in Nepal.

Biogas technology has proved to be very successful in an agro-based country like Nepal since it not only produces gas for heating, cooking and lighting purposes, but also provides good fertilizer in the form of digested slurry with a significant nutrient value. Biogas, a methane rich gas, is produced by the anaerobic digestion of cattle dung, buffalo dung and human excreta. There are two types of biogas plants (digesters) installed in Nepal, they are floating drum type (Indian model) and fixed dome type (Chinese model). However, the fixed dome model has become more popular (about 80 percent of the biogas plants), as these are easier to construct, operate and maintain using indigenous raw materials (Van Nes and Lam, 1997; Gongal and Shrestha, 1998). In Nepal, the overall livestock density per hectare of cultivated land has been estimated at 7.4. More than 83 percent of livestock holdings have cattle (HMG, 1998a). Livestock (cattle and buffaloes) population in Nepal has been estimated as over 10 million. These animals produce a huge quantity of dung, which can be utilised for biogas production. It is estimated that 12 million tones of animal dung (dry basis) were produced during the year 1997/1998 (WECS, 1999).

In Nepal, the interest in biogas began only in 1975/76 and the Gobar Gas Company was established in the year 1977 to promote biogas technology (Amatya and Shrestha, 1998). In 1992 the Biogas Support Programme (BSP) was set up to support the biogas programme through subsidies, quality control, training, etc. Due to the success of biogas development programmes and the availability of a government subsidy, as well as the interest and involvement of a number of INGOs and donor agencies, a number of private biogas companies started coming up (Khandelwal, 1996). So far, more than 41 new biogas companies have been established as a result of government privatisation policy. The potential number of biogas plants in the country is estimated at about 1.5 million based on the annual production of animal dung. If this capacity could be

TABLE 4.
Number of biogas installed households in Nepal by development regions and physiographic zones up to the beginning of 1997.

Development region	Physiographic zone			Total	%
	Mountain	Hill	Terai		
Eastern	37	1010	5097	6144	19.13
Central	277	3506	4891	8674	27.01
Western	0	10633	2855	1348	41.99
Mid-western	0	836	1831	2667	8.30
Far-western	4	43	1099	1146	3.57
Total	318	16028	15773	32119	100

Source: Van Nes and Lam (1997).

utilized in an effective manner, it can fulfil about 10 percent of the country's total energy requirement without affecting the productivity of the agriculture (WECS, 1994, HMG, 1998a). However, it has been estimated that there are over 32,000 biogas installed households in Nepal up to the beginning of 1997 (Van Nes and Lam, 1997). Now, this number has reached over 50 thousand in 61 districts out of 75 districts of the country. On a regional basis, the Western Development Region has a maximum number of installations (41.99 percent) followed by Central, Eastern, Mid-western and Far-western regions (Table 4). Similarly, among the physiographic zones the hill region is taking the leading position (49.9 percent) followed by Terai (49.11) and mountain (0.99).

Improved Cooking Stove (ICS) and Briquetting Technology

In view of the rapidly dwindling forests, the development and promotion of the Improved Cooking Stove (ICS), also known as Improved Fuel-wood Stove, has become one of the most important components of the fuel-wood management programme. ICS is a simple, low-cost technology that offers multiple benefits to those cooking with biomass energy. It conserves heat and reduces heat dissipation with minimum waste, makes kitchens smokeless and thereby creates a safer and healthier home environment and improved health conditions. It can save 18 to 42 percent of fuel-wood as compared to traditional stoves (Sulpya, 1991). Since the1980s, government of Nepal has played an important role in promoting and disseminating the ICS programme. A number of NGOs and INGOs have included ICS as an integral part of their overall development objectives and programmes (Amatya and Shrestha, 1998). There are over 44 various non-government and private sector organisations involved in ICS development in Nepal (WECS, 1995). There are several improved stove designs in Nepal (mud, metal, brick and ceramic insert stoves). In Nepal, about 44,000 insert-type stoves were distributed, of which 35 percent are believed to be in operation (Amatya and Shresta, 1998). At present, the ceramic insert-type design has been abandoned in favour of stoves built on site using locally available materials (mud, straw, animal dung) and skills (WECS, 1995; Amatya and Shrestha, 1998). It is believed that about 44,000 such mud stoves have been installed, of which more than 90 percent are still in operation.

Briquetting technology was introduced in Nepal in 1982 using rice husk as the main raw materials (WECS, 1995). The use of rice husk has been on the increase for industrial heating and process heating purposes. Briquettes are also being produced using sawdust as the raw material. Two different types of densification technology are currently in use. The first, called pyrolyzing technology, was developed in India and relies on the partial pyrolysis of rice husk, which is mixed with binder (mollase) and then made into pellets by casting and pressing. The second technology, imported from Taiwan, is the direct extrusion type, where the rice husk is dried and directly compacted. Although several private manufacturing industries have been established in Nepal for the production of briquettes, there have been no government initiatives for its promotion. However, financial support has been provided by the Agricultural Development Bank of Nepal (ADB/N), a rural development bank, and the National Industrial Development Corporation (NIDC), a semigovenment organisation, in terms of loans and consultancy services (WECS, 1995).

Solar Energy

No proper solar mapping of Nepal has been undertaken by any agency. However, it is practically established that the solar energy resource in Nepal is high and evenly distributed over the country and over the seasons, owing to the country's location close to the tropic of cancer. The average insulation is around 4.5 KWh per sq meter per day. Depending upon the location, solar average insulation varies from 10-15 percent and seasonal radiation is from 50-80 percent. The annual sunshine duration ranges from 1900 to 2500 hours. The potential of solar energy is an estimated 26.6 million MW (WECS, 1995). However, solar energy utilisation is still in the early stages of development in Nepal, although efforts have been initiated to develop devices for harnessing solar energy more efficiently.

Mainly, two types of solar energy technologies are available in the country: solar thermal system, and solar photovoltaic (SPV) system. There are two types of solar thermal devices used in the country; they are solar water heater and the solar drier. Solar water heaters of the thermo-syphon type are produced in a substantial number in urban areas, which are used for heating water in households as well as in hospitals, schools, hotels, and lodges. Another use of solar water heaters is in preheating water for industries requiring process heating. Currently they are being used to preheat water for some carpet industries (Shrestha and Bajracharya, 1998). The demand for the solar water heater in Nepal is growing sharply. This is also indicated from the fact that the number of solar water heater manufacturers has increased from a few in 1986/87 to more than 40 in 1991/92. The solar dryer is another solar thermal device, which has found limited use in Nepal, particularly used for drying spices, fruits, vegetables and herbs.

Another important area of solar energy use has been electricity generation from solar photovoltaic (SPV) systems. The electricity generated through SPV systems are used for generating residential electricity for less energy intensive uses such as lighting and operating appliances, operating small irrigation pumpsets for drinking and irrigation purposes, operating communication equipment in remote areas, etc (WECS, 1995). Nepal produces about 1 MW of electricity through solar energy, out of that 70 percent is used in telecommunication sector, and only 8 percent is used in the domestic sector (Kantipur, 1999). The Department of Civil Aviation and Nepal Telecommunication Corporation have been using SPV generators since 1963 to power their radio communication and navigational aids in remote areas. Twenty-eight airports use 365 solar modules with an approximate installed peak wattage of 14 kW (WECS, 1995). According to an estimate, more than 6000 units of 50W module SPV systems are in use in different parts of the country for powering communication equipment under the Nepal Telecommunications' Corporation (WLG and WECS, 1995). Besides, SPV systems are also used by a large number of other government and non-government organisations to power communication equipment in remote parts of the country.

Wind Energy

Wind energy is not used to any practical extent in Nepal. The few studies that have been conducted so far have indicated some potential for wind energy in Nepal (WECS, 1995). Because of Nepal's diverse topography and the consequent variation in meteorological conditions, it is difficult to generate wind energy in the country. Specific areas have been identified as favourable for viable wind generation. Surveys conducted so far indicate that wind could be used for electricity generation in the hills and for irrigation and pumping of drinking water in the Terai (WECS, 1995). According to preliminary estimates, Kagbeni and the surrounding area in Mustang District of the north-central part of Nepal is one of the potential sites to generate 200 MW of electric power with an annual energy production of 500 GWH from wind resources (Dangrid Consult, 1992). The prevailing average wind speed of 6 m/s has been recorded in these areas. Similarly, other highly potential areas for developing wind energy are Palpa, Ramechhap, Khumbu, Karnali-Chisapani and Jumla. Nepal Electricity Authority (NEA) had made efforts in installing a 30 kW wind power plant in Kagbeni, Mustang. After failure of this project, no further efforts were made in this direction. At present NEA is implementing a Wind Power Development Project to tap the huge wind energy in the country. Some agencies are working on the development of windmills and wind turbines to be used in the remote regions. Experts say that the wind energy sector lacks reliable data and shows the need for further studies to confirm whether that potential can be really materealised. In this response, the government, in its Ninth Five-Year Plan (1997-2001), has proposed for wind data collection and mapping in five potential districts.

POLICIES AND PROGRAMS

HMG policies, programmes and strategies regarding the development of the energy sector have been documented in its five years' plan. However, energy sector planning has gone through major changes, mainly due to international influence and the emergence of a focus on a market-based and liberalized economy. The government of Nepal has considered the development of Nepal's water resource to be a key instrument in economic development, and thus a key component for poverty alleviation. Therefore, hydropower development has become a major element in government policies and strategies and received high priority in all five years' plan. The Fifth Five Year Plan (1975-80), for the first time, considered the role of micro-hydro (up to 100 kW) plants in rural electrification. Sixth Plan (1980-85) covered the energy policy for hydropower and forestry sectors. Seventh Plan (1985-1990) has given due importance to the development of alternative energy for socioeconomic development and provided subsidies for the installation of biogas plants and micro-hydro turbines. Eighth Plan (1992-1997) was more specific and embraced a market-oriented, liberal economic policy for the development of the country. The focus on socio-economic and environmental concerns specifically related to poverty alleviation was given great consideration in planning. Only in the Eighth Plan has the importance of alternative energy been strongly felt at the policy and decision making level. The Ninth Plan (1997-2001) has followed the basic conceptual principles of objective and policy goals, which are, more or less, in line with the same patterns of the previous plan periods. The major objective of energy plans is to gradually replace imported fuels with indigenous energy sources that can be locally exploited.

In the Eighth and Ninth Plans energy development has been considered as an industry and has been given more stress on the involvement of private sector for the development and promotion of energy sector. There is a provision for external, internal and joint collaboration with the private sector in the development and financing of the energy sector. Detailed policy and programmes have been provided including provision for an institutional set-up for the development of alternative energy in Nepal. Subsequent changes were made in strategies to attract the private sector. For this purpose different acts, regulations and policies came into existence. The major acts, regulations and policies related to the promotion of energy resource development projects are: Hydropower Development Policy 1992, Water Resource Act 1992, Electricity Act 1992, Foreign Investment and Technology—Transfer Act 1992, Industrial Enterprises Act 1992, Irrigation Policy 1992, Electricity Regulations 1993, Water Resources Regulations 1993, Electricity Tariff Control Rules 1993, Forest Act 1993, Forest Regulations 1995, Environmental Protection Act 1997, Environmental Protection Regulations 1997 and Local Self-governance Act 1998. In these legal provisions the facilities, subsidies and cooperation to be provided to the private sector were made legal. A single door policy had been adopted for those types of procedures and facilities.

Policies adopted in the Ninth Five Year Plan (1997-2001) are in the following areas:

- Increase supply of traditional sources of energy and minimize negative environmental impact from their use.
- Involvement and encouragement of private sector, national and international non-government organisations to conduct research; promote, disseminate, establish, operate and develop the sources and technologies of energy

including alternative energy.

- Reduction in absorption of energy by increasing the existing bio energy utilisation efficiency while considering the replacement time for traditional energy.
- Replacement of imported energy by the application of hydro, solar and wind energy through adaptive study, research and development and commercialization of energy technology.
- Development of Alternative Energy Promotion Centre as a main agency for the promotion and development of alternative energy sources in Nepal.
- Purchasing of energy produced from rural alternative energy at a reasonable price.
- Provision of subsidy and formulation of necessary legal provisions.

As a result of the government's policy in the development of the energy sector, the private sector's activity has increased over the last few years. But a deeper look into their role for acting as a catalyst in local development is required. The existing policies are that no licenses are required to operate a small hydropower projects having up to 1000 KW capacity. For the hydropower projects with more than 1,000 KW capacity, it is required to seek permission from the government. Similarly, great efforts have been initiated by HMG Nepal to promote several rural electrification projects in different parts of the country. Such processes are exercised with the financial assistance of foreign donor agencies such as ADB, FINIDA, NORDIC, JICA, etc. The Ninth Five Year Plan aims to construct 1,024 km of transmission lines all over the country that will facilitate in extending electricity facility to an additional 1.1 million people of the kingdom by the year 2003 AD.

HMG Nepal has provided incentives and subsidy policy in the alternative energy sector. It is in the form of capital subsidy. The capital subsidy is being channeled through the Agriculture Development Bank of Nepal. In the construction of micro-hydro plants (up to 100 kW), the amount of subsidy is 75 percent in remote mountainous districts and 50 percent for the remaining districts (WECS, 1995). The amount of investment subsidy in biogas has been fixed at Rs. 7,000 and 10,000 for each plant installed in the Terai and the hills respectively independent of the size of the plant (WECS, 1995). Similarly, government of Nepal has been giving 40 to 45 percent subsidy in Terai and hills respectively from 1999 for solar domestic electricity installation. This programme is becoming popular as government has received over 10,000 applications for this programme (Kantipur, 1999). In the process, investment of NRs 105.564 million has been made as subsidy in the year 1997/98 in the alternative energy sector (HMG, 1998b) (Table 5). HMG has a plan to invest NRs 30 million as subsidy in solar and wind energy for the year 1999/2000. However the subsidy for alternative energy has been referred to as being inappropriate, inadequate, and inconsistent.

TABLE 5.
Total investment and subsidy provided in alternative energy sector during 1997/98.

Alternative energy	Physical target	Investment (in millions NRs)	Subsidy (in millions NRs)
Microhydro	425 kW	6.347	11.500
Biogas	9,869 biogas plants	146.139	87.064
Solar and wind energy	462 solar energy stations	25.387	7.000*
	Total	177.873	105.564

Source: HMG (1998b); *subsidy for solar energy has increased to NRs 15000 from 1999.

INTERMEDIATION AND PRIVATE SECTOR PARTICIPATION

Government line agencies, private entrepreneurs, INGOs and NGOs have been showing interest in the promotion and dissemination of alternative energy. The Agricultural Development Bank of Nepal (ADBN) has not only played the significant role of promoter of alternative energy, but has also taken on the responsibility of financial intermediation and is now engaged in channeling the government subsidy. Besides financial institutions, private sector entities, especially manufacturers and NGOs, have been playing the role of financial intermediaries to varying degrees (Amatya and Shrestha, 1998). Apart from domestic financial sources, grant and loan assistance is being received from bilateral and multilateral donor agencies such as United States Agency for International Development (USAID), United Nations Capital Development Fund (UNCDF), United Nations International Children Education Fund (UNICEF), Asian Development Bank (ADB) and SNV/Nepal (Amatya and Shrestha, 1998; Shrestha and Amatya, 1998).

The role of technical intermediary is being carried out by investors, manufacturers, technology suppliers and financial agencies (Amatya and Shrestha, 1998). The role of technical intermediary is to provide necessary information, research and development, technical services, training, etc. Various national and international institutions have been taking the role of technical intermediary (Table 6).

Water and Energy Commission was set up by His Majesty's Government of Nepal in 1976 to accomplish studies, surveys and investigations leading to policy recommendations relating to the development of the nation's water and energy resources. In 1981, a formal Secretariat for the Commission was established. The Commission and its Secretariat have a broad mandate to analyse and advise HMG of Nepal on national and regional water and energy policy, planning, and development issues. Similarly, Alternative Energy Promotion Centre, under the Ministry of Science and Technology, was established in 1997 to coordinate related agencies and move their activities. Similarly, in the semi-government side, Nepal Electricity Authority (NEA) is the only public utility responsible for the generation and distribution of electric power throughout the country and supplies electricity to isolated areas, mainly district headquarters, through a number of small hydropower plants.

The agencies ınstitutionally organised to operate national level research and development activities in the alternative energy sector are National Science and Technology Council (NSTC), Royal Nepal Academy for Science and Technology (RONAST), and Research Centre for Applied Science and Technology (RECAST). Some sporadic research activities have also been carried out by the Development Consulting Service (DCS), Intermediate Technology Development Group (ITDG), German Technical Cooperation (GTZ), Centre for Rural Technology (CRT), etc.

In the hydropower sector a number of public and private companies have emerged which are playing an important role in the development of mini/micro hydro projects in various capabilities (Thapa, 1995; WECS, 1995). The private sector participation, particularly in the micro hydropower plants, is important for manufacturers as well. There are about nine private manufacturers within Nepal for the construction of micro-hydro plants with a total manufacturing capacity of more than 2 MW of microhydro per year (WECS, 1994; Shrestha and Amatya, 1998). Various kinds of micro-hydropower devices, e.g. propeller turbines, cross-flow turbines, pelton wheels, multipurpose power units, Peltric -sets, and improved ghattas have been developed and are being successfully operated by the private sector in the country. Among solar thermal devices, solar heaters are being produced in substantial numbers in Nepal. However, there are only a few SPV manufacturers in Nepal producing solar photovoltaic modules and solar balance of systems, such as charge controllers, inverters and lanterns (Shrestha and Bajracharya, 1998). Besides these, consumer electrical items, such as bulbs, switch gear, cables solar panels for water heating systems, improved cooling stoves (ICS), used for domestic purpose and commercial establishments, are also produced by the private sector.

ENVIRONMENTAL IMPACTS

Almost 90 percent of energy in Nepal is supplied by biomass, and among biomass energy fuel-wood accounts for about 78.6 percent of the total energy use. The country's annual sustainable yield of fuel-wood amounts to approximately 7 million tonnes while annual consumption amounts to approximately 11 million tonnes. There stands a huge gap between demand and supply of the energy needs. So far, the commercial source of energy could not fill this gap. Instead, rapid population growth has created even higher demands for energy in its cheapest and most accessible form, i.e. fuel-wood. The demand for fuel-wood has created deforestation and soil erosion problems at many places. Deforestation rate was high (1.7 percent per annum) during the last two decades in general in Nepal and

TABLE 6.
Major institutions involved in the alternative energy sector in Nepal

Institution/organization	Major objectives
Agricultural Development Bank (ADB)	Financial intermediation
Alternative Energy Promotion Centre (AEPC), HMG	Coordination, R&D, promotion training in AE sector
Asian Development Bank (ADB)	Channeling subsidy, support on R&D efforts, training
Association for Context Appropriate Technology (FAKT)	Support the development of end use technology
Association of Micro-Hydro Manufacturers (AMHM)	Coordination among manufacturing companies
Balaju Technical School	Training
Balaju Yantra Shala (BYS)	Micro-hydro, solar water heater manufacturer
Biogas Support Programme (BSP)	Subsidies, quality control training in biogas sector
Butwal Technical School (BTS)	Training
Centre for Renewable Energy (CRE)	Promotion of solar technology in rural electrification
Centre for Rural Technology (CRT)	Promotion of AE technologies
Community Forestry Development Programme (CFDP)	Development and dissemination of ICS
Dept. of Soil Conservation and Watershed Management, HMG	Promotion and installation of solar devices
Development and Consulting Services (DCS)	Micro-hydro manufacturer
Food and Agriculture Organisation (FAO)	Support workshops and training on biogas, ICS
German Technical Cooperation (GTZ)	Support micro-hydro sector, Improved Cooking Stoves
Gobar Gas Company (GGC)	Construction, promotion & management of Biogas
Intermediate Technology Development Group (ITDG)	Strengthen the micro-hydro & appropriate technology
International Centre for Integrated Mountain Development (ICIMOD)	Planning and policy formulations, capability enhancement of implementing agencies in mini & micro-hydro
Kathmandu Metal Industries (KWI)	Micro-hydro manufacturer
Kishan Gobar Gas	Construction and installation of biogas plants
Lotus Energy (LE)	Manufacture of solar energy devices
Ministry of Finance (MF)	Formulation of economic policy, financial flow
Ministry of Local Development (MLD)	Promotion of ICS
Ministry of Water Resources (MWR)	Formulation of policy and development programmes
National Industrial Development Corporation (NIDC)	Provide loans and consultancy service
National Power Producers (NPP)	Micro-hydro manufacturer
National Structure and Engineering Co. (NECO)	Micro hydro manufacturer
Nepal-Australia Forestry Project	Dissemination of Improved Cooking Stoves
Nepal Electricity Authority (NEA)	Generation and distribution of electric power

(*continued*)

TABLE 6. (*Continued*)
Major institutions involved in the alternative energy sector in Nepal

Institution/organization	Major objectives
Nepal Gobar Gas Bistar	Construction and installation of biogas plants
Nepal Hydro and Electric Company P.Ltd. (NHE)	Micro-hydro manufacturer
Nepal Machine and Steel Structures (NMSS)	Micro-hydro manufacturer
Nepal Telecommunication Corporation (NTC)	Operate radio communication from SPV system
Nepal Yantra Shala and Engineering (NYSE)	Micro-hydro manufacturer
Production Credit for Rural Women (PCRW)	Promotion and dissemination of ICS
Rastriya Gobargas	Construction and installation of biogas plants
Research Ceatre for Applied Science and Technology (RECAST)	Research and development on AE sectors
Royal Nepal Academy of Science and Technology (RONAST)	Research and development on AE sectors
Save the Children USA	Promotion of ICS
Small Farmer Development Programme (SFDP)	Promote Improved Cooking Stoves
SNV Nepal	Grant assistance to micro-hydro, biogas
Sun Works Nepal (SWN)	Manufacture solar energy devices
Swiss Association for Technical Association (SATA)	Support the development of micro hydro, ICS
Swiss Federal Institute of Technology (SKAT)	Promotion of appropriate micro-hydro technology
Thapa Engineering Industry (TEI)	Micro-hydro manufacturer
United Mission to Nepal (UMN)	Promotion of mini and micro-hydro, biogas, ICS
Unfed Nations Capital Developmentt Fund (UNCDF)	Capital subsidy for installing micro-hydro, biogas
United Nations Development Programme (UNDP)	Support and promote ICS technology
United Nations International Children Education Fund (UNICEF)	Capital subsidy for installing biogas, promotion of ICS
United States Agency for International Development (USAID)	Support the micro-hydro, biogas projects.
Water and Energy Commission Secretariat (WECS)	Planning and policy formulation
Wisdom Light Group (WLG)	Manufacture of SPV technologies

2.3 percent in the hills of Nepal (HMG, 1998a). At present, officially, Nepal has 29 percent of forest and 10.7 percent of shrubland (good forest land is about 15 percent). The dwindling supplies of fuel-wood have forced people to burn increasing quantities of cattle dung and agricultural residues for fuel. There are grave consequences related to dung burning, as it reduces the organic manure input to the fields thereby depriving the soil of valuable nutrients and organic conditioning materials which are very vital to agricultural productivity. However, assessment of the environmental impacts from fuelwood/biomass energy use requires detailed analysis of physical, chemical and biological processes that are altered through energy into the production and consumption system as well as the institutional arrangements, policies, acts and regulations, that has a direct bearing upon the behavior of the fuel-wood/biomass users (Bhadra, 1993).

Since biomass energy has been used to light stoves, particularly in the rural areas, the health status of women and children has adversely been affected by indoor air pollution caused by the smoke of fuel-wood. A long distance walk to collect fuel-wood and more time needed to fetch drinking water in hills have emerged due to the drying up of water sources. The use of other commercial sources of energy, such as petroleum products, coal, and electricity, could reduce the growing pressure on fuel-wood demands. Many protected areas started setting kerosene depots to provide kerosene at subsidized rates. But, the import of petroleum products and coal has on the one hand led to an increasing demand on Nepal in foreign exchange and on the other the use of these types of energy, primarily for transport and industries, is concentrated in urban centers and cause pollution problems. The ambient air quality is deteriorating in big cities like Kathmandu, costing a huge price in the public health sector with an ever-increasing number of patients with respiratory complaints.

Alternative energy technologies (AETs) are gaining popularity among rural households due to the reduction in fuel-wood use as well as the implement of health and sanitation. The benefits are further extended towards reducing pressure on dwindling forest resources, health, and other socioeconomic benefits. AETs usually cause very little negative effect on the environment. However, in the case of the micro-hydro sector, the most noticeable change observed is related to the diversion mechanism, which may dry up the river between the intake and the tail-race. In the smallest schemes, the change may be barely noticeable as the river will essentially pass through the powerhouse with effectively no diversion. However, there is some possibility that the migratory fishes will be affected. Many schemes involve the introduction of electricity in the area for the first time. This can cause safety problems and it can also result in changed socioeconomic relationships caused primarily by the introduction of machines.

CONSTRAINTS AND GAPS IN THE ALTERNATIVE ENERGY SECTOR

Development of an alternative sources of energy programme in Nepal faces several constraints. Although the social cost of deforestation is obviously very high for the individual, fuel-wood is the cheapest and simplest source of energy. Utilisation of other alternative energy sources like solar, biogas, and micro-hydro involves high capital investment and thus, always faces the problem of economic constraints. The average per capita income of the people is very low, which restricts the capacity of the individuals to invest in alternative energy technology.

"Smaller the size, higher the costs" is a real constraint with hydropower projects. The initial investment in small hydro projects is relatively high with higher per KW generation cost than large hydro projects. Due to poor economic base and low purchasing capacity of the rural people, there is insufficient demand of electricity in the productive sectors of the economy. Hence, low capacity utilisation, higher tariff, and insufficient income to the plant are the major constraints to the micro-hydro sector. In the management aspects, the public sector mini-and micro-hydropower plants are more systematic since they are being managed by NEA. But, micro-hydro projects in the private sector face managerial constraints. In the private sector micro-hydro projects are managed and operated by the village level entrepreneurs who are seldom trained and lack technical knowledge of the proper operation, maintenance and repair of the plants. They also lack training in the management principles required to run the project efficiently and promote their sales.

As Nepal is a land-locked and mountainous country there are constraints in terms of transport and accessibility. The plains of Terai have moderate infrastructure facilities in comparison to the hilly and mountainous regions. Most of the hilly districts of the country are deprived of a road network. This hinders the supply of equipment and materials for installation and operation of energy devices and systems and also hinders the dissemination of the knowledge and skill of alternative energy technologies. Similarly, maintenance of the installed units face problems as repair workshops are virtually non-existent and spare parts are difficult to obtain. Furthermore, dissemination of new energy technology and propagation of energy conservation ideas are sometimes difficult due to the prevailing social structure and cultural tradition.

Low level of present technology base is a handicap towards better management of traditional energy resources like the forest and also towards further development of new and alternative energy resources. Suitable new energy technologies are required to exploit the most abundant domestic resources. The available technologies are not always suitable in the local context. For example, the major problem in family-size biogas plants is the low gas yield, particularly during the winter and rainy periods when the atmospheric temperature falls below the optimum value and causes occasional blockage and leakage of gas in the pipelines (WECS, 1995). Lack of appropriate technical manpower in adequate number is another major constraint in the area of alternative energy development. Thus, education and training placed in general within the broader framework of energy development and transfer of appropriate information and technology are necessary. In Nepal, there is a gap in coordination and interorganizational communication between the agencies dealing with the planning and implementation of alternative energy, which adversely affects the production and promotion of alternative energy technology devices.

Although the initial investments in alternative energy technology are relatively high at the community level, establishment of such technology in the rural hilly areas plays a crucial role in increasing production and productivity of the agricultural sector, including the processing of agricultural produce, forest conservation, creation of self employment opportunities and promotion of tourism. Thus, alternative energy resource needs to be developed at the local level for the benefit of the local inhabitants. In addition, to ensure that implemented schemes are financially and economically viable, the local governing bodies must get financial benefit from the established schemes. Any financial and technical interventions must emphasize the need to sustainably develop alternative energy resources at the local level. As mentioned above, a vast majority of the Nepalese population is still deprived of electricity. Although the private sector has been involved in power generation, the schemes they are involved in will contribute negligibly to the expansion of the rural networks, nor will it provide any direct financial benefit to the localities in which they are situated. The private sector and the attention of HMG have been focused on developing medium and large hydropower schemes (>5 MW) to increase the grid capacity or to explore the possibility of earning revenues by exporting electricity to neighboring countries. But high priority has to be given in the development of schemes that can be developed at the local level by national entrepreneurs keeping the local people in mind.

REFERENCES

Amatya, V B. and G.R. Shrestha. 1998. Review on policies and their implications on renewable energy technologies in Nepal. *In:* Rijal, K. (ed.) *Renewable energy technologies: a brighter future. Policy options for mountain communities in the HKH and agenda for action in Nepal.* International Centre for Integrated Mountain Development (ICIMOD), Kathmandu, Nepal. pp. 85-116.

Bhadra, B. 1993. Evaluation of environmental impacts of fuel-wood/biomass use. *Spotlight*, January 1999, pp 29-31.

CBS, 1998. *Agricultural statistics, Nepal, 1996/97.* His Majesty's Government of Nepal, National Planning Commission, Central Bureau of Statistics, Kathmandu, Nepal.

Dangrid Consult. 1992. *Wind power in Nepal.* Final report and recommendation, part one evaluation. His Majesty's Government of Nepal/Nepal Electricity Authority (NEA)/United Nations Development Programme (UNDP), Kathmandu, Nepal.

GATE, 1983. *Development and improvement of Nepalese water mill "ghatta".*

GATE, Germany.

Gongal, R.N. and S. Shrestha. 1998. A case study of the biogas programme in Nepal. *In*: Rijal, K (ed.) *Renewable energy technologies: a brighter future. Policy options for mountain communities in the HKH and agenda for action in Nepal.* International Centre for Integrated Mountain Development (ICIMOD), Kathmandu, Nepal pp. 157-169.

HMG, 1988. *Mater plan for the forestry sector, Nepal.* HMG/ADB/FINIDA, Kathmandu, Nepal.

HMG, 1998a. *A Compendium on Environmental Statistics, Nepal.* His Majesty's Government of Nepal, National Planning Commission, Central Bureau of Statistics, Kathmandu, Nepal.

HMG, 1998b. *Economic Survey.* His Majesty's Government of Nepal National Planning Commission, Kathmandu, Nepal.

Jha, H.B. 1995 Small hydropower development: cost effectiveness and private sector participation. *In*: Jha, H.B. (ed.) *Sustainable development of small hydropower in Nepal.* Centre for Economic and Technical Studies, Kathmandu, Nepal pp. 1-11.

Kantipur, 1999 Solar energy. *Kantipur* (daily news paper), Kantipur Publication, Kathmandu, Nepal, December 7, 1999, page 10.

Khandelwal, K.C. 1996. *Report on national policy framework for biogas development.* Food and Agriculture Organisation (FAO), Rome.

NEA. 1999. *Nepal Electricity Authority: a year in review FY 1998/99.* Nepal Electricity Authority, Kathmandu, Nepal.

Shrestha, D.B. 1999. Storage project for the next millennium. *Vidyut* (semi-annual magazine), Nepal Electricity Authority, Kathmandu, Nepal. 10(1):31-33.

Shrestha, S. and V.B. Amatya. 1998. A case study of micro-hydropower in Nepal *In*: Rijal, K. (ed.) *Renewable energy technologies: a brighter future. Policy options for mountain communities in the HKH and agenda for action in Nepal.* International Centre for Integrated Mountain Development (ICIMOD), Kathmandu, Nepal pp. 141-156.

Shrestha, S. and K. Bajracharya. 1998. A case study of solar photovoltaic technology in Nepal. *In*: Rijal K (ed.) *Renewable energy technologies: a brighter future. Policy options for mountain communities in the HKH and agenda for action in Nepal.* International Centre for Integrated Mountain Development (ICIMOD), Kathmandu, Nepal pp. 181-188.

Sulpya, K.M. 1991. *Biomass production and conservation of energy through improved cooking stoves in Nepal.* MIT Discussion Paper Series No. 12. International Centre for Integrated Mountain Development (ICIMOD), Kathmandu, Nepal.

Thapa, I.S. 1995. Development of small hydro electric power and its impact on the small and rural industries. *In*: Jha, H.B. (ed.) *Sustainable development of small hydropower in Nepal.* Centre for Economic and Technical Studies, Kathmandu, Nepal. pp. 35-56.

Van Nes, W.J. and J. Lam 1997. *Final report on the Biogas Support Programme phase I and phase II*. Biogas Support Programme, SNV, Kathmandu, Nepal.

WECS, 1992. *Energy sector synopsis report 1990/91*. Water and Energy Comnission Secretariat, His Majesty's Government of Nepal, Ministry of Water Resources, Kathmandu, Nepal.

WECS, 1994. *Energy sector synopsis report, 1992/93*. Water and Energy Commission Secretariat, His Majesty's Govermnent of Nepal, Ministry of Water Resources, Kathmandu, Nepal.

WECS, 1995. *Alternative energy technology: an overview and assessment*. Supporting Docunent No. 3, Perspective Plan. Water and Energy Commission Secretariat, His Majesty's Government of Nepal, Ministry of Water Resources, Kathmandu, Nepal.

WECS, 1996. *Energy Synopsis Report* 1994/95. Water and Energy Commission Secretariat, His Majesty's Government of Nepal, Ministry of Water Resources, Kathmandu, Nepal.

WECS, 1999. *Annual Report 1999. Report no. Adm./060799/1/1 Seq. No. 041*. His Majesty's Government of Nepal, Ministry of Water Resources, Water and Energy Commission Secretariat (WECS), Kathmandu, Nepal.

WLG and WECS, 1995. *Assessment of the socio-economic impact and technology of solar photovoltaic in Nepal*. Wisdom Light Group and Water and Energy Commission Secretariat, Kathmandu, Nepal.

Subject Index

A

B

C

D

E

F

G

H

I

J

K

L

M

N

O

P

Q

R

S

T

U

V

W

X

Y